**WITHDRAWN
NDSU**

HANDBOOK OF PHILOSOPHICAL LOGIC

VOLUME I

SYNTHESE LIBRARY

STUDIES IN EPISTEMOLOGY,

LOGIC, METHODOLOGY, AND PHILOSOPHY OF SCIENCE

Managing Editor:

JAAKKO HINTIKKA, *Florida State University, Tallahassee*

Editors:

DONALD DAVIDSON, *University of California, Berkeley*
GABRIËL NUCHELMANS, *University of Leyden*
WESLEY C. SALMON, *University of Pittsburgh*

VOLUME 164

HANDBOOK OF PHILOSOPHICAL LOGIC

Volume I:
Elements of Classical Logic

Edited by

D. GABBAY

Department of Mathematics and Computer Science, Bar-Ilan University, Israel

and

F. GUENTHNER

Neuphilologische Fakultaet, University of Tuebingen, West Germany

D. REIDEL PUBLISHING COMPANY

A MEMBER OF THE KLUWER ACADEMIC PUBLISHERS GROUP

DORDRECHT / BOSTON / LANCASTER

Library of Congress Cataloging in Publication Data
Main entry under title:

Elements of classical logic.

(Handbook of philosophical logic; v. 1) (Synthese library; v. 164)
Bibliography: p.
Includes indexes.
1. Logic—Addresses, essays, lectures. I. Gabbay, Dov M.,
1945– . II. Guenthner, Franz. III. Series.
BC6.H36 1983 vol. 1 [BC71] 160s [160]   83-4276
ISBN 90-277-1542-4

Published by D. Reidel Publishing Company,
P.O. Box 17, 3300 AA Dordrecht, Holland.

Sold and distributed in the U.S.A. and Canada
by Kluwer Academic Publishers,
190 Old Derby Street, Hingham, MA 02043, U.S.A.

In all other countries, sold and distributed
by Kluwer Academic Publishers Group,
P.O. Box 322, 3300 AH Dordrecht, Holland.

All Rights Reserved
© 1983 by D. Reidel Publishing Company
No part of the material protected by this copyright notice may be reproduced or
utilized in any form or by any means, electronic or mechanical,
including photocopying, recording or by any information storage and
retrieval system, without written permission from the copyright owner

Printed in The Netherlands

CONTENTS TO VOLUME I

ACKNOWLEDGEMENTS vii

PREFACE ix

A NOTE ON NOTATION xiii

I.1. WILFRID HODGES / Elementary Predicate Logic 1
I.2. GÖRAN SUNDHOLM / Systems of Deduction 133
I.3. HUGUES LEBLANC / Alternatives to Standard First-order Semantics 189
I.4. JOHAN VAN BENTHEM and KEES DOETS / Higher-order Logic 275
I.5. ALLEN HAZEN / Predicative Logics 331
I.6. DIRK VAN DALEN / Algorithms and Decision Problems: A Crash Course in Recursion Theory 409

NAME INDEX 479

SUBJECT INDEX 485

TABLE OF CONTENTS TO VOLUMES II, III, AND IV 495

ACKNOWLEDGEMENTS

The preparation of the *Handbook of Philosophical Logic* was generously supported by the Lady Davis Fund at Bar-Ilan University, Ramat-Gan, Israel and the Werner-Reimers-Stiftung, Bad Homburg, West Germany, which provided us with the chance of discussing the chapters in the *Handbook* at various workshops with the contributors. It is a great pleasure to acknowledge the assistance of these institutions during the preparation of this collection. We benefitted further from the editorial and personal advice and help from the publisher.

Most important of all, we would like to express our thanks to all the contributors to the *Handbook* for their unlimited goodwill, their professional counsel, as well as their friendly support. They have made the preparation of this collection a stimulating and gratifying enterprise.

D. GABBAY (*Bar-Ilan University*)
F. GUENTHNER (*University of Tuebingen*)

PREFACE

The aim of the first volume of the present *Handbook of Philosophical Logic* is essentially two-fold: First of all, the chapters in this volume should provide a concise overview of the main parts of classical logic. Second, these chapters are intended to present all the relevant background material necessary for the understanding of the contributions which are to follow in the next three volumes. We have thought it to be of importance that the connections between classical logic and its 'extensions' (covered in Volume II) as well as its most important 'alternatives' (covered in Volume III) be brought out clearly from the start.

The first chapter presents a clear and detailed picture of the range of what is generally taken to be the standard logical framework, namely, *predicate* (or *first-order quantificational*) logic. On the one hand, this chapter surveys both *propositional* logic and *first-order predicate* logic and, on the other hand, presents the main metalogical results obtained for them. Chapter I.1 also contains a discussion of the limits of first-order logic, i.e. it presents an answer to the question: Why has predicate logic played such a formidable role in the formalization of mathematics and in the many areas of philosophical and linguistic applications?

Chapter I.1 is prerequisite for just about all the other chapters in the entire *Handbook*, while the other chapters in Volume I provide more detailed discussions of material developed or hinted at in the first chapter.

Chapters I.2 centers on proof-theoretic questions and outlines several well-known frameworks for the development of a proof-theory for predicate logic.

The semantic account for the logical systems presented in Chapter I.1 is essentially model-theoretic; Chapter I.3 surveys several alternatives to the standard model-theoretic semantics for predicate logic.

Chapter I.4 takes a big step beyond predicate logic by introducing languages where quantification is not only over individuals but also over higher-order entities, such as sets, functions, sets of sets, etc. Various results concerning second-order and higher-order logic as well as relevant material from abstract model theory are pulled together here.

Chapter I.5 is essentially an introduction to ramified type theory and its philosophical and logical significance. The reason why it is included in the present volume is due to the fact that it covers material which overlaps greatly with the contents of all of the volumes of the *Handbook*. Being most closely connected to the chapter on higher-order logic, it was decided to make it part of Volume I.

Finally, Chapter I.6 is a concise guide to a number of concepts from the theory of algorithms and decidability which will turn up in many later chapters in the discussion of the meta-mathematical properties of the logical systems presented here.

In preparing the *Handbook of Philosophical Logic*, the editors have tried to follow a number of guidelines, the most important of which are the following. First, the field of philosophical logic being so vast and heterogeneous with roots in general philosophy, mathematical logic and theoretical linguistics to cite just a few, it was never our intention to aim for an exhaustive coverage of all the work done under the term philosophical logic. Most of the domains of investigation not covered in the present volumes were thus excluded on purpose as they seemed to belong more to such related disciplines like the philosophy of science or the philosophy of mathematics. Second, our preference has gone from the beginning towards the form of substantial self-contained survey essays as opposed to a large number of short articles. The reason for this is twofold: there exist already many short expository surveys of practically all the topics covered in the *Handbook*, on the other hand, relatively detailed expositions of the main results and problems in the various fields simply cannot be accommodated in the space of a few pages. Third, the division of the *Handbook* into the present four volumes is not intended to imply a strict and theoretical division of work in philosophical logic into exactly these subareas. On the contrary, the four volumes should rather be regarded as a whole. Neither the place, order of appearance nor the relative length of the various contributions should be taken as indications of the general structure of the entire field of philosophical logic. The *Handbook* was conceived to be a guide to the many areas of philosophical logic without imposing a particular picture of either their relative merits or status within the field. Fourth and finally, the survey articles in the *Handbook* should be accessible and useful to researchers and students in a variety of fields. It is hoped that philosophers, logicians, mathematicians, linguists, psychologists, computer scientists and many others can fruitfully draw upon the material presented in these volumes and arrive perhaps at the conclusion that these fields are not quite as independent as they might previously have thought. Although in some cases new results are included in the presentation, the emphasis in every essay is on providing a general survey of the subject in question. The level of sophistication and the required prerequisites are on the whole rather uniform. As pointed out above, it is in addition the essential function of Volume I to put the necessary background for later papers at the reader's disposition.

D. M. GABBAY
F. GUENTHNER

A NOTE ON NOTATION

Writings in the field of philosophical logic abound with logical symbols and every writer seems to have avowed or non-avowed (strong) preferences for one or the other system of notation. It had at one point been our intention to adopt a completely uniform set of symbols and notational conventions for the *Handbook*. For various reasons, we have left the choice of minor bits of notation (e.g. sentential symbols and the like) to the particular authors, unifying only whenever confusion might arise. The reader is invited to translate whatever seems to go against his or her notational taste into his or her favorite variant.

CHAPTER I.1

ELEMENTARY PREDICATE LOGIC
by WILFRID HODGES

Introduction	2

I: PROPOSITIONAL LOGIC

1.	Truth-functors	5
2.	Propositional arguments	8
3.	Why truth-tables work	11
4.	Some points of notation	14
5.	Properties of $\models$	17
6.	Decision methods	21
7.	Formal proof calculi	26

II: PREDICATE LOGIC

8.	Between propositional logic and predicate logic	33
9.	Quantifiers	35
10.	Satisfaction	39
11.	Quantifier notation	42
12.	Ambiguous constants	45
13.	First-order syntax formalized	49
14.	First-order semantics formalized	52
15.	First-order implications	56
16.	Creating models	60
17.	Consequences of the construction of models	65
18.	Identity	68
19.	Axioms as definitions	72
20.	Axioms with intended models	75
21.	Noun phrases	78

III: THE EXPRESSIVE POWER OF FIRST-ORDER LOGIC

22.	After all that, what is first-order logic?	83
23.	Set theory	86
24.	Encoding syntax	88
25.	Skolem functions	92
26.	Back-and-forth equivalence	97
27.	Lindström's theorem	101

D. Gabbay and F. Guenthner (eds.), Handbook of Philosophical Logic, Vol. I, 1–131.
© *1983 by D. Reidel Publishing Company.*

IV: APPENDICES

A.	A formal proof system	107
B.	Arithmetic	110
C.	Set theory	116

References   123

INTRODUCTION

Elementary (first-order) predicate logic is a child of many parents. At least three different groups of thinkers played their part in its conception, with three quite distinct motives. Maybe the mixture gave it hybrid strength. But whatever the reason, first-order logic is both the simplest, the most powerful and the most applicable branch of modern logic.

The first group who can claim paternity are the *Traditional Logicians*. For these scholars the central aim of logic was to schematize valid arguments. For present purposes an argument consists of a string of sentences called *premises*, followed by the word '*Therefore*', followed by a single sentence called the *conclusion*. An argument is called *valid* when its premises *entail* its conclusion, in other words, if the premises can't be true without the conclusion also being true.

A typical valid argument schema might be:

(1) a is more X than b. b is more X than c.
        Therefore a is more X than c.

This becomes a valid argument whenever we substitute names for a, b, c respectively and an adjective for X; as for example

(2)   Oslo is more clean than Ydstebøhavn. Ydstebøhavn is more clean than Trondheim. *Therefore* Oslo is more clean than Trondheim.

Arguments like (2) which result from such substitutions are called *instances* of the schema (1). Traditional logicians collected valid argument schemas such as (1). This activity used to be known as *formal logic* on the grounds that it was concerned with the forms of arguments. (Today we more often speak of formal versus informal logic, just as formal versus informal semantics, meaning mathematically precise versus mathematically imprecise.)

The ancients and the medievals had concerned themselves with small numbers of argument schemas gathered more or less *ad hoc*. Aristotle's syllogisms give twenty-four schemas, of which Aristotle himself mentions nineteen. The watershed between classical and modern logic lies in 1847, when George Boole (1815–1864) published a calculus which yielded infinitely many valid argument schemas of arbitrarily high complexity (Boole [1847, 1854]). Today we know Boole's calculus as *propositional logic*. Other early researchers who belong among the Traditionals are Augustus de Morgan (1806–1871) and C. S. Peirce (1839–1914). Their writings are lively with examples of people i being enemies to people j at time k, and other people overdrawing their bank accounts.

The second group of originators were the *Proof Theorists*. Among these should be included Gottlob Frege (1848–1925), Giuseppe Peano (1858–1932), David Hilbert (1862–1943), Bertrand Russell (1872–1970), Jacques Herbrand (1908–1931) and Gerhard Gentzen (1909–1945). Their aim was to systematize mathematical reasoning so that all assumptions were made explicit and all steps rigorous. For Frege this was a matter of integrity and mental hygiene. For Hilbert the aim was to make mathematical reasoning itself the object of mathematical study, partly in order to justify infinitary mathematics but partly also as a new method of mathematical research. This group devised both the notation and the proof theory of first-order logic. The earliest calculus adequate for first-order logic was the system which Frege published in his *Begriffschrift* [1879]. This was also the first work to discuss quantifiers.

With a slight anachronism I call the third group the *Model Theorists*. Their aim was to study mathematical structures from the point of view of the laws which these structures obey. The group includes Ernst Schröder (1841–1902), Leopold Löwenheim (1878–1957), Thoralf Skolem (1887–1963), C. H. Langford (1895?–1964), Kurt Gödel (1906–1978) and Alfred Tarski (1901–). The notion of a first-order property is already clear in Schröder's work [1895], though the earliest use I could find of the term 'first-order' in the modern sense is in Langford [1927]. (Langford quotes the slightly different use of the term in *Principia Mathematica*, Whitehead and Russell [1910].)

Our present understanding of what first-order logic is about was painstakingly built up by this group of workers during the years 1915 to 1935. The progress was conceptual as much as technical; a historian of logic feels his fingers tingle as he watches it. Increasing precision was an important part of it. But it is worth reflecting that by 1935 a logician could safely say 'The formal sentence S is true in the structure A' *and mean it*. Frege [1906] had found such language morally reprehensible (cf. Section 12 below). Skolem [1922] talked of formal axioms 'holding in a domain', but he felt obliged to add that this was 'only a manner of speaking, which can lead only to purely formal propositions – perhaps made up of very beautiful *words* . . . '. (On taking truth literally, see above all Kurt Gödel's letters to Hao Wang, quoted on p. 8ff of Wang [1974]. R. L. Vaught's historical paper [1974] is also valuable.)

Other groups with other aims have arisen more recently and found first-order logic helpful for their purposes. Let me mention two. One group are the computer scientists who are concerned with semantics and verification of computer programmes. I have to plead total ignorance of this field, but the reader can find explanations and references in Goldblatt [1982], Harel [1979] and Manna [1974]. Hajnal and Németi [1979] have a bibliography of over 400 items.

Last but in no way least come the semanticists. These workers start from the presumption that the meaning of a sentence in a natural language is built up from the meanings of the component words in a way which is part like and part unlike the grammatical structure of the sentence. The problem then is to describe the structures of meanings. One can see the beginnings of this enterprise in Bertrand Russell's theory of propositions and the 'logical forms' beloved of English philosophers earlier in this century; but the aims of these early investigations were not often clearly articulated. Two developments in the late 1960s have led to a burst of research in this area by both linguists and philosophers. First, the school of *generative semantics* (we may cite G. Lakoff and J. D. McCawley) claimed that every sentence of a natural language has an underlying structure which represents its meaning. The suggested underlying structures were justified in linguistic terms, but often they looked very much like the kind of analyses which an up-to-date Traditional Logician might make in the course of knocking arguments into tractable forms. Second, Montague [1974] used tools from logic to give extremely precise analyses of both the grammar and the semantics of some fragments of English. (Cf. Dowty *et al.* [1981] for an introduction and references to recent work in Montague grammer.) I should add that many researchers in this field, from Montague onwards, have found that they needed logical devices which go far beyond first-order logic.

Logicians like to debate over coffee when 'real' first-order logic first appeared in print. The earliest textbook account was in the *Grundzüge der theoretischen Logik* of Hilbert and Ackermann [1928], based on Hilbert's lectures of 1917–1922. Skolem's paper [1920] is undeniably about first-order logic. But Whitehead and Russell's *Principia Mathematica* [1910] belongs to an earlier era. It contains notation, axioms and theorems which we now regard as part of first-order logic, and for this reason it was quoted as a reference by Post, Langford, Herbrand and Gödel up to 1931, when it figured in the title of Gödel's famous paper on incompleteness, Gödel [1931].

But the first-order part of *Principia* is not distinguished from the rest; and more important, its authors had no notion of a precise syntax or the interpretation of formulas in structures.

I: PROPOSITIONAL LOGIC

1. TRUTH FUNCTORS

In propositional logic we use six artificial symbols $\neg, \wedge, \vee, \rightarrow, \leftrightarrow, \bot$, called *truth-functors*. These symbols all have agreed meanings. They can be used in English, or they can have an artificial language built around them.

Let me explain one of these symbols, $\wedge$, quite carefully. The remainder will then be easy.

We use $\wedge$ between sentences ϕ, ψ to form a new sentence

(1.1) $\quad (\phi \wedge \psi)$.

The brackets are an essential part of the notation. Here and below, 'sentence' means 'indicative sentence'. If ϕ and ψ are sentences, then in any situation,

(1.2) $\quad (\phi \wedge \psi)$ is true iff ϕ is true and ψ is true; otherwise it is false.

('Iff' means 'if and only if'.) This defines the meaning of $\wedge$.

Several points about this definition call for comment. First, we had to mention the situation, because a sentence can be true in one situation and not true in another. For example, the sentence may contain demonstrative pronouns or other indexicals that need to be given a reference, or words that need to be disambiguated. (The situation is not necessarily the 'context of utterance' – a sentence can be true in situations where it is never uttered.)

In propositional logic we assume that in every situation, each sentence under discussion is determinately either true or false and not both. This assumption is completely innocent. We can make it correct by adopting either or both of the following conventions. First, we can agree that although we intend to use the word 'true' as it is normally used, we shall take 'false' to mean simply 'not true'. And second, we can take it as understood that the term 'situation' covers only situations in which the relevant sentences are either true or false and not both. (We may also wish to put an embargo on nonsensical sentences, but this is not necessary.) There are of course several ways of being not true, but propositional logic doesn't distinguish between them.

Logicians always make one further assumption here: they assume that truth and falsehood – T and F for short – are objects. Then they say that the *truth-value* of a sentence is T if the sentence is true, and F otherwise. (Frege [1912]: '... in logic we have only two objects, in the first place: the two truth-values.') But I think in fact even the most scrupulous sceptic could follow the literature if he *defined* the truth-value of all true sentences to be his left big toe and that of false sentences to be his right. Many writers take truth to be the number 1, which they identify with the set $\{0\}$, and falsehood to be the number 0 which is identified with the empty set. Nobody is obliged to follow these choices, but technically they are very convenient. For example (1.2) says that if the truth-value of ϕ is x and the truth-value of ψ is y, then that of $(\phi \wedge \psi)$ is xy.

With this notation, the definition (1.2) of the meaning of $\wedge$ can be written in a self-explanatory chart:

(1.3)

ϕ	ψ	$(\phi \wedge \psi)$
T	T	T
T	F	F
F	T	F
F	F	F

The diagram (1.3) is called the *truth-table* of $\wedge$. Truth-tables were first introduced by C. S. Peirce in [1902].

Does (1.3) really define the meaning of $\wedge$? Couldn't there be two symbols $\wedge_1$ and $\wedge_2$ with different meanings, which both satisfied (1.3)?

The answer is that there certainly can be. For example, if $\wedge_1$ is any symbol whose meaning agrees with (1.3), then we can introduce another such symbol $\wedge_2$ by declaring that $(\phi \wedge_2 \psi)$ shall mean the same as the sentence

(1.4) $(\phi \wedge_1 \psi)$ and the number π is irrational.

(Wittgenstein [1961] said that $\wedge_1$ and $\wedge_2$ then mean the same! *Tractatus* 4.46ff, 4.465 in particular.) But this is the wrong way to read (1.3). Diagram (1.3) should be read as stating *what one has to check in order to determine that $(\phi \wedge \psi)$ is true*. One can verify that $(\phi \wedge \psi)$ is true without knowing that π is irrational, but not without verifying that ϕ and ψ are true. (See Michael Dummett [1958/9] and [1975] on the relation between meaning and truth-conditions.)

Some logicians have claimed that the sentence $(\phi \wedge \psi)$ means the same as the sentence

(1.5) ϕ and ψ.

Is this correct? Obviously the meanings are very close. But there are some apparent differences. For example, consider Mr Slippery who said in a court of law:

(1.6)   I heard a shot and I saw the girl fall.

when the facts are that he saw the girl fall and *then* heard the shot. Under these circumstances

(1.7)   (I heard a shot ∧ I saw the girl fall)

was true, but Mr Slippery could still get himself locked up for perjury. One might maintain that (1.6) does mean the same as (1.7) and was equally true, but that the conventions of normal discourse would have led Mr Slippery to choose a different sentence from (1.6) if he had not wanted to mislead the jury. (See Grice [1975] for these conventions; Cohen [1971] discusses the connection with truth-tables.)

Assuming, then, that the truth-table (1.3) does adequately define the meaning of ∧, we can define the meanings of the remaining truth-functors in the same way. For convenience I repeat the table for ∧.

(1.8)

ϕ	ψ	$\neg\phi$	$\phi \wedge \psi$	$\phi \vee \psi$	$\phi \to \psi$	$\phi \leftrightarrow \psi$	$\bot$
T	T	F	T	T	T	T	F
T	F		F	T	F	F	
F	T	T	F	T	T	F	
F	F		F	F	T	T	

$\neg\phi$ is read 'Not ϕ' and called the *negation* of ϕ. $(\phi \wedge \psi)$ is read 'ϕ and ψ' and called the *conjunction* of ϕ and ψ, with *conjuncts* ϕ and ψ. $(\phi \vee \psi)$ is read 'ϕ or ψ' and called the *disjunction* of ϕ and ψ, with *disjuncts* ϕ and ψ. $(\phi \to \psi)$ is read 'If ϕ then ψ' or 'ϕ arrow ψ'; it is called a *material implication* with *antecedent* ϕ and *consequent* ψ. $(\phi \leftrightarrow \psi)$ is read 'ϕ if and only if ψ', and is called the *biconditional* of ϕ and ψ. The symbol $\bot$ is read as 'absurdity', and it forms a sentence by itself; this sentence is false in all situations.

There are some alternative notations in common use; for example

(1.9) $-\phi$ or $\sim\phi$ for $\neg\phi$.
    $(\phi \mathbin{\&} \psi)$ for $(\phi \wedge \psi)$.
    $(\phi \supset \psi)$ for $(\phi \to \psi)$.
    $(\phi \equiv \psi)$ for $(\phi \leftrightarrow \psi)$.

Also the truth-functor symbols are often used for other purposes. For example the intuitionists use the symbols $\neg, \wedge, \vee, \to, \leftrightarrow$ but not with the meanings

given in (1.8); cf. van Dalen [Chapter III.4]. Some writers use the symbol →
for other kinds of implication, or even as a shorthand for the English words
'If ... then'.

A remark on metavariables. The symbols 'ϕ' and 'ψ' are not themselves
sentences and are not the names of particular sentences. They are used as
above, for making statements about any and all sentences. Symbols used in
this way are called *(sentence) metavariables*. They are part of the *metalanguage*,
i.e. the language we use for talking about formulas. I follow the convention
that when we talk about a formula, symbols which are not metavariables are
used as names for themselves. So for example the expression in line (1.1)
means the same as: the formula consisting of '(' followed by ϕ followed by
'∧' followed by ψ followed by ')'. I use quotation marks only when clarity
or style demand them. These conventions, which are normal in mathematical
writing, cut down the clutter but put some obligation on reader and writer to
watch for ambiguities and be sensible about them. Sometimes a more rigorous
convention is needed. Quine's corners ⌐ ¬ supply one; see Quine [1940]
Section 6. There are some more remarks about notation in Section 4 below.

2. PROPOSITIONAL ARGUMENTS

Besides the truth-functors, propositional logic uses a second kind of symbol,
namely the *sentence letters*

(2.1) $p, q, r, \ldots, p_1, p_2, \ldots$.

These letters have no fixed meaning. They serve to mark spaces where English
sentences can be written. We can combine them with the truth-functors to
produce expressions called *formulas*, which become sentences when the
sentence letters are replaced by sentences.

For example, from the sentence letters p, q and r we can built up the
formula

(2.2) $(p \wedge ((p \vee q) \rightarrow r))$

as follows:

(2.3)

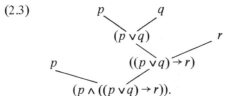

I.1: ELEMENTARY PREDICATE LOGIC

We call (2.3) the *formation tree* of the formula (2.2). Sentence letters themselves are reckoned to be *atomic formulas*, while formulas which use truth-functors are called *compound formulas*. In a compound formula there is always a truth-functor which was added last in the formation tree; this occurrence of the truth-functor is called the *main connective* of the formula. In (2.2) the main connective is the occurrence of ∧. The main connective of ⊥ is reckoned to be ⊥ itself.

Suppose ϕ is a formula. An *instance* of ϕ is a sentence which is got from ϕ by replacing each sentence letter in ϕ by an English sentence, in such a way that no sentence letter gets replaced by different sentences at different occurrences. (Henceforth, the symbols 'ϕ', 'ψ' are metavariables for formulas as well as sentences. The letters 'p', 'q' etc. are not metavariables; they are the actual symbols of propositional logic.)

Now if we know the truth-values of the inserted sentences in an instance of ϕ, then we can work out by table (1.8) what the truth-value of the whole instance must be. Taking (2.2) as an example, consider the following table:

(2.4)

	p	q	r	$(p \wedge ((p \vee q) \rightarrow r))$
(i)	T	T	T	T T T T T T T
(ii)	T	T	F	T F T T T F F
(iii)	T	F	T	T T T T F T T
(iv)	T	F	F	T F T T F F F
(v)	F	T	T	F F F T T T T
(vi)	F	T	F	F F F T T F F
(vii)	F	F	T	F F F F F T T
(viii)	F	F	F	F F F F F T F
				1 7 2 5 3 6 4

The rows (i)–(viii) on the left list all the possible ways in which the sentences put for p and q can have truth-values. The columns on the right are computed in the order shown by the numbers at the bottom. (The numbers at left and bottom are not normally written – I put them in to help the explanation.) Columns 1, 2, 3, 4 just repeat the columns on the left. Column 5 shows the truth-value of $(p \vee q)$, and is calculated from columns 2 and 3 by means of table (1.8). Then column 6 is worked out from columns 5 and 4, using the truth-table for $(\phi \rightarrow \psi)$ in (1.8). Finally, column 7 comes from columns 1 and 6 by the table for $(\phi \wedge \psi)$. Column 7 is written under the main connective of (2.2) and shows the truth-value of the whole instance of (2.2) under each of the eight possibilities listed on the left.

Table (2.4) is called the *truth-table* of the formula (2.2). As we constructed it, we were working out truth-tables for all the formulas shown in the formation tree (2.3), starting at the top and working downwards.

We are now equipped to use propositional logic to prove the validity of an argument. Consider:

(2.5)　That was a hornet, and soda only makes hornet and wasp stings worse. So you don't want to use soda.

This contains an argument along the following lines:

(2.6)　(You were stung by a hornet ∧ ((you were stung by a hornet ∨ you were stung by a wasp) → soda will make the sting worse)). *Therefore* soda will make the sting worse.

We replace the component sentences by letters according to the scheme:

(2.7)　p: You were stung by a hornet.
　　　q: You were stung by a wasp.
　　　r: Soda will make the sting worse.

The result is:

(2.8)　$(p \wedge ((p \vee q) \to r))$. *Therefore r.*

Then we calculate truth-tables for both premise and conclusion of (2.8) at the same time. Only the main columns are shown below:

(2.9)

	p	q	r	$(p \wedge ((p \vee q) \to r))$. Therefore r.	
(i)	T	T	T	T	T
(ii)	T	T	F	F	F
(iii)	T	F	T	T	T
(iv)	T	F	F	F	F
(v)	F	T	T	F	T
(vi)	F	T	F	F	F
(vii)	F	F	T	F	T
(viii)	F	F	F	F	F

Table (2.9) shows that if the premise of (2.6) is true then so is the conclusion. For if the premise is true, then the column under the premise shows that we are in row (i) or row (iii). In both of these rows, the last column in (2.9) shows that the conclusion is true. There is no row which has a T below $(p \wedge ((p \vee q) \to r))$ and an F below r. Hence, (2.6) is valid.

In the language of the traditional logician, these calculations showed that (2.8) is a valid argument schema. Every instance of (2.8) is a valid argument.

Note how the proof of the validity of an argument falls into two parts. The first is to translate the argument into the symbols of propositional logic. This involves no calculation, though a gauche translation can frustrate the second part. I say no more about this first part – the elementary textbooks give hundreds of examples (Kalish and Montague [1964], Mates [1965], Thomason [1970], Hodges [1977]). The second part of the proof is pure mechanical calculation using the truth-table definitions of the truth-functors. What remains to discuss below is the theory behind this mechanical part.

First and foremost, why does it work?

3. WHY TRUTH-TABLES WORK

If ϕ is any formula of propositional logic, then any assignment of truth-values to the sentence letters which occur in ϕ can be extended, by means of the truth-table definitions of the truth-functors, to give a truth-value to ϕ; this truth-value assigned to ϕ is uniquely determined and it can be computed mechanically.

This is the central thesis of propositional logic. In Section 2 I showed how the assignment to ϕ is calculated, with an example. But we shouldn't rest satisfied until we see, first, that this procedure *must always work*, and second, that the outcome is *uniquely determined by the truth-table definitions*. Now there are infinitely many formulas ϕ to be considered. Hence we have no hope of setting out all the possibilities on a page; we need to invoke some abstract principle to see why the thesis is true.

There is no doubt what principle has to be invoked. It is the principle of *induction on the natural numbers*, otherwise called *mathematical induction*. This principle says the following:

(3.1) Suppose that the number 0 has a certain property, and suppose also that whenever all numbers from 0 to n inclusive have the property, $n + 1$ must also have the property. Then all natural numbers from 0 upwards have the property.

This principle can be put in several forms; the form above is called *course-of-values induction*. (See Appendix B below.) For the moment we shall only be using one or two highly specific instances of it, where the property in question is a mechanically checkable property of arrays of symbols. Several writers have maintained that one knows the truth of any such instance of

(3.1) by a kind of inspection (*Anschauung*). (See for example Hilbert [1923], Herbrand [1930], Introduction. There is a discussion of the point in Steiner [1975].)

Essentially what we have to do is to tie a number n to each formula ϕ, calling n the *complexity* of ϕ, so that we can then use induction to prove:

(3.2)   For each number n from 0 upwards, the thesis stated at the beginning of this section is true for all formulas of complexity n.

There are several ways of carrying this through, but they all rest on the same idea, namely this: *all formulas are generated from atomic formulas in a finite number of steps and in a unique way; therefore each formula can be assigned a complexity which is greater than the complexities assigned to any formulas that went into the making of it.* It was Emil Post, one of the founders of formal language theory, who first showed the importance of this idea in his paper on truth-tables:

(3.3)   "It is desirable in what follows to have before us the vision of the totality of these [formulas] streaming out from the unmodified [sentence letters] through forms of ever-growing complexity . . ." (Post [1921], p. 266 of van Heijenoort [1967]).

For an exact definition of formulas and their complexities, we need to say precisely what sentence letters we are using. But it would be a pity to lumber ourselves with a set of letters that was inconvenient for some future purposes. So we adopt a compromise. Let X be any set of symbols to be used as sentence letters. Then we shall define the *propositional language of similarity type X*, in symbols $L(X)$. The set X is not fixed in advance; but as soon as it is fixed, the definition of $L(X)$ becomes completely precise. This is the usual modern practice.

The notions 'formula of similarity type X' (we say 'formula' for short) and 'complexity of a formula' are defined as follows.

(a)   Every symbol in X is a formula of complexity 0. $\bot$ is a formula of complexity 1.

(b)   If ϕ and ψ are formulas of complexities m and n respectively, then $\neg\phi$ is a formula with complexity $m + 1$, and $(\phi \wedge \psi), (\phi \vee \psi), (\phi \rightarrow \psi)$ and $(\phi \leftrightarrow \psi)$ are formulas of complexity $m + n + 1$.

(c)   Nothing is a formula except as required by (a) and (b).

For definiteness the *language of similarity type X*, $L(X)$, can be defined as the ordered pair $\langle X, F \rangle$ where F is the set of all formulas of similarity type X.

I.1: ELEMENTARY PREDICATE LOGIC

A *propositional language* is a language L(X) where X is a set of symbols; the *formulas of* L(X) are the formulas of similarity type X.

Frege would have asked: How do we know there is a unique notion 'formula of similarity type X' with the properties (a)–(c)? A full answer to this question lies in the theory of inductive definitions; cf. Appendix B below. But for the present it will be enough to note that by (a) and (b), every formation tree has a formula as its bottom line, and conversely by (c) every formula is the bottom line of a formation tree. We can prove rigorously by induction that if a formula has complexity n by definition (a)–(c) then it can't also have complexity m where $m \neq n$. This is actually not trivial. It depends on showing that the main connective in a compound formula is uniquely determined, and – ignoring $\neg$ and $\bot$ for simplicity – we can do that by showing that the main connective is the only truth-functor occurrence which has one more '(' than ')' to the left of it. (Cf. Kleene [1952] pp. 21ff.) The proof shows at the same time that every formula has a unique formation tree.

The atomic formulas are those which have complexity 0. A formula is called *basic* if it is either atomic or the negation of an atomic formula.

Now that the language has been adequately formulated, we come back to truth-tables. Let L be a propositional language with similarity type X. Then we define an L-*structure* to be a function from the set X to the set $\{T, F\}$ of truth-values. (Set-theoretic notions such as 'function' are defined in Appendix C below, or in any elementary textbook of set theory.) So an L-structure assigns a truth-value to each sentence letter of L. For each sentence letter ϕ we write $I_{\mathfrak{A}}(\phi)$ for the truth-value assigned to ϕ by the L-structure $\mathfrak{A}$. In a truth-table where the sentence letters of L are listed at top left, each row on the left will describe an L-structure, and every L-structure corresponds to just one row of the table.

Now we shall define when a formula ϕ of L is *true in* an L-structure $\mathfrak{A}$, or in symbols

(3.4) $\mathfrak{A} \models \phi$.

The definition of (3.4) will be by induction of the complexity of ϕ. This means that when ϕ has low complexity, the truth or falsity of (3.4) will be determined outright; when ϕ has higher complexity the truth of (3.4) depends in an unambiguous way on the truth of statements '$\mathfrak{A} \models \psi$' for formulas ψ of lower complexity than ϕ. (Cf. Appendix B.) We can prove by induction on the natural numbers that this definition determines exactly when (3.4) is true, and in fact that the truth or otherwise of (3.4) can be calculated mechanically once we know what $\mathfrak{A}$ and ϕ are. The definition is as follows:

(3.5) For each sentence letter ϕ, $\mathfrak{A} \models \phi$ iff $I_{\mathfrak{A}}(\phi) = T$.
It is false that $\mathfrak{A} \models \bot$.
For all formulas ϕ, ψ of L,
$\mathfrak{A} \models \neg \phi$ iff it is not true that $\mathfrak{A} \models \phi$;
$\mathfrak{A} \models (\phi \wedge \psi)$ iff $\mathfrak{A} \models \phi$ and $\mathfrak{A} \models \psi$;
$\mathfrak{A} \models (\phi \vee \psi)$ iff either $\mathfrak{A} \models \phi$ or $\mathfrak{A} \models \psi$ or both;
$\mathfrak{A} \models (\phi \rightarrow \psi)$ iff not: $\mathfrak{A} \models \phi$ but not $\mathfrak{A} \models \psi$.
$\mathfrak{A} \models (\phi \leftrightarrow \psi)$ iff either $\mathfrak{A} \models \phi$ and $\mathfrak{A} \models \psi$, or neither $\mathfrak{A} \models \phi$ nor $\mathfrak{A} \models \psi$.

Definition (3.5) is known as the *truth definition* for the language L. The statement '$\mathfrak{A} \models \phi$' is sometimes read as: $\mathfrak{A}$ is a *model of* ϕ.

The reader can verify that (3.5) matches the truth-table definitions of the truth-functors, in the following sense. The left-hand part of any row of a truth-table for ϕ describes an L-structure $\mathfrak{A}$ (for some appropriate language L). The truth-table gives ϕ the value T in this row if and only if $\mathfrak{A} \models \phi$; moreover the steps by which we calculated this value for ϕ in the table exactly match the steps by which the definition (3.5) above determines whether $\mathfrak{A} \models \phi$. In this sense, and only in this sense, (3.5) is a correct 'definition of truth for L'. Nobody claims that (3.5) explains what is meant by the word 'true'.

I should mention a useful piece of notation. We can write $\|\phi\|_{\mathfrak{A}}$ for the truth-value assigned to the formula ϕ by the structure $\mathfrak{A}$. Then $\|\phi\|_{\mathfrak{A}}$ can be defined in terms of $\models$ by:

$$(3.6) \quad \|\phi\|_{\mathfrak{A}} = \begin{cases} T & \text{if } \mathfrak{A} \models \phi, \\ F & \text{otherwise.} \end{cases}$$

Some writers prefer to define $\|\ \|_{\mathfrak{A}}$ directly, and then $\models$ in terms of $\|\ \|_{\mathfrak{A}}$. If we write 1 for T and 0 for F, an inductive definition of $\|\ \|_{\mathfrak{A}}$ will contain clauses such as

$$(3.7) \quad \|\neg \phi\|_{\mathfrak{A}} = 1 - \|\phi\|_{\mathfrak{A}}; \qquad \|(\phi \vee \psi)\|_{\mathfrak{A}} = \max\{\|\phi\|_{\mathfrak{A}}, \|\psi\|_{\mathfrak{A}}\}.$$

4. SOME POINTS OF NOTATION

In Section 3 we put the truth-table method onto a more solid footing. We extended it a little too, because we made no assumption that the language L had just finitely many sentence letters. The original purpose of the exercise was to prove valid argument schemas, and we can now redefine these in sharper terms too.

Let L be a fixed proportional language and $\phi_1, \ldots, \phi_n, \psi$ any formulas of L. Then the statement

(4.1) $\phi_1, \ldots, \phi_n \models \psi$

will mean: for every L-structure $\mathfrak{A}$, if $\mathfrak{A} \models \phi_1$ and ... and $\mathfrak{A} \models \phi_n$, then $\mathfrak{A} \models \psi$. We allow n to be zero; thus

(4.2) $\models \psi$

means that for every L-structure $\mathfrak{A}$, $\mathfrak{A} \models \psi$. To say that (4.1) is false, we write

(4.3) $\phi_1, \ldots, \phi_n \not\models \psi$.

Note that (4.1)–(4.3) are statements about formulas of L and not themselves formulas of L.

It is a misfortune that custom requires us to use the same symbol $\models$ both in '$\mathfrak{A} \models \phi$' (cf. (3.4) above) and in '$\phi_1, \ldots, \phi_n \models \psi$'. It means quite different things in the two cases. But one can always see which is meant, because in the first case a structure $\mathfrak{A}$ is mentioned immediately to the left of $\models$, and in the second usage $\models$ follows either a formula or an empty space. $\models$ can be pronounced 'double turnstile' or 'semantic turnstile', to contrast it with the symbol $\vdash$ ('turnstile' or 'syntactic turnstile') which occurs in the study of formal proof calculi (cf. Section 7 below).

The point of definition (4.1) should be clear. It says in effect that if we make any consistent replacement of the sentence letters by sentences of English, then in any situation where the sentences resulting from $\phi_1, \ldots, \phi_n$ are true, the sentence got from ψ will be true too. In short (4.1) says that

(4.4) $\phi_1, \ldots, \phi_n$. Therefore ψ.

is a valid argument schema. What's more, it says it without mentioning either English sentences or possible situations. Statements of form (4.1) or (4.2) are called *sequents* (= 'things that follow' in Latin). When (4.1) is true, $\phi_1, \ldots, \phi_n$ are said to *logically imply* ψ. When (4.2) is true, ψ is said to be a *tautology*; for a language with a finite number of sentence letters, this means that the truth-table of ψ has T all the way down its main column. Some elementary texts give long lists of tautologies (e.g. Kalish and Montague [1964] pp. 80–84).

While we are introducing notation, let me mention some useful abbreviations. Too many brackets can make a formula hard to read. So we shall agree that when naming formulas we can leave out some of the brackets. First, we can leave off the brackets at the two ends of an occurrence of $(\phi \wedge \psi)$ or

$(\phi \vee \psi)$ provided that the only truth-functor which occurs immediately outside them is either $\rightarrow$ or $\leftrightarrow$. For example we can abbreviate

(4.5) $\quad (p \leftrightarrow (q \wedge r))$ and $((p \wedge q) \rightarrow (r \vee s))$

to

(4.6) $\quad (p \leftrightarrow q \wedge r)$ and $(p \wedge q \rightarrow r \vee s)$

respectively; but we can *not* abbreviate

(4.7) $\quad (\neg(p \wedge q) \rightarrow r)$ and $((p \leftrightarrow q) \wedge r)$

to

(4.8) $\quad (\neg p \wedge q \rightarrow r)$ and $(p \leftrightarrow q \wedge r)$

respectively.

Second, we can leave off brackets at the ends of a formula. So the formulas in (4.6) can also be written

(4.9) $\quad p \leftrightarrow q \wedge r$ and $p \wedge q \rightarrow r \vee s$

respectively.

Third, if we have a string of $\wedge$'s with their associated brackets bunched up towards the left end of the formula, as in

(4.10) $\quad (((q \wedge r) \wedge s) \wedge t)$,

then we can omit all but the outermost brackets:

(4.11) $\quad (q \wedge r \wedge s \wedge t)$.

Formula (4.11) is called a *conjunction* whose *conjuncts* are q, r, s, t. Likewise we can abbreviate $(((q \vee r) \vee s) \vee t)$ to the *disjunction* $(q \vee r \vee s \vee t)$ with *disjuncts* q, r, s, t. (But the corresponding move with $\rightarrow$ or $\leftrightarrow$ is not allowed.)

All these conventions can be applied together, as when we write

(4.12) $\quad p \wedge q \wedge r \rightarrow s$

for

(4.13) $\quad (((p \wedge q) \wedge r) \rightarrow s)$.

When only these abbreviations are used, it is always possible to work out exactly which brackets have been omitted, so that there is no loss of information.

1.1: ELEMENTARY PREDICATE LOGIC

Jan Łukasiewicz pointed out that if we always write connectives to the left of the formulas they connect, then there is no need for any brackets at all. In this style the second formula of (4.5) could be written

(4.14) $\to \wedge pq \vee rs$,   or in Łukasiewicz's notation, $CKpqArs$.

Prior [1962] uses Łukasiewicz's notation throughout.

Note that the abbreviations described above only affect the way we talk about formulas of L — the formulas themselves remain untouched. The definition of 'formula of similarity type X' given in Section 3 stands without alteration. Some early writers were rather carefree about this point, making it difficult to follow what language L they really had in mind. If anybody wants to do calculations *in* L but still take advantage of our abbreviations, there is an easy way he can do it. He simply writes down abbreviated *names* of formulas instead of the formulas themselves. In other words, he works always in the metalanguage and never in the object language. This cheap trick will allow him the best of both worlds: a rigorously defined language and a relaxed and generous notation. Practising logicians do it all the time.

5. PROPERTIES OF $\models$

This section gathers up some properties of $\models$ which can be proved directly from the definitions in Sections 3 and 4 above. They are rather a ragbag, but there are some common themes.

THEOREM 5.1. *If $\mathfrak{A}$ and $\mathfrak{B}$ are structures which assign the same truth-values as each other to each sentence letter occurring in ϕ, then $\mathfrak{A} \models \phi$ iff $\mathfrak{B} \models \phi$.*

This is obvious from (3.5), but it can also be proved rigorously by induction on the complexity of ϕ. The most important consequence of Theorem 5.1 is:

THEOREM 5.2. *The truth of the sequent '$\phi_1, \ldots, \phi_n \models \psi$' doesn't depend on what language L the formulas $\phi_1, \ldots, \phi_n$ and ψ come from.*

In other words, although the definition of '$\phi_1, \ldots, \phi_n \models \psi$' was stated in terms of one language L containing $\phi_1, \ldots, \phi_n$ and ψ, any two such languages would give the same outcome. At first sight Theorem 5.2 seems a reasonable property to expect of any decent notion of entailment. But in other logics, notions of entailment which violate Theorem 5.2 have sometimes been proposed. (There is an example in Dunn and Belnap [1968], and another in Section 15 below.)

The next result turns all problems about sequents into problems about tautologies.

THEOREM 5.3. *Deduction Theorem*: $\phi_1, \ldots, \phi_n \models \psi$ *iff* $\phi_1, \ldots, \phi_{n-1} \models \phi_n \rightarrow \psi$.

Theorem 5.3 moves formulas to the right of $\models$. It has a twin that does the opposite:

THEOREM 5.4. $\phi_1, \ldots, \phi_n \models \psi$ *iff* $\phi_1, \ldots, \phi_n, \neg \psi \models \bot$.

We say that the formula ϕ is *logically equivalent* to the formula ψ if $\phi \models \psi$ and $\psi \models \phi$. This is equivalent to saying that $\models \phi \leftrightarrow \psi$. Intuitively speaking, logically equivalent formulas are formulas which behave in exactly the same way inside arguments. Theorem 5.5 makes this more precise:

THEOREM 5.5. *If* $\phi_1, \ldots, \phi_n \models \psi$, *and we take an occurrence of a formula* χ *inside one of* $\phi_1, \ldots, \phi_n, \psi$ *and replace it by an occurrence of a formula which is logically equivalent to* χ, *then the resulting sequent holds too.*

For example, $\neg p \vee q$ is logically equivalent to $p \rightarrow q$ (as truth-tables will confirm). Also we can easily check that

(5.1) $\quad r \rightarrow (\neg p \vee q), p \models r \rightarrow q$.

Then Theorem 5.5 tells us that the following sequent holds too:

(5.2) $\quad r \rightarrow (p \rightarrow q), p \models r \rightarrow q$.

An interesting consequence of Theorem 5.5 is:

THEOREM 5.6. *Every formula ϕ is logically equivalent to a formula which uses the same sentence letters as ϕ, but no truth-functors except* $\bot, \neg$ *and* $\rightarrow$.

Proof is as follows. Truth-tables will quickly show that

(5.3) $\quad \psi \wedge \chi$ is logically equivalent to $\neg(\psi \rightarrow \neg \chi)$,
$\psi \vee \chi$ is logically equivalent to $(\neg \psi \rightarrow \chi)$, and
$\psi \leftrightarrow \chi$ is logically equivalent to $\neg((\psi \rightarrow \chi) \rightarrow \neg(\chi \rightarrow \psi))$.

But then by Theorem 5.5, if we replace a part of ϕ of form $(\psi \wedge \chi)$ by $\neg(\psi \rightarrow \neg\chi)$, the resulting formula will be logically equivalent to ϕ. By replacements of this kind we can eliminate in turn all the occurrences of $\wedge$, $\vee$ and $\leftrightarrow$ in ϕ, and be left with a formula which is logically equivalent to ϕ. This proves Theorem 5.6. Noting that

(5.4) $\neg\phi$ is logically equivalent to $\phi \rightarrow \bot$,

we can eliminate $\neg$ too, at the cost of introducing some more occurrences of $\bot$. □

An argument just like the proof of Theorem 5.6 shows that every formula is logically equivalent to one whose only truth-functors are $\neg$ and $\wedge$, and to one whose only truth-functors are $\neg$ and $\vee$. But there are some limits to this style of reduction: there is no way of eliminating $\neg$ and $\bot$ in favour of $\wedge$, $\vee$, $\rightarrow$ and $\leftrightarrow$.

The next result is a useful theorem of Post [1921]. In Section 2 we found a truth-table for each formula. Now we go the opposite way and find a formula for each truth-table.

THEOREM 5.7. *Let P be a truth-table which writes either T or F against each possible assignment of truth-values to the sentence letters $p_1, \ldots, p_n$. Then P is the truth-table of some formula using no sentence letters apart from $p_1, \ldots, p_n$.*

I sketch the proof. Consider the jth row of the table, and write ϕ_j for the formula $p'_1 \wedge \cdots \wedge p'_n$, where each p'_i is p_i if the jth row makes p_i true, and $\neg p_i$ if the jth row makes p_i false. Then ϕ_j is a formula which is true at just the jth row of the table. Suppose the rows to which the table gives the value T are rows $j_1, \ldots, j_k$. Then take ϕ to be $\phi_{j_1} \vee \cdots \vee \phi_{j_k}$. If the table has F all the way down, take ϕ to be $\bot$. P is the truth-table of ϕ. □

Theorem 5.7 says in effect that we could never get a more expressive logic by inventing new truth-functors. Anything we could say with the new truth-functors could also be said using the ones we already have.

A formula is said to be in *disjunctive normal form* if it is either $\bot$ or a disjunction of conjunctions of basic formulas (basic = atomic or negated atomic). The proof of Theorem 5.7 actually shows that P is the truth-table of some formula in disjunctive normal form. Suppose now that we take any formula ψ, work out its truth-table P, and find a formula ϕ in disjunctive normal form with truth-table P. Then ψ and ϕ are logically equivalent, because they have the same truth-table. So we have proved:

THEOREM 5.8. *Every formula is logically equivalent to a formula in disjunctive normal form.*

One can also show that every formula is logically equivalent to one in *conjunctive normal form*, i.e. either $\neg\bot$ or a conjunction of disjunctions of basic formulas.

LEMMA 5.9. *Craig's Interpolation Lemma for propositional logic: If $\psi \models \chi$ then there exists a formula ϕ such that $\psi \models \phi$ and $\phi \models \chi$, and every sentence letter which occurs in ϕ occurs both in ψ and in χ.*

Lemma 5.9 is proved as follows. Let L be the language whose sentence letters are those which occur both in ψ and in χ, and L^+ the language whose sentence letters are those in either ψ or χ. Write out a truth-table for the letters in L, putting T against a row if and only if the assignment of truth-values in that row can be expanded to form a model of ψ. By Theorem 5.7, this table is the truth-table of some formula ϕ of L. Now we show $\phi \models \chi$. Let $\mathfrak{A}$ be any L^+-structure such that $\mathfrak{A} \models \phi$. Let $\mathfrak{C}$ be the L-structure which agrees with $\mathfrak{A}$ on all letters in L. Then $\mathfrak{C} \models \phi$ by Theorem 5.1. By the definition of ϕ it follows that some model $\mathfrak{B}$ of ψ agrees with $\mathfrak{C}$ on all letters in L. Now we can put together an L^+-structure $\mathfrak{D}$ which agrees with $\mathfrak{B}$ on all letters occurring in ψ, and with $\mathfrak{A}$ on all letters occurring in χ. (The overlap was L, but $\mathfrak{A}$ and $\mathfrak{B}$ both agree with $\mathfrak{C}$ and hence with each other on all letters in L.) Then $\mathfrak{D} \models \psi$ and hence $\mathfrak{D} \models \chi$ since $\psi \models \chi$; but then $\mathfrak{A} \models \chi$ too. The proof that $\psi \models \phi$ is easier and I leave it to the reader. □

Craig's Lemma is the most recent fundamental discovery in propositional logic. It is easy to state and to prove, but it was first published over a hundred years after propositional logic was invented (Craig [1957]). The corresponding lemma holds for full first-order logic too; this is much harder to prove. (See Lemma 27.4 below.)

Most of the topics in this section are taken further in Hilbert and Bernays [1934], Kleene [1952], Rasiowa and Sikorski [1963] and Bell and Machover [1977].

6. DECISION METHODS

Propositional logic owes much of its flavour to the fact that all interesting problems within it can be solved by scribbled calculations on the back of an envelope. If somebody presents you with formulas $\phi_1, \ldots, \phi_n$ and ψ, and asks you whether $\phi_1, \ldots, \phi_n$ logically imply ψ, then you can calculate the answer as follows. First choose a language L whose sentence letters are just those which occur in the formulas $\phi_1, \ldots, \phi_n, \psi$. If L has k sentence letters then there are just 2^k different L-structures. For each such structure you can check in a finite time whether it is a model of ϕ_1 and ... and ϕ_n but not of ψ. If you find an L-structure with these properties, then $\phi_1, \ldots, \phi_n$ don't logically imply ψ; if you don't, they do. This is the truth-table *decision method for logical implication.*

The question I want to consider next is whether this decision method can be improved. This is not a precise question. Some alternatives to truth-tables are very fast for very short sequents but get quite unwieldy for long ones. Other alternatives grow steadily more efficient as we progress to longer sequents. Some methods are easy to run on a computer but messy on paper; some are as handy one way as the other.

Let me sketch one alternative to truth-tables. An example will show the gist. We want to determine whether the following sequent holds:

(6.1) $\quad p \wedge q, \neg(p \wedge r) \models \neg r.$

By Theorem 5.4, (6.1) holds if and only if

(6.2) $\quad p \wedge q, \neg(p \wedge r), \neg\neg r \models \bot.$

Now (6.2) says that any structure in which all the formulas on the left of $\models$ are true is a model of $\bot$. But $\bot$ has no models; so (6.2) says *there is no model of $p \wedge q$, $\neg(p \wedge r)$ and $\neg\neg r$ simultaneously.* We try to refute this by constructing such a model. At each stage we ask: what must be true in the model for it to be a model of these sentences? For example, $\neg\neg r$ is true in a structure $\mathfrak{A}$ if and only if r is true in $\mathfrak{A}$; so we can simplify (6.2) by replacing $\neg\neg r$ by r:

(6.3) $\quad p \wedge q, \neg(p \wedge r), \neg\neg r \models \bot$
$\quad\quad\quad\quad\quad\quad |$
$\quad\quad p \wedge q, \neg(p \wedge r), r \models \bot.$

Likewise a structure is a model of $p \wedge q$ if and only if it is both a model of p and a model of q; so we can replace $p \wedge q$ by the two formulas p and q:

(6.4) $\quad p \wedge q, \neg(p \wedge r), \neg\neg r \models \bot$
$\quad\quad\quad\quad\quad\quad |$
$\quad\quad p \wedge q, \neg(p \wedge r), r \models \bot$
$\quad\quad\quad\quad\quad\quad |$
$\quad\quad p, q, \neg(p \wedge r), r \models \bot.$

Now there are just two ways of making $\neg(p \wedge r)$ true, namely to make $\neg p$ true and to make $\neg r$ true. (Of course these ways are not mutually exclusive.) So in our attempt to refute (6.2) we have two possible options to try, and the diagram accordingly branches in two directions:

(6.5) $\quad\quad\quad\quad\quad\quad p \wedge q, \neg(p \wedge r), \neg\neg r \models \bot$
$\quad\quad\quad\quad\quad\quad\quad\quad\quad\quad |$
$\quad\quad\quad\quad\quad\quad p \wedge q, \neg(p \wedge r), r \models \bot$
$\quad\quad\quad\quad\quad\quad\quad\quad\quad\quad |$
$\quad\quad\quad\quad\quad\quad p, q, \neg(p \wedge r), r \models \bot$
$\quad\quad\quad\quad\swarrow \quad\quad\quad\quad\quad\searrow$
$\quad p, q, \neg p, r \models \bot \quad\quad\quad p, q, \neg r, r \models \bot$

But there is no chance of having both p and $\neg p$ true in the same structure. So the left-hand fork is a non-starter, and we block it off with a line. Likewise the right-hand fork expects a structure in which $\neg r$ and r are both true, so it must be blocked off:

(6.6) $\quad\quad\quad\quad\quad\quad p \wedge q, \neg(p \wedge r), \neg\neg r \models \bot$
$\quad\quad\quad\quad\quad\quad\quad\quad\quad\quad |$
$\quad\quad\quad\quad\quad\quad p \wedge q, \neg(p \wedge r), r \models \bot$
$\quad\quad\quad\quad\quad\quad\quad\quad\quad\quad |$
$\quad\quad\quad\quad\quad\quad p, q, \neg(p \wedge r), r \models \bot$
$\quad\quad\quad\quad\swarrow \quad\quad\quad\quad\quad\searrow$
$\quad \underline{p, q, \neg p, r \models \bot} \quad\quad\quad \underline{p, q, \neg r, r \models \bot}.$

Since every possibility has been explored and closed off, we conclude that there is no possible way of refuting (6.2), and so (6.2) is correct.

What happens if we apply the same technique to an incorrect sequent? Here is an example:

(6.7) $\quad p \vee \neg(q \rightarrow r), q \rightarrow r \not\models q.$

I leave it to the reader to check the reasons for the steps below – he should note that $q \rightarrow r$ is true if and only if either $\neg q$ is true or r is true:

(6.8)
$$p \vee \neg(q \to r), q \to r, \neg q \models \bot$$

$p, q \to r, \neg q \models \bot$ $\qquad$ $\neg(q \to r), q \to r, \neg q \models \bot$

$p, \neg q, \neg q \models \bot$ $\quad$ $p, r, \neg q \models \bot$

Here two branches remain open, and since all the formulas in them have been decomposed into atomic formulas or negations of atomic formulas, there is nothing more we can do with them. In every such case it turns out that each open branch describes a structure which refutes the original sequent. For example, take the leftmost branch in (6.8). The formulas on the left side of the bottom sequent describe a structure $\mathfrak{A}$ in which p is true and q is false. The sequent says nothing about r, so we can make an arbitrary choice: let r be false in $\mathfrak{A}$. Then $\mathfrak{A}$ is a structure in which the two formulas on the left in (6.7) are true but that on the right is false.

This method always leads in a finite time *either* to a tree diagram with all branches closed off, in which case the beginning sequent was correct; *or* to a diagram in which at least one branch remains resolutely open, in which case this branch describes a structure which shows that the sequent was incorrect.

Diagrams constructed along the lines of (6.6) or (6.8) above are known as *semantic tableaux*. They were first invented, upside-down and with a different explanation, by Gentzen [1934]. The explanation given above is from Beth [1955] and Hintikka [1955].

We can cut out a lot of unnecessary writing by omitting the '$\models \bot$' at the end of each sequent. Also in all sequents below the top one, we need only write the new formulas. In this abbreviated style the diagrams are called *truth-trees*. Written as truth-trees, (6.6) looks like this:

(6.9) $\quad p \wedge q, \neg(p \wedge r), \neg\neg r$

r

p

q

$\neg p \qquad \neg r$

and (6.8) becomes

(6.10) $p \vee \neg(q \rightarrow r), q \rightarrow r, \neg q$

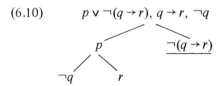

The rules for breaking down formulas in truth-trees can be worked out straight from the truth-table definitions of the truth-functors, but for the reader's convenience I list them:

(6.11)
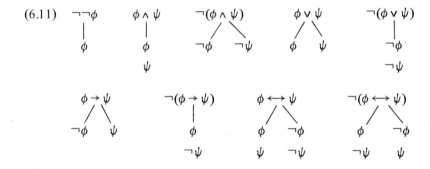

One is allowed to close off a branch as soon as either $\bot$ or any outright contradiction ϕ, $\neg \phi$ appears among the formulas in a branch. (Truth-trees are used in Jeffrey [1967]; see Smullyan [1968] and Bell and Machover [1977] for mathematical analyses.)

Truth-trees are one dialect of semantic tableaux. Here is another. We shall understand the generalized sequent

(6.12) $\phi_1, \ldots, \phi_n \vDash \psi_1, \ldots, \psi_m$

to mean that there is no structure which makes $\phi_1, \ldots, \phi_n$ all true and $\psi_1, \ldots, \psi_m$ all false. A structure in which $\phi_1, \ldots, \phi_n$ are true and $\psi_1, \ldots, \psi_m$ are false is called a *counterexample* to (6.12). When there is only one formula to the right of $\vDash$, (6.12) means just the same as our previous sequents (4.1).

Generalized sequents have the following two symmetrical properties:

(6.13) $\phi_1, \ldots, \phi_n, \neg \chi \vDash \psi_1, \ldots, \psi_m$ iff $\phi_1, \ldots, \phi_n \vDash \psi_1, \ldots, \psi_m, \chi$.

(6.14) $\phi_1, \ldots, \phi_n \vDash \psi_1, \ldots, \psi_m, \neg \chi$ iff $\phi_1, \ldots, \phi_n, \chi \vDash \psi_1, \ldots, \psi_m$.

I.1: ELEMENTARY PREDICATE LOGIC

Suppose now that we construct semantic tableaux as first described above, but using *generalized* sequents instead of sequents. The effect of (6.13) and (6.14) is that we handle $\neg$ *by itself*; as (6.11) shows, our previous tableaux could only tackle $\neg$ two at a time or in combination with another truth-functor.

Using generalized sequents, a proof of (6.1) goes as follows:

(6.15)
$$p \wedge q, \neg(p \wedge r) \models \neg r$$
(i)
$$p \wedge q, \neg(p \wedge r), r \models$$
(ii)
$$p \wedge q, r \models p \wedge r$$
(iii)
$$p, q, r \models p \wedge r$$
(iv)
$$p, q, r \models p \qquad p, q, r \models r$$

Steps (i) and (ii) are by (6.14) and (6.13) respectively. Step (iv) is justified as follows. We are trying to build a structure in which p, q and r are true but $p \wedge r$ is false, as a counterexample to the sequent '$p, q, r \models p \wedge r$'. By the truth-table for $\wedge$, it is necessary and sufficient to build *either* a structure in which p, q, r are true and p is false, *or* a structure in which p, q, r are true and r is false. We can close off under the bottom left sequent '$p, q, r \models p$' because a formula p occurs both on the right and on the left of $\models$, so that in a counterexample it would have to be both false and true, which is impossible. Likewise at bottom right.

Proofs with generalized sequents are virtually identical with the *cut-free sequent proofs* of Gentzen [1934], except that he wrote them upside down. Beth [1955, 1962] used them as a method for testing sequents. He wrote them in a form where, after the first sequent, one only needs to mention the new formulas.

Quine [1950] presents another quite fast decision method which he calls *fell swoop* (to be contrasted with the 'full sweep' of truth-tables).

I turn to the question how fast a decision method of testing sequents can be in the long run, i.e. as the number and lengths of the formulas increase. At the time of writing, this is one of the major unsolved problems of computation theory. A function $p(n)$ of the number n is said to be *polynomial* if it is calculated from n and some other fixed numbers by adding, subtracting and multiplying. (So for example $n^2 + 3$ and $2n^3 - n$ are polynomial functions of n but 3^n, $n!$ and $1/(n^2 + 1)$ are not.) It is not known whether there exist a decision method M for sequents of propositional logic, and a polynomial function $p(n)$, such that for every sequent S, if n is the number of symbols in S then M can determine in less than $p(n)$ steps whether or not S is correct. If the answer is Yes there are such M and $p(n)$, then we say that the decision problem for propositional logic is *solvable in polynomial time*. Cook [1971] showed that a large number of other interesting computational problems will be solvable in polynomial time if this one is. (See Garey and Johnson [1979].) I have the impression that everybody working in the field expects the answer to be No. This would mean in effect that for longer sequents the problem is too hard to be solved efficiently by a deterministic computer.

7. FORMAL PROOF CALCULI

During the first third of this century, a good deal of effort was put into constructing various formal proof calculi for logic. The purpose of this work was to reduce reasoning – or at least a sizeable part of mathematical reasoning – to precise mechanical rules. I should explain at once what a *formal proof calculus* (or *formal system*) is.

A formal proof calculus, call it Σ, is a device for proving sequents in a language L. First, Σ gives us a set of rules for writing down arrays of symbols on a page. An array which is written down according to the rules is called a *formal proof* in Σ. The rules must be such that one can check by inspection and calculation whether or not an array is a formal proof. Second, the calculus contains a rule to tell us how we can mechanically work out what are the *premises* and the *conclusion* of each formal proof.

We write

(7.1) $\phi_1, \ldots, \phi_n \vdash_\Sigma \psi$ or more briefly $\phi_1, \ldots, \phi_n \vdash \psi$

to mean that there is a formal proof in the calculus Σ whose premises all appear in the list $\phi_1, \ldots, \phi_n$, and whose conclusion is ψ. Some other ways of expressing (7.1) are:

> '$\phi_1, \ldots, \phi_n \vdash \psi$' is a *derivable sequent* of Σ;
> ψ is *deducible from* $\phi_1, \ldots, \phi_n$ in Σ;
> $\phi_1, \ldots, \phi_n$ *yield* ψ in Σ.

We call ψ a *derivable formula* of Σ if there is a formal proof in Σ with conclusion ψ and no premises. The symbol $\vdash$ is called *turnstile* or *syntactic turnstile*.

We say that the calculus Σ is:

> *sound* if $\phi_1, \ldots, \phi_n \vdash_\Sigma \psi$ implies $\phi_1, \ldots, \phi_n \models \psi$
> *strongly complete* if $\phi_1, \ldots, \phi_n \models \psi$ implies $\phi_1, \ldots, \phi_n \vdash_\Sigma \psi$,
> *weakly complete* if $\models \psi$ implies $\vdash_\Sigma \psi$,

where $\phi_1, \ldots, \phi_n, \psi$ range over the formulas of L. These definitions also make sense when $\models$ is defined in terms of other logics, not necessarily first-order. In this chapter, 'complete' will always mean 'strongly complete'.

The formal proofs in a calculus Σ are in general meaningless arrays of symbols. They need not be genuine proofs, that is, demonstrations that something is the case. But if we know that Σ is sound, then the fact that a certain sequent is derivable in Σ will prove that the corresponding sequent with $\models$ is correct. In some proof calculi the formal proofs are made to look as much as possible like intuitively correct reasoning, so that soundness can be checked easily.

We already have the makings of one formal proof calculus in Section 6 above: the *cut-free sequent proofs* using generalized sequents. As proofs, these are usually written the other way up, with $\vdash$ in place of $\models$, and with horizontal lines separating the sequents. Also there is no need to put in the lines which mark the branches that are closed, because every branch is closed.

For example, here is a cut-free sequent proof of the sequent '$p \wedge q, \neg(p \wedge r) \vdash \neg r$'; compare it with (6.15):

(7.2)
$$\frac{p \vdash p}{p, q, r \vdash p} \qquad \frac{r \vdash r}{p, q, r \vdash r}$$
$$\frac{p, q, r \vdash p \wedge r}{p \wedge q, r \vdash p \wedge r}$$
$$\frac{p \wedge q, \neg(p \wedge r), r \vdash}{p \wedge q, \neg(p \wedge r) \vdash \neg r}$$

To justify this proof we would show, working upwards from the bottom, that if there is a counterexample to the bottom sequent then at least one of the top sequents has a counterexample, which is impossible. Or equivalently, we could start by noting that the top sequents are correct, and then work *down* the tree, showing that each of the sequents must also be correct. By this kind of argument we can show that the cut-free sequent calculus is sound.

To prove that the calculus is complete, we borrow another argument from Section 6 above. Assuming that a sequent S is not derivable, we have to prove that it is not correct. To do this, we try to construct a cut-free sequent proof, working upwards from S. After a finite number of steps we shall have broken down the formulas as much as possible, but the resulting diagram can't be a proof of S because we assumed there isn't one. So at least one branch must still be 'open' in the sense that it hasn't revealed any immediate contradiction. Let B be such a branch. Let B_L be the set of all formulas which occur to the left of $\vdash$ in some generalized sequent in B, and let B_R be the same with 'right' for 'left'. We can define a structure $\mathfrak{A}$ by

(7.3) $\quad I_{\mathfrak{A}}(\phi) = \begin{cases} T & \text{if } \phi \text{ is a sentence letter which is in } B_L, \\ F & \text{if } \phi \text{ is a sentence letter not in } B_L. \end{cases}$

Then we can prove, by induction on the complexity of the formula ψ, that if ψ is any formula in B_L then $\mathfrak{A} \models \psi$, and if ψ is any formula in B_R then $\mathfrak{A} \models \neg \psi$. It follows that $\mathfrak{A}$ is a counterexample to the bottom sequent S, so that S is not correct.

The cut-free sequent calculus itself consists of a set of mechanical rules for constructing proofs, and it could be operated by somebody who had not the least idea what $\vdash$ or any of the other symbols mean. These rules are listed in Sundholm [Chapter I.2 below].

Gentzen [1934] had another formal proof calculus, known simply as the *sequent calculus*. This was the same as the cut-free sequent calculus, except that it allowed a further rule called the *cut rule* (because it cuts out a formula):

(7.4) $\quad \dfrac{\ldots \vdash {***}, \chi \qquad \ldots, \chi \vdash {***}}{\ldots \vdash {***}}$

This rule often permits much shorter proofs. Gentzen justified it by showing that any proof which uses the cut rule can be converted into a cut-free proof of the same sequent. This *cut elimination theorem* is easily the best mathematical theorem about proofs. Gentzen himself adapted it to give a proof of the consistency of first-order Peano arithmetic. By analyzing Gentzen's argument we can get sharp information about the degree to which different parts of mathematics rely on infinite sets. (Cf. Schütte [1960]. Gentzen's results on cut-elimination were closely related to deep but undigested work on quantifier logic which Jacques Herbrand had done before his death in a mountaineering accident at the age of 23; see Herbrand [1930] and the Introduction to Herbrand [1971].) Further details of Gentzen's sequent calculi, including the intuitionistic versions, are given in Kleene [1952] Chapter XV and Sundholm [Chapter I.2 below].

In the same paper [1934], Gentzen described yet a third formal proof calculus. This is known as the *natural deduction calculus* because proofs in this calculus start with their premises and finish at their conclusions (unlike sequent calculi and semantic tableaux), and all the steps between are intuitively natural (unlike the Hilbert-style calculi to be described below).

A proof in the natural deduction calculus is a tree of formulas, with a single formula at the bottom. The formulas at the tops of the branches are called the *assumptions* of the proof. Some of the assumptions may be *discharged* or *cancelled* by having square brackets [ ] written around them. The *premises* of the proof are its uncancelled assumptions, and the *conclusion* of the proof is the formula at the bottom.

Sundholm [Chapter I.2 below] gives the full rules of the natural deduction calculus. Here are a few illustrations. Leaving aside $\neg$ and $\bot$ for the moment, there are two rules for each truth-functor, namely an *introduction* rule and an *elimination* rule. The introduction rule for $\wedge$ is:

$$(7.6) \quad \frac{\phi \qquad \psi}{\phi \wedge \psi}$$

i.e. from ϕ and ψ deduce $\phi \wedge \psi$. The elimination rule for $\wedge$ comes in a left-hand version and a right-hand version:

$$(7.7) \quad \frac{\phi \wedge \psi}{\phi} \qquad \frac{\phi \wedge \psi}{\psi}.$$

The introduction rule for $\rightarrow$ says that if we have a proof of ψ from certain assumptions, then we can deduce $\phi \rightarrow \psi$ from those assumptions less ϕ:

$$(7.8) \quad \begin{array}{c} [\phi] \\ \vdots \\ \dfrac{\psi}{\phi \rightarrow \psi} \end{array}$$

The elimination rule for $\rightarrow$ is the *modus ponens* of the medievals:

$$(7.9) \quad \frac{\phi \qquad \phi \rightarrow \psi}{\psi}.$$

For example, to prove

(7.10) $\quad q, p \wedge q \to r \models p \to r$

in the natural deduction calculus we write:

(7.11) $\quad \dfrac{\dfrac{[p] \qquad q}{p \wedge q} \qquad p \wedge q \to r}{\dfrac{r}{p \to r}}$

Note that the assumption p is discharged at the last step when $p \to r$ is introduced.

The calculus reads $\neg \phi$ as a shorthand for $\phi \to \bot$. So for example, from ϕ and $\neg \phi$ we deduce $\bot$ by (7.9). There is an elimination rule for $\bot$. It says: given a proof of $\bot$ from certain assumptions, derive ϕ from the same assumptions less $\phi \to \bot$:

(7.12) $\quad \begin{array}{c} [\phi \to \bot] \\ \vdots \\ \dfrac{\bot}{\phi} \end{array}$

This is a form of *reductio ad absurdum*.

The rule about canceling assumptions in (7.8) should be understood as follows. When we make the deduction, we are *allowed* to cancel ϕ wherever it occurs as an assumption. But we are not obliged to; we can cancel some occurrences of ϕ and not others, or we can leave it completely uncanceled. The formula ϕ may not occur as an assumption anyway, in which case we can forget about canceling it. The same applies to $\phi \to \bot$ in (7.12). So (7.12) implies the following weaker rule in which we make no cancellations:

(7.13) $\quad \dfrac{\bot}{\phi}$

('Anything follows from a contradiction'.) Intuitionist logic accepts (7.13) but rejects the stronger rule (7.12) (cf. van Dalen [Chapter III.4]).

Belnap [1962] and Prawitz [1965] have explained the idea behind the natural deduction calculus in an interesting way. For each truth-functor the rules are of two sorts, the introduction rules and the elimination rules. In every case the elimination rules *only allow us to infer from a formula what we had to know in order to introduce the formula*. For example we can remove $\phi \rightarrow \psi$ only by rule (7.9), i.e. by using it to deduce ψ from ϕ; but $\phi \rightarrow \psi$ can only be introduced either as an explicit assumption or (by (7.8)) when we already know that ψ can be deduced from ϕ. (Rule (7.12) is in a special category. It expresses (1) that everything is deducible from $\bot$, and (2) that for each formula ϕ, at least one of ϕ and $\phi \rightarrow \bot$ is true.)

Popper ([1946/7], particularly p. 284) rashly claimed that he could define truth-functors just by writing down natural deduction rules for them. Prior [1960] gave a neat example to show that this led to absurdities. He invented the new truth-functor *tonk*, which is defined by the rules

(7.14)
$$\frac{\phi}{\phi \text{ tonk } \psi} \qquad \frac{\phi \text{ tonk } \psi}{\psi}$$

and then proceeded to infer everything from anything. Belnap [1962] points out that Prior's example works because its introduction and elimination rules fail to match up in the way described above. Popper should at least have imposed a requirement that the rules must match up. (Cf. Prawitz [1979], Tennant [1978], p. 74ff, Sundholm [Chapter III.8 below].)

Natural deduction calculi, all of them variants of Gentzen's, are given by Anderson and Johnstone [1962], Fitch [1952], Kalish and Montague [1964], Lemmon [1965], Prawitz [1965], Quine [1950], Suppes [1957], Tennant [1978], Thomason [1970] and van Dalen [1980]. Fitch (followed e.g. by Thomason) makes the trees branch to the right. Some versions (e.g. Quine's) disguise the pattern by writing the formulas in a vertical column. So they have to supply some other way of marking which formulas depend on which assumptions; different versions do this in different ways.

Just as a semantic tableau with its branches closed is at heart the same thing as a cut-free sequent proof written upside down, Prawitz [1965] has shown that after removing redundant steps, a natural deduction proof is really the same thing as a cut-free sequent proof written sideways. The relationship becomes clearer if we adapt the natural deduction calculus so as to allow a proof to have several alternative conclusions, just as it has several premises. Details of such calculi have been worked out by Kneale [1956] and more fully by Shoesmith and Smiley [1978].

A proof of $p \vee \neg p$ in Gentzen's natural deduction calculus takes several lines. This is a pity, because formulas of form $\phi \vee \neg \phi$ are useful halfway steps in proofs of other formulas. So some versions of natural deduction allow us to quote a few tautologies such as $\phi \vee \neg \phi$ whenever we need them in a proof. These tautologies are then called *axioms*. Technically they are formulas deduced from no assumptions, so we draw a line across the top of them, as at top right in (7.15) below.

If we wanted to undermine the whole idea of natural deduction proofs, we could introduce axioms which replace all the natural deduction rules except modus ponens. For example we can put (7.6) out of a job by using the axiom $\phi \to (\psi \to \phi \wedge \psi)$. Whenever Gentzen used (7.6) in a proof, we can replace it by

(7.15)
$$\frac{\psi \qquad \dfrac{\phi \qquad \overline{\phi \to (\psi \to \phi \wedge \psi)}}{\psi \to \phi \wedge \psi}}{\phi \wedge \psi}$$

using (7.9) twice. Likewise (7.7) become redundant if we use the axioms $\phi \wedge \psi \to \phi$ and $\phi \wedge \psi \to \psi$. Rule (7.8) is a little harder to dislodge, but it can be done, using the axioms $\phi \to (\psi \to \phi)$ and $(\phi \to \psi) \to ((\phi \to (\psi \to \chi)) \to (\phi \to \chi))$.

At the end of these manipulations we have what is called a *Hilbert-style* proof calculus. A Hilbert-style calculus consists of a set of formulas called *axioms*, together with one or two *derivation rules* for getting new formulas out of given ones. To prove $\phi_1, \ldots, \phi_n \models \psi$ in such a calculus, we apply the derivation rules as many times as we like to $\phi_1, \ldots, \phi_n$ and the axioms, until they give us ψ.

One Hilbert-style system is described in Appendix A below. Mates [1965] works out another such system in detail. Hilbert-style calculi for propositional logic were given by Frege [1879, 1893], Peirce [1885], Hilbert [1923] and Łukasiewicz (see Łukasiewicz and Tarski [1930]). (Cf. Sundholm [Chapter I.2, below].)

The typical Hilbert-style calculus is inefficient and barbarously unintuitive. But they do have two merits. The first is that their mechanics are usually very simple to describe – many Hilbert-style calculi for propositional logic have only one derivation rule, namely modus ponens. This makes them suitable for encoding into arithmetic (Section 24 below). The second merit is that we can strengthen or weaken them quite straightforwardly by tampering with the axioms, and this commends them to researchers in non-classical logics.

Soundness for these calculi is usually easy to prove: one shows (a) that the axioms are true in every structure and (b) that the derivation rules never lead from truth to falsehood. One way of proving completeness is to show that every natural deduction proof can be converted into a Hilbert-style proof of the same sequent, as hinted above. (Kleene [1952] Section 77 shows how to convert sequent proofs into Hilbert-style proofs and *vice versa*; see Sundholm [Chapter I.2, below].)

Alternatively we can prove their completeness directly, using maximal consistent sets. Since this is a very un-proof-theoretic approach, and this section is already too long, let me promise to come back to the matter at the end of Section 16 below. (Kalmár [1934/5] and Kleene independently found a neat proof of the weak completeness of Hilbert-style calculi, by converting a truth-table into a formal proof; cf. Kleene [1952] p. 132ff or Mendelson [1964] p. 36.)

II: PREDICATE LOGIC

8. BETWEEN PROPOSITIONAL LOGIC AND PREDICATE LOGIC

If we asked a Proof Theorist to explain what it means to say

(8.1) $\phi_1, \ldots, \phi_n$ logically imply ψ,

where $\phi_1, \ldots, \phi_n$ and ψ are formulas from propositional logic, he would explain that it means this: there is a proof of ψ from $\phi_1, \ldots, \phi_n$ in one of the standard proof calculi. A Model Theorist would prefer to use the definition we gave in Section 4 above, and say that (8.1) means: whenever $\phi_1, \ldots, \phi_n$ are true in a structure, then ψ is true in that structure too. The Traditional Logician for his part would explain it thus: every argument of the form '$\phi_1, \ldots, \phi_n$. *Therefore* ψ' is valid. There need be no fight between these three honest scholars, because it is elementary to show that (8.1) is true under any one of these definitions if and only if it is true under any other.

In the next few sections we shall turn from propositional logic to predicate logic, and the correct interpretation of (8.1) will become more contentious.

When $\phi_1, \ldots, \phi_n$ and ψ are sentences from predicate logic, the Proof Theorist has a definition of (8.1) which is a straightforward extension of his definition for propositional logic, so he at any rate is happy.

But the Traditional Logician will be in difficulties, because the quantifier expressions of predicate logic have a quite different grammar from all locutions of normal English; so he is hard put to say what would count as an argument of form '$\phi_1, \ldots, \phi_n$. *Therefore* ψ'. He will be tempted to say that really we should look at sentences whose deep structures (which he may call logical forms) are like the formulas $\phi_1, \ldots, \phi_n, \psi$. This may satisfy him, but it will hardly impress people who know that in the present state of the linguistic art one can find experts to mount convincing arguments for any one of seventeen deep structures for a single sentence. A more objective but admittedly vague option would be for him to say that (8.1) means that any argument which can be *paraphrased* into this form, using the apparatus of first-order logic, is valid.

But the man in the worst trouble in the Model Theorist. On the surface all is well – he has a good notion of 'structure', which he took over from the algebraists, and he can say just what it means for a formula of predicate logic to be 'true in' a structure. So he can say, just as he did for propositional logic, that (8.1) means that whenever $\phi_1, \ldots, \phi_n$ are true in a structure, then ψ is true in that structure too. His problems start as soon as he asks himself what a structure really is, and how he knows that they exist.

Structures, as they are presented in any textbook of model theory, are abstract set-theoretic objects. There are uncountably many of them and most of them are infinite. They can't be inspected on a page (like proofs in a formal calculus) or heard at Hyde Park Corner (like valid arguments). True, several writers have claimed that the only structures which exist are those which somebody constructs. (E.g. Putnam [1980] p. 482: "Models are ... constructions within our theory itself, and they have names from birth.") Unfortunately this claim is in flat contradiction to about half the major theorems of model theory (such as the Upward Löwenheim–Skolem Theorem, Theorem 17.5 in Section 17 below).

Anybody who wants to share in present-day model theory has to accept that structures are as disparate and intangible as sets are. One must handle them by set-theoretic principles and not by explicit calculation. Many model theorists have wider horizons even than this. They regard the whole universe V of sets as a structure, and they claim that first-order formulas in the language of set theory are true or false in this structure by just the same criteria as in smaller structures. The axioms of Zermelo–Fraenkel set theory, they claim, are simply true in V.

It is actually a theorem of set theory that a notion of truth adequate to cope with the whole universe of sets *cannot be formalized within set theory*. (We prove this in Section 24 below.) So a model theorist with this wider horizon is strictly not entitled to use formal set-theoretic principles either, and he is forced back onto his intuitive understanding of words like 'true', 'and', 'there is' and so forth. In mathematical practice this causes no problems whatever. The problems arise when one tries to justify what the mathematicians are doing.

In any event it is a major exercise to show that these three interpretations of (8.1) in predicate logic – or four if we allow the Model Theorist his wider and narrower options – agree with each other. But logicians pride themselves that it can be done. Section 17 will show how.

9. QUANTIFIERS

First-order predicate logic comes from propositional logic by adding the words 'every' and 'some'.

Let me open with some remarks about the meaning of the word 'every'. There is no space here to rebut rival views. (Cf. Leblanc [Chapter I.3, below]; on substitutional quantification see Dunn and Belnap [1968], Kripke [1976], Stevenson [1973].) But anybody who puts a significantly different interpretation on 'every' from the one presented below will have to see first-order logic in a different light too.

A person who understands the words 'every', 'Pole', the sentence

(9.1)   Richard is a Catholic.

and the principles of English sentence construction must also understand the sentence

(9.2)   Every Pole is a Catholic.

How?

First, (9.1) is true if and only if Richard satisfies a certain condition, namely that

(9.3)   He is a Catholic.

I underline the pronoun that stands for whatever does or does not satisfy the condition. Note that the condition expressed by (9.3) is one which people either satisfy or do not satisfy, regardless of how or whether we can identify them. Understanding the condition is a necessary part of understanding (9.1). In Michael Dummett's words ([1973] p. 517):

...given that we understand a sentence from which a predicate has been formed by omission of certain occurrences of a name, we are capable of recognizing what concept that predicate stands for in the sense of knowing what it is for it to be true of or false of any arbitrary object, whether or not the language contains a name for that object.

Second, the truth or otherwise of (9.2) in a situation depends on what class of Poles is on the agenda. Maybe only Poles at this end of town are under discussion, maybe Poles anywhere in the world; maybe only Poles alive now, maybe Poles for the last hundred years or so. Possibly the speaker was a little vague about which Poles he meant to include. I count the specification of the relevant class of Poles as part of the situation in which (9.2) has a truth-value. This class of Poles is called the *domain of quantification* for the phrase 'every Pole' in (9.2). The word 'Pole' is called the *restriction term*, because it restricts us to Poles; any further restrictions on the domain of quantification are called *contextual restrictions*.

So when (9.2) is used in a context, the word 'Pole' contributes a domain of quantification and the words 'is a Catholic' contribute a condition. The contribution of the word 'Every' is as follows: *In any situation, (9.2) is true iff every individual in the domain of quantification satisfies the condition.*

This analysis applies equally well to other simple sentences containing 'Every', such as:

(9.4)   She ate every flower in the garden.

For (9.4), the situation must determine what the garden is, and hence what is the class of flowers that were in the garden. This class is the domain of quantification; 'flower in the garden' is the restriction term. The sentence

(9.5)   She ate it.

expresses a condition which things do or do not satisfy, once the situation has determined who 'she' refers to. So in this example the condition varies with the situation. The passage from condition and domain of quantification to truth-value is exactly as before.

The analysis of

(9.6)   Some Pole is a Catholic

(9.7)   She ate some flower (that was) in the garden,

is the same as that of (9.2), (9.4) respectively, except at the last step. For (9.6) or (9.7) to be true we require that *at least one individual in the domain of quantification satisfies the condition.*

In the light of these analyses we can introduce some notation from first-order logic. In place of the underlined pronoun in (9.3) and (9.5) we shall use an *individual variable*, i.e. (usually) a lower-case letter from near the end of the alphabet, possibly with a subscript. Thus:

(9.8) x is a Catholic.

Generalizing (9.8), we use the phrase *1-place predicate* to mean a string consisting of words and one individual variable (which may be repeated), such that if the variable is understood as a pronoun referring to a certain person or object, then the string becomes a sentence which expresses that the person or object referred to satisfies a certain condition. The condition may depend on the situation into which the sentence is put.

For an example in which a variable occurs twice,

(9.9) x handed the melon to Schmidt, who gave it back to x.

is a 1-place predicate. It expresses the condition which Braun satisfies if and only if Braun handed the melon to Schmidt and Schmidt gave it back to Braun.

To return to (9.2), 'Every Pole is a Catholic': we have now analyzed this sentence into (a) a *quantifier word* 'Every', (b) the restriction term 'Pole', and (c) the predicate 'x is a Catholic'.

The separating out of the predicate (by Frege [1879], see also Mitchell [1883]) was vital for the development of modern logic. Predicates have the grammatical form of sentences, so that they can be combined by truth-functors. For example

(9.10)   (x is a Catholic $\wedge$ x is a philatelist)

is a predicate which is got by conjoining two other predicates with $\wedge$. It expresses the condition which a person satisfies if he is both a Catholic and a philatelist. Incidentally I have seen it suggested that the symbol $\wedge$ must have a different meaning in (9.10) from its meaning in propositional logic, because in (9.10) it stands between predicates which do not have truth-values. The answer is that predicates *do* gain truth-values when their variables are either replaced by or interpreted as names. The truth-value gained in this way by the compound predicate (9.10) is related to the truth-values gained by its two conjuncts in exactly the way the truth-table for $\wedge$ describes.

(A historical aside: Peirce [1885] points out that by separating off the predicate we can combine quantifiers with propositional logic; he says that all attempts to do this were 'more or less complete failures until Mr Mitchell showed how it was to be effected'. Mitchell published in a volume of essays by students of Peirce at Johns Hopkins (Members [1883]). Christine Ladd's paper in the same volume mentions both Frege's *Begriffschrift* [1879] and Schröder's review of it. It is abundantly clear that nobody in Peirce's group had read either. The same happens today.)

The account of quantifiers given above agrees with what Frege said in his *Funktion und Begriff* [1891] and *Grundgesetze* [1893], except in one point. Frege required that all conditions on possible values of the variable should be stated in the predicate. In other words, he allowed only one domain of quantification, namely absolutely everything. For example, if someone were to say, à propos of Poles in New York, 'Every Pole is a Catholic', Frege would take this to mean that absolutely everything satisfies the condition

(9.11) If x is a Pole in New York City then x is a Catholic.

If a person were to say

(9.12) Somebody has stolen my lipstick.

Frege's first move would be to interpret this as saying that at least one thing satisfies the condition expressed by

(9.13) x is a person and x has stolen my lipstick.

Thus Frege removed the restriction term, barred all contextual restrictions, and hence trivialized the domain of quantification.

There are two obvious advantages in getting rid of the restriction term: we have fewer separate expressions to deal with, and everything is thrown into the predicate where it can be analyzed by way of truth-functors.

However, it is often useful to keep the restriction terms, if only because it makes formulas easier to read. (There are solid technical dividends too, see Feferman [1968a, 1974].) Most logicians who do this follow the advice of Peirce [1885] and use a special style of variable to indicate the restriction. For example set theorists use Greek variables when the restriction is to ordinals. Variables that indicate a special restriction are said to be *sorted* or *sortal*. Two variables marked with the same restriction are said to be *of the same sort*. Logics which use this device are said to be *many-sorted*.

One can also go halfway with Frege and convert the restriction term into another predicate. In this style, 'Every Pole is a Catholic' comes out as a combination of three units: the quantifier word 'Every', the predicate 'x is a Catholic', and a second *relativisation predicate* 'x is a Pole'. The mathematical literature is full of *ad hoc* examples of this approach. See for example the bounded quantifiers of number theory in Section 24 below.

When people started to look seriously at other quantifier words besides 'every' and 'some', it became clear that Frege's method of eliminating the restriction term won't always work. For example, the sentence 'Most judges are freemasons' can't be understood as saying that most things satisfy a certain condition. (For a proof of this, and many other examples, see the study of natural language quantifiers by Barwise and Cooper [1981].) For this reason Neil Tennant (Altham and Tennant [1975]) and Barwise [1974] proposed very general formalisms which keep the relativization predicate separate from the main predicate.

Frege also avoided contextual restrictions. Given his aim, which was to make everything in mathematical reasoning fully explicit, this might seem natural. But it was a bad move. Contextual restrictions do occur, and a logician ought to be prepared to operate with them. In any case various writers have raised philosophical objections to Frege's habit of talking about just everything. Do we really have an undefinable notion of 'object', as Frege supposed? Is it determinate what objects there are? Don't we falsify the meanings of English sentences if we suppose that they state something about everything there is, when on the face of it they are only about Poles?

For a historical study of quantifiers in first-order logic, consult Goldfarb [1979].

10. SATISFACTION

As a convenient and well-known shorthand, we shall say that a person or thing *satisfies* the 1-place predicate ϕ if he or it satisfies the condition which the predicate ϕ expresses. (Notice that we are now allowing the metavariables 'ϕ', 'ψ' etc. to range over predicates as well as sentences and formulas. This shouldn't cause any confusion.)

Many writers put it a little differently. They say that a person or thing satisfies ϕ if the result of putting a name of the person or thing in place of every occurrence of the variable in ϕ is a true sentence. This way of phrasing matters is fine as a first approximation, but it runs into two hazards.

The first hazard is that not everything has a name, even if we allow phrases of the form 'the such-and-such' as names. For example there are uncountably many real numbers and only countably many names.

I can dispose of this objection quickly, as follows. I decree that for purposes of naming arbitrary objects, any ordered pair whose first term is an object and whose second term is the Ayatollah Khalkhali shall be a name of that object. There is a problem about using these names in sentences, but that's just a matter of finding an appropriate convention. So it is clear that if we have an abstract enough notion of what a name is, then every object can have a name.

More conscientious authors have tried to mount reasoned arguments to show that everything is in principle nameable. The results are not always a success. In one paper I recall, the author was apparently under the impression that the nub of the problem was to find a symbol that could be used for naming hitherto nameless objects. After quoting quite a lot of formulas from Quine's *Methods of Logic*, he eventually announced that lower-case italic *w* can always be used for the purpose. No doubt it can!

There is a second hazard in the 'inserted name' definition of satisfaction. If we allow phrases of the form 'the such-and-such' to count as names, it can happen that on the natural reading, a name means something different within the context of the sentence from what it means in isolation. For example, if my uncle is the mayor of Pinner, and in 1954 he fainted during the opening ceremony of the Pinner Fair, then the mayor of Pinner satisfies the predicate:

(10.1) In 1954 x fainted during the opening ceremony of the Pinner Fair.

But on the natural reading the sentence

(10.2) In 1954 the mayor of Pinner fainted during the opening ceremony of the Pinner Fair.

says something quite different and is probably false. One can avoid this phenomenon by sticking to names like 'the present mayor of Pinner' which automatically extract themselves from the scope of surrounding temporal operators (cf. Kamp [1971]). But other examples are less easily sorted out. If the programme note says simply 'Peter Warlock wrote this song', then Philip Heseltine, one of whose pen-names was 'Peter Warlock', surely satisfies the predicate

(10.3) The programme note attributes this song to x.

But my feeling is that on the natural reading, the sentence

(10.4) The programme note attributes this song to Philip Heseltine

is false. Examples like these should warn us to be careful in applying first-order formalisms to English discourse. (Cf. Bäuerle and Cresswell [Volume IV].)

I turn to some more technical points. We shall need to handle expressions like

(10.5) x was observed handing a marked envelope to y

which express a condition on *pairs* of people or things. It is, I think, quite obvious how to generalize the notion of a 1-place predicate to that of an *n-place predicate*, where *n* counts the number of distinct individual variables that stand in place of proper names. (Predicates with any positive number of places are also called *open sentences*.) Expression (10.5) is clearly a 2-place predicate. The only problem is to devise a convention for steering the right objects to the right variables. We do it as follows.

By the *free variables* of a predicate, we mean the individual variables which occur in proper name places in the predicate; so an *n*-place predicate has *n* free variables. (In Section 11 we shall have to revise this definition and exclude certain variables from being free.) We define an *assignment* g to a set of variables (in a situation) to be a function whose domain is that set of variables, with the stipulation that if x is a sorted variable then (in that situation) $g(x)$ meets the restriction which goes with the variable. So for example $g(y_{\text{raccoon}})$ has to be a raccoon.

We say that an assignment g is *suitable for* a predicate ϕ if every free variable of ϕ is in the domain of g. Using the inserted name definition of satisfaction as a temporary expedient, we define: if ϕ is a predicate and g is an assignment which is suitable for ϕ, then g *satisfies* ϕ (in a given situation) iff a true sentence results (in that situation) when we replace each variable x in ϕ by a name of the object $g(x)$.

We shall write

(10.6) α/x, β/y, γ/z, ...

to name the assignment g such that $g(x) = \alpha$, $g(y) = \beta$, $g(z) = \gamma$ etc. If $\mathfrak{A}$ is a situation, ϕ a predicate and g an assignment suitable for ϕ, then we write

(10.7) $\mathfrak{A} \models \phi[g]$

to mean that g satisfies ϕ in the situation $\mathfrak{A}$. The notation (10.7) is basic for all that follows, so let me give some examples. For simplicity I take $\mathfrak{A}$ to be the real world here and now. The following are true:

(10.8) $\mathfrak{A} \models$ In the year y, x was appointed Assistant Professor of Mathematics at w at the age of 19 years. [Dr Harvey Friedman/x, 1967/y, Stanford University California/w].

Example (10.8) asserts that in 1967 Dr Harvey Friedman was appointed Assistant Professor of Mathematics at Stanford University California at the age of 19 years; which must be true because the *Guinness Book of Records* says so.

(10.9) $\mathfrak{A} \models v$ is the smallest number which can be expressed in two different ways as the sum of two squares. [65/v].

(10.10) $\mathfrak{A} \models x$ wrote poems about the physical anatomy of x. [Walt Whitman/x].

This notation connects predicates with *objects*, not with names of objects. In (10.10) it is Mr Whitman himself who satisfies the predicate shown.

In the literature a slightly different and less formal convention is often used. The first time that a predicate ϕ is mentioned, it is referred to, say, as $\phi(y, t)$. This means that ϕ has at most the free variables y and t, and that these variables are to be considered *in that order*. To illustrate, let $\phi(w, x, y)$ be the predicate

(10.11) In the year y, x was appointed Assistant Professor of Mathematics at w at the age of 19 years.

Then (10.8) will be written simply as

(10.12) $\mathfrak{A} \models \phi$ [Stanford University California, Dr Harvey Friedman, 1967].

This handy convention can save us having to mention the variables again after the first time that a predicate is introduced.

There is another variant of (10.7) which is often used in the study of logics. Suppose that in situation $\mathfrak{A}$, g is an assignment which is suitable for the predicate ϕ, and S is a sentence which is got from ϕ by replacing each free variable x in ϕ by a name of $g(x)$. Then the truth-value of S is determined by $\mathfrak{A}, g$ and ϕ, and it can be written

(10.13) $g^*_\mathfrak{A}(\phi)$ or $\|\phi\|_{\mathfrak{A}, g}$.

So we have

(10.14) $\mathfrak{A} \models \phi[g]$ iff $g^*_\mathfrak{A}(\phi) = T$.

In (10.13), $g^*_\mathfrak{A}$ can be thought of as a function taking predicates to truth-values. Sometimes it is abbreviated to $g_\mathfrak{A}$ or even g, where this leads to no ambiguity.

11. QUANTIFIER NOTATION

Let us use the symbols $x_\text{boy}, y_\text{boy}$ etc. as sorted variables which are restricted to boys. We shall read the two sentences

(11.1) $\forall x_\text{boy} \, (x_\text{boy}$ has remembered to bring his woggle).

(11.2) $\exists x_\text{boy} \, (x_\text{boy}$ has remembered to bring his woggle).

I.1: ELEMENTARY PREDICATE LOGIC

as meaning exactly the same as (11.3) and (11.4) respectively:

(11.3) Every boy has remembered to bring his woggle.

(11.4) Some boy has remembered to bring his woggle.

In other words, (11.1) is true in a situation if and only if in that situation, every member of the domain of quantification of $\forall x_{boy}$ satisfies the predicate

(11.5) x_{boy} has remembered to bring his woggle.

Likewise (11.2) is true if and only if some member of the domain of quantification of $\exists x_{boy}$ satisfies (11.5). The situation has to determine what the domain of quantification is, i.e. what boys are being talked about.

The expression $\forall x_{boy}$ is called a *universal quantifier* and the expression $\exists x_{boy}$ is called an *existential quantifier*. Because of the restriction 'boy' on the variable, they are called *sorted* or *sortal* quantifiers. The symbols $\forall$, $\exists$ are called respectively the *universal* and *existential quantifier symbols*; $\forall$ is read 'for all', $\exists$ is read 'for some' or 'there is'.

For unsorted quantifiers using plain variables x, y, z etc., similar definitions apply, but now the domain of quantification for such a quantifier can be any class of things. Most uses of unsorted quantifiers are so remote from anything in ordinary language that we can't rely on the conventions of speech to locate a domain of quantification for us. So instead we have to assume that *each situation specifies a class which is to serve as the domain of quantification for all unsorted quantifiers.* Then

(11.6) $\forall x$ (if x is a boy then x has remembered to bring his woggle).

counts as true in a situation if and only if in that situation, every object in the domain of quantification satisfies the predicate

(11.7) if x is a boy then x has remembered to bring his woggle.

There is a corresponding criterion for the truth of a sentence starting with the unsorted existential quantifier $\exists x$; the reader can easily supply it.

The occurrences of the variable x_{boy} in (11.1) and (11.2), and of x in (11.6), are no longer doing duty for pronouns or marking places where names can be inserted. They are simply part of the quantifier notation. We express this by saying that these occurrences are *bound in* the respective sentences. We also say, for example, that the quantifier at the beginning of (11.1) *binds* the two occurrences of x_{boy} in that sentence. By contrast an occurrence of a variable in a predicate is called *free in* the predicate if it serves the role we discussed in Sections 9 and 10, of referring to whoever or whatever the predicate expresses a condition on. What we called the *free variables* of a predicate in Section 10 are simply those variables which have free occurrences in the predicate. Note that the concepts 'free' and 'bound' are relative: the occurrence of x_{boy} before 'has' in (11.1) is bound in (11.1) but free in (11.5). Consider also the predicate

(11.8) x_{boy} forgot his whistle, but $\forall x_{boy}$ (x_{boy} has remembered to bring his woggle).

Predicate (11.8) expresses the condition which Billy satisfies if Billy forgot his whistle but every boy has remembered to bring his woggle. So the first occurrence of x_{boy} in (11.8) is free in (11.8) but the other two occurrences are bound in (11.8).

I should recall here the well-known fact that in natural languages, a pronoun can be linked to a quantifier phrase that occurs much earlier, even in a different sentence:

(11.9)   HE: This evening I heard a nightingale in the pear tree.
SHE: It was a thrush – we don't get nightingales here.

In our notation this can't happen. *Our quantifiers bind only variables in themselves and the clause immediately following them.* We express this by saying that the *scope* of an occurrence of a quantifier consists of the quantifier itself and the clause immediately following it; a quantifier occurrence $\forall x$ or $\exists x$ binds all and only occurrences of the same variable x which lie within its scope.

It is worth digressing for a moment to ask why (11.9) makes life hard for logicians. The crucial question is: just when is the woman's remark 'It was a thrush' a true statement? We want to say that it's true if and only if the object referred to by 'It' is a thrush. But what is there for 'It' to refer to? Arguably the man hasn't referred to any nightingale, he has merely said that there was at least one that he heard in the pear tree. Also we want to say that if her remark is true, then it follows that he heard a thrush in the pear tree. But if this follows, why doesn't it also follow that the nightingale in the pear tree was a thrush? (which is absurd.) (There is a large literature on the problems of anaphora. See for example Chastain [1975], Partee [1978], Evans [1980], Kamp [1981].)

Returning to first-order logic, consider the sentence

(11.10) $\exists x_{boy}$ (x_{boy} kissed Brenda).

I.1: ELEMENTARY PREDICATE LOGIC

This sentence can be turned into a predicate by putting a variable in place of 'Brenda'. Naturally the variable we use has to be different from x_{boy}, or else it would get bound by the quantifier at the beginning. Apart from that constraint, any variable will do. For instance:

(11.11) $\exists x_{boy} (x_{boy}$ kissed $y_{girl\ with\ pigtails})$.

We need to describe the conditions in which Brenda satisfies (11.11). Brenda must of course be a girl with pigtails. She satisfies (11.11) if and only if there is a boy β such that the assignment

(11.12) β/x_{boy}, Brenda/$y_{girl\ with\ pigtails}$

satisfies the predicate 'x_{boy} kissed $y_{girl\ with\ pigtails}$'. Formal details will follow in Section 14 below.

12. AMBIGUOUS CONSTANTS

In his *Wissenschaftslehre* II ([1837] Section 147) Bernard Bolzano noted that we use demonstrative pronouns at different times and places to refer now to this, now to that. He continued:

> Since we do this anyhow, it is worth the effort to undertake this procedure with full consciousness and with the intention of gaining more precise knowledge about the nature of such propositions by observing their behaviour with respect to truth. Given a proposition, we could merely inquire whether it is true or false. But some very remarkable properties of propositions can be discovered if, in addition, we consider the truth values of all those propositions which can be generated from it, if we take some of its constituent ideas as variable and replace them by any other ideas whatever.

We can abandon to the nineteenth century the notion of 'variable ideas'. What Bolzano did in fact was to introduce *totally ambiguous symbols*. When a writer uses such a symbol, he has to indicate what it means, just as he has to make clear what his demonstrative pronouns refer to. In our terminology, the situation must fix the meanings of such symbols. Each totally ambiguous symbol has a certain grammatical type, and the meaning supplied must fit the grammatical type; but that apart, anything goes.

Let us refer to a sentence which contains totally ambiguous symbols as a *sentence schema*. Then an *argument schema* will consist of a string of sentence schemas called *premises*, followed by the word '*Therefore*', followed by a sentence schema called the *conclusion*. A typical argument schema might be:

(12.1) *a* is more *X* than *b*. *b* is more *X* than *c*. *Therefore a* is more *X* than *c*.

A traditional logician would have said that (12.1) is a valid argument schema if and only if all its instances are valid arguments (cf. (1) in the Introduction above). Bolzano said something different. Following him, we shall say that (12.1) is *Bolzano-valid* if for every situation in which a, b, c are interpreted as names and X is interpreted as an adjective, either one or more of the premises are not true, or the conclusion is true. We say that the premises in (12.1) *Bolzano-entail* the conclusion if (12.1) is Bolzano-valid.

Note the differences. For the traditional logician entailment is from sentences to sentences, not from sentence schemas to sentence schemas. Bolzano's entailment is between schemas, not sentences, and moreover he defines it without mentioning entailment between sentences. The schemas become sentences of a sort when their symbols are interpreted, but Bolzano never asks whether these sentences "can't be true without certain other sentences being true" (to recall our definition of entailment in the Introduction) – he merely asks when they *are* true. The crucial relationship between Bolzano's ideas and the traditional ones is that *every instance of a Bolzano-valid argument schema is a valid argument.* If an argument is an instance of a Bolzano-valid argument schema, then that fact itself is a reason why the premises can't be true without the conclusion also being true, and so the argument is valid.

In first-order logic we follow Bolzano and study entailments between schemas. We use two kinds of totally ambiguous constants. The first kind are the *individual constants*, which are normally chosen from lower-case letters near the beginning of the alphabet: a, b, c etc. These behave grammatically as singular proper names, and are taken to stand for objects. The other kind are the *predicate* (or *relation*) *constants*. These are usually chosen from the letters P, Q, R etc. They behave as verbs or predicates, in the following way. To specify a meaning for the predicate constant P, we could write

(12.2) $Pxyz$ means x aimed at y and hit z.

The choice of variables here is quite arbitrary, so (12.2) says the same as:

(12.3) $Pyst$ means y aimed at s and hit t.

We shall say that under the interpretation (12.2), an ordered 3-tuple $\langle \alpha, \beta, \gamma \rangle$ of objects *satisfies P* if and only if the assignment

(12.4) $\alpha/x, \beta/y, \gamma/z$

satisfies the predicate 'x aimed at y and hit z'. So for example the ordered 3-tuple ⟨Bert, Angelo, Chen⟩ satisfies P under the interpretation (12.2) or

(12.3) if and only if Bert aimed at Angelo and hit Chen. (We take P to be satisfied by ordered 3-tuples rather than by assignments because, unlike a predicate, the symbol P comes without benefit of variables.) The collection of all ordered 3-tuples which satisfy P in a situation where P has the interpretation (12.2) is called the *extension* of P in that situation. In general a collection of ordered n-tuples is called an *n-place relation*.

Since P is followed by three variables in (12.2), we say that P in (12.2) is serving as a *3-place predicate constant*. One can have n-place predicate constants for any positive integer n; the extension of such a constant in a situation is always an n-place relation. In theory a predicate constant could be used both as a 3-place and as a 5-place predicate constant in the same setting without causing mishap, but in practice logicians try to avoid doing this.

Now consider the sentence

(12.5) $\forall x$ (if Rxc then x is red).

with 2-place predicate constant R and individual constant c. What do we need to be told about a situation $\mathfrak{A}$ in order to determine whether (12.5) is true or false in $\mathfrak{A}$? The relevant items in $\mathfrak{A}$ seem to be:

(a) the domain of quantification for $\forall x$.
(b) the object named by the constant c. (Note: it is irrelevant what meaning c has over and above naming this object, because R will be interpreted by a predicate.) We call this object $I_{\mathfrak{A}}(c)$.
(c) the extension of the constant R. (Note: it is irrelevant what predicate is used to give R this extension; the extension contains all relevant information.) We call this extension $I_{\mathfrak{A}}(R)$.
(d) the class of red things.

In Section 14 we shall define the important notion of a *structure* by extracting what is essential from (a)–(d). Logicians normally put into the definition of 'structure' some requirements that are designed to make them simpler to handle. Before matters get buried under symbolism, let me say what these requirements amount to in terms of $\mathfrak{A}$. (See Appendix C below for the set-theoretic notions used.)

(1) There is to be a collection of objects called the *domain* of $\mathfrak{A}$, in symbols $|\mathfrak{A}|$.
(2) $|\mathfrak{A}|$ is the domain of quantification for all unsorted quantifiers. Two sorted quantifiers with variables of the same sort (if there are any)

always have the same domain of quantification, which is included in $|\mathfrak{A}|$.
(3) For every individual constant c, the interpretation $I_\mathfrak{A}(c)$ is a member of $|\mathfrak{A}|$; for every predicate constant R, the relation $I_\mathfrak{A}(R)$ is a relation on $|\mathfrak{A}|$.
(4) Some authors require $|\mathfrak{A}|$ to be a pure set. Most authors require it to have at least one member. A very few authors (e.g. Carnap [1956], Hintikka [1955]) require it to be at most countable.

Requirements (1)–(3) mean in effect that first-order logicians abandon any pretence of following the way that domains of quantification are fixed in natural languages. Frege's device of Section 9 (e.g. (9.11)) shows how we can meet these requirements and still say what we wanted to say, though at greater length. Requirements (4) are an odd bunch; I shall study their reasons and justifications in due course below.

Logicians also allow one important relaxation of (a)–(d). They permit an n-place predicate symbol to be interpreted by *any* n-place relation on the domain, not just one that comes from a predicate. Likewise they permit an individual constant to stand for any member of the domain, regardless of whether we can identify that member. The point is that the question whether *we* can describe the extension or the member is totally irrelevant to the question what is true in the structure.

Note here the 3-way analogy

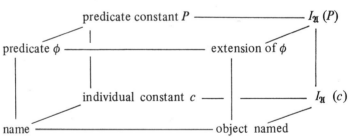

The front face of this cube is essentially due to Frege. Would he have accepted the back?

No, he would not. In 1899 Hilbert published a study of the axioms of geometry. Among other things, he asked questions of the form 'Do axioms A, B, C together entail axiom D?' (The famous problem of the independence of Euclid's parallel postulate is a question of this sort.) Hilbert answered these questions by regarding the axioms as schemas containing ambiguous signs, and then giving number-theoretic interpretations which made the premises A, B and C true but the conclusion D false. Frege read the book (Hilbert [1899]) and reacted angrily. After a brief correspondence with Hilbert (Frege and Hilbert [1899/1900]), he published a detailed critique [1906], declaring (Frege [1971] p. 66): "Indeed, if it were a matter of deceiving oneself and others, there would be no better means than ambiguous signs."

Part of Frege's complaint was that Hilbert had merely shown that certain argument schemas were not Bolzano-valid; he had not shown that axioms A, B and C, taken literally as statements about points, lines etc. in real space, do not entail axiom D taken literally. This is true and need not detain us – Hilbert had answered the questions he wanted to answer. Much more seriously, Frege asserted that Hilbert's propositions, being ambiguous, did not express determinate thoughts and hence could not serve as the premises or conclusions of inferences. In short, Frege refused to consider Bolzano-valid argument schemas as any kind of valid argument. So adamant was he about this that he undertook to translate the core of Hilbert's reasoning into what he considered an acceptable form which never mentioned schematic sentences. This is not difficult to do – it is a matter of replacing statements of the form 'Axiom A entails axiom B' by statements of the form 'For all relations P and R, if P and R do this then they do that'. But the resulting translation is quite unreadable, so good mathematics is thrown away and all for no purpose.

Frege's rejection of ambiguous symbols is part and parcel of his refusal to handle indexical expressions; see Perry [1977] for some discussion of the issue. It is sad to learn that the grand architect of modern logic fiercely rejected the one last advance which was needed to make his ideas fruitful.

13. FIRST-ORDER SYNTAX FORMALIZED

The main purpose of this section and the next is to extract the formal content of Sections 9–12 above. I give the definitions first under the assumption that there are no sorted variables. Also I ignore for the moment the fact that some first-order logicians use = and function symbols. Section 18 below will be more broad-minded.

A *similarity type* is defined to be a set of individual constants together with a set of predicate constants; each predicate constant is assumed to be labelled somehow to indicate that it is an n-place predicate constant, for some positive integer n. Some writers include the n as a superscript: R^{133} is a 133-place predicate constant.

We shall define the *first-order language* L of *similarity type X*. For definiteness, L shall be an ordered triple $\langle X, T(X), F(X) \rangle$ where X is the similarity type, and $T(X)$ and $F(X)$ are respectively the set of all terms and formulas of similarity type X (known more briefly as the terms and formulas of L). Grammatically speaking, the terms of L are its noun phrases and the formulas are its sentences. Metavariables σ, τ will range over terms, and metavariables ϕ, ψ, χ will range over formulas.

We start the definition by defining the *variables* to be the countably many symbols

(13.1) $\quad x_0, x_1, x_2, \ldots$.

Unofficially everybody uses the symbols x, y, z etc. as variables. But in the spirit of Section 4 above, these can be understood as metavariables ranging over variables. The *terms* of L are defined to be the variables of L and the individual constants in X.

An *atomic formula* of L is an expression of form $P(\sigma_1, \ldots, \sigma_n)$ where P is an n-place predicate constant in X and $\sigma_1, \ldots, \sigma_n$ are terms of L. The class of *formulas* of L is defined inductively, and as the induction proceeds we shall define also the set of subformulas of the formula ϕ, and the set $FV(\phi)$ of free variables of ϕ:

(a) Every atomic formula ϕ of L is a formula of L; it is its only subformula, and $FV(\phi)$ is the set of all variables which occur in ϕ. $\perp$ is a formula of L; it is its only subformula, and $FV(\perp)$ is empty.

(b) Suppose ϕ and ψ are formulas of L and x is a variable. Then: $\neg \phi$ is a formula of L; its subformulas are itself and the subformulas of ϕ; $FV(\neg \phi)$ is $FV(\phi)$. Also $(\phi \wedge \psi)$, $(\phi \vee \psi)$, $(\phi \rightarrow \psi)$ and $(\phi \leftrightarrow \psi)$ are formulas of L; the subformulas of each of these formulas are itself, the subformulas of ϕ and the subformulas of ψ; its free variables are those of ϕ together with those of ψ. Also $\forall x \phi$ and $\exists x \phi$ are formulas of L; for each of these, its subformulas are itself and the subformulas of ϕ; its free variables are those of ϕ excluding x.

(c) Nothing is a formula of L except as required by (a) and (b).

The *complexity* of a formula ϕ is defined to be the number of subformulas of ϕ. This definition disagrees with that in Section 3, but it retains the crucial property that every formula has a higher complexity than any of its proper subformulas. (The *proper subformulas* of ϕ are all the subformulas of ϕ except ϕ itself.) A formula is said to be *closed*, or to be a *sentence*, if it has no free variables. Closed formulas correspond to sentences of English, non-closed formulas to predicates or open sentences of English. Formulas of a formal language are sometimes called *well-formed formulas*, or *wffs* for short.

If ϕ is a formula, x is a variable and τ is a term, then there is a formula $\phi[\tau/x]$ which 'says the same thing about the object τ as ϕ says about the object x'. At a first approximation, $\phi[\tau/x]$ can be described as the formula which results if we put τ in place of each free occurrence of x in ϕ; when this description works, we say τ is *free for x in ϕ* or *substitutable* for x in ϕ. Here is an example where the approximation doesn't work: ϕ is $\exists y\, R(x,y)$ and τ is y. If we put y for x in ϕ, the resulting formula $\exists y\, R(y,y)$ says nothing at all about 'the object y', because the inserted y becomes bound by the quantifier $\exists y$ — a phenomenon known as *clash of variables*. In such cases we have to define $\phi[\tau/x]$ to be $\exists z\, R(y,z)$ where z is some other variable. (There is a good account of this messy matter in Bell and Machover [1977] Chapter 2, Section 3.)

Note the useful shorthand: if ϕ is described at its first occurrence as $\phi(x)$, then $\phi(\tau)$ means $\phi[\tau/x]$. Likewise if ϕ is introduced as $\phi(y_1, \ldots, y_n)$ then $\phi(\tau_1, \ldots, \tau_n)$ means the formula which says about the objects $\tau_1, \ldots, \tau_n$ the same thing as ϕ says about the objects $y_1, \ldots, y_n$.

Not much in the definitions above needs to be changed if you want a system with sorted variables. You must start by deciding what kind of sortal system you want. There will be a set S of sorts s, t etc., and for each sort s there will be sorted variables x_0^s, x_1^s, x_2^s etc. But then (a) do you want every object to belong to some sort? If so, the similarity type must assign each individual constant to at least one sort. (b) Do you want the sorts to be mutually exclusive? Then the similarity type must assign each individual constant to at most one sort. (c) Do you want to be able to say 'everything', rather than just 'everything of such-and-such a sort'? If not then the unsorted variables (13.1) should be struck out.

Some formal languages allow restricted quantification. For example, in languages designed for talking about numbers, we have formulas $(\forall x < y)\phi$ and $(\exists x < y)\phi$, read respectively as 'For all numbers x less than y, ϕ' and 'There is a number x less than y such that ϕ'. These expressions can be regarded as metalanguage abbreviations for $\forall x(x < y \rightarrow \phi)$ and $\exists x(x < y \wedge \phi)$ respectively (where '$x < y$' in turn is an abbreviation for '$<(x,y)$'). Or we can alter the definition of 'formula of L' to allow restricted quantifiers in L itself.

One often sees expressions such as '$\forall xy\phi$' or '$\exists \vec{z}\phi$'. These are metalanguage abbreviations. $\forall xy$ is short for $\forall x \forall y$. $\vec{z}$ means a finite sequence $z_1, \ldots, z_n$. Furthermore, the abbreviations of Section 4 remain in force.

All the syntactic notions described in this section can be defined using only concrete instances of the induction axiom as in Section 3 above.

14. FIRST-ORDER SEMANTICS FORMALIZED

We turn to the definition of structures. (They are also known as *models* – but it is better to reserve this term for the context 'model of ϕ'.) Let L be a language with similarity type X. Then an L-*structure* $\mathfrak{A}$ is defined to be an ordered pair $\langle A, I \rangle$ where:

(1) A is a class called the *domain* of $\mathfrak{A}$, in symbols $|\mathfrak{A}|$. The elements of A are called the *elements of* $\mathfrak{A}$, and the cardinality of A is called the *cardinality of* $\mathfrak{A}$. So for example we call $\mathfrak{A}$ *finite* or *empty* if A is finite or empty. Many writers use the convention that A, B and C are the domains of $\mathfrak{A}, \mathfrak{B}$ and $\mathfrak{C}$ respectively.

(2) I is a function which assigns to each individual constant c of X an element $I(c)$ of A, and to each n-place predicate symbol R of X an n-place relation $I(R)$ on A. I is referred to as $I_\mathfrak{A}$.

Structure means: L-structure for some language L.

If Z is a set of variables, then an *assignment* to Z in $\mathfrak{A}$ is defined to be a function from Z to A. If g is an assignment to Z in $\mathfrak{A}$, x is a variable not in Z and α is an element of $\mathfrak{A}$, then we write

(14.1) $g, \alpha/x$

for the assignment h got from g by adding x to g's domain and putting $h(x) = \alpha$. (Some writers call assignments *valuations.*)

For each assignment g in $\mathfrak{A}$ and each individual constant c we define $c[g]$ to be the element $I_\mathfrak{A}(c)$. For each variable x and assignment g whose domain contains x, we define $x[g]$ to be the element $g(x)$. Then $\tau[g]$ is 'the element named by the term τ under the assignment g'.

For each formula ϕ of L and each assignment g to the free variables of ϕ in $\mathfrak{A}$, we shall now define the conditions under which $\mathfrak{A} \models \phi[g]$ (cf. (10.7) above). The definition is by induction on the complexity of ϕ.

(a) If R is an n-place predicate constant in X and $\tau_1, \ldots, \tau_n$ are terms, then $\mathfrak{A} \models R(\tau_1, \ldots, \tau_n)$ iff the ordered n-tuple $\langle \tau_1[g], \ldots, \tau_n[g] \rangle$ is in $I_\mathfrak{A}(R)$.

(b) It is never true that $\mathfrak{A} \models \bot$.

(c) $\mathfrak{A} \models \neg\phi[g]$ iff it is not true that $\mathfrak{A} \models \phi[g]$.

$\mathfrak{A} \models \phi \wedge \psi[g]$ iff $\mathfrak{A} \models \phi[g_1]$ and $\mathfrak{A} \models \psi[g_2]$, where g_1 and g_2 are the results of restricting g to the free variables of ϕ and ψ respectively. Etc. as in (3.5).

(d) If x is a free variable of ϕ, then:
$\mathfrak{A} \models \forall x \phi[g]$ iff for every element α of A, $\mathfrak{A} \models \phi[g, \alpha/x]$;
$\mathfrak{A} \models \exists x \phi[g]$ iff for at least one element α of A, $\mathfrak{A} \models \phi[g, \alpha/x]$.
If x is not a free variable of ϕ, then $\mathfrak{A} \models \forall x \phi[g]$ iff $\mathfrak{A} \models \phi[g]$, and $\mathfrak{A} \models \exists x \phi[g]$ iff $\mathfrak{A} \models \phi[g]$.

We say an assignment g in $\mathfrak{A}$ is *suitable for* the formula ϕ if every free variable of ϕ is in the domain of g. If g is suitable for ϕ, we say that $\mathfrak{A} \models \phi[g]$ if and only if $\mathfrak{A} \models \phi[h]$, where h comes from g by throwing out of the domain of g those variables which are not free variables of ϕ.

If ϕ is a sentence, then ϕ has no free variables and we can write just $\mathfrak{A} \models \phi$ in place of $\mathfrak{A} \models \phi[\ ]$. This notation agrees with (3.4) above. When $\mathfrak{A} \models \phi$, we say that $\mathfrak{A}$ is a *model of* ϕ, or that ϕ is *true in* $\mathfrak{A}$. '$\mathfrak{A} \models \phi[g]$' can be pronounced '$g$ *satisfies* ϕ *in* $\mathfrak{A}$'.

To anybody who has mastered the symbolism it should be obvious that clauses (a)–(d) really do determine whether or not $\mathfrak{A} \models \phi$, for every L-structure $\mathfrak{A}$ and every sentence ϕ of L. *If $\mathfrak{A}$ is a set* then we can formalize the definition in the language of set theory and prove that it determines $\models$ uniquely, using only quite weak set-theoretic axioms (cf. Barwise [1975] Chapter III). Set structures are adequate for most applications of first-order logic in mathematics, so that many textbooks simply state without apology that a structure has to be a set. We shall return to this point in Section 17 below.

The definition of $\models$ given above is called the *truth-definition*, because it specifies exactly when a symbolic formula is to count as 'true in' a structure. It solves no substantive problems about what is true – we are just as much in the dark about Fermat's Last Theorem or the Reichstag fire after writing it down as we were before. But it has attracted a lot of attention as a possible answer to the question of what is Truth. Many variants of it have appeared in the literature, which can cause anguish to people anxious to get to the heart of the matter. Let me briefly describe three of these variants; they are all mathematically equivalent to the version given above. (Cf. Leblanc [Chapter I.3, below].)

In the first variant, *assignments are sequences*. More precisely an assignment in $\mathfrak{A}$ is defined to be a function g from the natural numbers N to the domain A of $\mathfrak{A}$. Such a function can be thought of as an infinite sequence $\langle g(0), g(1), g(2), \ldots \rangle$. The element $g(i)$ is assigned to the ith variable x_i, so that $x_i[g]$ is defined to be $g(i)$. In (c) and (d) we have to make some changes for the purely technical reason that g assigns elements to *every* variable and not just those free in ϕ. In (c) the clause for $\phi \wedge \psi$ becomes

$$\mathfrak{A} \models \phi \wedge \psi[g] \quad \text{iff} \quad \mathfrak{A} \models \phi[g] \text{ and } \mathfrak{A} \models \psi[g],$$

which is an improvement (and similarly with $(\phi \vee \psi)$, $(\phi \rightarrow \psi)$ and $(\phi \leftrightarrow \psi)$). But (d) becomes distorted, because g already makes an assignment to the quantified variable x; this assignment is irrelevant to the truth of $\mathfrak{A} \models \forall x \phi[g]$, so we have to discard it as follows. For each number i and element α of $\mathfrak{A}$, let $g(\alpha/i)$ be the assignment h which is exactly like g except that $h(i) = \alpha$. Then (d) is replaced by:

(d') For each variable x_i: $\mathfrak{A} \models \forall x_i \phi[g]$ iff for every element α of A, $\mathfrak{A} \models \phi[g(\alpha/i)]$.

together with a similar clause for $\exists x_i \phi$.

In the second variant, we copy (3.6) and define the *truth-value* of ϕ in $\mathfrak{A}$, $\|\phi\|_\mathfrak{A}$, to be the set of all assignments g to the free variables of ϕ such that $\mathfrak{A} \models \phi[g]$. When ϕ is a sentence, there is only one assignment to the free variables of ϕ, namely the empty function 0; so $\|\phi\|_\mathfrak{A}$ is $\{0\}$ if ϕ is true in $\mathfrak{A}$, and the empty set (again 0) if ϕ is false in $\mathfrak{A}$. This variant is barely more than a change of notation. Instead of '$\mathfrak{A} \models \phi[g]$' we write '$g \in \|\phi\|_\mathfrak{A}$'. The clauses (a)–(d) can be translated easily into the new notation.

Some writers combine our first and second variants, taking $\|\phi\|_\mathfrak{A}$ to be the set of all sequences g such that $\mathfrak{A} \models \phi[g]$. In this style, the clause for $\phi \wedge \psi$ in (c) becomes rather elegant:

$$\|\phi \wedge \psi\|_\mathfrak{A} = \|\phi\|_\mathfrak{A} \cap \|\psi\|_\mathfrak{A}.$$

However, when ϕ is a sentence the definition of 'ϕ is true in $\mathfrak{A}$' becomes 'every sequence is in $\|\phi\|_\mathfrak{A}$', or equivalently 'at least one sequence is in $\|\phi\|_\mathfrak{A}$'. I have heard students repeat this definition with baffled awe as if they learned it in the Eleusinian Mysteries.

The third variant dispenses with assignments altogether and adds new constant names to the language L. Write $L(c)$ for the language got from L by adding c as an extra individual constant. If $\mathfrak{A}$ is an L-structure and α is an element of $\mathfrak{A}$, write $(\mathfrak{A}, \alpha)$ for the $L(c)$-structure $\mathfrak{B}$ which is the same as $\mathfrak{A}$ except that $I_\mathfrak{B}(c) = \alpha$. If ϕ is a formula of L with just the free variable x, one can prove by induction on the complexity of ϕ that

(14.2) $\quad (\mathfrak{A}, \alpha) \models \phi[c/x] \quad \text{iff} \quad \mathfrak{A} \models \phi[\alpha/x].$

(Warning: $[c/x]$ on the left is a substitution in the formula ϕ; α/x on the right is an assignment to the variable x.) The two sides in (14.2) are just different ways of expressing that α satisfies ϕ in $\mathfrak{A}$. Hence we have

(14.3) $\mathfrak{A} \models \forall x \phi$ iff for every element α of $\mathfrak{A}$, $(\mathfrak{A}, \alpha) \models \phi[c/x]$,

and a similar clause for $\exists x \phi$. In our third variant, (14.3) is taken as the *definition* of $\models$ for sentences of form $\forall x \phi$. This trick sidesteps assignments. Its disadvantage is that we have to alter the language and the structure each time we come to apply clause (d). The great merit of assignments is that they enable us to keep the structure fixed while we wiggle around elements in order to handle the quantifiers.

There are L-structures whose elements are all named by individual constants of L. For example, the natural numbers are sometimes understood as a structure in which every number n is named by a numeral constant $\bar{n}$ of the language. For such structures, *and only for such structures*, (14.3) can be replaced by

(14.4) $\mathfrak{A} \models \forall x \phi$ iff for every individual constant c of L, $\mathfrak{A} \models \phi[c/x]$.

Some writers confine themselves to structures for which (14.4) applies.

Alfred Tarski's famous paper on the concept of truth in formalized languages [1935/6] was the first paper to present anything like our definition of $\models$. Readers should be aware of one vital difference between his notion and ours. His languages have no ambiguous constants. True, Tarski says they have constants. But he explains that by 'constants' he means negation signs, quantifier symbols and suchlike, together with symbols of fixed meaning such as the inclusion sign $\subseteq$ in set theory. (See Section 20 below on symbols with an 'intended interpretation'.) The only concession that Tarski makes to the notion of an L-structure is that he allows the domain of elements to be any class, not necessarily the class of everything. Even then he says that relativizing to a particular class is 'not essential for the understanding of the main theme of this work'! (Cf. pages 199, 212 of the English translation of Tarski [1935/6].) Carnap's truth-definition [1935] is also a little sideways from modern versions.

There is no problem about adapting Tarski's definition to our setting. It can be done in several ways. Probably the simplest is to allow some of his constants to turn ambiguous; then his definition becomes our first variant.

Finally I should mention structures for many-sorted languages, if only to say that no new issues of principle arise. If the language L has a set S of sorts, then for each sort s in S, an L-structure $\mathfrak{A}$ must carry a class $s(\mathfrak{A})$ of *elements of sort s*. In accordance with Section 12, $s(\mathfrak{A})$ must be included in $|\mathfrak{A}|$. If the individual constant c is of sort s, then $I_{\mathfrak{A}}(c)$ must be an element of $s(\mathfrak{A})$. If we have required that every element should be of at least one sort, then $|\mathfrak{A}|$ must be the union of the classes $s(\mathfrak{A})$.

15. FIRST-ORDER IMPLICATIONS

Let me make a leap that will seem absurd to the Traditional Logician, and define sequents with infinitely many premises.

Suppose L is a first-order language. By a *theory in* L we shall mean a set of sentences of L – it can be finite or infinite. The metavariables $\Delta, \Gamma, \Theta, \Lambda$ will range over theories. If Δ is a theory in L and $\mathfrak{A}$ is an L-structure, we say that $\mathfrak{A}$ is a *model of* Δ if $\mathfrak{A}$ is a model of every sentence in Δ.

For any theory Δ in L and sentence ϕ of L, we define

(15.1) $\Delta \models \phi$ ('Δ *logically implies* ϕ', 'ϕ is a *logical consequence* of Δ')

to mean that every L-structure which is a model of Δ is also a model of ϕ. If Δ has no models, (15.1) is reckoned to be true by default. A *counterexample* to (15.1) is an L-structure which is a model of Δ but not of ϕ. We write

(15.2) $\models \phi$ ('ϕ is *logically valid*')

to mean that every L-structure is a model of ϕ; a *counterexample* to (15.2) is an L-structure which is not a model of ϕ. The expressions (15.1) and (15.2) are called *sequents*. This definition of logical implication was first set down by Tarski [1936], though it only makes precise what Bolzano [1837] Section 155 and Hilbert [1899] already understood.

If the language L has at least one individual constant c, then every L-structure must have an element $I_{\mathfrak{A}}(c)$, so the domain of $\mathfrak{A}$ can't be empty. It follows that in this language the sentence $\exists x \neg \bot$ must be logically valid, so we can 'prove' that at least one thing exists.

On the other hand if L has no individual constants, then there is an L-structure whose domain is empty. This is not just a quirk of our conventions: one can quite easily think of English sentences uttered in contexts where the natural domain of quantification happens to be empty. In such a language L, $\exists x \neg \bot$ is not logically valid.

This odd state of affairs deserves some analysis. Suppose L does have an individual constant c. By the Bolzano–Tarski definition (15.1), when we consider logical implication in L we are only concerned with structures in which c names something. In other words, the Bolzano–Tarski definition slips into every argument a tacit premise that *every name does in fact name something*.

If we wanted to, we could adapt the Traditional Logician's notion of a valid argument in just the same way. For a traditional example, consider

(15.3) Every man runs. *Therefore* Socrates, if he is a man, runs.

On the traditional view, (15.3) is not a valid argument – it could happen that every man runs and yet there is no such entity as Socrates. On the Bolzano–Tarski view we must consider only situations in which 'Socrates' names something or someone, and on that reckoning, (15.3) is valid. (According to Walter Burleigh in the fourteenth century, (15.3) is not valid outright but it is valid at the times when Socrates exists. Cf. Bocheński [1961] p. 193; I have slightly altered Burleigh's example. I don't know how one and the same argument can be valid at 4 p.m. and invalid at 5 p.m.)

Once this much is clear, we can decide whether we want to do anything about it. From the Traditional Logician's point of view it might seem sensible to amend the Bolzano–Tarski definition. This is the direction which *free logic* has taken. (Cf. Bencivenga [Chapter III.6].)

The mainstream has gone the other way. Non-referring constants are anathema in most mathematics. Besides, Hilbert-style calculi with identity always have $\exists x(x = x)$ as a provable formula. (See Remark 6 in Appendix A below. On the other hand semantic tableau systems which allow empty structures, such as Hodges [1977], are arguably a little simpler and more natural than versions which exclude them.) If $\exists x \neg \bot$ is logically valid in some languages and not in others, the easiest remedy is to make it logically valid in all languages, and we can do that by *requiring all structures to have non-empty domains*. Henceforth we shall do so (after pausing to note that Schröder [1895] p. 5 required all structures to have at least two elements).

Let us review some properties of $\models$. Analogues of Theorems 5.1–5.4 (allowing infinitely many premises!) and Theorem 5.5 now hold. The relevant notion of logical equivalence is this: the formula ϕ is *logically equivalent* to the formula ψ if for every structure $\mathfrak{A}$ and every assignment g in $\mathfrak{A}$ which is suitable for both ϕ and ψ, $\mathfrak{A} \models \phi[g]$ if and only if $\mathfrak{A} \models \psi[g]$. For example

(15.4) $\forall x \phi$ is logically equivalent to $\neg \exists x \neg \phi$,
$\exists x \phi$ is logically equivalent to $\neg \forall x \neg \phi$.

A formula is said to be *basic* if it is either atomic or the negation of an atomic formula. A formula is in *disjunctive normal form* if it is either $\bot$ or a disjunction of conjunctions of basic formulas. One can show:

(15.5) *Every formula of L is logically equivalent to a formula of L with the same free variables, in which all quantifiers are at the left-hand end, and the part after the quantifiers is in disjunctive normal form.*

A formula with its quantifiers all at the front is said to be in *prenex form*. (In Section 25 below we meet Skolem normal forms, which are different from (15.5) but also prenex.)

Proof calculi for propositional logic are generally quite easy to adapt to predicate logic. Sundholm [Chapter I.2, below] surveys the possibilities. Usually in predicate logic one allows arbitrary formulas to occur in a proof, not just sentences, and this can make it a little tricky to say exactly what is the informal idea expressed by a proof. (This applies particularly to Hilbert-style calculi; cf. Remarks 4 and 5 in Appendix A below. Some calculi paper over the difficulty by writing the free variables as constants.) When one speaks of a formal calculus for predicate logic as being *sound* or *complete* (cf. Section 7 above), one always ignores formulas which have free variables.

Gentzen's natural deduction calculus can be adapted to predicate logic simply by adding four rules, namely introduction and elimination rules for $\forall$ and $\exists$. The *introduction rule* for $\exists$ says:

(15.6) From $\phi[\tau/x]$ infer $\exists x \phi$.

(If the object τ satisfies ϕ, then at least one thing satisfies ϕ.) The *elimination rule* for $\exists$ says:

(15.7) Given a proof of ψ from $\phi[y/x]$ and assumptions $\chi_1, \ldots, \chi_n$, where y is not free in any of $\exists x \phi, \psi, \chi_1, \ldots, \chi_n$, deduce ψ from $\exists x \phi$ and $\chi_1, \ldots, \chi_n$.

The justification of (15.7) is of some philosophical interest, as the following example will show. We want to deduce an absurdity from the assumption that there is a greatest integer. So we let y be a greatest integer, we get a contradiction $y < y + 1 \leq y$, whence $\bot$. Then by (15.7) we deduce $\bot$ from $\exists x$ (x is a greatest integer). Now the problem is: How can we possibly 'let y be a greatest integer', since there aren't any? Some logicians exhort us to '*imagine* that y is a greatest integer', but I always found that this one defeats my powers of imagination.

The Bolzano–Tarski definition of logical implication is a real help here, because it steers us away from matters of 'If it were the case that ...' towards questions about what actually is the case in structures which do exist.

We have to decide how natural deduction proofs are supposed to match the Bolzano–Tarski definition, bearing in mind that formulas with free variables may occur. The following interpretation is the right one: the existence of a natural deduction proof with conclusion ψ and premises $\chi_1, \ldots, \chi_n$ should tell us that for every structure $\mathfrak{A}$ and every assignment g in $\mathfrak{A}$ which is suitable for all of $\psi, \chi_1, \ldots, \chi_n$, we have $\mathfrak{A} \models (\chi_1 \wedge \cdots \wedge \chi_n \to \psi)[g]$. (This is *not* obvious – for Hilbert-style calculi one has to supply a quite different rationale, cf. Remark 5 on Hilbert-style calculi in Appendix A.)

Now we can justify (15.7). Let $\mathfrak{A}$ be a structure and g an assignment in $\mathfrak{A}$ which is suitable for $\exists x \phi, \chi_1, \ldots, \chi_n$ and ψ. We wish to show that

(15.8) $\mathfrak{A} \models (\exists x \phi \wedge \chi_1 \wedge \cdots \wedge \chi_n \to \psi)[g]$.

By the truth-definition in Section 14 we can assume that the domain of g is just the set of variables free in the formulas listed, so that in particular y is not in the domain of g. There are now two cases. The first is that $\mathfrak{A} \models \neg(\exists x \phi \wedge \chi_1 \wedge \cdots \wedge \chi_n)[g]$. Then truth-tables show that (15.8) holds. The second case is that $\mathfrak{A} \models (\exists x \phi \wedge \chi_1 \wedge \cdots \wedge \chi_n)[g]$, so there is an element α of $\mathfrak{A}$ such that $\mathfrak{A} \models (\phi[y/x] \wedge \chi_1 \wedge \cdots \wedge \chi_n)[g, \alpha/y]$. By the proof which (15.7) assumes to be given, $\mathfrak{A} \models (\phi[y/x] \wedge \chi_1 \wedge \cdots \wedge \chi_n \to \psi)[g, \alpha/y]$, so $\mathfrak{A} \models \psi[g, \alpha/y]$. But then since y is not free in ψ, $\mathfrak{A} \models \psi[g]$, which again implies (15.8).

I do not think this solves all the philosophical problems raised by (15.7). Wiredu [1973] seems relevant.

The references given for the proof calculi discussed in Section 7 remain relevant, except Łukasiewicz and Tarski [1930] which is only about propositional logic. The various theorems of Gentzen [1934], including the cut-elimination theorem, all apply to predicate logic. From the point of view of these calculi, the difference between propositional and predicate logic is relatively slight and has to do with checking that certain symbols don't occur in the wrong places in proofs.

Proof calculi for many-sorted languages are also not hard to come by. See Schmidt [1938], Wang [1952], Feferman [1968].

Quantifiers did provoke one quite new proof-theoretic contrivance. In the 1920s a number of logicians (notably Skolem, Hilbert, Herbrand) regarded quantifiers as an intrusion of infinity into the finite-minded world of propositional logic, and they tried various ways of – so to say – deactivating quantifiers. Hilbert proposed the following: replace $\exists x \phi$ everywhere by the sentence $\phi[\epsilon x \phi / x]$, where '$\epsilon x \phi$' is interpreted as 'the element I choose among those that satisfy ϕ'. The interpretation is of course outrageous, but

Hilbert showed that his ε-calculus proved exactly the same sequents as more conventional calculi. See Hilbert and Bernays [1939] and Leisenring [1969].

It can easily be checked that any sequent which can be proved by the natural deduction calculus sketched above (cf. Sundholm [Chapter I.2, below] for details) is correct. But nobody could claim to see, just by staring at it, that this calculus can prove *every* correct sequent of predicate logic. Nevertheless it can, as the next section will show.

16. CREATING MODELS

The natural deduction calculus for first-order logic is *complete* in the sense that if $\Delta \models \psi$ then the calculus gives a proof of ψ from assumptions in Δ. This result, or rather the same result for an equivalent Hilbert-style calculus, was first proved by Kurt Gödel in his doctoral dissertation [1930]. Strictly Thoralf Skolem has already proved it in his brilliant papers [1922, 1928, 1929], but he was blissfully unaware that he had done so. (See Vaught [1974] and Wang [1970]; Skolem's finitist philosophical leanings seem to have blinded him to some mathematical implications of his work.)

A theory Δ in the language L is said to be *consistent* for a particular proof calculus if the calculus gives no proof of $\bot$ from assumptions in Δ. (Some writers say instead: 'gives no proof of a contradiction $\phi \wedge \neg \phi$ from assumptions in Δ'. For the calculi we are considering, this amounts to the same thing.) We shall demonstrate that *if Δ is consistent for the natural deduction calculus then Δ has a model*. This implies that the calculus is complete, as follows. Suppose $\Delta \models \psi$. Then $\Delta, \psi \to \bot \models \bot$ (cf. Theorem 5.4 in Section 5), hence Δ together with $\psi \to \bot$ has no model. But then the theory consisting of Δ together with $\psi \to \bot$ is not consistent for the natural deduction calculus, so we have a proof of $\bot$ from $\psi \to \bot$ and sentences in Δ. One can then quickly construct a proof of ψ from sentences in Δ by the rule (7.12) for $\bot$.

So the main problem is to show that every consistent theory has a model. This involves constructing a model – but out of what? Spontaneous creation is not allowed in mathematics; the pieces must come from somewhere. Skolem [1922] and Gödel [1930] made their models out of natural numbers, using an informal induction to define the relations. A much more direct source of materials was noticed by Henkin [1949] and independently by Rasiowa and Sikorski [1950]: they constructed the model of Δ out of the theory Δ itself. (Their proof was closely related to Kronecker's [1882] method of constructing extension fields of a field K out of polynomials over K. Both he and they factored out a maximal ideal in a ring.)

I.1: ELEMENTARY PREDICATE LOGIC

Hintikka [1955] and Schütte [1956] extracted the essentials of the Henkin–Rasiowa–Sikorski proof in an elegant form, and what follows is based on their account. For simplicity we assume that the language L has infinitely many individual constants but its only truth-functors are $\neg$ and $\wedge$ and its only quantifier symbol is $\exists$. A theory Δ in L is called a *Hintikka set* if it satisfies these seven conditions:

(a) $\bot$ is not in Δ.
(b) If ϕ is an atomic formula in Δ then $\neg \phi$ is not in Δ.
(c) If $\neg\neg \psi$ is in Δ then ψ is in Δ.
(d) If $\psi \wedge \chi$ is in Δ then ψ and χ are both in Δ.
(e) If $\neg(\psi \wedge \chi)$ is in Δ then either $\neg \psi$ is in Δ or $\neg \chi$ is in Δ.
(f) If $\exists x \psi$ is in Δ then $\psi[c/x]$ is in Δ for some individual constant c.
(g) If $\neg \exists x \psi$ is in Δ then $\neg \psi[c/x]$ is in Δ for each individual constant c.

We can construct an L-structure $\mathfrak{A}$ out of a theory Δ as follows. The elements of $\mathfrak{A}$ are the individual constants of L. For each constant c, $I_\mathfrak{A}(c)$ is c itself. For each n-place predicate constant R of L the relation $I_\mathfrak{A}(R)$ is defined to be the set of all ordered n-tuples $\langle c_1, \ldots, c_n \rangle$ such that the sentence $R(c_1, \ldots, c_n)$ is in Δ.

Let Δ be a Hintikka set. We claim that the structure $\mathfrak{A}$ built out of Δ is a model of Δ. It suffices to show the following, by induction on the complexity of ϕ: if ϕ is in Δ then ϕ is true in $\mathfrak{A}$, and if $\neg \phi$ is in Δ then $\neg \phi$ is true in $\mathfrak{A}$. I consider two sample cases. First let ϕ be atomic. If ϕ is in Δ then the construction of $\mathfrak{A}$ guarantees that $\mathfrak{A} \models \phi$. If $\neg \phi$ is in Δ, then by clause (b), ϕ is not in Δ; so by the construction of $\mathfrak{A}$ again, $\mathfrak{A}$ is not a model of ϕ and hence $\mathfrak{A} \models \neg \phi$. Next suppose ϕ is $\psi \wedge \chi$. If ϕ is in Δ, then by clause (d), both ψ and χ are in Δ; since they have lower complexities than ϕ, we infer that $\mathfrak{A} \models \psi$ and $\mathfrak{A} \models \chi$; so again $\mathfrak{A} \models \phi$. If $\neg \phi$ is in Δ then by clause (e) either $\neg \psi$ is in Δ or $\neg \chi$ is in Δ; suppose the former. Since ψ has lower complexity than ϕ, we have $\mathfrak{A} \models \neg \psi$; it follows again that $\mathfrak{A} \models \neg \phi$. The remaining cases are similar. So *every Hintikka set has a model*.

It remains to show that if Δ is consistent, then by adding sentences to Δ we can get a Hintikka set Δ^+; Δ^+ will then have a model, which must also be a model of Δ because Δ^+ includes Δ. The strategy is as follows.

STEP 1. Extend the language L of T to a language L^+ which has infinitely many new individual constants $c_0, c_1, c_2, \ldots$. These new constants are known as the *witnesses* (because in (f) above they will serve as witnesses to the truth of $\exists x \psi$).

STEP 2. List all the sentences of L^+ at $\phi_0, \phi_1, \ldots$ in an infinite list so that every sentence occurs infinitely often in the list. This can be done by some kind of zigzagging back and forth.

STEP 3. At this very last step there is a parting of the ways. Three different arguments will lead us home. Let me describe them and then compare them.

The first argument we may call the *direct* argument: we simply add sentences to Δ as required by (c)–(g), making sure as we do so that (a) and (b) are not violated. To spell out the details, we define by induction theories $\Delta_0, \Delta_1, \ldots$ in the language L^+ so that (i) every theory Δ_i is consistent; (ii) for all i, Δ_{i+1} includes Δ_i; (iii) for each i, only finitely many of the witnesses appear in the sentences in Δ_i; (iv) Δ_0 is Δ; and (v) for each i, if ϕ_i is in Δ_i then:

(c′) if ϕ_i is of form $\neg\neg\psi$ then Δ_{i+1} is Δ_i together with ψ;
(d′) if ϕ_i is of form $\psi \wedge \chi$ then Δ_{i+1} is Δ_i together with ψ and χ;
(e′) if ϕ_i is of form $\neg(\psi \wedge \chi)$ then Δ_{i+1} is Δ_i together with at least one of $\neg\psi, \neg\chi$;
(f′) if ϕ_i is of form $\exists x \psi$ then Δ_{i+1} is Δ_i together with $\psi[c/x]$ for some witness c which doesn't occur in Δ_i;
(g′) if ϕ_i is of form $\neg\exists x \psi$ then Δ_{i+1} is Δ_i together with $\neg\psi[c/x]$ for the first witness c such that $\neg\psi[c/x]$ is not already in Δ_i.

It has to be shown that theories Δ_i exist meeting conditions (i)–(v). The proof is by induction. We satisfy (i)–(v) for Δ_0 by putting $\Delta_0 = \Delta$ (and this is the point where we use the assumption that Δ is consistent for natural deduction). Then we must show that if we have got as far as Δ_i safely, Δ_{i+1} can be constructed too. Conditions (ii) and (iii) are actually implied by the others and (iv) is guaranteed from the beginning. So we merely need show that

(16.1) assuming Δ_i is consistent, Δ_{i+1} can be chosen so that it is consistent and satisfies the appropriate one of (c′)–(g′).

There are five cases to consider. Let me take the hardest, which is (f′). It is assumed that ϕ_i is $\exists x \psi$ and is in Δ_i. By (iii) so far, some witness has not yet been used; let c be the first such witness and let Δ_{i+1} be Δ_i together with $\psi[c/x]$. If by misfortune Δ_{i+1} was inconsistent, then since c never occurs in Δ_i or ϕ_i, the elimination rule for $\exists$ (Section 15 or Sundholm [Chapter I.2, below] shows that we can prove $\bot$ already from $\exists x \psi$ and assumptions in Δ_i. But $\exists x \psi$ was in Δ_i, so we have a contradiction to our assumption that Δ_i was consistent. Hence Δ_{i+1} is consistent as required.

I.1: ELEMENTARY PREDICATE LOGIC

When the theories Δ_i have been constructed, let Δ^+ be the set of all sentences which are in at least one theory Δ_i. Since each Δ_i was consistent, Δ^+ satisfies conditions (a) and (b) for a Hintikka set. The requirements (c')–(g'), and the fact that in the listing $\phi_0, \phi_1, \ldots$ we keep coming round to each sentence infinitely often, ensure that Δ^+ satisfies conditions (c)–(g) as well. So Δ^+ is a Hintikka set and has a model, which completes the construction of a model of Δ.

The second argument we may call the *tree* argument. A hint of it is in Skolem [1929]. We imagine a man constructing the theories Δ_i as in the direct argument above. When he faces clauses (c'), (d'), (f') or (g'), he knows at once how he should construct Δ_{i+1} out of Δ_i; the hardest thing he has to do is to work out which is the first witness not yet used in Δ_i in the case of clause (f'). But in (e') we can only prove for him that at least one of $\neg\psi$ and $\neg\chi$ can consistently be added to Δ_i, so he must check for himself whether Δ_i together with $\neg\psi$ is in fact consistent. Let us imagine that he is allergic to consistency calculations. Then the best he can do is to make *two alternative suggestions* for Δ_{i+1}, viz. Δ_i with $\neg\psi$, and Δ_i with $\neg\chi$. Thus he will make not a chain of theories $\Delta_0, \Delta_1, \ldots$ but a branching tree of theories:

(16.2)
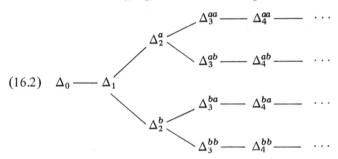

Now he no longer knows which of these theories are consistent. So he forgets about consistency and looks directly at conditions (a) and (b) in the definition of a Hintikka set. At least he can tell by inspection whether a theory violates these. So he prunes off the tree all theories which fail (a) or (b) – he can do this as he goes along. Some theories in the tree will become dead ends. But the argument we gave for the earlier direct approach shows that at every level in the tree there must be some theory which can be extended to the next level.

Now a combinatorial theorem known as *König's tree lemma* says that if a tree has a positive but finite number of items at the nth level, for every natural number n, then the tree has a branch which runs up through all these levels. So we know that (16.2) has an infinite branch. Let $\Delta_0, \Delta_1, \Delta_2, \ldots$ be such a branch and let Δ^+ be the set of all sentences which occur in at least one theory Δ_i in the branch. The previous argument shows that Δ^+ satisfies (c)–(g), and we know that Δ^+ satisfies (a) and (b) because otherwise it would have been pruned off at some finite stage. So again Δ^+ is a Hintikka set.

The third argument is the *maximizing* argument, sometimes known as the *Henkin-style* argument, though Skolem's argument in [1922] seems to be of this type. This argument is an opposite to the second kind of argument: instead of using (a)–(g) in the construction and forgetting consistency, we exploit consistency and leave (a)–(g) on one side until the very end. We define by induction theories $\Delta_0, \Delta_1, \ldots$ in the language L^+ so that (i) every theory Δ_i is consistent; (ii) for all i, Δ_{i+1} includes Δ_i; (iii) for each i, only finitely many of the witnesses appear in the sentences in Δ_i; (iv) Δ_0 is Δ; and (v) for each i,

(α) if Δ_i together with ϕ_i is consistent then Δ_{i+1} contains ϕ_i;
(β) if ϕ_i is in Δ_{i+1} and is of form $\exists x\psi$, then for some witness c which doesn't occur in Δ_i or in ϕ_i, $\psi[c/x]$ is in Δ_{i+1}.

The argument to justify this construction is the same as for the direct argument, except that (c'), (d'), (e') and (g') are now irrelevant. As before, let Δ^+ be the set of sentences which occur in at least one theory Δ_i. Clause (α) in the construction guarantees that

(16.3) for every sentence ϕ of L^+, if Δ^+ together with ϕ is consistent, then ϕ is in Δ^+.

From (16.3) and properties of natural deduction we infer

(16.4) for every sentence ϕ of L^+, if ϕ is provable from assumptions in Δ^+ then ϕ is in Δ^+.

Knowing (16.3) and (16.4), we can show that Δ^+ satisfies (c)–(g). For example, take (e) and suppose that $\neg(\psi \wedge \chi)$ is in Δ^+ but $\neg\psi$ is not in Δ^+. Then by (16.3) there is a proof of $\bot$ from Δ^+ and $\neg\psi$. Using the natural deduction rules we can adapt this proof to get a proof of $\neg\chi$ from Δ^+, and it follows by (16.4) that $\neg\chi$ is in Δ^+. Since the Δ_i are all consistent, Δ^+ also satisfies (a) and (b). So once again Δ^+ is a Hintikka set.

Some authors take care of clause (β) before the main construction. They can do it by adding to Δ a collection of sentences of form $\exists x\psi \to \psi[c/x]$. The argument which justified (f') will justify this too.

The first and third arguments above are very closely related. I gave both of them in the form that would serve for a countable language, but they adapt to first-order languages of any cardinality. The merit of the maximizing argument is that the construction is easy to describe. (For example, the listing $\phi_0, \phi_1, \ldots$ need not repeat any formulas.)

The first and second arguments have one advantage over the third. Suppose Δ is a finite set of prenex sentences of form $\exists \vec{x} \forall \vec{y} \psi$, with no quantifiers in ψ. Then these two arguments find Δ^+ after only a finite number of steps in the construction. So Δ^+ is finite and has a finite model, and it follows that we can compute whether or not a sentence of this form has a model. (This is no longer true if function symbols are added to the language as in Section 18 below.) The decidability of propositional logic is a special case of this. So also are various theorems about finite models for modal logics.

When Δ_0 is finite, closer inspection of the trees (16.2) shows that they are just the natural extension to predicate logic of the semantic tableaux of propositional logic. If Δ_0 has no models then every branch comes to a dead end after a finite number of steps. If Δ_0 has a model, then the tree has a branch which never closes, and we can read this branch as a description of a model. So the tree argument has given us a complete proof calculus for predicate logic. (Cf. Beth [1955, 1962], Jeffrey [1967], Smullyan [1968], Bell and Machover [1977] for predicate logic semantic tableaux.) Incidentally it is most unpleasant to prove the completeness of semantic tableaux by the direct or maximizing arguments. One needs facts of the form: if $\Delta \vdash \psi$ and $\Delta, \psi \vdash \chi$ then $\Delta \vdash \chi$. To prove these is to prove Gentzen's cut-elimination theorem.

Notice that even when Δ_0 is finite, semantic tableaux no longer provide a method for deciding whether Δ_0 has a model. If it does have a model, the tree may simply go on branching for ever, and we may never know whether it is going to close off in the next minute or the next century. In Section 24 below we prove a theorem of Church [1936] which says that there is not and cannot be any mechanical method for deciding which sentences of predicate logic have models.

17. CONSEQUENCES OF THE CONSTRUCTION OF MODELS

Many of the most important consequences of the construction in the previous section are got by making some changes in the details. For example, instead of using the individual constants of the language as elements, we can number these constants as $b_0, b_1, \ldots$, and use the number n in place of the constant b_n. Since numbers can be thought of as pure sets (Mendelson [1964] pp. 170ff. or Appendix C below), the structure which emerges at the end will be a pure set structure. Hence, for any standard proof calculus for a language L of predicate logic:

THEOREM 17.1. *Suppose T is a theory and ψ a sentence of* L, *such that the calculus doesn't prove ψ from T. Then there is a pure set structure which is a model of T and not of ψ.*

In terms of the discussion in Section 8 above, this shows that the Proof Theorist's notion of logical implication agrees with the Model Theorist's, whether or not the Model Theorist restricts himself to pure set structures.

We can take matters one step further by encoding all symbols and formulas of L as natural numbers. So a theory in L will be a set of numbers. Suppose the theory T is in fact the set of all numbers which satisfy the first-order formula ϕ in the language of arithmetic; then by analyzing the proof of Theorem 17.1 we can find another first-order formula χ in the language of arithmetic, which defines a structure with natural numbers as its elements, so that:

THEOREM 17.2. *In first-order Peano arithmetic we can prove that if some standard proof calculus doesn't prove T is inconsistent, then the structure defined by χ is a model of T.*

(Cf. Kleene [1952] p. 394 and Hasenjaeger [1953] for a sharper result.)

Theorem 17.2 is philosophically very interesting. Suppose T is a finite theory, and proof-theoretically T doesn't imply ψ. Applying Theorem 17.2 to the theory $T \cup \{\neg\psi\}$, we get a formula χ which defines a natural number structure $\mathfrak{A}$ in which T is true and ψ is false. By means of χ, the formulas of T and ψ can be read as meaningful statements about $\mathfrak{A}$ and hence about the natural numbers. The statements in T are true but ψ is false, so we have found an invalid argument of the form 'T. Therefore ψ'. It follows that if a first-order sequent is correct by the Traditional Logician's definition, then it is correct by the Proof Theorist's too. Since the converse is straightforward to prove, we have a demonstration that *the Traditional Logician's notion of validity exactly coincides with the Proof Theorist's.* The proof of this result uses nothing stronger than the assumption that the axioms of first-order Peano arithmetic have a model.

The Traditional Logician's notion of logical implication is quite informal – on any version it involves the imprecise notion of a 'valid English argument'. Nevertheless we have now proved that it agrees exactly with the mathematically precise notion of logical implication given by the Proof Theorist. (Cf. Kreisel [1967].) People are apt to say that it is impossible to prove that an informal notion and a formal one agree exactly. Since we have just done the impossible, maybe I should add a comment. Although the notion of a valid argument is vague, there is no doubt that (i) if there is a formal proof of a sequent, then any argument with the form of that sequent must be valid, and (ii) if there is an explicitly definable counterexample to the sequent, then there is an invalid argument of that form. We have shown, by strict mathematics, that every finite sequent has either a formal proof or an explicitly definable counterexample. So we have trapped the informal notion between two formal ones. Contrast *Church's thesis*, that the effectively computable functions (informal notion) are exactly the recursive ones (formal). There is no doubt that the existence of a recursive definition for a function makes the function effectively computable. But nobody has yet thought of any kind of mathematical object whose existence undeniably implies that a function is *not* effectively computable. So Church's thesis remains unproved. (Van Dalen [Chapter I.5, below] discusses Church's thesis.)

I return to the completeness proof. By coding all expressions of L into numbers or sets, we made it completely irrelevant that the symbols of L can be written on a page, or even that there are at most countably many of them. *So let us now allow arbitrary sets to serve instead of symbols.* Languages of this abstract type can be called *set languages*. They are in common use today even among proof theorists. Of course to use these languages we have to rely either on our intuitions about sets or on proofs in axiomatic set theory; there is no question of checking by inspection. Henkin's [1949] completeness proof was given in this setting. In fact he proved:

THEOREM 17.3. *If* L *is a first-order set language and T a theory in* L *whose cardinality is at most the infinite cardinal* κ, *then either a finite part of T can be proved inconsistent by a proof calculus, or T has a model with at most* κ *elements.*

Theorem 17.3 has several important mathematical consequences. For example the *Compactness Theorem* says:

THEOREM 17.4. *Let T be a first-order theory (in a set language). If every finite set of sentences in T has a model, then T has a model.*

Theorem 17.4 for countable languages was proved by Gödel in [1930]. For propositional logic with arbitrarily many symbols it was proved by Gödel [1931a], in answer to a question of Menger. The first proof of Theorem 17.4 was sketched rather inadequately by Anatolii Mal'cev in [1936] (see the review of Mal'cev [1941] by Henkin and Mostowski [1959]). But in [1941] Mal'cev showed that Theorem 17.4 has interesting and far from trivial consequences in group theory, thus beginning one of the most important lines of application of first-order logic in mathematics.

The last consequence I shall draw from Theorem 17.3 is not really interesting until identity is added to the language (see the next section); but this is a convenient place to state it. It is the *Upward and Downward Löwenheim-Skolem Theorem*:

THEOREM 17.5. *Let T be a first-order theory in a language with λ formulas, and κ an infinite cardinal at least as great as λ. If T has a model with infinitely many elements then T has one with exactly κ elements.*

Theorem 17.4 was proved in successively stronger versions by Löwenheim [1915], Skolem [1920, 1922], Tarski in unpublished lectures in 1928, Mal'cev [1936] and Tarski and Vaught [1957]; see Vaught [1974] for a thorough history of this and Theorems 17.3, 17.4. The texts of Bell and Slomson [1969] and Chang and Keisler [1973] develop these theorems, and Sacks [1972] and Cherlin [1976] study some of their applications in algebra. Skolem [1955] expressly dissociated himself from the Upward version of Theorem 17.5, which he regarded as nonsense.

18. IDENTITY

The symbol '=' is reserved for use as a 2-place predicate symbol with the intended meaning

(18.1) $a = b$ iff a and b are one and the same thing.

When $\mathfrak{A}$ is a structure for a language containing '=', we say that $\mathfrak{A}$ has *standard identity* if the relation $I_{\mathfrak{A}}(=)$ holds between elements α and β of $\mathfrak{A}$ precisely when α and β are the same element.

'$x = y$' is read as 'x equals y', rather misleadingly – all men may be created equal but they are not created one and the same man. Another reading is 'x is identical with y'. As far as English usage goes, this is not much improvement on 'equals': there are two identical birds feeding outside my window, but they aren't the same bird (and think of identical twins). Be that as it may, '=' is called the *identity* sign and the relation it expresses in (18.1) is called *identity*.

Let L be a language containing the symbol '='. It would be pleasant if we could find a theory Δ in L whose models are exactly the L-structures with standard identity. Alas, there is no such theory. *For every L-structure $\mathfrak{A}$ with standard identity there is an L-structure $\mathfrak{B}$ which is a model of the same sentences of L as $\mathfrak{A}$ but doesn't have standard identity.* Let us prove this.

Take an L-structure $\mathfrak{A}$ with standard identity and let $\delta_1, \ldots, \delta_{2,000,000}$ be two million objects which are not in the domain of $\mathfrak{A}$. Let β be an element of $\mathfrak{A}$. We construct the L-structure $\mathfrak{B}$ thus. The elements of $\mathfrak{B}$ are those of $\mathfrak{A}$ together with $\delta_1, \ldots, \delta_{2,000,000}$. For each individual constant c we put $I_\mathfrak{B}(c) = I_\mathfrak{A}(c)$. For each element α of $\mathfrak{B}$ we define an element $\hat{\alpha}$ of $\mathfrak{A}$ as follows: if α is in the domain of $\mathfrak{A}$ then $\hat{\alpha}$ is α, and if α is one of the δ_j's then $\hat{\alpha}$ is β. For every n-place predicate constant R we choose $I_\mathfrak{B}(R)$ so that if $\langle \alpha_1, \ldots, \alpha_n \rangle$ is any n-tuple of elements of $\mathfrak{B}$, then:

(18.2) $\langle \alpha_1, \ldots, \alpha_n \rangle$ is in $I_\mathfrak{B}(R)$ iff $\langle \hat{\alpha}_1, \ldots, \hat{\alpha}_n \rangle$ is in $I_\mathfrak{A}(R)$.

This defines $\mathfrak{B}$. By induction on the complexity of ϕ we can prove that for every formula $\phi(x_1, \ldots, x_n)$ of L and every n-tuple $\langle \alpha_1, \ldots, \alpha_n \rangle$ of elements of $\mathfrak{B}$,

(18.3) $\mathfrak{B} \models \phi[\alpha_1/x_1, \ldots, \alpha_n/x_n]$ iff $\mathfrak{A} \models \phi[\hat{\alpha}_1/x_1, \ldots, \hat{\alpha}_n/x_n]$.

In particular, $\mathfrak{A}$ and $\mathfrak{B}$ are models of exactly the same sentences of L. Since $\mathfrak{A}$ has standard identity, $\mathfrak{A} \models (x = x)[\beta/x]$. Then from (18.2) it follows that the relation $I_\mathfrak{B}(=)$ holds between any two of the elements $\delta_1, \ldots, \delta_{2,000,000}$, and so $I_\mathfrak{B}(=)$ is vastly different from standard identity.

So we look for a second best. Is there a theory Δ which is true in all L-structures with standard identity, and which logically implies every sentence of L that is true in all such L-structures? This time the answer is positive. The following theory will do the job:

(18.4) $\forall x \; x = x$.

(18.5)   All sentences of the form $\forall \vec{z} x y \, (x = y \to (\phi \to \phi[y/x]))$.

Formula (18.4) is known as the *law of reflexivity of identity*. (18.5) is not a single sentence but an infinite family of sentences, namely all those which can be got by putting any formula ϕ of L into the expression in (18.5);

$\vec{z}$ are all the free variables of ϕ except for x and y. These sentences (18.5) are collectively known as *Leibniz' Law*. They are the nearest we can get within L to saying that if $a = b$ then anything which is true of a is true of b too.

By inspection it is clear that every L-structure with standard identity is a model of (18.4) and (18.5). To show that (18.4) and (18.5) logically imply every sentence true in all structures with standard identity, let me prove something stronger, namely: *For every L-structure $\mathfrak{B}$ which is a model of* (18.4) *and* (18.5) *there is an L-structure $\mathfrak{A}$ which is a model of exactly the same sentences of* L *as $\mathfrak{B}$ and has standard identity*. Supposing this has been proved, let Δ be the theory consisting of (18.4) and (18.5), and let ψ be a sentence of L which is not logically implied by Δ. Then some L-structure $\mathfrak{B}$ is a model of Δ and $\neg\psi$; so some structure $\mathfrak{A}$ with standard identity is also a model of $\neg\psi$. It follows that ψ is not true in all structures with standard identity.

To prove what I undertook to prove, let $\mathfrak{B}$ be a model of Δ. Then we can show that the following hold, where we write $=_\mathfrak{B}$ for $I_\mathfrak{B}(=)$:

(18.6) the relation $I_\mathfrak{B}(=)$ is an equivalence relation;

(18.7) for every n-place predicate constant R of L, if $\alpha_1 =_\mathfrak{B} \beta_1, \ldots, \alpha_n =_\mathfrak{B} \beta_n$ and $\langle \alpha_1, \ldots, \alpha_n \rangle$ is in $I_\mathfrak{B}(R)$ then $\langle \beta_1, \ldots, \beta_n \rangle$ is in $I_\mathfrak{B}(R)$.

Statement (18.7) can be proved by applying Leibniz' Law n times. Then (18.6) follows from (18.7) and reflexivity of identity, taking '=' for R. Statements (18.6) and (18.7) together are summarized by saying that the relation $=_\mathfrak{B}$ is a *congruence* for L. For each element α of $\mathfrak{B}$, we write $\alpha^=$ for the equivalence class of α under the relation $=_\mathfrak{B}$.

Now we define the L-structure $\mathfrak{A}$ as follows. The domain of $\mathfrak{A}$ is the class of all equivalence classes $\alpha^=$ of elements α of $\mathfrak{B}$. For each individual constant c we define $I_\mathfrak{A}(c)$ to be $I_\mathfrak{B}(c)^=$. For each n-place predicate symbol R of L we define $I_\mathfrak{A}(R)$ by:

(18.8) $\langle \alpha_1^=, \ldots, \alpha_n^= \rangle$ is in $I_\mathfrak{A}(R)$ iff $\langle \alpha_1, \ldots, \alpha_n \rangle$ is in $I_\mathfrak{B}(R)$.

Definition (18.8) presupposes that the right-hand side of (18.8) is true or false depending only on the equivalence classes of $\alpha_1, \ldots, \alpha_n$; but (18.7) assured this.

In particular, $\alpha^= =_\mathfrak{A} \beta^=$ if and only if $\alpha =_\mathfrak{B} \beta$, in other words, if and only if $\alpha^=$ equals $\beta^=$. Hence, $\mathfrak{A}$ has standard identity. It remains only to show that for every formula $\phi(x_1, \ldots, x_n)$ of L and all elements $\alpha_1, \ldots, \alpha_n$ of $\mathfrak{B}$,

(18.9) $\mathfrak{A} \models \phi[\alpha_1^=/x_1, \ldots, \alpha_n^=/x_n]$ iff $\mathfrak{B} \models \phi[\alpha_1/x_1, \ldots, \alpha_n/x_n]$.

I.1: ELEMENTARY PREDICATE LOGIC 71

Statement (18.9) is proved by induction on the complexity of ϕ.

Most logicians include '=' as a part of the vocabulary of every language for predicate logic, and interpret it always to mean standard identity. Since it is in every language, it is usually not mentioned in the similarity type. The proof calculi have to be extended to accommodate '='. One way to extend the natural deduction calculus is to add two new rules:

$$(18.10) \quad \frac{}{x = x} \qquad \frac{x = y \quad \phi}{\phi[y/x]}$$

The first rule deduces $x = x$ from no premises.

Identity is needed for virtually all mathematical applications of logic. It also makes it possible to express in formulas the meanings of various English phrases such as 'the', 'only', 'at least one', 'at most eight' etc. (see e.g. Section 21 below).

Many mathematical applications of logic need symbols of another kind, called *function symbols*. The definitions given above can be stretched to allow function symbols as follows. Symbols f, g, h etc., with or without subscripts, are called *function constants*. A similarity type may contain function constants, each of which is labelled as an *n-place function constant* for some positive integer n. If the language L has an n-place function constant f and $\mathfrak{A}$ is an L-structure, then f is interpreted by $\mathfrak{A}$ as an *n-place function* $I_{\mathfrak{A}}(f)$ which assigns one element of $\mathfrak{A}$ to each ordered n-tuple of elements of $\mathfrak{A}$. For example the 2-place function constant '+' may be interpreted as a function which assigns 5 to $\langle 2, 3 \rangle$, 18 to $\langle 9, 9 \rangle$ and so forth – though of course it can also be interpreted as some quite different function.

There are various ways of writing functions, such as

$$(18.11) \quad \sin x, \quad \sqrt{x}, \quad x^2, \quad \hat{x}, \quad x^y, \quad x+y, \quad \langle x, y \rangle.$$

But the general style is '$f(x_1, \ldots, x_n)$', and logicians' notation tends to follow this style. The details of syntax and proof theory with function symbols are rather messy, so I omit them and refer the reader to Hilbert and Bernays [1934] for details.

One rarely needs function symbols outside mathematical contexts. In any case, provided we have '=' in our language, everything that can be said with function symbols can also be said without them. Briefly, the idea is to use a predicate constant R in such a way that '$R(x_1, \ldots, x_{n+1})$' means '$f(x_1, \ldots, x_n) = x_{n+1}$'. When the function symbol f is in the language, it is true in all structures – and hence logically valid – that for all $x_1, \ldots, x_n$

there is a unique x_{n+1} such that $f(x_1, \ldots, x_n) = x_{n+1}$. Translating f into R, this becomes

(18.12) $\forall x_1 \cdots x_n zt \exists y((R(x_1, \ldots, x_n, z) \wedge R(x_1, \ldots, x_n, t) \to z = t) \wedge R(x_1, \ldots, x_n, y))$.

Since (18.12) is not logically valid, it may have to be assumed as an extra premise when we translate arguments involving f into arguments involving R.

19. AXIOMS AS DEFINITIONS

Axioms are, roughly speaking, the statements which one writes down at the beginning of a book in order to define the subject-matter of the book and provide a basis for deductions made in the book. For example any textbook of group theory will start by telling you that a group is a triple $\langle G, *, e \rangle$ where $*$ is a binary operation in the set G and e is an element of G such that

(19.1) $*$ is associative, i.e. for all x, y and z, $x * (y * z) = (x * y) * z$,

(19.2) e is an identity, i.e. for all x, $x * e = e * x = x$,

(19.3) every element x has an inverse, i.e. an element y such that $x * y = y * x = e$.

Statements (19.1)–(19.3) are known as the *axioms for groups*. I could have chosen examples from physics, economics or even ethics.

It is often said that in an 'axiomatic theory' such as group theory, the axioms are 'assumed' and the remaining results are 'deduced from the axioms'. This is completely wrong. W. R. Scott's textbook *Group Theory* [1964] contains 457 pages of facts about groups, and the last fact which can by any stretch of the imagination be described as being 'deduced from (19.1)–(19.3)' occurs on page 8. We could indeed rewrite Scott's book as a set of deductions from assumed axioms, but the axioms would be those of set theory, not (19.1)–(19.3). These three group axioms would appear, not as assumptions but as *part of the definition of 'group'*.

The definition of a group can be paraphrased as follows. First we can recast the triple $\langle G, *, e \rangle$ as an L-structure $\mathfrak{G} = \langle G, I_{\mathfrak{G}} \rangle$ in a first-order language L with one 2-place function symbol $*$ and one individual constant e. Then $\mathfrak{G}$ is a group if and only if $\mathfrak{G}$ is a model of the following three sentences:

(19.4) $\forall xyz\ x * (y * z) = (x * y) * z$,

(19.5) $\forall x(x * e = x \land e * x = x)$,

(19.6) $\forall x \exists y(x * y = e \land y * x = e)$.

Generalizing this, let Δ be any theory in a first-order language L. Let **K** be a class of L-structures. Then Δ is said to *axiomatize* **K**, and **K** is called $Mod(\Delta)$, if **K** is the class of all L-structures which are models of Δ. The sentences in Δ are called *axioms* for **K**. Classes of form $Mod(\{\phi\})$ for a single first-order sentence ϕ are said to be *first-order definable*. Classes of form $Mod(\Delta)$ for a first-order theory Δ are said to be *generalized first-order definable*. The class of groups is first-order definable – we can use the conjunction of the three sentences (19.4)–(19.6).

Many other classes of structure which appear in pure or applied mathematics are (generalized) first-order definable. To give examples I need only list the axioms. First, *equivalence relations*:

(19.7) $\forall x\ R(x,x)$ 'R is reflexive'

(19.8) $\forall xy\ (R(x,y) \rightarrow R(y,x))$ 'R is symmetric'

(19.9) $\forall xyz\ (R(x,y) \land R(y,z) \rightarrow R(x,z))$ 'R is transitive'.

Next, *partial orderings*:

(19.10) $\forall x\ x \leq x$ '$\leq$ is reflexive'

(19.11) $\forall xyz(x \leq y \land y \leq z \rightarrow x \leq z)$ '$\leq$ is transitive'

(19.12) $\forall xy(x \leq y \land y \leq x \rightarrow x = y)$ '$\leq$ is antisymmetric'.

Then *total* or *linear orderings* are axiomatized by (19.11) and (19.12) and

(19.13) $\forall xy(x \leq y \lor y \leq x)$ '$\leq$ is connected'.

Total orderings can also be axiomatized as follows, using $<$ instead of $\leq$:

(19.14) $\forall xyz(x < y \land y < z \rightarrow x < z)$

(19.15) $\forall x\ \neg x < x$

(19.16) $\forall xy(x < y \lor y < x \lor x = y)$.

A total ordering in the second style can be converted into a total ordering in the first style by reading $x \leq y$ as meaning $x < y \lor x = y$. There is a similar conversion from the first style to the second. We can express various conditions on linear orderings by adding further axioms to (19.11)–(19.13):

(19.17) $\exists x \forall y\, y \leqslant x$ 'there is a last element'

(19.18) $\forall x \exists y (\neg x = y \wedge \forall z (x \leqslant z \leftrightarrow x = z \vee y \leqslant z))$
'every element has an immediate successor'

Algebra is particularly rich in first-order or generalized first-order definable classes, for example rings, fields, lattices, categories, toposes, algebraically closed fields, vector spaces over a given field. *Commutative groups* are axiomatized by adding to (19.4)–(19.6) the axiom

(19.19) $\forall xy\, x * y = y * x$.

All the examples mentioned so far are first-order definable except for algebraically closed fields and vector spaces over an infinite field, which need infinitely many sentences to define them.

The notion of first-order definable classes was first made explicit in a paper of Tarski [1954]. If we know that a class of structures is generalized first-order definable then we immediately know various other facts about it, for example that it is closed under taking ultraproducts (cf. Chang and Keisler [1973] or Bell and Slomson [1969] – they are defined in Appendix C below) and that implicit definitions in the class can all be made explicit ('Beth's theorem' – Theorem 27.5 in Section 27 below). On the other hand, if one is not interested in model-theoretic facts like these, the informal style of (19.1)–(19.3) makes just as good a definition of a class as any set of first-order formulas. (In the philosophy of science, structuralists have given reasons for preferring the informal set-theoretic style; see Sneed [1971] and Stegmüller [1976].)

It was Hilbert and his school who first exploited axioms, higher-order as well as first-order, as a means of defining classes of structures. Hilbert was horrifically inaccurate in describing what he was doing. When he set up geometric axioms, he said that they defined what was meant by a point. Frege then caustically asked how he could use this definition to determine whether his pocket watch was a point (Frege and Hilbert [1899/1900]). Hilbert had simply confused defining a class of structures with defining the component relations and elements of a single structure. (Cf. the comments of Bernays [1942].) In this matter Hilbert was a spokesman for a confusion which many people shared. Even today one meets hopeful souls who believe that the axioms of set theory define what is meant by 'set'.

Hilbert spoke only for himself when he added the lunatic remark that 'If ... arbitrarily posited axioms together with all their consequences do not contradict one another, then they are true and the things defined by these axioms exist'. (Frege and Hilbert [1899/1900].) For example, one infers,

I.1: ELEMENTARY PREDICATE LOGIC

if the axioms which say there is a measurable cardinal are consistent, then there is a measurable cardinal. If the axioms which say there is no measurable cardinal are consistent, then there is no measurable cardinal. If both sets of axioms are consistent . . . . As I read him, Hilbert had at the back of his mind an informal version of the completeness theorem which Gödel proved in [1930]: if a set of first-order axioms is proof-theoretically consistent then it has a model. This result by itself has no philosophical implications about the existence of abstract entities. Readers of Hilbert's philosophical remarks should always bear in mind his slogan *'Wir sind Mathematiker'* ([1925]).

20. AXIOMS WITH INTENDED MODELS

Axioms are not always intended to define a class of structures as in Section 19 above. Often they are written down *in order to set on record certain facts about a particular structure*. The structure in question is then called the *intended interpretation* or *standard model* of the axioms. The best known example is probably the axioms of Peano arithmetic, which were set down by Dedekind [1888, 1890] as a statement of the 'fundamental properties' of the natural number sequence (the first-order formalization is due to Gödel [1931], cf. Appendix B below). Euclid's axioms and postulates of geometry are another example, since he undoubtedly had space in mind as the intended interpretation.

The object in both Dedekind's case and Euclid's was to write down some elementary facts about the standard model so that further information could be got by making deductions from these facts. With this aim it becomes very natural to write the axioms in a first-order language, because we understand first-order deducibility well and so we shall know exactly what we are entitled to deduce from the axioms.

However, there is no hope at all of *defining* the natural numbers, even up to isomorphism, by means of any first-order axioms. Let me sketch a proof of this – it will be useful later. Suppose L is the first-order language of arithmetic, with symbols to represent plus and times, a 2-place predicate constant < ('less than'), and a name n^* for each natural number n. Let L^+ be L with a new individual constant c added. Let Δ be the set of all sentences of L which are true in the standard model. Let Δ^+ be Δ together with the sentences

(20.1) $0^* < c, \quad 1^* < c, \quad 2^* < c, \quad \ldots$.

Now if Γ is any finite set of sentences from Δ^+ then Γ has a model: take the standard model of Δ and let c stand for some natural number which is greater than every number mentioned in Γ. So by the Compactness Theorem (Theorem 17.4 in Section 17 above), Δ^+ has a model $\mathfrak{A}$. Since Δ^+ includes Δ, $\mathfrak{A}$ is a model of Δ and hence is a model of exactly the same sentences of L as the standard model. But $\mathfrak{A}$ also has an element $I_\mathfrak{A}(c)$ which by (20.1) is 'greater than' $I_\mathfrak{A}(0^*)$, $I_\mathfrak{A}(1^*)$, $I_\mathfrak{A}(2^*)$ and all the 'natural numbers' of $\mathfrak{A}$. So $\mathfrak{A}$ is a model of Δ with an 'infinite element'. Such models of Δ are called *non-standard models of arithmetic*. They were first constructed by Skolem [1934], and today people hold conferences on them.

But one can reasonably ask whether, say, the first-order Peano axioms (cf. Appendix B) imply all first-order sentences which are true in the standard model. This is equivalent to asking whether the axioms are a *complete* theory in the sense that if ϕ is any sentence of their language, then either ϕ or $\neg\phi$ is a consequence of the axioms. Gödel's epoch-making paper [1931] showed that the first-order Peano axioms are not complete; in fact no mechanically describable theory in this language is both complete and true in the standard model. In Section 24 below I shall sketch a proof of this.

There is a halfway house between the use of axioms to define a class and their use to say things about a standard model. Often we want to work with a class **K** of L-structures which may not be generalized first-order definable. In such cases we say that a theory Δ is a *set of axioms* for **K** if every structure in **K** is a model of Δ; we call it a *complete* set of axioms for **K** if moreover every sentence of L which is true in all structures in **K** is a logical consequence of Δ.

Let me give three examples. (i) For the first, paraphrasing Carnap ([1956] p. 222ff) I consider the class of all structures which represent possible worlds, with domain the set of all people, '*Bx*' interpreted as '*x* is a bachelor' and '*Mx*' as '*x* is married'. Obviously this class is not generalized first-order definable. But the following sentence is a complete set of axioms:

(20.2) $\forall x(Bx \rightarrow \neg Mx)$.

In Carnap's terminology, when **K** is the class of all structures in which certain symbols have certain fixed meanings, axioms for **K** are called *meaning postulates*. (Lakoff [1972] discusses some trade-offs between meaning postulates and deep structure analysis in linguistics.)

(ii) For a second sample, consider second-order logic (cf. [Chapter I.4, below]). In this logic we are able to say 'for all subsets P of the domain, . . . ', using second-order quantifiers '$\forall P$'. For reasons explained in Chapter I.4 below, there is no hope of constructing a complete proof calculus for second-order logic. But we do have some incomplete calculi which are good for most practical purposes. They prove, among other things, the formula

(20.3) $\forall PQ(\forall z(P(z) \leftrightarrow Q(z)) \to P = Q)$

which is the second-order logician's version of the axiom of extensionality.

Second-order logic can be translated wholesale into a kind of two-sorted first-order logic by the following device. Let L be any (first-order) language. Form a two-sorted language $L^\downarrow$ with the same predicate and individual constants as L, together with one new 2-place predicate constant ϵ. For each L-structure $\mathfrak{A}$, form the $L^\downarrow$-structure $\mathfrak{A}^\downarrow$ as follows. The domain of $\mathfrak{A}^\downarrow$ is $|\mathfrak{A}| \cup \mathscr{P}|\mathfrak{A}|$, $|\mathfrak{A}|$ is the domain for the first sort and $\mathscr{P}|\mathfrak{A}|$ is the domain for the second. ($\mathscr{P}X$ = the set of all subsets of X.) If α and β are elements of $\mathfrak{A}^\downarrow$, then

(20.4) $\langle \alpha, \beta \rangle$ is in $I_{\mathfrak{A}^\downarrow}(\epsilon)$ iff α is an element of the first sort, β of the second sort, and $\alpha \in \beta$.

The constants of L are interpreted in the first sort of $\mathfrak{A}^\downarrow$ just as they were in $\mathfrak{A}$. Now each second-order statement ϕ about L-structures $\mathfrak{A}$ is equivalent to a first-order statement $\phi^\downarrow$ about $L^\downarrow$-structures $\mathfrak{A}^\downarrow$. For example, if we use number superscripts to distinguish the first and second sorts of variables, the axiom of extensionality (20.3) translates into

(20.5) $\forall x^2 y^2 (\forall z^1 (z^1 \epsilon x^2 \leftrightarrow z^1 \epsilon y^2) \to x^2 = y^2)$.

Axiom (20.5) is a first-order sentence in $L^\downarrow$.

Let **K** be the class of all $L^\downarrow$-structures of form $\mathfrak{A}^\downarrow$ for some L-structure $\mathfrak{A}$. Let **QC**^{2} be some standard proof calculus for second-order logic, and let Δ be the set of all sentences $\phi^\downarrow$ such that ϕ is provable by **QC**^{2}. Then Δ is a set of axioms for **K**, though not a complete one. The $L^\downarrow$-structures in **K** are known as the *standard* models of Δ. There will be plenty of non-standard models of Δ too, but because of (20.5) they can all be seen as 'parts of' standard models in the following way. For each element β of the second sort in the model $\mathfrak{B}$ of Δ, let β^+ be the set of elements α such that $\langle \alpha, \beta \rangle \in I_\mathfrak{B}(\epsilon)$. By (20.5), $\beta^+ = \gamma^+$ implies $\beta = \gamma$. So in $\mathfrak{B}$ we can replace each element β of the second sort by β^+. Then the second sort consists of subsets of the domain of the first sort, but not necessarily all the subsets. All the subsets are in the second domain if and only if this doctored version of $\mathfrak{B}$ is a standard model. (Models of Δ, standard or non-standard, are known as *Henkin models of second-order logic*, in view of Henkin [1950].)

How can one distinguish between a proof calculus for second-order logic on the one hand, and on the other hand a first-order proof calculus which also proves the sentences in Δ? The answer is easy: one can't. In our notation above, the proof calculus for second-order logic has '$P(z)$' where the first-order calculus has '$z^1 \epsilon x^2$', but this is no more than a difference of notation. Take away this difference and the two calculi become exactly the same thing. Don't be misled by texts like Church [1956] which present 'calculi of first order' in one chapter and 'calculi of second order' in another. The latter calculi are certainly different from the former, because they incorporate a certain amount of set theory. But what makes them second-order calculi, as opposed to two-sorted first-order calculi with extra non-logical axioms, is *solely their intended interpretation.*

It follows, incidentally, that it is quite meaningless to ask whether the proof theory of actual mathematics is first-order or higher-order. (I recently saw this question asked. The questioner concluded that the problem is 'not easy'.)

Where then can one meaningfully distinguish second-order from first-order? One place is the *classification of structures*. The class **K** of standard models of Δ is not a first-order definable class of $L^{\downarrow}$-structures, but it is second-order definable.

More controversially, we can distinguish between *first-order and second-order statements about a specific structure*, even when there is no question of classification. For example the sentence (20.3) says about an L-structure $\mathfrak{A}$ something which can't be expressed in the first-order language of $\mathfrak{A}$. This is not a matter of classification, because (20.3) is true in *all* L-structures.

(iii) In Section 18 we studied the class of all L-structures with standard identity. Quine [1970] p. 63f studies them too, and I admire his nerve. He first demonstrates that in any language L with finite similarity type there is a formula ϕ which defines a congruence relation in every L-structure. From Section 18 we know that ϕ cannot always express identity. Never mind, says Quine, let us *redefine* identity by the formula ϕ. This happy redefinition instantly makes identity first-order definable, at least when the similarity type is finite. It also has the consequence, not mentioned by Quine, that for any two different things there is some language in which they are the same thing. (Excuse me for a moment while I redefine exams as things that I don't have to set.)

21. NOUN PHRASES

In this section I want to consider whether we can make any headway by adding to first-order logic some symbols for various types of noun phrase. Some types of noun phrase, such as 'most Xs', are not really fit for formalizing because their meanings are too vague or too shifting. Of those which can be formalized, some never give us anything new, in the sense that any formula using a symbol for them is logically equivalent to a formula of first-order logic (with =); to express this we say that these formalizations give *conservative extensions* of first-order logic. Conservative extensions are not necessarily a waste of time. Sometimes they enable us to say quickly something that can only be said lengthily in first-order symbols, sometimes they behave more like natural languages than first-order logic does. So they may be useful to linguists or to logicians in a hurry.

Many (perhaps most) English noun phrases have to be symbolized as *quantifiers and not as terms*. For example the English sentence

(21.1) I have inspected every batch.

finds itself symbolized by something of form

(21.2) For every batch x, I have inspected x.

Let me recall the reason for this. If we copied English and simply put the noun phrase in place of the variable x, there would be no way of distinguishing between (i) the negation of 'I have inspected every batch' and (ii) the sentence which asserts, of every batch, that I have not inspected it. In style (21.2) there is no confusion between (i), viz.

(21.3) $\neg$ For every batch x, I have inspected x.

and (ii), viz.

(21.4) For every batch x, $\neg$ I have inspected x.

Confusions like that between (i) and (ii) are so disastrous in logic that it constantly amazes logicians to see that natural languages, using style (21.1), have not yet collapsed into total anarchy.

In the logician's terminology, the *scope* of the quantifier 'For every batch x' in (21.4) is the whole sentence, while in (21.3) it is only the part after the negation sign. Unlike its English counterpart, the quantifier doesn't *replace* the free occurrences of x in the predicate, it *binds* them. (More precisely, an occurrence of a quantifier with variable x binds all occurrences of x which are within its scope and not already bound.) This terminology carries over at once to the other kinds of quantifier that we shall consider, for example

(21.5) $\neg$ For one in every three men x, x is colourblind.

The quantifier 'For one in every three men x' binds both occurrences of the variable, and doesn't include the negation in its scope.

I shall consider three groups of noun phrases. The first yield conservative extensions of first-order logic and are quite unproblematic. The second again give conservative extensions and are awkward. The third don't yield conservative extensions – we shall prove this. In all cases I assume that we start with a first-order language L with identity.

The *first* group are noun phrases such as 'At least n things', 'At most n things', 'Exactly n things'. We begin by finding, for each natural number n, a formula $\exists_{\geq n} x\phi$ of L which expresses 'There are at least n things x such that ϕ'. We do it recursively:

(21.6) $\quad \exists_{\geq 0} x\phi \quad$ is $\quad \neg \bot; \quad \exists_{\geq 1} x\phi \quad$ is $\quad \exists x\phi$.

(21.7) $\quad \exists_{\geq n+1} x\phi \quad$ is $\quad \exists y(\phi[y/x] \wedge \exists_{\geq n} x(\neg x = y \wedge \phi)) \quad$ when $n \geq 1$.

To these definitions we add:

(21.8) $\quad \exists_{\leq n} x\phi \quad$ is $\quad \neg \exists_{\geq n+1} x\phi$.

(21.9) $\quad \exists_{= n} x\phi \quad$ is $\quad \exists_{\geq n} x\phi \wedge \exists_{\leq n} x\phi$.

$\exists_{=1} x\phi$ is sometimes written $\exists ! x\phi$.

Definitions (21.6)–(21.9) are in the metalanguage; they simply select formulas of L. But there is no difficulty at all in adding the symbols $\exists_{\geq n}$, $\exists_{\leq n}$ and $\exists_{=n}$ for each natural number to the language L, and supplying the needed extra clauses in the definition of $\models$, together with a complete formal calculus.

The *second* group are singular noun phrases of the form 'The such-and-such'. These are known as *definite descriptions*. Verbal variants of definite descriptions, such as 'My father's beard' for 'The beard of my father', are generally allowed to be definite descriptions too.

According to Bertrand Russell (Russell [1905], Whitehead and Russell [1910] Introduction, Chapter III), the sentence

(21.10) The author of 'Slawkenburgius on Noses' was a poet.

can be paraphrased as stating three things: (1) at least one person wrote 'Slawkenburgius on Noses'; (2) at most one person wrote 'Slawkenburgius on Noses'; (3) some person who did write 'Slawkenburgius on Noses' was a poet. I happily leave to Bencivenga [III.6] and Salmon [IV.7] the question whether Russell was right about this. But assuming he was, his theory calls for the following symbolization. We write '$\{\imath x\psi\}$' to represent 'the person or thing x such that ψ', and we define

(21.11) $\quad \{\imath x\psi\}\phi \quad$ to mean $\quad \exists_{=1} x\psi \wedge \exists x(\psi \wedge \phi)$.

I.1: ELEMENTARY PREDICATE LOGIC

Expression (21.11) can be read either as a metalinguistic definition of a formula of L, or as a shorthand explanation of how the expressions $\{\imath x\psi\}$ can be added to L. In the latter case the definition of $\models$ has to sprout one extra clause:

(21.12) $\mathfrak{A} \models \{\imath x\psi\}\phi[g]$ iff there is a unique element α of $\mathfrak{A}$ such that $\mathfrak{A} \models \psi[g, \alpha/x]$, and for this α, $\mathfrak{A} \models \phi[g, \alpha/x]$.

There is something quite strongly counterintuitive about the formulas on either side in (21.11). It seems in a way obvious that when there is a unique such-and-such, we can refer to it by saying 'the such-and-such'. But Russell's paraphrase never allows us to use the expression $\{\imath x\psi\}$ this way. For example if we want to say that the such-and-such equals 5, Russell will not allow us to render this as '$\{\imath x\psi\} = 5$'. The expression $\{\imath x\psi\}$ has the wrong grammatical type, and the semantical explanation in (21.12) doesn't make it work like a name. On the right-hand side in (21.11) the position is even worse – the definition description has vanished without trace.

Leaving intuition on one side, there are any number of places in the course of formal calculation where one wants to be able to say 'the such-and-such', and then operate with this expression *as a term*. For example formal number theorists would be in dire straits if they were forbidden use of the term

(21.13) $\mu x\psi$, i.e. the least number x such that ψ.

Likewise formal set theorists need a term

(21.14) $\{x \mid \psi\}$, i.e. the set of all sets x such that ψ.

Less urgently, there are a number of mathematical terms which bind variables, for example the integral $\int_b^a f(x)\, dx$ with bound variable x, which are naturally defined as 'the number λ such that ... (here follows half a page of calculus)'. If we are concerned to formalize mathematics, the straightforward way to formalize such an integral is by a definite description term.

Necessity breeds invention, and in the event it is quite easy to extend the first-order language L by adding *terms* $\imath x\psi$. (The definitions of 'term' and 'formula' in Section 13 above have to be rewritten so that the classes are defined by simultaneous induction, because now we can form terms out of formulas as well as forming formulas out of terms.) There are two ways to proceed. One is to take $\imath x\psi$ as a name of the unique element satisfying ψ, if there is such a unique element, and as undefined otherwise; then to reckon an atomic formula false whenever it contains an undefined term. This is equivalent to giving each occurrence of $\imath x\psi$ the smallest possible scope, so that the notation need not indicate any scope. (Cf. Kleene [1952] p. 327; Kalish and Montague [1964] Chapter VII.) The second is to note that questions of scope only arise if there is not a unique such-and-such. So we can choose a constant of the language, say 0, and read $\imath x\psi$ as

(21.15) the element which is equal to the unique x such that ψ if there is such a unique x, and is equal to 0 if there is not.

(Cf. Montague and Vaught [1959], Suppes [1972].)

Russell himself claimed to believe that definite descriptions 'do not name'. So it is curious to note (as Kaplan does in his illuminating paper [1966] on Russell's theory of descriptions) that Russell himself didn't use the notation (21.11) which makes definite descriptions into quantifiers. What he did instead was to invent the notation $\imath x\psi$ *and then use it both as a quantifier and as a term*, even though this makes for a contorted syntax. Kaplan detects in this 'a lingering ambivalence' in the mind of the noble lord.

The *third* group of noun phrases express things which can't be said with first-order formulas. Peirce [1885] invented the *two-thirds* quantifier which enables us to say 'At least $\frac{2}{3}$ of the company have white neckties'. (His example.) Peirce's quantifier was unrestricted. It seems more natural, and changes nothing in principle, if we allow a relativization predicate and write $\frac{2}{3}x(\psi, \phi)$ to mean 'At least $\frac{2}{3}$ of the things x which satisfy ψ satisfy ϕ'.

Can this quantifier be defined away in the spirit of (21.6)–(21.9)? Unfortunately not. Let me prove this. By a *functional* I shall mean an expression which is a first-order formula except that formula metavariables may occur in it, and it has no constant symbols except perhaps =. By substituting actual formulas for the metavariables, we get a first-order formula. Two functionals will be reckoned *logically equivalent* if whenever the same formulas are substituted for the metavariables in both functionals, the resulting first-order formulas are logically equivalent. For example the expression $\exists_{\geq 2} x\phi$, viz.

(21.16) $\exists y(\phi[y/x] \wedge \exists x(\neg x = y \wedge \phi))$,

is a functional which is logically equivalent to $\exists_{\geq 3} x\phi \vee \exists_{=2} x\phi$. Notice that we allow the functional to change some variables which it binds, so as to avoid clash of variables.

A theorem of Skolem [1919] and Behmann [1922] (cf. Ackermann [1962] pp. 41–47) states that *if a functional binds only one variable in each inserted formula, then it is logically equivalent to a combination by $\neg$, $\wedge$ and $\vee$ of equations $y = z$ and functionals of the form $\exists_{=n} x\chi$ where χ is a functional*

I.1: ELEMENTARY PREDICATE LOGIC

without quantifiers. Suppose now that we could define away the quantifier $\frac{2}{3}x(,)$. The result would be a functional binding just the variable x in ψ and ϕ, so by the Skolem–Behmann theorem we could rewrite it as a propositional compound of a finite number of functionals of form $\exists_{=n}x\chi$, and some equations. (The equations we can forget, because the meaning of $\frac{2}{3}x(\psi,\phi)$ shows that it has no significant free variables beyond those in ψ or ϕ.) If n is the greatest integer for which $\exists_{=n}x$ occurs in the functional, then the functional is incapable of distinguishing any two numbers greater than n, so that it can't possibly express that one of them is at least $\frac{2}{3}$ times the other.

A harder example is

(21.17) The average Briton speaks at least two-thirds of a foreign language.

I take this to mean that if we add up the number of foreign languages spoken by each Briton, and divide the sum total by the number of Britons, then the answer is at least $\frac{2}{3}$. Putting $\psi(x)$ for 'x is a Briton' and $\phi(x,y)$ for 'y is a foreign language spoken by x', this can be symbolized as $\{Av\frac{2}{3}xy\}(\psi,\phi)$. Can the quantifier $\{Av\frac{2}{3}xy\}$ be defined away in a first-order language? Again the answer is no. This time the Skolem–Behmann result won't apply directly, because $\{Av\frac{2}{3}xy\}$ binds two variables, x and y, in the second formula ϕ. But indirectly the same argument will work. $\frac{2}{3}x(\psi,\phi)$ expresses just the same thing as $\forall z(\psi[z/x] \to \{Av\frac{2}{3}xy\}(\psi, z = x \wedge \phi[y/x] \wedge \psi[y/x]))$. Hence if $\{Av\frac{2}{3}xy\}$ could be defined away, then so could $\frac{2}{3}x$, and we have seen that this is impossible.

Barwise and Cooper [1981] made a thorough study of the logical properties of natural language noun phrases. See also Montague [1970] and [1973], particularly his discussion of 'the'. Van Benthem and Doets [Chapter I.4 below] have a fuller discussion of things not expressible in first-order language.

III: THE EXPRESSIVE POWER OF FIRST-ORDER LOGIC

22. AFTER ALL THAT, WHAT IS FIRST-ORDER LOGIC?

It may seem perverse to write twenty-one sections of a chapter about elementary (i.e. first-order) logic without ever saying what elementary logic is. But the easiest definition is ostensive: elementary logic is the logic that we have been doing in Sections 1–18 above. But then, why set *that* logic apart from any other? What particular virtues and vices does it have?

At first sight the Traditional Logician might well prefer a stronger logic. After all, the more valid argument schemas you can find him the happier he is. But in fact Traditional Logicians tend to draw a line between what is 'genuinely logic' and what is really mathematics. The 'genuine logic' usually turns out to be a version of first-order logic.

One argument often put forward for this choice of 'genuine logic' runs along the following lines. In English we can group the parts of speech into two groups. The first group consists of *open classes* such as nouns, verbs, adjectives. These classes expand and contract as people absorb new technology or abandon old-fashioned morality. Every word in these classes carries its own meaning and subject-matter. In the second group are the *closed classes* such as pronouns and conjunctions. Each of these classes contains a fixed, small stock of words; these words have no subject-matter, and their meaning lies in the way they combine with open-class words to form phrases. Quirk and Greenbaum [1973] p. 18 list the following examples of closed-class words: the, a, that, this, he, they, anybody, one, which, of, at, in, without, in spite of, and, that, when, although, oh, ah, ugh, phew.

The Traditional Logicians' claim is essentially this: 'genuine logic' is the logic which assembles those valid argument schemas in which open-class words are replaced by schematic letters and closed-class words are not. Quirk and Greenbaum's list already gives us $\wedge$ 'and', $\neg$ 'without', $\forall$ 'anybody', $\exists$ 'a', and of course the words 'not', 'if', 'then', 'or' are also closed-class words. The presence of 'at', 'in spite of' and 'phew' in their list doesn't imply we ought to have added any such items to our logic, because these words don't play any distinctive role in arguments. (The presence of 'when' is suggestive though.) Arguably it is impossible to express second-order conditions in English without using open-class words such as 'set' or 'concept'.

It's a pretty theory. Related ideas run through Quine's [1970]. But for myself I can't see why features of the surface grammar of a few languages that we know and love should be considered relevant to the question what is 'genuine logic'.

We turn to the Proof Theorist. His views are not very helpful to us here. As we saw in Section 20 above, there is in principle no difference between a first-order proof calculus and a non-first-order one. Still, he is likely to make the following comment, which is worth passing on. For certain kinds of application of logic in mathematics, a stronger logic may lead to weaker results. To quote one example among thousands: in a famous paper [1965], Ax and Kochen showed that for each positive integer d there are only finitely many primes which contradict a conjecture of Artin about d. Their proof

I.1: ELEMENTARY PREDICATE LOGIC

used heavy set theory and gave no indication what these primes were. Then Cohen [1969] found a proof of the same result using no set-theoretic assumptions at all. From his proof one can calculate, for each d, what the bad primes are. By using the heavy guns, Ax and Kochen had gained intuition but lost information. The moral is that we should think twice before strengthening our logic. The mere fact that a thing is provable in a weaker logic may lead us to further information.

We turn to the Model Theorist. He was probably taught that 'first-order' means we only quantify over elements, not over subsets of the domain of a structure. By now he will have learned (Section 21 above) that some kinds of quantification over elements are not first-order either.

What really matters to a Model Theorist in his language is the interplay of strength and weakness. Suppose he finds a language which is so weak that it can't tell a Montagu from a Capulet. Then at once he will try to use it to prove things about Capulets, as follows. First he shows that something is true for all Montagus, and then he shows that this thing is expressible in his weak language L. Then this thing must be true for at least one Capulet too, otherwise he could use it to distinguish Montagus from Capulets in L. If L is bad enough at telling Montagus and Capulets apart, he may even be able to deduce that *all* Capulets have the feature in question. These methods, which are variously known as *overspill* or *transfer* methods, can be extremely useful if Montagus are easier to study than Capulets.

It happens that first-order languages are excellent for encoding finite combinatorial information (e.g. about finite sequences or syntax), but hopelessly bad at distinguishing one infinite cardinal or infinite ordering from another infinite cardinal or infinite ordering. This particular combination makes first-order model theory very rich in transfer arguments. For example, the whole of Abraham Robinson's nonstandard analysis (Robinson [1967]) is one vast transfer argument. The Model Theorist will not lightly give up a language which is as splendidly weak as the Upward and Downward Löwenheim–Skolem Theorem and the Compactness Theorem (Section 17 above) show first-order languages to be.

This is the setting into which Per Lindström's theorem came (Section 27 below). He showed that any language which has as much coding power as first-order languages, but also the same weaknesses which have just been mentioned, must actually be a first-order language in the sense that each of its sentences has exactly the same models as some first-order sentence.

23. SET THEORY

In 1922 Skolem described a set of first-order sentences which have become accepted, with slight variations, as the definitive axiomatization of set theory and hence in some sense a foundation for mathematics. Skolem's axioms were in fact a first-order version of the informal axioms which Zermelo [1908] had given, together with one extra axiom (Replacement) which Fraenkel [1922] had also seen was necessary. The axioms are known as ZFC – Zermelo–Fraenkel set theory with Choice. They are listed in Appendix C below and developed in detail in Suppes [1972] and Levy [1979].

When these axioms are used as a foundation for set theory or any other part of mathematics, they are read as being about a particular collection V, the class of all sets. Mathematicians differ about whether we have any access to this collection V independently of the axioms. Some writers (Gödel [1947]) believe V is the standard model of the axioms, while others (von Neumann [1925]) regard the symbol 'V' as having no literal meaning at all. But everybody agrees that the axioms have a standard reading, namely as being about V. In this the axioms of ZFC differ from, say, the axioms for group theory, which are never read as being about The Group, but simply as being true in any group.

These axioms form a foundation for mathematics in two different ways. First, some parts of mathematics are directly about sets, so that all their theorems can be phrased quite naturally as statements about V. For example the natural numbers are now often taken to be sets. If they are sets, then the integers, the rationals, the reals, the complex numbers and various vector spaces over the complex numbers are sets too. Thus the whole of real and complex analysis is now recognized as being part of set theory and can be developed from the axioms of ZFC.

Some other parts of mathematics are not about sets, but can be *encoded* in V. We already have an example in Section 17 above, where we converted languages into sets. There are two parts to an encoding. First the entities under discussion are replaced by sets, and we check that all the relations between the original entities go over into relations in V that can be defined within the language of first-order set theory. In the case of our encoded languages, it was enough to note that any finite sequence of sets $a_1, \ldots, a_n$ can be coded into an ordered n-tuple $\langle a_1, \ldots, a_n \rangle$, and that lengths of sequences, concatenations of sequences and the result of altering one term of a sequence can all be defined. (Cf. Gandy [1974].)

The second part of an encoding is to check that all the theorems one wants to prove can be deduced from the axioms of ZFC. Most theorems of elementary syntax can be proved using only the much weaker axioms of Kripke–Platek set theory (cf. Barwise [1975]); these axioms plus the axiom of infinity suffice for most elementary model theory too.

Thus the possibility of encoding pieces of mathematics in set theory rests on two things: first the expressive power of the first-order language for talking about sets, and second the proving power of the set-theoretic axioms. Most of modern mathematics lies within V or can be encoded within it in the way just described. Not all the encodings can be done in a uniform way; see for example Feferman [1969] for a way of handling tricky items from category theory, and the next section below for a trickier item from set theory itself. I think it is fair to say that all of modern mathematics can be encoded in set theory, but it has to be done locally and not all at once, and sometimes there is a perceptible loss of meaning in the encoding. (Incidentally the rival system of *Principia Mathematica*, using a higher-order logic, came nowhere near this goal. As Gödel says of *Principia* in his [1951]: 'it is clear that the theory of real numbers in its present form cannot be obtained'.)

One naturally asks how much of the credit for this universality lies with first-order logic. Might a weaker logic suffice? The answer is unknown. A good deal of mathematics can be encoded in an elementary topos (cf. Johnstone [1977]). All the axioms defining an elementary topos are Horn sentences (see Chang and Keisler [1973] for these); the language of Horn sentences is a proper part of first-order language and it is quite well behaved, to the point that it satisfies a form of Craig's Interpolation Lemma. But it seems one needs arbitrary first-order formulas to encode things in a topos.

Amazingly, Skolem's purpose in writing down the axioms of ZFC was to debunk the enterprise: 'But in recent times I have seen to my surprise that so many mathematicians think that these axioms of set theory provide the ideal foundation for mathematics; therefore it seemed to me that the time had come to publish a critique' ([1922]).

In fact Skolem showed that, since the axioms form a countable first-order theory, they have a countable model $\mathfrak{A}$. In $\mathfrak{A}$ there are 'sets' which satisfy the predicate 'x is uncountable', but since $\mathfrak{A}$ is countable, these 'sets' have only countably many 'members'. This has become known as Skolem's Paradox, though in fact there is no paradox. The set-theoretic predicate 'x is uncountable' is written so as to catch the uncountable elements of V, and there is no reason at all to expect it to distinguish the uncountable elements of other models of set theory. More precisely, this predicate says 'there is no

1-1 function from x to the set ω'. In a model $\mathfrak{A}$ which is different from V, this only expresses that there is no function which is an element of $\mathfrak{A}$ and which is 1-1 from x to ω.

According to several writers, the real moral of Skolem's Paradox is that there is no standard model of ZFC, since for any model $\mathfrak{A}$ of ZFC there is another model $\mathfrak{B}$ which is not isomorphic to $\mathfrak{A}$ but is indistinguishable from $\mathfrak{A}$ by first-order sentences. If you have already convinced yourself that the only things we can say about an abstract structure $\mathfrak{A}$ are of the form 'Such-and-such first-order sentences are true in $\mathfrak{A}$', then you should find this argument persuasive. (See Klenk [1976] and Putnam [1980] for further discussion.)

Skolem's own explanation of why his argument debunks axiomatic set-theoretic foundations is very obscure. He says in several places that the conclusion is that the meaning of 'uncountable' is relative to the axioms of set theory. I have no idea what this means. The obvious conclusion, surely, is that the meaning of 'uncountable' is relative to the *model*. But Skolem said that he didn't believe in the existence of uncountable sets anyway, and we learn he found it disagreeable to review the articles of people who did (Skolem [1955]).

Contemporary set theorists make free use of nonstandard – especially countable – models of ZFC. One usually requires the models to be well-founded, i.e. to have no elements which descend in an infinite sequence

(23.1) $\quad \cdots \in a_2 \in a_1 \in a_0$.

It is easy to see that this is not a first-order condition on models (for example, Hodges [1972] constructs models of full first-order set theory with arbitrarily long descending sequences of ordinals but no uncountable increasing well-ordered sequences – these models are almost inversely well-founded.) However, if we restrict ourselves to models which are subsets of V, then the statement that such a model contains no sequence (23.1) can be written as a first-order formula in the language of V. The moral is that it is simply meaningless to classify mathematical statements absolutely as 'first-order' or 'not first-order'. One and the same statement can perfectly well express a second-order condition on structure $\mathfrak{A}$ but a first-order condition on structure $\mathfrak{B}$. (Cf. Section 20 above.)

24. ENCODING SYNTAX

I begin by showing that the definition of truth in the class V of all sets is not itself expressible in V by a first-order formula. This will demonstrate

I.1: ELEMENTARY PREDICATE LOGIC

that there is at least one piece of mathematics which can't be encoded in set theory without serious change of meaning.

As we saw in the previous section, there is no problem about encoding the first-order language L of set theory into V. Without going into details, let me add that we can go one stage further and add to the language L a name for each set; the resulting language L^+ can still be encoded in V as a definable proper class. Let us assume this has been done, so that every formula of L^+ is in fact a set. For each set b, we write $\ulcorner b \urcorner$ for the constant of L^+ which names b. (This is nothing to do with Quine's corners $\ulcorner \ \urcorner$.) When we speak of sentences of L^+ being true in V, we mean that they are true in the structure whose domain is V where '$\in$' is interpreted as set membership and each constant $\ulcorner b \urcorner$ is taken as a name of b.

A class X of sets is said to be *definable* by the formula ψ if for every set α,

(24.1) $V \models \psi[\alpha/x]$ iff $\alpha \in X$.

Since every set α has a name $\ulcorner \alpha \urcorner$, (24.1) is equivalent to:

(24.2) $V \models \psi(\ulcorner \alpha \urcorner/x)$ iff $\alpha \in X$

where I now write $\psi(\ulcorner \alpha \urcorner/x)$ for the result of putting $\ulcorner \alpha \urcorner$ in place of free occurrences of x in ψ.

Suppose now that the class of true sentences of L^+ can be defined by a formula *True* of L^+ with the free variable x. Then for every sentence ϕ of L^+, according to (24.2),

(24.3) $V \models \mathit{True}(\ulcorner \phi \urcorner/x)$ iff $V \models \phi$.

But since the syntax of L^+ is definable in V, there is a formula χ of L^+ with just x free, such that for every formula ϕ of L^+ with just x free, if $\ulcorner \phi \urcorner = b$ then

(24.4) $V \models \chi(\ulcorner b \urcorner/x)$ iff $V \models \neg \mathit{True}(\ulcorner \phi(\ulcorner b \urcorner/x) \urcorner/x)$.

Now put $b = \ulcorner \chi \urcorner$. Then by (24.3) and (24.4),

(24.5) $V \models \chi(\ulcorner b \urcorner/x)$ iff $V \models \mathit{True}(\ulcorner \chi(\ulcorner b \urcorner/x) \urcorner/x)$ iff $V \models \neg \chi(\ulcorner b \urcorner/x)$.

Evidently the two ends of (24.5) make a contradiction. Hence the class of true sentences of L can't be defined by any formula of L. Thus we have shown that

THEOREM 24.1. *The class of pairs $\langle \phi, g \rangle$ where ϕ is a formula of the language L of set theory, g is an assignment in V and $V \models \phi[g]$, is not definable in V by any formula of the language L^+ of set theory with names for arbitrary sets.*

This is one version of Tarski's [1935/6] *theorem on the undefinability of truth*. Another version, with essentially the same proof, is:

THEOREM 24.2. *The class of sentences ϕ of L which are true in V is not definable in V by any formula of L.*

Of course, the set b of all true sentences of L would be definable in V if we allowed ourselves a name for b. Hence the difference between Theorems 24.1 and 24.2. These two theorems mean that the matter of truth in V has to be handled either informally or not at all.

Lévy [1965] gives several refined theorems about definability of truth in V. He shows that truth for certain limited classes of sentences of L^+ can be defined in V; in fact each sentence of L^+ lies in one of his classes. As I remarked earlier, everything can be encoded but not all at once.

Tarski's argument was based on a famous paper of Gödel [1931], to which I now turn. When formalizing the language of arithmetic it is common to include two restricted quantifiers $(\forall x < y)$ and $(\exists x < y)$, meaning respectively 'for all x which are less than y' and 'there is an x which is less than y, such that'. A formula in which every quantifier is restricted is called a Δ_0 formula. Formulas of form $\forall \vec{x} \phi$ and $\exists \vec{x} \phi$, where ϕ is a Δ_0 formula, are said to be Π_1 and Σ_1 formulas respectively. (See under 'Arithmetical hierarchy' in Van Dalen [Chapter I.6, below].)

N shall be the structure whose elements are the natural numbers; each natural number is named by an individual constant $\ulcorner n \urcorner$, and there are relations or functions giving 'plus' and 'times'. A relation on the domain of N which is defined by a Π_1 or Σ_1 formula is said to be a Π_1 or Σ_1 relation respectively. Some relations can be defined both ways; these are said to be Δ_1 relations. The interest of these classifications lies in a theorem of Kleene [1943]:

THEOREM 24.3. *An n-place relation R on the natural numbers is Δ_1 iff there is a computational test which decides whether any given n-tuple is in R; an n-tuple relation R on the natural numbers is Σ_1 iff a computer can be programmed to print out all and only the n-tuples in R.*

Hilbert in [1925], the paper that started this whole line of enquiry, had laid great stress on the fact that we can test the truth of a Δ_0 sentence in a finite number of steps, because each time we meet a restricted quantifier we have only to check a finite number of numbers. This is the central idea of the proofs from left to right in Kleene's equivalences. The other directions

I.1: ELEMENTARY PREDICATE LOGIC

are proved by encoding computers into N; see Theorems 2.5 and 2.14 in Van Dalen [Chapter I.6, below].

Now all grammatical properties of a sentence can be checked by mechanical computation. So we can encode the language of first-order Peano arithmetic into N in such a way that all the grammatical notions are expressed by Δ_1 relations. (This follows from Theorem 24.3, but Gödel [1931] wrote out an encoding explicitly.) We shall suppose that this has been done, so that from now on every formula or symbol of the language of arithmetic is simply a number. Thus every formula ϕ is a number which is named by the individual constant $\ulcorner\phi\urcorner$. Here $\ulcorner\phi\urcorner$ is also a number, but generally a different number from ϕ; $\ulcorner\phi\urcorner$ is called the *Gödel number* of ϕ. Note that if T is any mechanically describable theory in the language of arithmetic, then a suitably programmed computer can spew out all the consequences of T one by one, so that by Kleene's equivalences (Theorem 24.3), the set of all sentences ϕ such that $T \vdash \phi$ is a Σ_1 set.

We need one other piece of general theory. Tarski *et al.* [1953] describe a sentence Q in the language of arithmetic which is true in N and has the remarkable property that for every Σ_1 sentence ϕ,

(24.6) $Q \vdash \phi$ iff $N \models \phi$.

We shall use these facts to show that the set of numbers n which are not sentences deducible from Q is not a Σ_1 set. Suppose it were a Σ_1 set, defined by the Σ_1 formula ψ. Then for every number n we would have

(24.7) $N \models \psi(\ulcorner n\urcorner/x)$ iff not$(Q \vdash n)$.

Now since all syntactic notions are Δ_1, with a little care one can find a Σ_1 formula χ with just x free, such that for every formula ϕ with just x free, if $\ulcorner\phi\urcorner = n$ then

(24.8) $N \models \chi(\ulcorner n\urcorner/x)$ iff $N \models \psi(\ulcorner\phi(\ulcorner n\urcorner/x)\urcorner/x)$.

Putting $n = \ulcorner\chi\urcorner$ we get by (24.6), (24.7) and (24.8):

(24.9) $N \models \chi(\ulcorner n\urcorner/x)$ iff $N \models \psi(\ulcorner\chi(\ulcorner n\urcorner/x)\urcorner/x)$ iff not $(Q \vdash \chi(\ulcorner n\urcorner/x))$
  iff not $(N \models \chi(\ulcorner n\urcorner/x))$

where the last equivalence is because $\chi(\ulcorner n\urcorner/x)$ is a Σ_1 sentence. The two ends of (24.9) make a contradiction; so we have proved that the set of numbers n which are not sentences deducible from Q is not Σ_1. Hence the set of numbers which *are* deducible is not Δ_1, and therefore by Theorem 24.3 there is no mechanical test for what numbers belong to it. We have proved:

there is no mechanical test which determines, for any given sentence ϕ of the language of arithmetic, whether or not $\vdash (Q \to \phi)$. This immediately implies Church's theorem [1936]:

THEOREM 24.4. *There is no mechanical test to determine which sentences of first-order languages are logically valid.*

Now we can very easily prove a weak version of Gödel's [1931] incompleteness theorem too. Let P be first-order Peano arithmetic. Then it can be shown that $P \vdash Q$. Hence from (24.6) we can infer that (24.6) holds with P in place of Q. So the same argument as above shows that the set of non-consequences of P is not Σ_1. If P had as consequences all the sentences true in N, then the non-consequences of P would consist of (i) the sentences ϕ such that $P \vdash \neg \phi$, and (ii) the numbers which are not sentences. But these together form a Σ_1 set. Hence, as Gödel proved,

THEOREM 24.5. *There are sentences which are true in N but not deducible from P.*

Finally Tarski's theorem (Theorems 24.1, 24.2) on the undefinability of truth applies to arithmetic just as well as to set theory. A set of numbers which is definable in N by a first-order formula is said to be *arithmetical*. Tarski's theorem on the undefinability of truth in N states:

THEOREM 24.6. *The class of first-order sentences which are true in N is not arithmetical.*

Van Benthem and Doets [Chapter I.4, below] show why Theorem 24.6 implies that there can be no complete formal proof calculus for second-order logic.

For recent work connecting Gödel's argument with modal logic, see Boolos [1979] and Smoryński [Chapter II.9].

25. SKOLEM FUNCTIONS

When Hilbert interpreted $\exists x \phi$ as saying in effect 'The element x which I choose satisfies ϕ' (cf. Section 15 above), Brouwer accused him of 'causing mathematics to degenerate into a game' (Hilbert [1928]). Hilbert was delighted with this description, as well he might have been, since games which are closely related to Hilbert's idea have turned out to be an extremely powerful tool for understanding quantifiers.

I.1: ELEMENTARY PREDICATE LOGIC

Before the technicalities, here is an example. Take the sentence

(25.1) Everybody in Croydon owns a dog.

Imagine a game G: you make the first move by producing someone who lives in Croydon, and I have to reply by producing a dog. I win if and only if the dog I produce belongs to the person you produced. Assuming that I have free access to other people's dogs, (25.1) is true if and only if I can always win the game G. This can be rephrased: (25.1) is true if and only if there is a function F assigning a dog to each person living in Croydon, such that whenever we play G, whatever person x you produce, if I retaliate with dog $F(x)$ then I win. A function F with this property is called a *winning strategy* for me in the game G. By translating (25.1) into a statement about winning strategies, we have turned a statement of form $\forall x \exists y \phi$ into one of form $\exists F \forall x \psi$.

Now come the technicalities. For simplicity I shall assume that our language L doesn't contain $\bot$, $\to$ or $\leftrightarrow$, and that all occurrences of $\neg$ are immediately in front of atomic formulas. The arguments of Sections 5 and 15 show that every first-order formula is logically equivalent to one in this form, so the theorems proved below hold without this restriction on L. $\mathfrak{A}$ shall be a fixed L-structure. For each formula ϕ of L and assignment g in $\mathfrak{A}$ to the free variables of ϕ, we shall define a game $G(\mathfrak{A}, \phi; g)$ to be played by two players $\forall$ and $\exists$. The definition of $G(\mathfrak{A}, \phi; g)$ is by induction on the complexity of ϕ, and it very closely follows the definition of $\models$ in Section 14:

(a) If ϕ is atomic then neither player makes any move in $G(\mathfrak{A}, \phi; g)$ or $G(\mathfrak{A}, \neg \phi; g)$; player $\exists$ wins $G(\mathfrak{A}, \phi; g)$ if $\mathfrak{A} \models \phi[g]$, and he wins $G(\mathfrak{A}, \neg \phi; g)$ if $\mathfrak{A} \models \neg \phi[g]$; player $\forall$ wins iff player $\exists$ doesn't win.

(b) Suppose ϕ is $\psi \wedge \chi$, and g_1 and g_2 are respectively the restrictions of g to the free variables of ψ, χ; then player $\forall$ has first move in $G(\mathfrak{A}, \phi; g)$, and the move consists of deciding whether the game shall proceed as $G(\mathfrak{A}, \psi; g_1)$ or as $G(\mathfrak{A}, \chi; g_2)$.

(c) Suppose ϕ is $\psi \vee \chi$, and g_1, g_2 are as in (b); then player $\exists$ moves by deciding whether the game shall continue as $G(\mathfrak{A}, \psi; g_1)$ or $G(\mathfrak{A}, \chi; g_2)$.

(d) If ϕ is $\forall x \psi$ then player $\forall$ chooses an element α of $\mathfrak{A}$, and the game proceeds as $G(\mathfrak{A}, \psi; g, \alpha/x)$.

(e) If ϕ is $\exists x \psi$ then player $\exists$ chooses an element α of $\mathfrak{A}$, and the game proceeds as $G(\mathfrak{A}, \psi; g, \alpha/x)$.

If g is an assignment suitable for ϕ, and h is the restriction of g to the free variables of ϕ, then $G(\mathfrak{A}, \phi; g)$ shall be $G(\mathfrak{A}, \phi; h)$. When ϕ is a sentence, h is empty and we write the game simply as $G(\mathfrak{A}, \phi)$.

The quantifier clauses for these games were introduced in Henkin [1961]. It is then clear how to handle the other clauses; see Hintikka [1973] Chapter V. Lorenzen [1961, 1962] (cf. also Lorenzen and Schwemmer [1975]) described similar games, but in his versions the winning player had to *prove* a sentence, so that his games turned out to define intuitionistic provability where ours will define truth. (Cf. Felscher [III.5].) In Lorenzen [1962] one sees a clear link with cut-free sequent proofs.

A *strategy* for a player in a game is a set of rules that tell him how he should play, in terms of the previous moves of the other player. The strategy is called *winning* if the player wins every time he uses it, regardless of how the other player moves. Leaving aside the game-theoretic setting, the next result probably ought to be credited to Skolem [1920]:

THEOREM 25.1. *Assume the axiom of choice* (cf. Appendix C). *Then for every L-structure* $\mathfrak{A}$, *every formula* ϕ *of L and every assignment g in* $\mathfrak{A}$ *which is suitable for* ϕ, $\mathfrak{A} \models \phi[g]$ *iff player* $\exists$ *has a winning strategy for the game* $G(\mathfrak{A}, \phi; g)$.

Theorem 25.1 is proved by induction on the complexity of ϕ. I consider only clause (d), which is the one that needs the axiom of choice. The 'if' direction is not hard to prove. For the 'only if', suppose that $\mathfrak{A} \models \forall x \psi[g]$, where g is an assignment to the free variables of $\forall x \psi$. Then $\mathfrak{A} \models \psi[g, \alpha/x]$ for every element α; so by the induction assumption, player $\exists$ has a winning strategy for each $G(\mathfrak{A}, \psi; g, \alpha/x)$. Now *choose* a winning strategy S_α for player $\exists$ in each game $G(\mathfrak{A}, \psi; g, \alpha/x)$. Player $\exists$'s winning strategy for $G(\mathfrak{A}, \phi; g)$ shall be as follows: wait to see what element α player $\forall$ chooses, and then follow S_α for the rest of the game.

Theorem 25.1 has a wide range of consequences. First, it shows that games can be used to give a definition of truth in structures. In fact this was Henkin's purpose in introducing them. See Chapter III of Hintikka [1973] for some phenomenological reflections on this kind of truth-definition.

For the next applications we should bear in mind that *every first-order formula can be converted into a logically equivalent first-order formula which is prenex, i.e. with all its quantifiers at the left-hand end.* (Cf. (15.5).) When ϕ is prenex, a strategy for player $\exists$ takes a particularly simple form. It consists of a set of functions, one for each existential quantifier in ϕ, which tell player $\exists$ what element to choose, depending on what elements were chosen by player $\forall$ at earlier universal quantifiers.

I.1: ELEMENTARY PREDICATE LOGIC

For example if ϕ is $\forall x \exists y \forall z \exists t R(x,y,z,t)$, then a strategy for player $\exists$ in $G(\mathfrak{A}, \phi)$ will consist of two functions, a 1-place function F_y and a 2-place function F_t. This strategy will be winning if and only if

(25.2)   for all elements α and γ, $\mathfrak{A} \models R(x,y,z,t)[\alpha/x, F_y(\alpha)/y, \gamma/z, F_t(\alpha,\gamma)/t]$.

Statement (25.2) can be paraphrased as follows. Introduce new function symbols f_y and f_t. Write $\phi\hat{}$ for the sentence got from ϕ by removing the existential quantifiers and then putting $f_y(x), f_t(x,z)$ in place of y, t respectively. So $\phi\hat{}$ is $\forall x \forall z R(x, f_y(x), z, f_t(x,z))$. We expand $\mathfrak{A}$ to a structure $\mathfrak{A}\hat{}$ by adding interpretations $I_{\mathfrak{A}\hat{}}(f_y)$ and $I_{\mathfrak{A}\hat{}}(f_t)$ for the new function symbols; let F_y and F_t be these interpretations. Then by (25.2),

(25.3) F_y, F_t are a winning strategy for player $\exists$ in $G(\mathfrak{A}, \phi)$ iff $\mathfrak{A}\hat{} \models \phi\hat{}$.

Functions F_y, F_t which do satisfy either side of (25.3) are called *Skolem functions for* ϕ. Putting together (25.3) and Theorem 25.1, we get:

(25.4) $\mathfrak{A} \models \phi$ iff by adding functions to $\mathfrak{A}$ we can get a structure $\mathfrak{A}\hat{}$ such that $\mathfrak{A}\hat{} \models \phi\hat{}$.

A sentence $\phi\hat{}$ can be defined in the same way whenever ϕ is any prenex sentence; (25.4) will still apply. Note that $\phi\hat{}$ is of form $\forall \vec{x} \psi$ where ψ has no quantifiers; a formula of this form is said to be *universal*.

From (25.4) we can deduce:

THEOREM 25.2. *Every prenex first-order sentence ϕ is logically equivalent to a second-order sentence $\exists \vec{f} \phi\hat{}$ in which $\phi\hat{}$ is universal.*

In other words, we can always push existential quantifiers to the left of universal quantifiers, provided that we convert the existential quantifiers into second-order function quantifiers $\exists \vec{f}$. Another consequence of (25.4) is:

LEMMA 25.3. *For every prenex first-order sentence ϕ we can effectively find a universal sentence $\phi\hat{}$ which has a model iff ϕ has a model.*

Because of Lemma 25.3, $\phi\hat{}$ is known as the *Skolem normal form of ϕ for satisfiability*.

Lemma 25.3 is handy for simplifying various logical problems. But it would be handier still if no function symbols were involved. At the end of Section 18 we saw that anything that can be said with a function constant can also be said with a relation constant. However, in order to make the implication from right to left in (25.4) still hold when relations are used instead of functions, we have to require that the relations really do represent functions, in other words some sentences of form (18.12) must hold. These sentences are $\forall\exists$ sentences, i.e. they have form $\forall\vec{x}\exists\vec{y}\,\psi$ where ψ has no quantifiers. The upshot is that for every prenex first-order sentence ϕ *without function symbols* we can effectively find an $\forall\exists$ first-order sentence $\phi_\wedge$ *without function symbols but with extra relation symbols*, such that ϕ has a model if and only if $\phi_\wedge$ has a model. The sentence $\phi_\wedge$ is also known as the *Skolem normal form of ϕ for satisfiability*.

For more on Skolem normal forms see Chapter 2 of Kreisel and Krivine [1967].

Skolem also applied Theorem 25.1 to prove his part of the Löwenheim–Skolem Theorem 17.5. We say that L-structures $\mathfrak{A}$ and $\mathfrak{B}$ are *elementarily equivalent* to each other if exactly the same sentences of L are true in $\mathfrak{A}$ as in $\mathfrak{B}$. Skolem showed:

THEOREM 25.4. *If L is a language with at most countably many formulas and $\mathfrak{A}$ is an infinite L-structure, then by choosing countably many elements of $\mathfrak{A}$ and throwing out the rest, we can get a countable L-structure $\mathfrak{B}$ which is elementarily equivalent to $\mathfrak{A}$.*

This is proved as follows. There are countably many sentences of L which are true in $\mathfrak{A}$. For each of these sentences ϕ, player $\exists$ has a winning strategy S_ϕ for $G(\mathfrak{A}, \phi)$. All we need do is find a countable set X of elements of $\mathfrak{A}$ such that if player $\forall$ chooses his elements from X, all the strategies S_ϕ tell player $\exists$ to pick elements which are in X too. Then X will serve as the domain of $\mathfrak{B}$, and player $\exists$ will win each $G(\mathfrak{B}, \phi)$ by playing the same strategy S_ϕ as for $G(\mathfrak{A}, \phi)$. Starting from any countable set X_0 of elements of $\mathfrak{A}$, let X_{n+1} be X_n together with all elements called forth by any of the strategies S_ϕ when player $\forall$ chooses from X_n; then X can be the set of all elements which occur in X_n for at least one natural number n.

In his paper [1920], Skolem noticed that the proof of Theorem 25.1 gives us information in a rather broader setting too. Let $\mathfrak{L}_{\omega_1\omega}$ be the logic we get if, starting from first-order logic, we allow formulas to contain conjunctions or disjunctions of countably many formulas at a time. For example, in $\mathfrak{L}_{\omega_1\omega}$ there is an infinite sentence

(25.5) $\quad \forall x(x = 0 \vee x = 1 \vee x = 2 \vee \cdots)$

which says 'Every element is a natural number'. If we add (25.5) to the axioms of first-order Peano arithmetic we get a theory whose only models are the natural number system and other structures which are exact copies of it. This implies that the Compactness Theorem (Theorem 17.4) and the Upward Löwenheim–Skolem Theorem (Theorem 17.5) both fail when we replace first-order logic by $\mathfrak{L}_{\omega_1\omega}$.

Skolem noticed that the proof of Theorem 25.1 tells us:

THEOREM 25.5. *If ϕ is a sentence of the logic $\mathfrak{L}_{\omega_1\omega}$ and $\mathfrak{A}$ is a model of ϕ, then by choosing at most countably many elements of $\mathfrak{A}$ we can get an at most countable structure $\mathfrak{B}$ which is also a model of ϕ.*

So a form of the Downward Löwenheim–Skolem Theorem (cf. Theorem 17.5) does hold in $\mathfrak{L}_{\omega_1\omega}$.

26. BACK-AND-FORTH EQUIVALENCE

In this section and the next, we shall prove that certain things are definable by first-order formulas. The original versions of the theorems we prove go back to the mid 1950s. But for us their interest lies in the proofs which Per Lindström gave in [1969]. He very cleverly used the facts (1) that first-order logic is good for encoding finite sequences, and (2) that first-order logic is bad for distinguishing infinite cardinals. His proofs showed that anything we can say using a logic which shares features (1) and (2) with first-order logic can also be said with a first-order sentence; so first-order logic is essentially the only logic with these features.

I should say what we mean by a logic. A *logic* $\mathfrak{L}$ is a family of languages, one for each similarity type, together with a definition of what it is for a sentence of a language L of $\mathfrak{L}$ to be true in an L-structure. Just as in first-order logic, an L-structure is a structure which has named relations and elements corresponding to the similarity type of L. We shall always assume that the analogue of Theorem 5.1 holds for $\mathfrak{L}$, i.e., that the truth-value of a sentence ϕ in a structure $\mathfrak{A}$ doesn't depend on how $\mathfrak{A}$ interprets constants which don't occur in ϕ.

We shall say that a logic $\mathfrak{L}$ is an *extension of first-order logic* if, roughly speaking, it can do everything that first-order logic can do and maybe a bit more. More precisely, it must satisfy three conditions. (i) Every first-order formula must be a formula of $\mathfrak{L}$. (ii) If ϕ and ψ are formulas of $\mathfrak{L}$ then so are $\neg\phi, \phi \wedge \psi, \phi \vee \psi, \phi \rightarrow \psi, \phi \leftrightarrow \psi, \forall x\phi, \exists x\phi$; we assume the symbols

¬ etc. keep their usual meanings. (iii) $\mathfrak{L}$ is *closed under relativization*. This means that for every sentence ϕ of $\mathfrak{L}$ and every 1-place predicate constant P not in ϕ, there is a sentence $\phi^{(P)}$ such that a structure $\mathfrak{A}$ is a model of $\phi^{(P)}$ if and only if the part of $\mathfrak{A}$ with domain $I_\mathfrak{A}(P)$ satisfies ϕ. For example, if $\mathfrak{L}$ can say 'Two-thirds of the elements satisfy $R(x)$', then it must also be able to say 'Two-thirds of the elements which satisfy $P(x)$ satisfy $R(x)$'. First-order logic itself is closed under relativization; although I haven't called attention to it earlier, it is a device which is constantly used in applications.

The logic $\mathfrak{L}_{\omega_1\omega}$ mentioned in the previous section is a logic in the sense defined above, and it is an extension of first-order logic. Another logic which extends first-order logic is $\mathfrak{L}_{\infty\omega}$; this is like first-order logic except that we are allowed to form conjunctions and disjunctions of arbitrary sets of formulas, never mind how large. Russell's logic, got by adding definite description operators to first-order logic, is another extension of first-order logic though it never enables us to say anything new.

We shall always require logics to obey one more condition, which needs some definitions. L-structures $\mathfrak{A}$ and $\mathfrak{B}$ are said to be *isomorphic* to each other if there is a function F from the domain of $\mathfrak{A}$ to the domain of $\mathfrak{B}$ which is bijective, and such that for all elements $\alpha_0, \alpha_1, \ldots$ of $\mathfrak{A}$ and every atomic formula ϕ of L,

(26.1) $\mathfrak{A} \models \phi[\alpha_0/x_0, \alpha_1/x_1, \ldots]$ iff $\mathfrak{B} \models \phi[F(\alpha_0)/x_0, F(\alpha_1)/x_1, \ldots]$.

It will be helpful in this section and the next if we omit the x_i's when writing conditions like (26.1); so (26.2) means the same as (26.1) but is briefer:

(26.2) $\mathfrak{A} \models \phi[\alpha_0, \alpha_1, \ldots]$ iff $\mathfrak{B} \models \phi[F(\alpha_0), F(\alpha_1), \ldots]$.

If (26.1) or equivalently (26.2) holds, where F is a bijection from the domain of $\mathfrak{A}$ to that of $\mathfrak{B}$, we say that F is an *isomorphism* from $\mathfrak{A}$ to $\mathfrak{B}$. Intuitively, $\mathfrak{A}$ is isomorphic to $\mathfrak{B}$ when $\mathfrak{B}$ is a perfect copy of $\mathfrak{A}$.

If $\mathfrak{L}$ is a logic, we say that structures $\mathfrak{A}$ and $\mathfrak{B}$ are $\mathfrak{L}$-*equivalent* to each other if every sentence of $\mathfrak{L}$ which is true in one is true in the other. Thus 'elementarily equivalent' means $\mathfrak{L}$-equivalent where $\mathfrak{L}$ is first-order logic. The further condition we impose on logics is this: structures which are isomorphic to each other must also be $\mathfrak{L}$-equivalent to each other. Obviously this is a reasonable requirement. Any logic you think of will meet it.

Now we shall introduce another kind of game. This one is used for comparing two structures. Let $\mathfrak{A}$ and $\mathfrak{B}$ be L-structures. The game $EF_\omega(\mathfrak{A};\mathfrak{B})$ is played by two players $\forall$ and $\exists$ as follows. There are infinitely many moves. At the ith move, player $\forall$ chooses one of $\mathfrak{A}$ and $\mathfrak{B}$ and then selects an element of the structure he has chosen; then player ∃ must pick an element from the other structure. The elements chosen from $\mathfrak{A}$ and $\mathfrak{B}$ at the ith move are written α_i and β_i respectively. Player ∃ wins the game if and only if for every atomic formula ϕ of L,

(26.3) $\quad \mathfrak{A} \models \phi[\alpha_0, \alpha_1, \ldots]$ iff $\mathfrak{B} \models \phi[\beta_0, \beta_1, \ldots]$.

We say that $\mathfrak{A}$ and $\mathfrak{B}$ are *back-and-forth equivalent* to each other if player ∃ has a winning strategy for this game.

The game $EF_\omega(\mathfrak{A};\mathfrak{B})$ is known as the *Ehrenfeucht–Fraïssé* game of length ω, for reasons that will appear in the next section. One feels that the more similar $\mathfrak{A}$ and $\mathfrak{B}$ are, the easier it ought to be for player ∃ to win the game. The rest of this section is devoted to turning this feeling into theorems. For an easy start:

THEOREM 26.1. *If $\mathfrak{A}$ is isomorphic to $\mathfrak{B}$ then $\mathfrak{A}$ is back-and-forth equivalent to $\mathfrak{B}$.*

Given an isomorphism F from $\mathfrak{A}$ to $\mathfrak{B}$, player ∃ should always choose so that for each natural number i, $\beta_i = F(\alpha_i)$. Then he wins. Warning: we are talking set theory now, so F may not be describable in terms which any human player could use, even if he could last out the game.

As a partial converse to Theorem 26.1:

THEOREM 26.2. *If $\mathfrak{A}$ is back-and-forth equivalent to $\mathfrak{B}$ and both $\mathfrak{A}$ and $\mathfrak{B}$ have at most countably many elements, then $\mathfrak{A}$ is isomorphic to $\mathfrak{B}$.*

For this, imagine that player ∀ chooses his moves so that he picks each element of $\mathfrak{A}$ or $\mathfrak{B}$ at least once during the game; he can do this if both structures are countable. Let player ∃ use his winning strategy. When all the α_i's and β_i's have been picked, define F by putting $F(\alpha_i) = \beta_i$ for each i. (The definition is possible because (26.3) holds for each atomic formula '$x_i = x_j$'.) Comparing (26.2) with (26.3), we see that F is an isomorphism. The idea of this proof was first stated by Hausdorff ([1914] p. 99) in his proof of a theorem of Cantor about dense linear orderings. Fraïssé [1954] noticed that the argument works just as well for structures as for orderings.

Now we are going to show that whether or not $\mathfrak{A}$ and $\mathfrak{B}$ have countably many elements, if $\mathfrak{A}$ and $\mathfrak{B}$ are back-and-forth equivalent then they are elementarily equivalent. This was known to Fraïssé [1955], and Karp [1965] gave direct proof of the stronger result that $\mathfrak{A}$ is back-and-forth equivalent to $\mathfrak{B}$ if and only if $\mathfrak{A}$ is $\mathfrak{L}_{\infty\omega}$-equivalent to $\mathfrak{B}$. The interest of our proof (which was extracted from Lindström [1969] by Barwise [1974]) is that it works for any extension of first-order logic which obeys the Downward Löwenheim–Skolem Theorem. To be precise:

THEOREM 26.3. *Suppose $\mathfrak{L}$ is an extension of first-order logic, and every structure of at most countable similarity type is $\mathfrak{L}$-equivalent to a structure with at most countably many elements. Suppose also that every sentence of $\mathfrak{L}$ has at most countably many distinct symbols. Then any two structures which are back-and-forth equivalent are $\mathfrak{L}$-equivalent to each other.*

Theorem 26.3 can be used to prove Karp's result too, by a piece of set-theoretic strong-arm tactics called 'collapsing cardinals' (as in Barwise [1973]). By Skolem's observation (Theorem 25.5), Theorem 26.3 applies almost directly to $\mathfrak{L}_{\omega_1\omega}$ (though one still has to use 'countable fragments' of $\mathfrak{L}_{\omega_1\omega}$ – I omit details).

Let me sketch the proof of Theorem 26.3. Assume all the assumptions of Theorem 26.3, and let $\mathfrak{A}$ and $\mathfrak{B}$ be L-structures which are back-and-forth equivalent. We have to show that $\mathfrak{A}$ and $\mathfrak{B}$ are $\mathfrak{L}$-equivalent. Replacing $\mathfrak{B}$ by an isomorphic copy if necessary, we can assume that $\mathfrak{A}$ and $\mathfrak{B}$ have no elements in common. Now we construct a jumbo structure:

(26.4) $\mathfrak{C}$: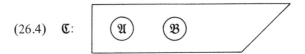

The language of $\mathfrak{C}$ shall contain two 1-place predicate constants $\partial^{\mathfrak{A}}$ and $\partial^{\mathfrak{B}}$. Also for each predicate constant R and individual constant c of L the language of $\mathfrak{C}$ shall contain two symbols $R^{\mathfrak{A}}$, $R^{\mathfrak{B}}$ and $c^{\mathfrak{A}}$, $c^{\mathfrak{B}}$. The elements in $I_{\mathfrak{C}}(\partial^{\mathfrak{A}})$ are precisely the elements of $\mathfrak{A}$, and each $I_{\mathfrak{C}}(R^{\mathfrak{A}})$ and $I_{\mathfrak{C}}(c^{\mathfrak{A}})$ is to be identical with $I_{\mathfrak{A}}(R)$ and $I_{\mathfrak{A}}(c)$ respectively. Thus $\mathfrak{C}$ contains an exact copy of $\mathfrak{A}$. Likewise with $\mathfrak{B}$ in place of $\mathfrak{A}$. The remaining pieces of $\mathfrak{C}$ outside $\mathfrak{A}$ and $\mathfrak{B}$ consist of enough set-theoretic apparatus to code up all finite sequences of elements of $\mathfrak{A}$ and $\mathfrak{B}$. Finally the language of $\mathfrak{C}$ shall have a 2-place predicate constant S which encodes the winning strategy of player $\exists$ in the game $EF_\omega(\mathfrak{A};\mathfrak{B})$ as follows:

(26.5) $I_{\mathfrak{C}}(S)$ contains exactly those ordered pairs $\langle\langle\gamma_0,\ldots,\gamma_{n-1}\rangle,\gamma_n\rangle$ such that γ_n is the element which player $\exists$'s winning strategy tells him to play if player $\forall$'s previous moves were $\gamma_0,\ldots,\gamma_{n-1}$.

Now we wish to show that any sentence of $\mathfrak{L}$ which is true in $\mathfrak{A}$ is true also in $\mathfrak{B}$, and *vice versa*. Since each sentence of $\mathfrak{L}$ contains at most countably many symbols, we can assume without any loss of generality that the similarity type of $\mathfrak{A}$ and $\mathfrak{B}$ has just countably many symbols; hence the same is true for $\mathfrak{C}$, and thus by the assumption in Theorem 26.3, $\mathfrak{C}$ is $\mathfrak{L}$-equivalent to a structure $\mathfrak{C}'$ with at most countably many elements. The sets $I_{\mathfrak{C}'}(\partial^{\mathfrak{A}})$ and $I_{\mathfrak{C}'}(\partial^{\mathfrak{B}})$ of $\mathfrak{C}'$ define L-structures $\mathfrak{A}'$ and $\mathfrak{B}'$ which are $\mathfrak{L}$-equivalent to $\mathfrak{A}$ and $\mathfrak{B}$ respectively, since everything we say in $\mathfrak{L}$ about $\mathfrak{A}$ can be rewritten as a statement about $\mathfrak{C}$ using $\partial^{\mathfrak{A}}$ and the $R^{\mathfrak{A}}$ and $c^{\mathfrak{A}}$. (Here we use the fact that $\mathfrak{L}$ allows relativization.)

Since $\mathfrak{L}$ contains all first-order logic, everything that we can say in a first-order language about $\mathfrak{C}$ must also be true in $\mathfrak{C}'$. For example we can say in first-order sentences that for every finite sequence $\gamma_0, \ldots, \gamma_{n-1}$ of elements of $\mathfrak{A}$ or $\mathfrak{B}$ there is a unique element γ_n such that $\langle\langle\gamma_0, \ldots, \gamma_{n-1}\rangle, \gamma_n\rangle$ is in $I_{\mathfrak{C}}(S)$; also that if player $\exists$ in $\mathrm{EF}_\omega(\mathfrak{A};\mathfrak{B})$ reads $I_{\mathfrak{C}}(S)$ as a strategy for him, then he wins. So all these things must be true also for $\mathfrak{A}', \mathfrak{B}'$ and $I_{\mathfrak{C}'}(S)$. (The reader can profitably check for himself that all this can be coded into first-order sentences, but if he gets stuck he can consult Barwise [1974] or Flum [1975].)

Therefore $\mathfrak{A}'$ is back-and-forth equivalent to $\mathfrak{B}'$. But both $\mathfrak{A}'$ and $\mathfrak{B}'$ are bits of $\mathfrak{C}'$, so they have at most countably many elements. Hence by Theorem 26.2, $\mathfrak{A}'$ is isomorphic to $\mathfrak{B}'$ and therefore $\mathfrak{A}'$ is $\mathfrak{L}$-equivalent to $\mathfrak{B}'$. But $\mathfrak{A}'$ was $\mathfrak{L}$-equivalent to $\mathfrak{A}$ and $\mathfrak{B}'$ was $\mathfrak{L}$-equivalent to $\mathfrak{B}$. So finally we deduce that $\mathfrak{A}$ and $\mathfrak{B}$ are $\mathfrak{L}$-equivalent.

27. LINDSTRÖM'S THEOREM

Theorem 26.3 showed that any extension of first-order logic which obeys a form of the Downward Löwenheim–Skolem Theorem is in a sense no stronger than the infinitary logic $\mathfrak{L}_{\infty\omega}$. This result is relatively shallow and not terribly useful; the logic $\mathfrak{L}_{\infty\omega}$ is quite powerful and not very well understood. (See Van Benthem and Doets [Chapter I.4, below].) Lindström [1969] found a stronger and more subtle result: he showed that if in addition $\mathfrak{L}$ obeys a form of the Compactness Theorem or the Upward Löwenheim–Skolem Theorem then every sentence of $\mathfrak{L}$ has exactly the same models as some first-order sentence. Since a first-order sentence contains only finitely many symbols, this result evidently needs some finiteness restriction on the sentences of $\mathfrak{L}$. So from now on we shall assume that *all similarity types are finite and have no function symbols.*

Lindström's argument relies on some detailed information about Ehrenfeucht–Fraïssé games. The Ehrenfeucht–Fraïssé game $EF_n(\mathfrak{A};\mathfrak{B})$ of length n, where n is a natural number, is fought and won exactly like $EF_\omega(\mathfrak{A};\mathfrak{B})$ except that the players stop after n moves. We say that the structures $\mathfrak{A}$ and $\mathfrak{B}$ are *n-equivalent* if player $\exists$ has a winning strategy for the game $EF_n(\mathfrak{A};\mathfrak{B})$. If $\mathfrak{A}$ and $\mathfrak{B}$ are back-and-forth equivalent then they are n-equivalent for all n; the converse is not true.

Ehrenfeucht–Fraïssé games of finite length were invented by Ehrenfeucht [1960] as a means of showing that two structures are elementarily equivalent. He showed that if two structures $\mathfrak{A}$ and $\mathfrak{B}$ are n-equivalent for all finite n then $\mathfrak{A}$ and $\mathfrak{B}$ are elementarily equivalent (which follows easily from Theorem 26.3), and that if the similarity type is finite and contains no function symbols, then the converse holds too. Fraïssé's definitions were different, but in his [1955] he proved close analogues of Ehrenfeucht's theorems, including an analogue of the following:

THEOREM 27.1. *Let L be a first-order language. Then for every natural number n there is a finite set of sentences $\sigma_{n,1}, \ldots, \sigma_{n,j_n}$ of L such that:*

(i) *every L-structure $\mathfrak{A}$ is a model of exactly one of $\sigma_{n,1}, \ldots, \sigma_{n,j_n}$; if $\mathfrak{A} \models \sigma_{n,i}$ we say that $\mathfrak{A}$ has n-type $\sigma_{n,i}$;*
(ii) *L-structures $\mathfrak{A}$ and $\mathfrak{B}$ are n-equivalent iff they have the same n-type.*

Theorem 27.1 is best proved by defining a more complicated game. Suppose $\gamma_0, \ldots, \gamma_{k-1}$ are elements of $\mathfrak{A}$ and $\delta_0, \ldots, \delta_{k-1}$ are elements of $\mathfrak{B}$. Then the game $EF_n(\mathfrak{A}, \gamma_0, \ldots, \gamma_{k-1}; \mathfrak{B}, \delta_0, \ldots, \delta_{k-1})$ shall be played exactly like $EF_n(\mathfrak{A};\mathfrak{B})$, but at the end when elements $\alpha_0, \ldots, \alpha_{n-1}$ of $\mathfrak{A}$ and $\beta_0, \ldots, \beta_{n-1}$ of $\mathfrak{B}$ have been chosen, player $\exists$ wins if and only if for every atomic formula ϕ,

(27.1) $\mathfrak{A} \models \phi[\gamma_0, \ldots, \gamma_{k-1}, \alpha_0, \ldots, \alpha_{n-1}]$
   iff $\mathfrak{B} \models \phi[\delta_0, \ldots, \delta_{k-1}, \beta_0, \ldots, \beta_{n-1}]$.

So this game is harder for player $\exists$ to win than $EF_n(\mathfrak{A};\mathfrak{B})$ was. We say that $\langle \mathfrak{A}, \gamma_0, \ldots, \gamma_{k-1} \rangle$ is *n-equivalent* to $\langle \mathfrak{B}, \delta_0, \ldots, \delta_{k-1} \rangle$ if player $\exists$ has a winning strategy for the game $EF_n(\mathfrak{A}, \gamma_0, \ldots, \gamma_{k-1}; \mathfrak{B}, \delta_0, \ldots, \delta_{k-1})$. We assert that for each finite k and n there is a finite set of formulas $\sigma^k_{n,1}, \sigma^k_{n,2}$ etc. of L such that

(i) for every L-structure $\mathfrak{A}$ and elements $\gamma_0, \ldots, \gamma_{k-1}$ of $\mathfrak{A}$ there is a unique i such that $\mathfrak{A} \models \sigma^k_{n,i}[\gamma_0, \ldots, \gamma_{k-1}]$; this $\sigma^k_{n,i}$ is called the *n-type* of $\langle \mathfrak{A}, \gamma_0, \ldots, \gamma_{k-1} \rangle$;

(ii) $\langle \mathfrak{A}, \gamma_0, \ldots, \gamma_{k-1} \rangle$ and $\langle \mathfrak{B}, \delta_0, \ldots, \delta_{k-1} \rangle$ are n-equivalent iff they have the same n-type.

Theorem 27.1 will then follow by taking k to be 0. We prove the assertion above for each k by induction on n.

When $n = 0$, for each k there are just finitely many sequences $\langle \mathfrak{A}, \gamma_0, \ldots, \gamma_{k-1} \rangle$ which can be distinguished by atomic formulas. (Here we use the fact that the similarity type is finite and there are no function symbols.) So we can write down finitely many formulas $\sigma_{0,1}^k$, $\sigma_{0,2}^k$ etc. which distinguish all the sequences that can be distinguished.

When the formulas have been constructed and (i), (ii) proved for the number n, we construct and prove them for $n + 1$ as follows. Player $\exists$ has a winning strategy for $\mathrm{EF}_{n+1}(\mathfrak{A}, \gamma_0, \ldots, \gamma_{k-1}; \mathfrak{B}, \delta_0, \ldots, \delta_{k-1})$ if and only if he can make his first move so that he has a winning strategy from that point onwards, i.e. if he can ensure that α_0 and β_0 are picked so that

$\langle \mathfrak{A}, \gamma_0, \ldots, \gamma_{k-1}, \alpha_0 \rangle$ is n-equivalent to $\langle \mathfrak{B}, \delta_0, \ldots, \delta_{k-1}, \beta_0 \rangle$.

In other words, using (iii) for n which we assume has already been proved, player $\exists$ has this winning strategy if and only if for every element α of $\mathfrak{A}$ there is an element β of $\mathfrak{B}$ so that

$\langle \mathfrak{A}, \gamma_0, \ldots, \gamma_{k-1}, \alpha \rangle$ has the same n-type as $\langle \mathfrak{B}, \delta_0, \ldots, \delta_{k-1}, \beta \rangle$,

and *vice versa* with $\mathfrak{A}$ and $\mathfrak{B}$ reversed. But this is equivalent to the condition:

for every i,

$\mathfrak{A} \models \exists x_k \sigma_{n,i}^{k+1}[\gamma_0, \ldots, \gamma_{k-1}]$ iff $\mathfrak{B} \models \exists x_k \sigma_{n,i}^{k+1}[\delta_0, \ldots, \delta_{k-1}]$.

It follows that we can build suitable formulas $\sigma_{n+1,i}^k$ by taking conjunctions of formulas of form $\exists x_k \sigma_{n,i}^{k+1}$ or $\neg \exists x_k \sigma_{n,i}^{k+1}$, running through all the possibilities.

When the formulas $\sigma_{n,i}^k$ have all been defined, we take $\sigma_{n,i}$ to be $\sigma_{n,i}^0$. Thus Theorem 27.1 is proved.

Barwise ([1975] Chapter VII.6) describes the formulas $\sigma_{n,i}^k$ in detail in a rather more general setting. The sentences $\sigma_{n,i}$ were first described by Hintikka [1953] (cf. also Chapter XI of his [1973]), but their meaning was mysterious until Ehrenfeucht's paper appeared. We shall call the sentences *Hintikka sentences*. Hintikka proved that every first-order sentence is logically equivalent to a (finite) disjunction of Hintikka sentences. We shall prove this too, but by Lindström's proof [1969] which assumes only some general facts about the expressive power of first-order logic; so the proof will show that any sentence in any logic with this expressive power has the same models as some first-order sentence, viz. a disjunction of Hintikka sentences. Lindström proved:

THEOREM 27.2. *Let $\mathfrak{L}$ be any extension of first-order logic with the two properties:*

(a) *(Downward Löwenheim–Skolem) If a sentence ϕ of $\mathfrak{L}$ has an infinite model then ϕ has a model with at most countably many elements.*

(b) *Either (Upward Löwenheim–Skolem) if a sentence of $\mathfrak{L}$ has an infinite model then it has one with uncountably many elements; or (Compactness) if Δ is a theory in $\mathfrak{L}$ such that every finite set of sentences from Δ has a model then Δ has a model.*

Then every sentence of $\mathfrak{L}$ has exactly the same models as some first-order sentence.

The proof is by the same kind of coding as the proof of Theorem 26.3. Instead of proving Theorem 27.2 directly, we shall show:

THEOREM 27.3. *Let $\mathfrak{L}$ be any extension of first-order logic obeying* (a) *and* (b) *as in Theorem 27.2, and let ϕ and ψ be sentences of $\mathfrak{L}$ such that no model of ϕ is also a model of ψ. Then for some integer n there is a disjunction σ of Hintikka sentences $\sigma_{n,i}$ such that $\phi \models \sigma$ and $\psi \models \neg\sigma$.*

To get Theorem 27.2 from Theorem 27.3, let ψ be $\neg\phi$.

Suppose then that Theorem 27.3 is false. This means that there exist sentences ϕ and ψ of $\mathfrak{L}$ with no models in common, and for every natural number n there is no disjunction of Hintikka sentences $\sigma_{n,i}$ which separates the models of ϕ from the models of ψ. So by Theorem 27.1 there are, for each n, n-equivalent structures $\mathfrak{A}_n$ and $\mathfrak{B}_n$ such that $\mathfrak{A}_n$ is a model of ϕ and $\mathfrak{B}_n$ is a model of ψ. By (a) we can assume that $\mathfrak{A}_n$ and $\mathfrak{B}_n$ have at most countably many elements (since the sentences $\sigma_{n,i} \wedge \phi$ and $\sigma_{n,i} \wedge \psi$ are both in $\mathfrak{L}$).

So now once again we build a mammoth model $\mathfrak{C}$:

1.1: ELEMENTARY PREDICATE LOGIC

The coding is more complicated this time. $\mathfrak{C}$ contains a copy of the natural numbers N, picked out by a predicate constant ∂^N. There are 2-place predicate constants $\partial^{\mathfrak{A}}$, $\partial^{\mathfrak{B}}$. $I_{\mathfrak{C}}(\partial^{\mathfrak{A}})$ contains just those pairs $\langle \alpha, n \rangle$ such that n is a natural number and α is an element of $\mathfrak{A}_n$. Similarly with the $\mathfrak{B}_n$. Also $\mathfrak{C}$ has constants which describe each $\mathfrak{A}_n$ and $\mathfrak{B}_n$ completely, and $\mathfrak{C}$ contains all finite sequences of elements taken from any $\mathfrak{A}_n$ or $\mathfrak{B}_n$, together with enough set theory to describe lengths of sequences etc. There is a relation $I_{\mathfrak{C}}(S)$ which encodes the winning strategies for player $\exists$ in all games $\mathrm{EF}_n(\mathfrak{A}_n; \mathfrak{B}_n)$. Finally $\mathfrak{C}$ can be assumed to have just countably many elements, so we can incorporate a relation which sets up a bijection between N and the whole of the domain of $\mathfrak{C}$.

We shall need the fact that everything salient about $\mathfrak{C}$ can be said in *one single sentence* χ of $\mathfrak{L}$. Since N is in $\mathfrak{C}$ and we can build in as much set-theoretic equipment as we please, this is no problem, bearing in mind that $\mathfrak{L}$ is an extension of first-order logic. Barwise [1974] and Flum [1975] give details.

Now by (b), the sentence χ has a model $\mathfrak{C}'$ in which some 'infinite' number ∞ comes after all the 'natural numbers' $I_{\mathfrak{C}'}(0), I_{\mathfrak{C}'}(1), I_{\mathfrak{C}'}(2), \ldots$ in $I_{\mathfrak{C}'}(\partial^N)$. If the Upward Löwenheim–Skolem property holds, then this is because the N-part of any uncountable model of χ must have the same cardinality as the whole model, in view of the bijection which we incorporated. If on the other hand the Compactness property holds, we follow the construction of non-standard models in Section 20 above.

By means of $I_{\mathfrak{C}'}(\partial^{\mathfrak{A}})$ and $I_{\mathfrak{C}'}(\partial^{\mathfrak{B}})$, the structure $\mathfrak{C}'$ encodes structures $\mathfrak{A}'_{\infty}$ and $\mathfrak{B}'_{\infty}$, and $I_{\mathfrak{C}'}(S)$ encodes a winning strategy for player $\exists$ in the game $\mathrm{EF}_{\infty}(\mathfrak{A}'_{\infty}; \mathfrak{B}'_{\infty})$. All this is implied by a suitable choice of χ. The game $\mathrm{EF}_{\infty}(\mathfrak{A}'_{\infty}; \mathfrak{B}'_{\infty})$ turns out to be bizarre and quite unplayable; but the important point is that if player $\exists$ has a winning strategy for this game, then he has one for the shorter and entirely playable game $\mathrm{EF}_{\omega}(\mathfrak{A}'_{\infty}; \mathfrak{B}'_{\infty})$. Hence $\mathfrak{A}'_{\infty}$ and $\mathfrak{B}'_{\infty}$ are back-and-forth equivalent.

But now χ records that all the structures encoded by $\partial^{\mathfrak{A}}$ are models of ϕ, while those encoded by $\partial^{\mathfrak{B}}$ are models of ψ. Hence $\mathfrak{A}'_{\infty} \models \phi$ but $\mathfrak{B}'_{\infty} \models \psi$. Since ϕ and ψ have no models in common, it follows that $\mathfrak{B}'_{\infty} \models \neg \phi$. The final step is to use assumption 27.2(a), the Downward Löwenheim–Skolem property, to prove a slightly sharpened version of Theorem 26.3. To be precise, since $\mathfrak{A}'_{\infty}$ and $\mathfrak{B}'_{\infty}$ are back-and-forth equivalent and $\mathfrak{A}'_{\infty}$ is a model of the sentence ϕ of $\mathfrak{L}$, $\mathfrak{B}'_{\infty}$ must also be a model of ϕ. (The proof is like that in Section 26, but we use the fact that the similarity type is finite and has no function symbols in order to boil down the essential properties of $\mathfrak{C}$ into a single sentence.) So we have reached a contradiction, and Theorem 27.3 is proved.

The proof of Theorem 27.3, less the last paragraph, adapts to give a proof of *Craig's Interpolation Lemma* for predicate logic:

LEMMA 27.4. *Let ϕ and ψ be sentences of first-order predicate logic such that $\phi \models \neg \psi$. Then there is a first-order sentence σ such that $\phi \models \sigma$, $\psi \models \neg \sigma$, and every constant symbol which occurs in σ occurs both in ϕ and in ψ.*

Let $\mathfrak{L}$ in the proof of Theorem 27.3 be first-order logic and let L be the first-order language whose constants are those which occur both in ϕ and in ψ. Using Section 18, we can assume that L has no function symbols. If $\mathfrak{A}$ is any model of ϕ, then we get an L-structure $\mathfrak{A}|L$ by discarding all constant symbols not in L, without changing the elements or the interpretations of the symbols which are in L. Likewise for every model $\mathfrak{B}$ of ψ. Now suppose that the conclusion of Lemma 27.4 fails. Then for each natural number n there is no disjunction σ of Hintikka sentences $\sigma_{n,i}$ *in the language* L such that $\phi \models \sigma$ and $\psi \models \neg \sigma$, and hence there are models $\mathfrak{A}_n, \mathfrak{B}_n$ of ϕ, ψ respectively, such that $\mathfrak{A}_n|L$ is n-equivalent to $\mathfrak{B}_n|L$. Proceed now as in the proof of Theorem 27.3, using the Compactness and Downward Löwenheim–Skolem Theorems to find a countable $\mathfrak{C}'$ with an infinite natural number ∞. Excavate models $\mathfrak{A}'_\infty, \mathfrak{B}'_\infty$ of ϕ, ψ from $\mathfrak{C}'$ as before, noting this time that $\mathfrak{A}'_\infty|L$ is back-and-forth equivalent to $\mathfrak{B}'_\infty|L$. Then by Theorem 26.2, since $\mathfrak{A}'_\infty|L$ and $\mathfrak{B}'_\infty|L$ are countable and back-and-forth equivalent, they are isomorphic. It follows that we can add to $\mathfrak{A}'_\infty$ interpretations of those symbols which are in ψ but not in L, using $\mathfrak{B}'_\infty$ as a template. Let $\mathfrak{D}$ be the resulting structure. Then $\mathfrak{D} \models \phi$ since $\mathfrak{A}'_\infty \models \phi$, and $\mathfrak{D} \models \psi$ since $\mathfrak{B}'_\infty \models \psi$. This contradicts the assumption that $\phi \models \neg \psi$. Hence Lemma 27.4 is proved.

Craig himself [1957a] used his interpolation result to give a proof of *Beth's Definability Theorem* (Beth [1953]):

THEOREM 27.5. *Let L be a first-order language and Δ a first-order theory which uses the language L together with one extra n-place predicate constant R. Suppose that for every L-structure $\mathfrak{A}$ there is at most one way of adding to $\mathfrak{A}$ an interpretation of R so that the resulting structure is a model of Δ. Then Δ has a consequence of form $\forall x_1 \ldots x_n (R(x_1, \ldots, x_n) \leftrightarrow \phi)$, where ϕ is a formula in the language L.*

Time's wingèd chariot prevents a proper discussion of implicit and explicit definitions here, but Beth's theorem is proved as Theorem 2.2.22 in Chang and Keisler [1973]. There is some useful background on implicit definitions in Chapter 8 of Suppes [1957]. Craig's and Beth's results have interested philosophers of science; see e.g. Sneed [1971].

IV: APPENDICES

These three appendices will show in outline how one can construct a formal calculus of set theory, which in some sense formalizes the whole of mathematics. I have put this material into appendices, first because it is turgid, and second because I should hate to resuscitate the dreadful notion that the business of logicians is to produce all-embracing formal systems.

A. A FORMAL PROOF SYSTEM

We shall define a formal proof system for predicate logic with identity. To cover propositional logic too, the language will have some sentence letters. The calculus is a Hilbert-style system.

First we define the *language* L, by describing its similarity type, its set of terms and its set of formulas (cf. Sections 3 and 13 above).

The *similarity type* of L is made up of the following sentence letters, individual constants, predicate constants and function constants. The *sentence letters* are the expressions p_n, where n is a natural number subscript. The *individual constants* are the expressions c_n, where n is a natural number subscript. The *predicate constants* are the expressions P_n^m, where n is a natural number subscript and m is a positive integer superscript. The *function constants* are the expressions f_n^m, where n is a natural number subscript and m is a positive integer superscript. A predicate or function constant is said to be *m-place* if its superscript is m.

The *terms* of L are defined inductively as follows: (i) Every variable is a term, where the *variables* are the expressions x_n with natural number subscript n. (ii) For each function symbol f_n^m, if $\tau_1, \ldots, \tau_m$ are terms then the expression $f_n^m(\tau_1, \ldots, \tau_m)$ is a term. (iii) Nothing is a term except as required by (i) and (ii).

The *formulas* of L are defined inductively as follows: (i) Every sentence letter is a formula. (ii) The expression $\bot$ is a formula. (iii) For each predicate constant R_n^m, if $\tau_1, \ldots, \tau_m$ are terms then the expression $R_n^m(\tau_1, \ldots, \tau_m)$ is a formula. (iv) If σ and τ are terms then the expression $(\sigma = \tau)$ is a formula. (v) If ϕ and ψ are formulas, then so are the expressions $\neg \phi$, $(\phi \wedge \psi)$, $(\phi \vee \psi)$, $(\phi \to \psi)$, $(\phi \leftrightarrow \psi)$. (vi) For each variable x_n, if ϕ is a formula then so are the expressions $\forall x_n \phi$ and $\exists x_n \phi$. (vii) Nothing is a formula except as required by (i)–(vi).

A full account would now define two further notions, $FV(\phi)$ (the set of variables with free occurrences in ϕ) and $\phi[\tau_1 \cdots \tau_k / x_{i_1} \cdots x_{i_k}]$ (the formula which results when we simultaneously replace all free occurrences of x_{i_j} in ϕ by τ_j, for each j, $1 \leq j \leq k$, avoiding clash of variables). Cf. Section 13 above.

Now that L has been defined, formulas occurring below should be read as metalinguistic names for formulas of L. Hence, we can make free use of the metalanguage abbreviations in Sections 4 and 13.

Now we define the proof system – let us call it **H**. We do this by describing the axioms, the derivations, and the way in which a sequent is to be read off from a derivation. (Sundholm [Chapter I.2, below] describes an alternative Hilbert-style system **CQC** which is equivalent to **H**.)

The *axioms* of **H** are all formulas of the following forms:

H1. $\phi \to (\psi \to \phi)$

H2. $(\phi \to \psi) \to ((\phi \to (\psi \to \chi)) \to (\phi \to \chi))$

H3. $(\neg \phi \to \psi) \to ((\neg \phi \to \neg \psi) \to \phi)$

H4. $((\phi \to \bot) \to \bot) \to \phi$

H5. $\phi \to (\psi \to \phi \wedge \psi)$

H6. $\phi \wedge \psi \to \phi, \quad \phi \wedge \psi \to \psi$

H7. $\phi \to \phi \vee \psi, \quad \psi \to \phi \vee \psi$

H8. $(\phi \to \chi) \to ((\psi \to \chi) \to (\phi \vee \psi \to \chi))$

H9. $(\phi \to \psi) \to ((\psi \to \phi) \to (\phi \leftrightarrow \psi))$

H10. $(\phi \leftrightarrow \psi) \to (\phi \to \psi), \quad (\phi \leftrightarrow \psi) \to (\psi \to \phi)$

H11. $\phi[\tau/x] \to \exists x \phi$

H12. $\forall x \phi \to \phi[\tau/x]$

H13. $x = x$

H14. $x = y \to (\phi \to \phi[y/x])$

I.1. ELEMENTARY PREDICATE LOGIC

A *derivation* (or *formal proof*) in **H** is defined to be a finite sequence

(A.1) $\langle\langle\phi_1, m_1\rangle, \ldots, \langle\phi_n, m_n\rangle\rangle$

such that $n \geq 1$, and for each i ($1 \leq i \leq n$) one of the five following conditions holds:

(1) $m_i = 1$ and ϕ_i is an axiom;
(2) $m_i = 2$ and ϕ_i is any formula of L;
(3) $m_i = 3$ and there are j and k in $\{1, \ldots, i-1\}$ such that ϕ_k is $\phi_j \to \phi_i$;
(4) $m_i = 4$ and there is j ($1 \leq j < i$) such that ϕ_j has the form $\psi \to \chi$, x is a variable not free in ψ, and ϕ_i is $\psi \to \forall x \chi$;
(5) $m_i = 5$ and there is j ($1 \leq j < i$) such that ϕ_j has the form $\psi \to \chi$, x is a variable not free in χ, and ϕ_i is $\exists x \psi \to \chi$.

Conditions (3)–(5) are called the *derivation rules* of the calculus. They tell us how we can add new formulas to the end of a derivation. Thus (3) says that if ψ and $\psi \to \chi$ occur in a derivation, then we can add χ at the end; this is the rule of *modus ponens*.

The *premises* of the derivation (A.1) are those formulas ϕ_i such that $m_i = 2$. Its *conclusion* is ϕ_n. We say that ψ is *derivable from* $\chi_1, \ldots, \chi_k$ in the calculus **H**, in symbols

(A.2) $\chi_1, \ldots, \chi_n \vdash_{\mathbf{H}} \psi$,

if there exists a derivation whose premises are all among $\chi_1, \ldots, \chi_n$ and whose conclusion is ψ.

REMARKS. 1. The calculus **H** is sound and strongly complete for propositional and predicate logic with identity. (Cf. Section 7; as in Section 15, this says nothing about provable sequents in which some variables occur free.)

2. In practice most logicians would write the formulas of a derivation as a column or a tree, and they would omit the numbers m_i.

3. To prove the completeness of **H** by either the first or the third method in Section 16, one needs to know for all sentences $\chi_1, \ldots, \chi_n$ and ψ,

(A.3)   if $\chi_1, \ldots, \chi_n \vdash_{\mathbf{H}} \psi$ then $\chi_1, \ldots, \chi_{n-1} \vdash_{\mathbf{H}} \chi_n \to \psi$.

Statement (A.3) is the *Deduction Theorem* for **H**. It remains true if we allow free variables to occur in the formulas, provided that they occur only in certain ways. See Kleene [1952] Sections 21–24 for details.

4. Completeness and soundness tell us that if $\chi_1, \ldots, \chi_n$ and ψ are sentences, then (A.2) holds if and only if $\chi_1, \ldots, \chi_n \models \psi$. This gives an intuitive

meaning to such sequents. But when $\chi_1, \ldots, \chi_n$ and ψ are allowed to be any formulas of L, then to the best of my knowledge there are no natural necessary and sufficient conditions for (A.2) to hold. So it seems impossible to explain what if anything (A.2) tells us, except by referring to the fine details of the calculus **H**. This is a general feature of Hilbert-style calculi for predicate logic, and I submit that it makes them thoroughly inappropriate for introducing undergraduates to logic.

5. If we are thinking of varying the rules of the calculus, or even if we just want a picture of what the calculus is about, it is helpful to have at least a *necessary* condition for (A.2) to hold. The following supplies one. The *universal closure* of ϕ is $\forall y_1 \ldots y_n \phi$, where $y_1, \ldots, y_n$ are the free variables of ϕ. Let ϕ_1 be the universal closure of $\chi_1 \wedge \cdots \wedge \chi_n$ and ϕ_2 the universal closure of ψ. Then one can show that

(A.4)   if $\chi_1, \ldots, \chi_n \vdash_{\mathbf{H}} \psi$ then $\phi_1 \models \phi_2$.

The proof of (A.4) is by induction on the lengths of derivations. Statement (A.4) is one way of showing that **H** is sound.

6. The following derivation shows that $\vdash_{\mathbf{H}} \exists x(x = x)$:

(A.5) $x = x$ (axiom H13)
        $x = x \to \exists x(x = x)$ (axiom H11)
        $\exists x(x = x)$ (from above by modus ponens)

Statement (A.4) shows the reason, namely:

(A.6) $\forall x(x = x \wedge (x = x \to \exists x(x = x))) \models \exists x(x = x)$.

On any reasonable semantic interpretation (cf. Section 14 above), the left-hand side in (A.6) is true in the empty structure but the right-hand side is false. Suppose now that we want to modify the calculus in order to allow empty structures. Then we must alter the derivation rule which took us from left to right in (A.6), and this is the rule of modus ponens. (Cf. Bencivenga [Chapter III.6].) It is important to note here that even if (A.4) was a tidy two-way implication, the modus ponens rule would *not* express 'ϕ and $\phi \to \psi$ imply ψ', but rather something of form '$\forall \vec{x}(\phi \wedge (\phi \to \psi))$ implies $\forall \vec{y} \psi$'. As it is, the meaning of modus ponens in **H** is quite obscure. (Cf. Section 24 in Kleene [1952].)

B. ARITHMETIC

I begin with naive arithmetic, not formal Peano arithmetic. One needs to have at least an intuitive grasp of naive arithmetic in order to understand what a formal system is. In any case Peano [1889] reached his axioms by throwing naive arithmetic into fancy symbols.

Naive arithmetic is adequately summed up by the following five axioms, which come from Dedekind [1888, 1890]. Here and below, 'number' means 'natural number', and I start with 0 (Dedekind's first number was 1).

NA1.   0 is a number.

NA2.   For every number n there is a next number after n; this next number is called Sn or the *successor* of n.

NA3.   Two different numbers never have the same successor.

NA4.   0 is not the successor of any number.

NA5.   (Induction axiom) Let K be any set with the properties (i) 0 is in K, (ii) for every number n in K, Sn is also in K. Then every number is in K.

These axioms miss one vital feature of numbers, viz. their order. So we define $<$ as follows. First we define an *initial segment* to be a set K of numbers such that if a number Sn is in K then n is also in K. We say:

(B.1) $m < n$ iff there is an initial segment which contains m but not n.

The definition (B.1) implies:

(B.2)   If $m < Sn$ then either $m < n$ or $m = n$.

For future reference I give a proof. Suppose $m < Sn$ but not $m = n$. Then there is an initial segment K such that m is in K and Sn is not in K. Now there are two cases. *Case 1*: n is not in K. Then by (B.1), $m < n$. *Case 2*: n is in K. Then let M be K with n omitted. Since $m \neq n$, M contains m but not n. Also M is an initial segment; for if Sk is in M but k is not, then by the definition of M we must have $k = n$, which implies that Sn is in M and hence in K; contradiction. So we can use M in (B.1) to show $m < n$.

(B.3)   For each number m it is false that $m < 0$.

(B.3) is proved 'by induction on m', using the induction axiom NA5. Proofs of this type are written in a standard style, as follows:

CASE 1. $m = 0$. Then $m < 0$ would imply by (B.1) that there was a set containing 0 but not 0, which is impossible.

CASE 2. $m = Sk$, assuming it proved when $m = k$. Suppose $Sk < 0$. Then by (B.1) there is an initial segment containing Sk and not 0. Since K is an initial segment containing Sk, k is also in K. So by (B.1) again, K shows that $k < 0$. But the induction hypothesis states that not $k < 0$; contradiction.

This is all one would normally say in the proof. To connect it with NA5, let M be the set of all numbers m such that not $m < 0$. The two cases show exactly what has to be shown, according to NA5, in order to prove that every number is in M.

Here are two more provable facts.

(B.4)   The relation $<$ is a linear ordering of the numbers (in the sense of (19.14)–(19.16) above).

(B.5)   Every non-empty set of numbers has a first element.

Fact (B.5) states that the numbers are *well-ordered*, and it is proved as follows. Let X be any set of numbers without a first element. Let Y be the set of numbers not in X. Then by induction on n we show that every number n is in Y. So X is empty.

Fact (B.5) is one way of justifying *course-of-values induction*. This is a style of argument like the proof of (B.3) above, except that in Case 2, instead of proving the result for Sk assuming it was true for k, we prove it for Sk assuming it was true *for all numbers* $\leq k$. In many theorems about logic, one shows that every formula has some property A by showing (i) that every atomic formula has property A and (ii) that if ϕ is a compound formula whose proper subformulas have A then ϕ has A. Arguments of this type are course-of-values inductions on the complexity of formulas.

In naive arithmetic we can justify two important types of definition. The first is sometimes called *recursive definition* and sometimes *definition by induction*. It is used for defining functions whose domain is the set of natural numbers. To define such a function F recursively, we first say outright what $F(0)$ is, and then we define $F(Sn)$ in terms of $F(n)$. A typical example is the recursive definition of addition:

(B.6) $m + 0 = m, \quad m + Sn = S(m + n).$

Here $F(n)$ is $m + n$; the definition says first that $F(0)$ is m and then that for each number n, $F(Sn)$ is $SF(n)$. To justify such a definition, we have to show that there is exactly one function F which satisfies the stated conditions. To show there is *at most* one such function, we suppose that F and G are two functions which meet the conditions, and we prove by induction on n that for every n, $F(n) = G(n)$; this is easy. To show that there is *at least* one is harder. For this we define an *n-approximation* to be a function whose domain is the set of all numbers $< n$, and which obeys the conditions in the recursive definition for all numbers in its domain. Then we show by induction on n (i) that there is at least one n-approximation, and (ii) that if $m < k < n$, f is a k-approximation and g is an n-approximation, then $f(m) = g(m)$. Then finally we define F explicitly by saying that $F(m)$ is the unique number h such that $f(m) = h$ whenever f is an n-approximation for some number n greater than m.

After defining $+$ by (B.6), we can go on to define $\cdot$ by:

(B.7) $\quad m \cdot 0 = 0, \quad m \cdot Sn = m \cdot n + m.$

The functions definable by a sequence of recursive definitions in this way, using equations and previously defined functions, are called *primitive recursive* functions. Van Dalen [Chapter I.5, below] discusses them further.

There is a course-of-values recursive definition too: in this we define $F(0)$ outright, and then $F(Sn)$ in terms of values $F(k)$ for numbers $k \leqslant n$. For example if $F(n)$ is the set of all formulas of complexity n, understood as in Section 3 above, then the definition of $F(n)$ will have to refer to the sets $F(k)$ for all $k < n$. Course-of-values definitions can be justified in the same way as straightforward recursive definitions.

The second important type of definition that can be justified in naive arithmetic is also known as *inductive definition*, though it is quite different from the 'definition by induction' above. Let H be a function and X a set. We say that X is *closed under H* if for every element x of X, if x is in the domain of H then $H(x)$ is also in X. We say that X is the *closure* of Y under H if (i) every element of Y is in X, (ii) X is closed under H, and (iii) if Z is any set which includes Y and is closed under H then Z also includes X. (Briefly, 'X is the smallest set which includes Y and is closed under H'.) Similar definitions apply if we have a family of functions $H_1, \ldots, H_k$ instead of the one function H; also the functions can be n-place functions with $n > 1$.

A set is said to be *inductively defined* if it is defined as being the closure of some specified set Y under some specified functions $H_1, \ldots, H_k$. A typical inductive definition is the definition of the set of terms of a language L. The usual form for such a definition is:

(i) Every variable and every individual constant is a term.
(ii) For each function constant f, if f is n-place and $\tau_1, \ldots, \tau_n$ are terms, then the expression $f(\tau_1, \ldots, \tau_n)$ is a term.
(iii) Nothing is a term except as required by (i) and (ii).

Here we are defining the set X of terms. The so-called *basic* clause (i) describes Y as the set of all variables and all individual constants. The *inductive* clause (ii) describes the functions H_i, one for each function constant. Finally the *extremal* clause (iii) says that X is the closure of Y under the H_i. (Many writers omit the extremal clause, because it is rather predictable.)

Frege [1884] may have been the first to argue that inductive definitions need to be justified. He kept asking: How do we know that there *is* a smallest set which includes Y and is closed under H? One possible justification runs as follows. We recursively define $F(n)$, for each positive integer n, to be the set of all sequences $\langle b_1, \ldots, b_n \rangle$ such that b_1 is in Y and for every i ($1 \leqslant i < n$), b_{i+1} is $H(b_i)$. Then we define X to be the set of all b such that for some number n there is a sequence in $F(n)$ whose last term is b. Clearly Y is included in X, and we can show that X is closed under H. If Z is any set which is closed under H and includes Y, then an induction on the lengths of sequences shows that every element of X is in Z.

Naive arithmetic, as described above, is an axiomatic system but not a formal one. Peano [1889] took the first step towards formalizing it, by inventing a good symbolism. But the arguments above use quite an amount of set theory, and Peano made no attempt to write down what he was assuming about sets. Skolem [1923] threw out the set theory and made his assumptions precise, but his system was rather weak. First-order Peano arithmetic, a formalization of the first-order part of Peano's axioms, was introduced in Gödel [1931].

P, or *first-order Peano arithmetic*, is the following formal system. The constants of the language are an individual constant $\bar{0}$, a 1-place function symbol S and 2-place function symbols $+$ and $\cdot$, forming terms of form Sx, $(x + y)$, $(x \cdot y)$. Write $\bar{n}$ as an abbreviation for $S \ldots (n \text{ times}) \ldots S\bar{0}$; the symbols $\bar{n}$ are called *numerals*. We use a standard proof calculus for first-order logic (e.g. the calculus **H** of Appendix A) together with the following axioms:

P1. $\forall xy(Sx = Sy \rightarrow x = y)$

P2. $\forall x \neg (Sx = 0)$

P3. (Axiom schema of induction) All sentences of the form
$\forall \vec{z}(\phi[\bar{0}/x] \wedge \forall x(\phi \rightarrow \phi[Sx/x]) \rightarrow \forall x\phi)$

P4. $\forall x(x + \bar{0} = x)$

P5. $\forall xy(x + Sy = S(x + y))$

I.1: ELEMENTARY PREDICATE LOGIC

P6. $\quad \forall x(x \cdot \bar{0} = \bar{0})$

P7. $\quad \forall xy(x \cdot Sy = (x \cdot y) + x)$

The axioms are read as being just about numbers, so that $\forall x$ is read as 'for all numbers x'. In this way the symbols $\bar{0}$ and S in the language take care of axioms NA1 and NA2 without further ado. Axioms NA3 and NA4 appear as **P1** and **P2**. Since we can refer only to numbers and not to sets, axiom NA5 has to be recast as a condition on those sets of numbers which are definable by first-order formulas; this accounts for the axiom schema of induction, **P3**.

P4–P7 are the recursive definitions of addition and multiplication, cf. **(B.6)** and **(B.7)** above. In naive arithmetic there was no need to assume these as axioms, because we could *prove* that there are unique functions meeting these conditions. However, the proof used some set-theoretic notions like 'function defined on the numbers $0, \ldots, n-1$', which can't be expressed in a first-order language using just $\bar{0}$ and S. So we have to put the symbols $+$, $\cdot$ into the language – in particular they occur in formulas in the axiom schema of induction – and we have to assume the definitions **P4–P7** as axioms.

Gödel showed that with the aid of first-order formulas involving only $\bar{0}$, S, $+$ and $\cdot$, he could explicitly define a number of other notions. For example

(B.8) $\quad x < y \quad \text{iff} \quad \exists z(x + Sz = y).$

Also by using a clever trick with prime numbers he could encode each finite sequence $\langle m_1, m_2, \ldots \rangle$ of numbers as a single number

(B.9) $\quad 2^{m_1+1} \cdot 3^{m_2+1} \cdot 5^{m_3+1} \cdot \ldots$

and he could express the relation 'x is the yth term of the sequence coded by z' by a first-order formula. But then he could carry out 'in **P**' all the parts of naive arithmetic which use only numbers, finite sequences of numbers, finite sequences of finite sequences of numbers, and so on. This includes the argument which justifies primitive recursive definitions. In fact:

(1) *For every recursive definition δ of a number function, using just first-order formulas, there is a formula $\phi(x, y)$ such that in **P** we can prove that ϕ defines a function obeying δ. (If δ is primitive recursive then ϕ can be chosen to be Σ_1, cf.* Section 24.)

(2) *For every inductive definition of a set, where a formula ψ defines the basic set Y and formulas χ define the functions H in the inductive clause, there is a formula $\phi(x)$ such that we can prove in*

P *that the numbers satisfying ϕ are those which can be reached in a finite number of steps from Y by H. (If ψ and χ are Σ_1 then ϕ can be chosen to be Σ_1.)*

These two facts state in summary form why the whole of elementary syntax can be formalized within **P**.

There are some things that can be said in the language of **P** but not proved or refuted from the axioms of **P**. For example the statement that **P** itself is consistent (i.e. doesn't yield ⊥) can be formalized in the language of **P**. In [1931] Gödel showed that this formalized statement is not deducible from **P**, although we all hope it is true.

There are some other things that can't even be said in the language of **P**. For example we can't say in this language that the set X defined by ϕ in (2) above really is the closure of Y under H, because that would involve us in saying that 'if Z is *any set* which includes Y and is closed under H then Z includes X'. In the first-order language of **P** there is no way of talking about 'all sets of numbers'. For the same reason, many statements about real numbers can't be expressed in the language of **P** – even though some can by clever use of rational approximations.

In second-order arithmetic we can talk about real numbers, because real numbers can be represented as sets of natural numbers. Actually the natural numbers themselves are definable up to isomorphism in second-order logic without special arithmetical axioms. In third-order logic we can talk about sets of real numbers, fourth-order logic can talk about sets of sets of real numbers, and so on. Most of the events that take place in any standard textbook of real analysis can be recorded in, say, fifth-order logic. See Van Benthem and Doets [Chapter I.4, below] for these higher-order logics.

C. SET THEORY

The efforts of various nineteenth-century mathematicians reduced all the concepts of real and complex number theory to one basic notion: classes. So when Frege, in his *Grundgesetze der Arithmetik* I [1893], attempted a formal system which was to be adequate for all of arithmetic and analysis, the backbone of his system was a theory of classes. One of his assumptions was that for every condition there is a corresponding class, namely the class of all the objects that satisfy the condition. Unfortunately this assumption leads to contradictions, as Russell and Zermelo showed. Frege's approach has now been abandoned.

Today the most commonly adopted theory of classes is Zermelo–Fraenkel set theory, ZF. This theory was first propounded by Zermelo [1908] as an informal axiomatic theory. It reached its present shape through contributions of Mirimanoff, Fraenkel, Skolem and von Neumann. (Cf. Fraenkel's historical introduction to Bernays and Fraenkel [1958].)

Officially ZF is a set of axioms in a first-order language whose only constant is the 2-place predicate symbol $\in$ ('is a member of'). But all set theorists make free use of symbols introduced by definition.

Let me illustrate how a set theorist introduces new symbols. The axiom of Extensionality says that no two different sets have the same members. The Pair-set axiom says that if x and y are sets then there is at least one set which has just x and y as members. Putting these two axioms together, we infer that there is exactly one set with just x and y as members. Introducing a new symbol, we call this set $\{x, y\}$. There are also some definitions which don't depend on the axioms. For example we say x is *included in* y, or a *subset of* y, if every member of x is a member of y. This prompts the definition

(C.1) $x \subseteq y$ iff $\forall t(t \in x \to t \in y)$.

The language with these extra defined symbols is in a sense impure, but it is much easier to read than the pure set language with only $\in$, and one can always paraphrase away the new symbols when necessary. In what follows I shall be relentlessly impure. (On introducing new terms by definition, cf. Section 21 above. Suppes [1972] and Levy [1979] are careful about it.)

The first three axioms of ZF are about what kind of things we choose to count as sets. The axiom of Extensionality says that sets will count as equal when they have the same members:

ZF1.   (Extensionality) $\forall xy(x \subseteq y \wedge y \subseteq x \to x = y)$

We think of sets as being built up by assembling their members, starting with the empty or null set 0 which has no members:

ZF2.   (Null-set) $\forall t \; t \notin 0$ ($x \notin y$ means $\neg(x \in y)$.)

In a formal calculus which proves $\exists x \; x = x$, the Null-set axiom is derivable from the Separation axiom below and can be omitted. The axiom of Regularity expresses – as well as one can express it with a first-order statement – that x will not count as a set unless each of the members of x could be assembled together at an earlier stage than x itself. (So for example there is no 'set' x such that $x \in x$.)

ZF3.   (Regularity) $\forall x(x = 0 \vee \exists y(y \in x \wedge \forall z(z \in y \to z \notin x)))$.

The next three axioms state that certain collections can be built up:

ZF4. (Pair-set) $\forall xyt(t \in \{x,y\} \leftrightarrow t=x \vee t=y)$

ZF5. (Union) $\forall xt(t \in \bigcup x \leftrightarrow \exists y(t \in y \wedge y \in x))$

ZF6. (Power-set) $\forall xt(t \in \mathscr{P}x \leftrightarrow t \subseteq x)$.

Axioms ZF3–ZF6 allow some constructions. We write $\{x\}$ for $\{x,x\}$, $x \cup y$ for $\bigcup\{x,y\}$, $\{x_1,x_2,x_3\}$ for $\{x_1,x_2\} \cup \{x_3\}$, $\{x_1,\ldots,x_4\}$ for $\{x_1,x_2,x_3\} \cup \{x_4\}$, and so on. Likewise we can form ordered pairs $\langle x,y \rangle = \{\{x\},\{x,y\}\}$, ordered triplets $\langle x,y,z \rangle = \langle \langle x,y \rangle,z \rangle$ and so on. Building up from 0 we can form $1 = \{0\}$, $2 = \{0,1\}$, $3 = \{0,1,2\}$ etc.; the axiom of Regularity implies that $0, 1, 2, \ldots$ are all distinct. We can regard $0, 1, 2, \ldots$ as the natural numbers.

We need to be able to express 'x is a natural number' in the language of set theory, without using informal notions like 'and so on'. It can be done as follows. First, following von Neumann, we define Ord(x), 'x is an ordinal', by:

(C.2) Ord(x) iff $\bigcup x \subseteq x \wedge \forall yz(y \in x \wedge z \in x \rightarrow y \in z \vee z \in y \vee y = z)$.

This somewhat technical definition implies that the ordinals are linearly ordered by $\in$, and that they are well-ordered (i.e. every non-empty set of them has a least element, cf. (B.5) above). We can prove that the first ordinals are $0, 1, 2, \ldots$. Greek letters α, β, γ are used for ordinals. For every ordinal α there is a first greater ordinal; it is written $\alpha + 1$ and defined as $\alpha \cup \{\alpha\}$. For every set X of ordinals there is a first ordinal β which is greater than or equal to every ordinal in X, viz. $\beta = \bigcup X$. Each ordinal β has just one of the following three forms: either $\beta = 0$, or β is a *successor* (i.e. of form $\alpha + 1$), or β is a *limit* (i.e. of form $\bigcup X$ for a non-empty set X of ordinals which has no greatest member). Now the natural numbers can be defined as follows:

(C.3) x is a natural number iff Ord(x) $\wedge \forall y(y \in x + 1 \rightarrow y = 0 \vee y$ is a successor).

The remaining four axioms, ZF7–ZF10, are needed for talking about infinite sets. Each of them says that sets exist with certain properties. Nothing in ZF1–ZF6 implies that there are any infinite sets. We fill the gap by decreeing that the set ω of all natural numbers exists:

ZF7. (Infinity) $\forall t(t \in \omega \leftrightarrow t$ is a natural number).

The next axiom says that within any given set x we can collect together those members w which satisfy the formula $\phi(\vec{z},w)$. Here ϕ is allowed to be any first-order formula in the language of set theory, and it can mention other sets $\vec{z}$. Strictly ZF8 is an axiom schema and not a single axiom.

ZF8. (Separation) $\forall \vec{z} \forall x \exists t (t \in \{w \in x \mid \phi\} \leftrightarrow t \in x \wedge \phi[t/w])$.

For example this tells us that for any sets x and y there is a set whose members are exactly those members w of x which satisfy the formula $w \in y$; in symbols this set is $\{w \in x \mid w \in y\}$. So we can introduce a new symbol for this set, and write $x \cap y = \{w \in x \mid w \in y\}$. Similarly we can define: $\cap x = \{w \in \bigcup x \mid \forall z (z \in x \to w \in z)\}$, $x \times y = \{t \in \mathscr{P} \mathscr{P}(x \cup y) \mid \exists z w (z \in x \wedge w \in y \wedge t = \langle z, w \rangle)\}$, $x^2 = x \times x$ and more generally $x^{n+1} = x^n \times x$. An *n-place relation* on the set x is a subset of x^n. We can define 'f is a function from x to y', in symbols $f: x \to y$, by:

(C.4) $f: x \to y$ iff $f \subseteq x \times y \wedge \forall w (w \in x \to \exists z \forall t (t = z \leftrightarrow \langle w, t \rangle \in f))$.

We say f is an *n-place function* from x to y if $f: x^n \to y$. When $f: x \to y$, we call x the *domain* of f, and we can define it in terms of f by: dom $f = \{w \in \bigcup \bigcup f \mid \exists z \langle w, z \rangle \in f\}$. We define the *value* of f for *argument* w, in symbols $f(w)$, as $\{t \in \bigcup \bigcup \bigcup f \mid \exists z (\langle w, z \rangle \in f \wedge t \in z)\}$. A *bijection* (or *one–one correspondence*) *from x to y* is a function f such that $f: x \to y$ and every element z of y is of form $f(w)$ for exactly one w in x. A *sequence of length* α is defined to be a function with domain α.

The system of axioms ZF1–ZF8 is sometimes known as *Zermelo set theory*, or *Z* for short. It is adequate for formalizing all of naive arithmetic, not just the finite parts that can be axiomatized in first-order Peano arithmetic. The Separation axiom is needed. For example in the proof of (B.2) we had to know that there is a set M whose members are all the members of K except n; M is $\{w \in K \mid w \neq n\}$.

First-order languages can be defined formally within Z. For example we can define a *similarity type* for predicate logic to be a set whose members each have one of the following forms: (i) $\langle 1, x \rangle$, (ii) $\langle 2, m, x \rangle$ where m is a positive natural number, (iii) $\langle 3, m, x \rangle$ where m is a positive natural number. The elements of form (i) are called *individual constants*, those of form (ii) are the *m-place predicate constants* and those of form (iii) are the *m-place function constants*. *Variables* can be defined as ordered pairs of form $\langle 4, n \rangle$ where n is a natural number. *Terms* can be defined inductively by: (a) Every variable or individual constant is a term. (b) If f is an *m*-place function constant and $\tau_1, \ldots, \tau_m$ are terms then $\langle 5, f, \tau_1, \ldots, \tau_m \rangle$ is a term. (c) Nothing is a term except as required by (a) and (b). By similar devices we can define the whole language L of a given similarity type X. *L-structures* can be defined to be ordered pairs $\langle A, I \rangle$ where A is a non-empty set and I is a function with domain X, such that for each individual constant c of X, $I(c) \in A$ (and so on as in Section 14). Likewise we can define $\models$ for L-structures.

The two remaining axioms of ZF are needed for various arguments in infinite arithmetic.

In Appendix B we saw how one can define functions with domain the natural numbers, by recursion. We want to be able to do the same in set theory, but with any ordinal as the domain. For example if the language L is not countable, then the proof of completeness in Section 16 above will need to be revised so that we build a chain of theories Δ_i for all $i \in \alpha$, where α is some ordinal greater than ω. One can try to justify recursive definitions on ordinals, just as we justified definitions in Appendix B. It turns out that one piece of information is missing. We need to know that if a formula defines a function f whose domain is an ordinal, then f is a set. The following axiom supplies this missing information. It says that if a formula ϕ defines a function with domain a set, then the image of this function is again a set:

ZF9. (Replacement) $\forall \vec{z} x (\forall y w t (y \in x \wedge \phi \wedge \phi[w/t] \to t = w) \to \exists u \forall t (t \in u \leftrightarrow \exists y (y \in x \wedge \phi)))$.

Like Separation, the Replacement axiom is really an axiom schema.

The final axiom is the axiom of Choice, which is needed for most kinds of counting argument. This axiom can be given in many forms, all equivalent in the sense that any one can be derived from any other using ZF1–ZF9. The form given below, Zermelo's Well-ordering principle, means intuitively that the elements of any set can be checked off one by one against the ordinals, and that the results of this checking can be gathered together into a set.

ZF10. (Well-ordering) $\forall x \exists f \alpha (\alpha$ is an ordinal and f is a bijection from α to $x)$.

Axiom ZF10 is unlike axioms ZF4–ZF9 in a curious way. These earlier axioms each said that there is a set with just such-and-such members. But ZF10 says that a certain set exists (the function f) without telling us what the members of the set are. So arguments which use the axiom of Choice have to be less explicit than arguments which use only ZF1–ZF9.

Using ZF10, the theory of cardinality proceeds as follows. The *cardinality* $|x|$ or $\bar{\bar{x}}$ of a set x is the first ordinal α such that there is a bijection from α to x. Ordinals which are the cardinalities of sets are called *cardinals*. Every cardinal is equal to its own cardinality. Every natural number is a cardinal.

I.1: ELEMENTARY PREDICATE LOGIC

A set is said to be *finite* if its cardinality is a natural number. The cardinals which are not natural numbers are said to be *infinite*. The infinite cardinals can be listed in increasing order as $\omega_0, \omega_1, \omega_2, \ldots$; ω_0 is ω. For every ordinal α there is an αth infinite cardinal ω_α, sometimes also written $\aleph_\alpha$. It can be proved that there is no greatest cardinal, using Cantor's theorem that for every set x, $\mathscr{P}(x)$ has greater cardinality than x.

Let me give an example of a principle equivalent to ZF10. If I is a set and for each $i \in I$ a set A_i is given, then $\Pi_I A_i$ is defined to be the set of all functions $f: I \to \cup \{A_i \mid i \in I\}$ such that for each $j \in I$, $f(j) \in A_j$. $\Pi_I A_i$ is called the *product* of the sets A_i. Then ZF10 is equivalent to the statement: If the sets A_i in a product are all non-empty then their product is also not empty.

The compactness theorem for propositional logic with any set of sentence letters is not provable from ZF1–ZF9. *A fortiori* neither is the compactness theorem for predicate logic. Logicians have dissected the steps between ZF10 and the compactness theorem, and the following notion is one of the results. (It arose in other parts of mathematics too.)

Let I be any set. Then an *ultrafilter* on I is defined to be a subset D of $\mathscr{P}(I)$ such that (i) if a and $b \in D$ then $a \cap b \in D$, (ii) if $a \in D$ and $a \subseteq b \subseteq I$ then $b \in D$, and (iii) for all subsets a of I, exactly one of I and $I-a$ is in D (where $I-a$ is the set of all elements of I which are not in a). For example if $i \in I$ and $D = \{a \in \mathscr{P}(I) \mid i \in a\}$ then D is an ultrafilter on I; ultrafilters of this form are called *principal* and they are uninteresting. From ZF1–ZF9 it is not even possible to show that there exist any non-principal ultrafilters at all. But using ZF10 one can prove the following principle:

THEOREM (C.5) *Let I be any infinite set. Then there exist an ultrafilter D on I and for each $i \in I$ an element $a_i \in D$, such that for every $j \in I$ the set $\{i \in I \mid j \in a_i\}$ is finite.*

An ultrafilter D with the property described in Theorem (C.5) is said to be *regular*. Regular ultrafilters are always non-principal.

To derive the compactness theorem from Theorem (C.5), we need to connect ultrafilters with structures. This is done as follows. For simplicity we can assume that the language L has just one constant symbol, the 2-place predicate constant R. Let D be an ultrafilter on the set I. For each $i \in I$, let $\mathfrak{A}_i$ be an L-structure with domain A_i. Define a relation $\sim$ on $\Pi_I A_i$ by:

(C.6) $f \sim g$ iff $\{i \in I \mid f(i) = g(i)\} \in D$.

Then since D is an ultrafilter, $\sim$ is an equivalence relation; write $f^\sim$ for the equivalence class containing f. Let B be $\{f^\sim \mid f \in \Pi_I A_i\}$. Define an L-structure $\mathfrak{B} = \langle B, I_\mathfrak{B} \rangle$ by putting

(C.7) $\quad \langle f^\sim, g^\sim \rangle \in I_\mathfrak{B}(R)$ iff $\{i \in I \mid \langle f(i), g(i) \rangle \in I_{\mathfrak{A}_i}(R)\} \in D$.

(Using the fact that D is an ultrafilter, this definition makes sense.) Then $\mathfrak{B}$ is called the *ultraproduct* of the $\mathfrak{A}_i$ by D, in symbols $\Pi_D \mathfrak{A}_i$ or D-prod $\mathfrak{A}_i$. By a theorem of Jerzy Łoś, if ϕ is any sentence of the first-order language L, then

(C.8) $\quad \Pi_D \mathfrak{A}_i \models \phi$ iff $\{i \in I \mid \mathfrak{A}_i \models \phi\} \in D$.

Using the facts above, we can give another proof of the compactness theorem for predicate logic. Suppose that Δ is a first-order theory, and every finite subset of Δ has a model. We have to show that Δ has a model. If Δ itself is finite, there is nothing to prove. So assume now that Δ is infinite, and let I in Theorem (C.5) be Δ. Let D and the sets a_ϕ ($\phi \in \Delta$) be as in Theorem (C.5). For each $i \in \Delta$, the set $\{\phi \mid i \in a_\phi\}$ is finite, so by assumption it has a model $\mathfrak{A}_i$. Let $\mathfrak{B}$ be $\Pi_D \mathfrak{A}_i$. For each sentence $\phi \in \Delta$, $a_\phi \subseteq \{i \in \Delta \mid \mathfrak{A}_i \models \phi\}$, so by (ii) in the definition of an ultrafilter, $\{i \in \Delta \mid \mathfrak{A}_i \models \phi\} \in D$. It follows by Łoś's theorem (C.8) that $\mathfrak{B} \models \phi$. Hence Δ has a model, namely $\mathfrak{B}$.

There are full accounts of ultraproducts in Bell and Slomson [1969] and Chang and Keisler [1973]. One principle which often turns up when ultraproducts are around is as follows. Let X be a set of subsets of a set I. We say that X has the *finite intersection property* if for every finite subset $\{a_1, \ldots, a_n\}$ of X, the set $a_1 \cap \cdots \cap a_n$ is not empty. The principle states that *if X has the finite intersection property then there is an ultrafilter D on I such that $X \subseteq D$*. This can be proved quite quickly from ZF10.

Some writers refer to ZF1–ZF9, without the axiom of Choice, as ZF; they write ZFC when Choice is included. There are a number of variants of ZF. For example the set-class theory of Gödel and Bernays (cf. Mendelson [1964]) allows one to talk about 'the class of all sets which satisfy the formula ϕ' provided that ϕ has no quantifiers ranging over classes. This extension of ZF is only a notational convenience. It enables one to replace axiom schemas by single axioms, so as to get a system with just finitely many axioms.

Another variant allows elements which are not sets – these elements are called *individuals*. Thus we can talk about the set {Geoffrey Boycott} without having to believe that Geoffrey Boycott is a set. In informal set theory of course one considers such sets all the time. But there seems to be no mathematical advantage in admitting individuals into formal set theory;

rather the contrary, we learn nothing new and the proofs are messier. A set is called a *pure set* if its members, its members' members, its members' members' members etc. are all of them sets. In ZF all sets are pure.

ACKNOWLEDGMENTS

I thank the many people who commented on the first version of this chapter, and especially Franz Guenthner, Hans Kamp and Dirk van Dalen for their detailed improvements and corrections. I am also very much in debt to Jan and Emilia Mycielski who allowed me to type the chapter in their idyllic house in the foothills of the Rockies.

Bedford College, London

REFERENCES

The text of Church [1956] is a reliable and thorough source of information on anything that happened in first-order logic before the mid 1950s. I regret that I only learned of the valuable historical survey by Gregory Moore (Moore [1980]) after this chapter was written.

Ackermann, W.: 1962, *Solvable Cases of the Decision Problem*, North-Holland, Amsterdam.
Altham, J. E. J. and Tennant, N. W.: 1975, 'Sortal quantification', in E. L. Keenan [1975] pp. 46–58.
Anderson, J. M. and Johnstone Jr., H. W.: 1962, *Natural Deduction: the Logical Basis of Axiom Systems*, Wadsworth, Belmont, California.
Ax, J. and Kochen, S.: 1965, 'Diophantine problems over local fields: I', *Amer. J. Math.* **87**, 605–630.
Bar-Hillel, Y.: 1971, *Pragmatics of Natural Languages*, Reidel, Dordrecht.
Barwise, J.: 1973, 'Abstract logics and $L_{\infty\omega}$', *Annals Math. Logic* **4**, 309–340.
Barwise, J.: 1974, 'Axioms for abstract model theory', *Annals Math. Logic* **7**, 221–265.
Barwise, J.: 1975, *Admissible Sets and Structures*, Springer, Berlin.
Barwise, J. and Cooper, R.: 1981, 'Generalized quantifiers and natural languages', *Linguistics and Philosophy* **4**, 159–219.
Behmann, H.: 1922, 'Beiträge zur Algebra der Logik, insbesondere zum Entscheidungsproblem', *Math. Annalen* **86**, 163–229.
Bell, J. L. and Machover, M.: 1977, *A Course in Mathematical Logic*, North-Holland, Amsterdam.
Bell, J. L. and Slomson, A. B.: 1969, *Models and Ultraproducts*, North-Holland, Amsterdam.
Belnap, N. D.: 1962, 'Tonk, plonk and plink', *Analysis* **22**, 130–134; reprinted in Strawson [1967] pp. 132–137.

Benacerraf, P. and Putnam, H. (eds.): 1964, *Philosophy of Mathematics: Selected Readings*, Prentice-Hall, Englewood Cliffs, NJ.

Bernays, P.: 1942, review of Max Steck, 'Ein unbekannter Brief von Gottlob Frege über Hilberts erste Vorlesung über die Grundlagen der Geometrie', *J. Symbolic Logic* **7**, 92f.

Bernays, P. and Fraenkel, A. A.: 1958, *Axiomatic Set Theory*, North-Holland, Amsterdam.

Beth, E. W.: 1953, 'On Padoa's method in the theory of definition', *Koninklijke Nederlandse Akad. van Wetensch.* **56**, ser. A, Math. Sciences, 330–339.

Beth, E. W.: 1955, 'Semantic entailment and formal derivability', *Mededelingen der Koninklijke Nederlandse Akad. van Wetensch.*, afd letterkunde, 18; reprinted in Hintikka [1969] pp. 9–41.

Beth, E. W.: 1962, *Formal Methods*, Reidel, Dordrecht.

Bocheński, I. M.: 1970, *A History of Formal Logic*, tr. I. Thomas, Chelsea Publ. Co., New York.

Bolzano, B.: 1837, *Wissenschaftslehre*, ed. and tr. R. George as *Theory of Science*, Univ. California Press, Berkeley and Los Angeles, 1972.

Boole, G.: 1847, *The Mathematical Analysis of Logic*, Macmillan, Barclay and Macmillan, Cambridge; pp. 45–124 of George Boole, *Studies in Logic and Probability*, Open Court, La Salle, Illinois, 1952.

Boole, G.: 1854, *An Investigation of the Laws of Thought*, Walton and Maberley, London; also Open Court, La Salle, Illinois, 1952.

Boolos, G.: 1979, *The Unprovability of Consistency: an Essay in Modal Logic*, Cambridge U.P.

Carnap, R.: 1935, 'Ein Gültigkeitskriterium für die Sätze der klassischen Mathematik', *Monatshefte Math. und Phys.* **42**, 163–190.

Carnap, R.: 1956, *Meaning and Necessity*, 2nd edn., Univ. Chicago Press.

Chang, C. C. and Keisler, H. J.: 1973, *Model Theory*, North-Holland, Amsterdam.

Chastain, C.: 1975, 'Reference and context', in K. Gunderson (ed.), *Minnesota Studies in the Philosophy of Science, VII, Language, Mind, and Knowledge*, Univ. Minnesota Press, Minneapolis, pp. 194–269.

Cherlin, G.: 1976, *Model Theoretic Algebra: Selected Topics*, Lecture Notes in Math. **521**, Springer, Berlin.

Church, A.: 1936, 'A note on the Entscheidungsproblem', *J. Symbolic Logic* **1**, 40f, 101f.

Church, A.: 1956, *Introduction to Mathematical Logic I*, Princeton U.P., Princeton, NJ.

Cohen, L. J.: 1971, 'Some remarks on Grice's views about the logical particles of natural language', in Bar-Hillel [1971] pp. 60–68.

Cohen, P. J.: 1969, 'Decision procedures for real and p-adic fields', *Comm. Pure Appl. Math.* **22**, 131–151.

Cook, S. A.: 1971, 'The complexity of theorem-proving procedures', in *Proc. 3rd Ann. ACM Symp. on Theory of Computing*, Assoc. for Computing Machinery, NY, pp. 151–158.

Craig, W.: 1957, 'Linear reasoning. A new form of the Herbrand–Gentzen theorem', *J. Symbolic Logic* **22**, 250–268.

Craig, W.: 1957a, 'Three uses of the Herbrand–Gentzen theorem in relating model theory and proof theory', *J. Symbolic Logic* **22**, 269–285.

I.1: ELEMENTARY PREDICATE LOGIC

Davidson, D. and Harman, G. (eds.): 1972, *Semantics of Natural Languages*, Reidel, Dordrecht.
Dedekind, R.: 1888, *Was sind und was sollen die Zahlen?*, Brunswick.
Dedekind, R.: 1890, Letter to Keferstein, in van Heijenoort [1967] pp. 90–103.
Dowty, D., Wall, R. and Peters, S.: 1981, *Introduction to Montague Semantics*, Reidel, Dordrecht.
Dummett, M. A. E.: 1958/9, 'Truth', *Proc. Aristotelian Soc.* **59**, 141–162; reprinted in Strawson [1967], pp. 49–68.
Dummett, M. A. E.: 1973, *Frege: Philosophy of Language*, Duckworth, London.
Dummett, M. A. E.: 1975, 'What is a theory of meaning?', in *Mind and Language*, ed. Samuel Guttenplan, Clarendon Press, Oxford.
Dunn, J. M. and Belnap, N. D.: 1968, 'The substitution interpretation of the quantifiers', *Noûs* **2**, 177–185.
Ehrenfeucht, A.: 1960, 'An application of games to the completeness problem for formalized theories', *Fundamenta Math.* **49**, 129–141.
Evans, G.: 1980, 'Pronouns', *Linguistic Inquiry* **11**, 337–362.
Evans, G. and McDowell, J. (eds.): 1976, *Truth and Meaning: Essays in Semantics*, Clarendon Press, Oxford.
Feferman, S.: 1968, 'Lectures on proof theory', in *Proc. Summer School of Logic, Leeds 1967*, Lecture Notes in Mathematics **70**, Springer, Berlin, pp. 1–109.
Feferman, S.: 1968a, 'Persistent and invariant formulas for outer extensions', *Compositio Math.* **20**, 29–52.
Feferman, S.: 1969, 'Set-theoretical foundations of category theory', in *Reports of the Midwest Category Seminar III*, Lecture Notes in Mathematics **106**, Springer, Berlin, pp. 201–247.
Feferman, S.: 1974, 'Applications of many-sorted interpolation theorems', in Henkin *et al.* (eds.) [1974], pp. 205–223.
Fitch, F. B.: 1952, *Symbolic Logic*, Ronald Press, New York.
Flum, J.: 1975, 'First-order logic and its extensions', in *ISILC Logic Conference*, Lecture Notes in Mathematics **499**, Springer, Berlin 1975, pp. 248–307.
Fraenkel, A.: 1922, 'Zu den Grundlagen der Cantor–Zermeloschen Mengenlehre', *Math. Annalen* **86**, 230–237.
Fraissé, R.: 1954, 'Sur l'extension aux relations de quelques propriétés des ordres', *Ann. Sci. École Norm. Sup.* **71**, 363–388.
Fraissé, R.: 1955, 'Sur quelques classifications des relations, basées sur des isomorphismes restreints', *Alger–Mathématiques* **2**, 16–60 and 273–295.
Frege, G.: 1879, *Begriffsschrift*, Halle; tr. in van Heijenoort [1967] pp. 1–82.
Frege, G.: 1884, *Die Grundlagen der Arithmetik*, Breslau; tr. by J. L. Austin, 'The Foundations of Arithmetic', 2nd edn., Blackwell, Oxford, 1953.
Frege, G.: 1891, *Funktion und Begriff*, Jena; in Frege [1967] pp. 125–142; tr. in Frege [1952].
Frege, G.: 1893, *Grundgesetze der Arithmetik I*, Jena; partial tr. with intro. by Montgomery Furth, 'The Basic Laws of Arithmetic', Univ. California Press, Berkeley, 1964.
Frege, G.: 1906, 'Über die Grundlagen der Geometrie', *Jahresbericht der Deutschen Mathematiker-Vereinigung* **15**, 293–309, 377–403 and 423–430; tr. in Frege [1971].

Frege, G.: 1912, Anmerkungen zu: Philip E. B. Jourdain, 'The development of the theories of mathematical logic and the principles of mathematics' (1912), in Frege [1967] pp. 334–341.

Frege, G.: 1952, *Translations from the Philosophical Writings of Gottlob Frege*, P. Geach *et al.* (eds.), Blackwell, Oxford.

Frege, G.: 1967, *Kleine Schriften*, I. Angelelli (ed.), Georg Olms Verlagsbuchhandlung, Hildesheim.

Frege, G.: 1971, *On the Foundations of Geometry and Formal Theories of Arithmetic*, tr. with introduction by E. W. Kluge, Yale U.P., New Haven.

Frege, G. and Hilbert, D.: 1899/1900, Correspondence leading to 'On the foundations of geometry', in Frege [1967] pp. 407–418; tr. in Frege [1971] pp. 6–21.

Gandy, R. O.: 1974, 'Set-theoretic functions for elementary syntax', in T. J. Jech (ed.), *Axiomatic Set Theory II*, Amer. Math. Soc., Providence RI, pp. 103–126.

Garey, M. R. and Johnson, D. S.: 1979, *Computers and Intractability*, W. H. Freeman, San Francisco.

Gentzen, G.: 1934, 'Untersuchungen über das logische Schliessen', *Math. Zeitschrift* **39**, 176–210, 405–431.

Gödel, K.: 1930, 'Die Vollständigkeit der Axiome des logischen Funktionenkalküls', *Monatshefte für Mathematik und Physik* **37**, 349–360; tr. in van Heijenoort [1967], pp. 582–591.

Gödel, K.: 1931, 'Über formal unentscheidbare Sätze der Principia Mathematica und verwandter Systeme I', *Monatshefte für Mathematik und Physik* **38**, 173–198; tr. in van Heijenoort [1967], pp. 596–616.

Gödel, K.: 1931a, 'Eine Eigenschaft der Realisierungen des Aussagenkalküls', *Ergebnisse Math. Kolloq.* **3**, 20–21.

Gödel, K.: 1947, 'What is Cantor's continuum problem?', *Amer. Math. Monthly* **54**, 515–525; revised and expanded version in Benacerraf *et al.* [1964] pp. 258–273.

Gödel, K.: 1951, 'Russell's mathematical logic', in P. A. Schilpp (ed.), *The Philosophy of Bertrand Russell*, Tudor Publ. Co., New York, pp. 123–153; also in Pears [1972], pp. 192–226, and Benacerraf *et al.* [1964] pp. 211–232.

Goldblatt, R.: 1982, *Axiomatizing the Logic of Programmes*, Lecture Notes in Computer Science **130**, Springer, Berlin.

Goldfarb, W. D.: 1979, 'Logic in the twenties: the nature of the quantifier', *J. Symbolic Logic* **44**, 351–368.

Grice, H. P.: 1975, 'Logic and conversation', in P. Cole *et al.* (eds.), *Syntax and Semantics 3, Speech Acts*, Academic Press, New York, pp. 41–58.

Hajnal, A. and Németi, I.: 1979, 'Applications of universal algebra, model theory, and categories in computer science', *Comput. Linguist. Comput. Lang.* **13**, 251–282.

Harel, D.: 1979, *First-order Dynamic Logic*, Lecture Notes in Computer Science **68**, Springer, Berlin.

Hasenjaeger, G.: 1953, 'Eine Bemerkung zu Henkins Beweis für die Vollständigkeit des Prädikatenkalküls der ersten Stufe, *J. Symbolic Logic* **18**, 42–48.

Hausdorff, F.: 1914, *Grundzüge der Mengenlehre*, Veit, Leipzig.

Henkin, L.: 1949, 'The completeness of the first-order functional calculus', *J. Symbolic Logic* **14**, 159–166; reprinted in Hintikka [1969].

Henkin, L.: 1950, 'Completeness in the theory of types', *J. Symbolic Logic* **15**, 81–91; reprinted in Hintikka [1969].

Henkin, L.: 1961, 'Some remarks on infinitely long formulas', in *Infinitistic Methods: Proc. Symp. on Foundations of Mathematics, Warsaw*, Pergamon, London, pp. 167-183.
Henkin, L. and Mostowski, A.: 1959, Review of Mal'cev [1941], *J. Symbolic Logic* **24**, 55-57.
Henkin, L. *et al.* (eds.): 1974, *Proceedings of the Tarski Symposium*, Proc. Symposia in Pure Math. XXV, Amer. Math. Soc., Providence, RI.
Herbrand, J.: 1930, *Recherches sur la théorie de la démonstration*, Thèse, University of Paris; tr. in Herbrand [1971] pp. 44-202.
Herbrand, J.: 1971, *Logical writings*, W. D. Goldfarb (ed.), Harvard Univ. Press, Cambridge, Mass.
Hilbert, D.: 1899, *Grundlagen der Geometrie*, Teubner, Leipzig.
Hilbert, D.: 1923, 'Die Logischen Grundlagen der Mathematik', *Math. Annalen* **88**, 151-165; in Hilbert [1970] pp. 178-195.
Hilbert, D.: 1924, 'Über das Unendliche', *Math. Annalen* **95**, 161-190; tr. in van Heijenoort [1967], pp. 367-392; partial tr. in Benacerraf *et al.* [1964].
Hilbert, D.: 1928, 'Die Grundlagen der Mathematik', *Abhandlungen aus dem Math. Seminar der Hamburgischen Universität* **6**, 65-85; tr. in van Heijenoort [1967], pp. 464-479.
Hilbert, D.: 1970, *Gesammelte Abhandlungen III: Analysis, Grundlagen der Mathematik, Physik, Verschiedenes*, Springer, Berlin.
Hilbert, D. and Ackermann, W.: 1928, *Grundzüge der theoretischen Logik*, Springer, Berlin.
Hilbert, D. and Bernays, P.: 1934, *Grundlagen der Mathematik I*, Springer, Berlin.
Hilbert, D. and Bernays, P.: 1939, *Grundlagen der Mathematik II*, Springer, Berlin.
Hintikka, J.: 1953, 'Distributive normal forms in the calculus of predicates', *Acta Philosophica Fennica* **6**.
Hintikka, J.: 1955, 'Form and content in quantification theory', *Acta Philosophica Fennica* **8**, 7-55.
Hintikka, J. (ed.): 1969, *The Philosophy of Mathematics*, Oxford U.P.
Hintikka, J.: 1973, *Logic, Language-games and Information*, Oxford U.P.
Hintikka, J., Niiniluoto, I. and Saarinen, E. (eds.): 1979, *Essays on Mathematical and Philosophical Logic*, Reidel, Dordrecht.
Hodges, W.: 1972, 'On order-types of models', *J. Symbolic Logic* **37**, 69f.
Hodges, W.: 1977, *Logic*, Penguin Books, Harmondsworth, Middx.
Jeffrey, R. C.: 1967, *Formal Logic: its Scope and Limits*, McGraw-Hill, New York.
Johnstone, P. T.: 1977, *Topos Theory*, Academic Press, London.
Kalish, D. and Montague, R.: 1964, *Logic: Techniques of Formal Reasoning*, Harcourt, Brace and World, New York.
Kalmár, L.: 1934/5, 'Über die Axiomatisierbarkeit des Aussagenkalküls', *Acta Scient. Math. Szeged* **7**, 222-243.
Kamp, H.: 1971, 'Formal properties of "Now"', *Theoria* **37**, 227-273.
Kamp, H.: 1981, 'A theory of truth and semantic representation', in Groenendijk, J. A. G. *et al.* (eds.), *Formal Methods in the Study of Language*, Math. Centrum, Amsterdam, pp. 277-322.
Kaplan, D.: 1966, 'What is Russell's theory of descriptions?', in *Proceedings of Internat. Colloq. on Logic, Physical Reality and History, Denver, 1966*, Plenum, New York; reprinted in Pears [1972] pp. 227-244.

Karp, C.: 1965, 'Finite quantifier equivalence', in J. Addison *et al.* (eds.), *The Theory of Models*, North-Holland, Amsterdam.
Keenan, E. L. (ed.): 1975, *Formal Semantics of Natural Language*, Cambridge U.P.
Kleene, S. C.: 1943, 'Recursive predicates and quantifiers', *Trans. Amer. Math. Soc.* **53**, 41–73.
Kleene, S. C.: 1952, *Introduction to Metamathematics*, North-Holland, Amsterdam.
Klenk, V.: 1976, 'Intended models and the Löwenheim–Skolem theorem', *J. Philos. Logic* **5**, 475–489.
Kneale, W.: 1956, 'The province of logic', in H. D. Lewis (ed.), *Contemporary British Philosophy*, 3rd series, George Allen and Unwin, London, pp. 237–261.
Kreisel, G.: 1967, 'Informal rigour and completeness proofs', in Lakatos (ed.), *Problems in the Philosophy of Mathematics*, North-Holland, Amsterdam, pp. 138–157; partially reprinted in Hintikka [1969] pp. 78–94.
Kreisel, G. and Krivine, J. L.: 1967, *Elements of Mathematical Logic (Model Theory)*, North-Holland, Amsterdam.
Kripke, S.: 1976, 'Is there a problem about substitutional quantification?', in Evans and McDowell [1976] pp. 325–419.
Kronecker, L.: 1882, 'Grundzüge einer arithmetischen Theorie der algebraischen Grössen', *Crelle's Journal* **92**, 1–122.
Lakoff, G.: 1972, 'Linguistics and natural logic', in Davidson and Harman [1972] pp. 545–665.
Langford, C. H.: 1927, 'Some theorems on deducibility', *Annals of Math.* **28**, 16–40.
Leisenring, A. C.: 1969, *Mathematical Logic and Hilbert's ϵ-symbol*, Gordon and Breach, New York.
Lemmon, E. J.: 1965, *Beginning Logic*, Nelson, London.
Levy, A.: 1965, 'A hierarchy of formulas in set theory', *Memoirs of Amer. Math. Soc.* **57**.
Levy, A.: 1979, *Basic Set Theory*, Springer, New York.
Lindström, P.: 1969, 'On extensions of elementary logic', *Theoria* **35**, 1–11.
Lorenzen, P.: 1961, 'Ein dialogisches Konstruktivitätskriterium', in *Infinitistic Methods, Proc. of a Symp. on Foundations of Mathematics, Warsaw*, Pergamon Press, London, pp. 193–200.
Lorenzen, P.: 1962, *Metamathematik*, Bibliographisches Institut, Mannheim.
Lorenzen, P. and Schwemmer, O.: 1975, *Konstruktive Logik, Ethik und Wissenschaftstheorie*, Bibliographisches Institut, Mannheim.
Löwenheim, L.: 1915, 'Über Möglichkeiten im Relativkalkül', *Math. Annalen* **76**, 447–470; tr. in van Heijenoort [1967] pp. 228–251.
Łukasiewicz, J. and Tarski, A.: 1930, 'Untersuchungen über den Aussagenkalkül', *Comptes Rendus des séances de la Société des Sciences et des Lettres de Varsovie*, 23 cl. iii, 30–50; tr. in Tarski [1956] pp. 38–59.
Mal'cev, A. I.: 1936, 'Untersuchungen aus dem Gebiete der mathematischen Logik', *Mat. Sbornik* **1**, 323–336; tr. in Mal'cev [1971] pp. 1–14.
Mal'cev, A. I.: 1936a, 'On a general method for obtaining local theorems in group theory' (Russian), *Ivanov Gos. Ped. Inst. Uč. Zap. Fiz.-Mat. Fak.* **1** No. 1, 3–9; tr. in Mal'cev [1971] pp. 15–21.
Mal'cev, A. I.: 1971, *The Metamathematics of Algebraic Systems; Collected Papers 1936–1967*, tr. and ed. B. F. Wells III, North-Holland, Amsterdam.

Manna, Z.: 1974, *Mathematical Theory of Computation*, McGraw-Hill, New York.
Mates, B.: 1965, *Elementary Logic*, Oxford U.P., New York.
Members of the Johns Hopkins University, Boston: 1883, *Studies in Logic*, Little, Brown and Co.
Mendelson, E.: 1964, *Introduction to Mathematical Logic*, Van Nostrand, Princeton, NJ.
Mitchell, O. H.: 1883, 'On a new algebra of logic', in Members [1883] pp. 72–106.
Montague, R.: 1970, 'English as a formal language', in B. Visentini *et al.* (eds.), *Linguaggi nella Società e nella Tecnica*, Milan 1970; in Montague [1974] pp. 188–221.
Montague, R.: 1973, 'The proper treatment of quantification in ordinary English', in J. Hintikka *et al.* (eds.), *Approaches to Natural Language*, Reidel, Dordrecht; in Montague [1974] pp. 247–270.
Montague, R.: 1974, *Formal Philosophy, Selected Papers of Richard Montague*, ed. R. H. Thomason, Yale U.P., New Haven.
Montague, R. and Vaught, R. L.: 1959, 'Natural models of set theory', *Fundamenta Math.* **47**, 219–242.
Moore, G. H.: 1980, 'Beyond first-order Logic: the historical interplay between mathematical logic and axiomatic set theory', *History and Philosophy of Logic* **1**, 95–137.
Partee, B.: 1978, 'Bound variables and other anaphors', in D. Waltz (ed.), *Tinlap-2, Theoretical Issues in Natural Language Processing*, Assoc. for Computing Machinery, New York, pp. 248–280.
Peano, G.: 1889, *Arithmetices Principia, Nova Methodo Exposita*, Turin; tr. in van Heijenoort [1967] pp. 20–55.
Pears, D. F. (ed.): 1972, *Bertrand Russell. A Collection of Critical Essays*, Anchor Books, Doubleday, New York.
Peirce, C. S.: 1883, 'A theory of probable inference. Note B. The logic of relatives', in Members [1883]; reprinted in Peirce [1933] vol. III, pp. 195–209.
Peirce, C. S.: 1885, 'On the algebra of logic', *Amer. J. Math.* **7**, 180–202; reprinted in Peirce [1933] vol. III, pp. 210–238.
Peirce, C. S.: 1902, *The Simplest Mathematics*, in Peirce [1933], vol. IV, pp. 189–262.
Peirce, C. S.: 1933, *Collected Papers of Charles Sanders Peirce*, ed. C. Hartshorne *et al.* Harvard U.P., Cambridge, Mass.
Perry, J.: 1977, 'Frege on demonstratives', *Philos. Review* **86**, 474–497.
Popper, K. R.: 1946/7, 'Logic without assumptions', *Proc. Aristot. Soc.* 1946/7, pp. 251–292.
Post, E.: 1921, 'Introduction to a general theory of elementary propositions', *Amer. J. Math.* **43**, 163–185; reprinted in van Heijenoort [1967] pp. 264–283.
Prawitz, D.: 1965, *Natural Deduction: a Proof-theoretical Study*, Almqvist and Wiksell, Stockholm.
Prawitz, D.: 1979, 'Proofs and the meaning and the completeness of the logical constants', in Hintikka *et al.* [1979] pp. 25–40.
Prior, A. N.: 1960, 'The runabout inference ticket', *Analysis* **21**, 38–39; reprinted in Strawson [1967] pp. 129–131.
Prior, A. N.: 1962, *Formal Logic*, Oxford U.P.
Putnam, H.: 1980, 'Models and reality', *J. Symbolic Logic* **45**, 464–482.
Quine, W. V.: 1940, *Mathematical Logic*, revised ed., Harvard U.P., Cambridge Mass., 1951.
Quine, W. V.: 1950, *Methods of Logic*, Holt, New York.

Quine, W. V.: 1970, *Philosophy of Logic*, Prentice-Hall, Englewood Cliffs, NJ.
Quirk, R. and Greenbaum, S.: 1973, *A University Grammar of English*, Longman, London.
Rasiowa, H. and Sikorski, R.: 1950, 'A proof of the completeness theorem of Gödel', *Fundamenta Math.* **37**, 193-200.
Rasiowa, H. and Sikorski, R.: 1963, *The Mathematics of Metamathematics*, Monografie Matematyczne, Polska Akad. Nauk.
Robinson, A.: 1967, 'The metaphysics of the calculus', in Lakatos (ed.), *Problems in the Philosophy of Mathematics*, North-Holland, Amsterdam, pp. 28-40; reprinted in Hintikka [1969] pp. 153-163.
Russell, B.: 1905, 'On denoting', *Mind* **14**, 479-493; reprinted in Russell [1956].
Russell, B.: 1956, *Logic and Knowledge, Essays 1901-1950*, ed. R. C. Marsh, George Allen and Unwin, London.
Sacks, G. E.: 1972, *Saturated Model Theory*, Benjamin, Reading, Mass.
Schmidt, H. A.: 1938, 'Über deduktive Theorien mit mehreren Sorten von Grunddingen', *Math. Annalen* **115**, 485-506.
Schröder, E.: 1895, *Vorlesungen über die Algebra der Logik*, Vol. 3, Leipzig.
Schütte, K.: 1956, 'Ein System des verknüpfenden Schliessens', *Arch. Math. Logik Grundlagenforschung* **2**, 55-67.
Schütte, K.: 1977, *Proof Theory*, tr. J. N. Crossley, Springer, Berlin.
Scott, W. R.: 1964, *Group Theory*, Prentice-Hall, Englewood Cliffs, NJ.
Shoesmith, D. J. and Smiley, T. J.: 1978, *Multiple-Conclusion Logic*, Cambridge U.P.
Skolem, T.: 1919, 'Untersuchungen über die Axiome des Klassenkalküls und über Produktations- und Summationsprobleme, welche gewisse von Aussagen betreffen', *Videnskapsselskapets Skrifter, I. Matem.-naturv. klasse, no. 3*; reprinted in Skolem [1970].
Skolem, T.: 1920, 'Logisch-kombinatorische Untersuchungen über die Erfüllbarkeit oder Beweisbarkeit mathematischer Sätze nebst einem Theoreme über dichte Mengen', *Videnskapsselskapets Skrifter, I. Matem.-Naturv. Klasse 4*; reprinted in Skolem [1970] pp. 103-136; partial tr. in van Heijenoort [1967] pp. 252-263.
Skolem, T.: 1922, 'Einige Bemerkungen zur axiomatischen Begründung der Mengenlehre', in *Matematikerkongressen i Helsingfors den 4-7 Juli 1922*, pp. 217-232; reprinted in Skolem [1970] pp. 137-152; tr. in van Heijenoort [1967] pp. 290-301.
Skolem, T.: 1923, 'Begründung der elementaren Arithmetik durch die rekurrierende Denkweise ohne Anwendung scheinbarer Veränderlichen mit unendlichem Ausdehnungsbereich, *Videnskapsselskapets Skrifter I, Matem.-naturv. Klasse 6*; tr. in van Heijenoort [1967] 303-333.
Skolem, T.: 1928, 'Über die mathematische Logik', *Norsk. Mat. Tidsk.* **10**, 125-142; reprinted in Skolem [1970] pp. 189-206; tr. in van Heijenoort [1967] pp. 513-524.
Skolem, T.: 1929, 'Über einige Grundlagenfragen der Mathematik', *Skr. Norsk. Akad. Oslo I Mat.-Natur. Kl. 4*, 1-49; reprinted in Skolem [1970] pp. 227-273.
Skolem, T.: 1934, 'Über die Nichtcharakterisierbarkeit der Zahlenreihe mittels endlich oder abzählbar unendlich vieler Aussagen mit ausschliesslich Zahlenvariablen', *Fundamenta Math.* **23**, 150-161; reprinted in Skolem [1970] pp. 355-366.
Skolem, T.: 1955, 'A critical remark on foundational research', *Kongelige Norsk. Vidensk. Forhand. Trondheim* **28**, 100-105; reprinted in Skolem [1970] pp. 581-586.

Skolem, T.: 1970, *Selected Works in Logic*, ed. J. E. Fenstad, Universitetsforlaget, Oslo.
Smullyan, R.: 1968, *First-order Logic*, Springer, Berlin.
Sneed, J. D.: 1971, *The Logical Structure of Mathematical Physics*, Reidel, Dordrecht.
Stegmüller, W.: 1976, *The Structure and Dynamics of Theories*, Springer, New York.
Steiner, M.: 1975, *Mathematical Knowledge*, Cornell Univ. Press, Ithaca.
Stevenson, L.: 1973, 'Frege's two definitions of quantification', *Philos. Quarterly* **23**, 207–223.
Strawson, P. F. (ed.): 1967, *Philosophical Logic*, Oxford U.P.
Suppes, P.: 1957, *Introduction to Logic*, Van Nostrand, Princeton, NJ.
Suppes, P.: 1972, *Axiomatic Set Theory*, Dover, NY.
Tarski, A.: 1935/6, 'Der Wahrheitsbegriff in den formalisierten Sprachen', based on a paper in *Ruch. Filozoficzny* xii (1930/1); tr. in Tarski [1956] pp. 152–278.
Tarski, A.: 1936, 'O pojciu wynikania logicznego', *Przegląd Filozoficzny* **39**, 58–68; tr. as 'On the concept of logical consequence' in Tarski [1956] pp. 409–420.
Tarski, A.: 1954, 'Contributions to the theory of models I, II', *Indag. Math.* **16**, 572–588.
Tarski, A.: 1956, *Logic, Semantics, Metamathematics, Papers from 1923 to 1938*, tr. J. H. Woodger, Oxford U.P.
Tarski, A., Mostowski, A., and Robinson, R. M.: 1953, *Undecidable Theories*, North-Holland, Amsterdam.
Tarski, A. and Vaught, R. L.: 1956, 'Arithmetical extensions of relational systems', *Compositio Math.* **13**, 81–102.
Tennant, N. W.: 1978, *Natural Logic*, Edinburgh U.P.
Thomason, R. H.: 1970, *Symbolic Logic, An Introduction*, Macmillan, London.
Van Dalen, D.: 1980, *Logic and Structure*, Springer, Berlin.
Van Heijenoort, J. (ed.): 1967, *From Frege to Gödel, A Source Book in Mathematical Logic, 1879–1931*, Harvard U.P., Cambridge, Mass.
Vaught, R. L.: 1974, 'Model theory before 1945', in Henkin *et al.* [1974] pp. 153–172.
Von Neumann, J.: 1925, 'Eine Axiomatisierung der Mengenlehre', *J. für die Reine und Angew. Math.* **154**, 219–240; tr. in van Heijenoort [1967] pp. 393–413.
Wang, H.: 1952, 'Logic of many-sorted theories', *J. Symbolic Logic* **17**, 105–116.
Wang, H.: 1970, 'A survey of Skolem's work in logic', in Skolem [1970] pp. 17–52.
Wang, H.: 1974, *From Mathematics to Philosophy*, Routledge and Kegan Paul, New York.
Whitehead, A. N. and Russell, B.: 1910, *Principia Mathematica* I, Cambridge U.P.; to *56, reprinted 1962.
Wiredu, J. E.: 1973, 'Deducibility and inferability', *Mind* **82**, 31–55.
Wittgenstein, L.: 1921, *Tractatus Logico-Philosophicus*, Annalen der Naturphilosophie; reprinted with tr. by D. F. Pears and B. F. McGuinness, Routledge and Kegan Paul, London, 1961.
Zermelo, E.: 1908, 'Untersuchungen über die Grundlagen der Mengenlehre I', *Math. Annalen* **65**, 261–281; tr. in van Heijenoort [1967] pp. 199–215.
Zucker, J.: 1974, 'The correspondence between cut-elimination and normalization', *Annals of Math. Logic* **7**, 1–156.

CHAPTER I.2

SYSTEMS OF DEDUCTION
by GÖRAN SUNDHOLM

Introduction	133
A remark about notation	134
1. Hilbert-Frege style systems	134
2. Natural deduction	148
3. Sequent calculi	168
References	187

INTRODUCTION

Formal calculi of deduction have proved useful in logic and in the foundations of mathematics, as well as in metamathematics. Examples of some of these uses are:

(a) The use of formal calculi in attempts to give a secure foundation for mathematics, as in the original work of Frege.
(b) To generate syntactically an already given semantical consequence relation, e.g. in some branches of technical modal logic.
(c) Formal calculi can serve as heuristic devices for finding metamathematical properties of the consequence relation, as was the case, e.g., in the early development of infinitary logic via the use of cut-free Gentzen sequent calculi.
(d) Formal calculi have served as the *objects* of mathematical study, as in traditional work on Hilbert's consistency programme.
(e) Certain versions of formal calculi have been used in attempts to formulate philosophical insights into the nature of reasoning.

It goes without saying that a particular type of calculus which serves admirably for one of the above uses, which are just a small indication of the many uses to which calculi of deduction have been put, does not have to be at all suitable for some of the other. Thus, a rich variety of different techniques have been developed for doing the 'book-keeping' of formal deduction, each one with its own advantages and disadvantages. It is the purpose of the present chapter to present a number of these techniques and to indicate some of the connections between the various versions.

More precisely, we shall concentrate on three main types of deductive systems known as (a) *Hilbert-Frege* style systems, (b) *Natural Deduction* systems and (c) *Sequent Calculi*. Under each of these headings we are going to study different variants of the main idea underlying the deductive technique in question. In particular, we shall relate the sequent calculus of Gentzen to currently fashionable *'tableaux'* systems of logic and show in what way they are essentially just a variant of the original Gentzen idea.

A REMARK ABOUT NOTATION

In the present chapter we largely follow the notation of Gentzen [1934], and in particular, we use '$\supset$' as implication and '&' as conjunction, as well as the *'falsum'* or *absurdity* symbol '$\perp$'. The arrow '$\rightarrow$' we reserve for the *sequent arrow*.

Various versions of predicate logic can be formulated more conveniently by the use of a special category of free individual variables, *'parameters'*. As individual variables, free or bound, we use

$$x_0, x_1, \ldots, y_0, y_1, \ldots \quad x, y, z, \ldots$$

and as parameters

$$a_0, a_1, \ldots, b_0, b_1, \ldots \quad a, b, \ldots.$$

Greek letters are used as schematic letters for formulae, cf. Hodges [Chapter I.1].

We shall sometimes use subscripts on our turnstiles, e.g. '$\vdash_N \varphi$' will be used to indicate that φ is a theorem of a Natural Deduction system.

1. HILBERT-FREGE STYLE SYSTEMS

We begin by considering the historically earliest of the three main types of systems, viz. the Hilbert-Frege style systems. One of the main advantages of such systems in their contemporary versions is a certain neatness of formulation and typographical ease. It is therefore somewhat ironical that the first such system in Frege [1879] was typographically most cumbersome. The sort of presentation used here has become standard since the codification given in Hilbert and Bernays [1934].

The central notion in Hilbert-Frege style systems is *provability* (other

I.2: SYSTEMS OF DEDUCTION

terms used include *derivability* and *theoremhood*). One lays down that certain wffs are *axioms* and defines the *theorems* as the least class of wffs which contains all axioms and is closed under certain *rules of proof*. The most familiar of such rules is unquestionably *modus ponens*, MP, which states: If $\varphi \supset \psi$ and φ are theorems, then so is ψ.

Sometimes the term 'rule of inference' is used where we use 'rule of proof'. This is a question of terminological choice, but in order to emphasize the fact that, say, the premises and the conclusion of MP are all *theorems*, we prefer the present usage and reserve 'rule of inference' for those situations where the premises are allowed to depend on assumptions. In present terminology, therefore, the theorems are inductively defined by the axioms and rules of proof.

Let us begin by considering a very simple system for **CPC** based on $\supset$ and $\neg$ as primitives and with other connectives, e.g. &, introduced by standard definitions, cf. Hodges [Chapter I.1].

The axioms are given in the form of *axiom schemata*, any instance of which is an axiom.

(A1) $\quad \varphi \supset (\psi \supset \varphi)$

(A2) $\quad (\varphi \supset (\psi \supset \theta)) \supset ((\varphi \supset \psi) \supset (\varphi \supset \theta))$

(A3) $\quad (\neg \psi \supset \neg \varphi) \supset (\varphi \supset \psi)$.

Hence

$$(p_0 \supset p_1) \supset (\neg p_3 \supset (p_0 \supset p_1))$$

is an instance of A1, and if φ and ψ are wffs then

$$(\varphi \supset (\varphi \supset \psi)) \supset ((\varphi \supset \varphi) \supset (\varphi \supset \psi))$$

is a schema which gives a subclass of the axioms which are instances of A2.

We now define the theorems of the system:

DEFINITION 1. *Every axiom is a theorem.*

DEFINITION 2. *If $\varphi \supset \psi$ and φ both are theorems, then so is ψ.*

The system so given may be called **HCPC** – 'H' for Hilbert – and is known to be complete for tautologies. Note that **HCPC** has only three axiom schemata but *infinitely many axioms*, and that MP is the only rule of proof. Sometimes one sees formulations using three *axioms* only – the relevant instances of (A1)–(A3) for sentence variables p_0, p_1 and p_2, but with an extra rule of proof instead. This is the

Substitution rule: The result of substituting a wff for a sentence variable in a theorem is a theorem.

In the case of predicate logic, the substitution rule becomes very complicated if one wants to formulate it in an exact way, cf. Church [1956] pp. 289–290, and formulations using axiom schemata have become almost universal.

We write '$\vdash \varphi$' to indicate that φ is a theorem of **HCPC**, and then the proper way to present MP becomes:

$$\frac{\vdash \varphi \supset \psi \quad \vdash \varphi}{\vdash \psi}(MP)$$

By the inductive generation of the theorems, to every theorem there corresponds a proof tree, or derivation, of φ. Such a tree D is a finite tree of wffs which is regulated by MP, has got φ at its root and axioms only as top formulae.

We are now going to give a proof tree for the schema $\varphi \supset \varphi$, thereby establishing that this is a *theorem schema*.

$$\frac{\dfrac{(\varphi \supset ((\varphi \supset \varphi) \supset \varphi)) \supset ((\varphi \supset (\varphi \supset \varphi)) \supset (\varphi \supset \varphi)) \quad \varphi \supset ((\varphi \supset \varphi) \supset \varphi)}{(\varphi \supset (\varphi \supset \varphi)) \supset (\varphi \supset \varphi) \quad \varphi \supset (\varphi \supset \varphi)}}{\varphi \supset \varphi}$$

This is a tree of the form

$$\frac{\dfrac{(1) \quad (2)}{(3)} \quad (4)}{(5)}$$

where (1) is a schematic instance of (A2), (2) is an instance of (A1) and (3) is a consequence by MP of (1) and (2). Likewise, (5) is an MP consequence of (3) and the (A1) instance (4). This is the shortest proof known to us of $\varphi \supset \varphi$ in **HCPC**, and amply brings out one of the drawbacks of the Hilber-Frege style systems. If one is interested in actually carrying out derivations in the systems, the work involved rapidly becomes enormous and quite unintuitive. If, on the other hand, we were allowed to use proofs from assumptions, then one could prove the above schema easily enough, provided that proofs from assumptions have the property that if ψ is provable from assumptions φ and $\varphi_1, \varphi_2, \ldots, \varphi_k$, then $\varphi \supset \psi$ is provable from assumptions

$\varphi_1, \ldots, \varphi_k$ only. We say that *D is a proof from assumptions* $\varphi_1, \ldots, \varphi_k$ of φ, if D is a finite tree of wffs regulated by MP and with φ as its end formula. All the top formulae of D are either axioms or one of $\varphi_1, \ldots, \varphi_k$. Thus, we may use the assumptions *as if they were axioms* in a proof from these assumptions. If there is a proof of φ from assumptions $\varphi_1, \ldots, \varphi_k$ we write

$$\varphi_1, \ldots, \varphi_k \vdash \varphi.$$

This notion is extended to schemata in the obvious way. We can also define a consequence relation between possibly infinite sets Γ of assumptions and wffs by putting

$$\Gamma \vdash \varphi \quad \text{iff} \quad \varphi_1, \ldots, \varphi_k \vdash \varphi, \quad \text{for some } \{\varphi_1, \ldots, \varphi_k\} \subseteq \Gamma.$$

For this notion of consequence from assumption by means of a proof tree from the assumptions, one is able to establish one of the central theorems of elementary metamathematics.

THE DEDUCTION THEOREM (Herbrand, Tarski). *If* $\Gamma, \varphi \vdash \psi$, *then* $\Gamma \vdash \varphi \supset \psi$.

Proof. (after Hilbert and Bernays [1934]).

By hypothesis, there is a proof tree D of ψ from the assumptions φ and Γ. Such a D must, in principle, look like

$$\begin{array}{ccc} \varphi, & \varphi_1, \ldots, \varphi_k, & \gamma_1, \ldots, \gamma_m \\ & \vdots & \\ & \dfrac{\theta \supset \delta \quad \theta}{\delta}\text{(MP)} & \\ & \vdots & \\ & \psi & \end{array}$$

where the top formulae $\varphi_1, \ldots, \varphi_k$ are all assumptions from the set Γ and $\gamma_1, \ldots, \gamma_m$ are all axioms.

We need to find a proof tree for $\varphi \supset \psi$ from assumption in Γ only. Consider first the '$\varphi \supset$' transformation of D; that is, in front of every wff in D we write '$\varphi \supset$'. This transformed tree, call it '$\varphi \supset D$', is no longer a proof tree from assumptions but looks like:

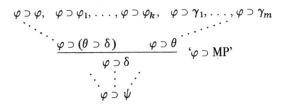

Our task is thus to show that at each step in this transformed tree '$\varphi \supset D$' we can restore provability from Γ. We begin by considering the three sorts of top formulae:

(a) The top formula is $\varphi \supset \varphi$. We have already seen how to prove this without assumptions.

(b) The top formula is one of the $\varphi \supset \varphi_i$, where φ_i is in Γ. Then we use (A1) – this is, in fact, the main *raison d'être* for the schema (A1). It is exactly what is needed to go from φ_i to $\varphi \supset \varphi_i$ – given *modus ponens* – to construct a proof of $\varphi \supset \varphi_i$ from Γ:

$$\frac{\varphi_i \supset (\varphi \supset \varphi_i) \quad \varphi_i}{\varphi \supset \varphi_i}.$$

This proof tree uses only one assumption φ_i which we assume is a member of Γ. Thus, provability from Γ is also restored here.

(c) The top formula is $\varphi \supset \gamma_j$ where γ_j is an axiom. Then, in this case,

$$\frac{\gamma_j \supset (\varphi \supset \gamma_j) \quad \gamma_j}{\varphi \supset \gamma_j}$$

is actually a proof of $\varphi \supset \gamma_j$ from no assumptions at all, and hence, *a fortiori*, provability from assumptions in Γ is also restored here.

This ends the discussion of the top formulae. It remains, however, to check that the transformation of instances of MP preserve derivability from assumptions in Γ. So assume that we are given proofs from assumptions in Γ of $\varphi \supset \theta$ and $\varphi \supset (\theta \supset \delta)$. Using these given proof trees from assumptions we continue via (A2):

$$(\text{MP}) \quad \frac{\dfrac{(\varphi \supset (\theta \supset \delta)) \supset ((\varphi \supset \theta) \supset (\varphi \supset \delta)) \quad \varphi \supset (\theta \supset \delta)}{(\varphi \supset \theta) \supset (\varphi \supset \delta)} \quad \varphi \supset \theta}{\varphi \supset \delta}(\text{MP})$$

But, by hypothesis, we have obtained proofs from assumptions in Γ of

$\varphi \supset (\theta \supset \delta)$ and $\varphi \supset \theta$, and the remaining top formula is an axiom. Therefore, here provability from Γ has also been secured and *the proof* of the Deduction Theorem *is completed.* □

REMARK. The proof is very general and, in fact, shows that the Deduction Theorem holds for any system where the notion of consequence from sets of assumptions is introduced via proofs from assumptions, *provided* that

(1) (A1) and (A2) are axiom schemata of the system, and
(2) MP is the only rule of proof.

In order to see the importance of the second of these two conditions, we will consider a case where the Deduction Theorem does not hold, or better, where the notion of consequence from sets of assumptions cannot be introduced via proofs from assumptions, and where the latter are straightforwardly introduced, as in **HCPC** above. When proof trees are extended to those from assumptions, what in effect takes place is that MP is converted into a *rule of inference* rather than a rule of proof, because it now licenses the step from $\Gamma \vdash \varphi \supset \theta$ and $\Delta \vdash \varphi$ to $\Gamma, \Delta \vdash \theta$. (Here we use '$\Gamma, \Delta$' as an abbreviation of '$\Gamma \cup \Delta$' and similarly for 'Γ, φ' and '$\Gamma \cup \{\varphi\}$'.)

Consider now the modal logic **S4**, cf. Bull and Segerberg [Chapter II.1], where we have a further primitive connective $\square$ (and & is defined from $\neg$ and $\supset$), with the additional axiom schemata and extra rule of proof:

(A4) $\quad \square \varphi \supset \varphi$

(A5) $\quad \square(\varphi \supset \psi) \supset (\square \varphi \supset \square \psi)$

(A6) $\quad \square \varphi \supset \square \square \varphi$

Necessitation (Nec) $\quad \dfrac{\vdash \varphi}{\vdash \square \varphi}$

If we were now to define proofs from assumptions in such a way that the proof trees have to be regulated by MP *and* Nec, then, as above, in the case of *modus ponens*, we would have converted *the rule of proof* (Nec) into a rule of inference, which licenses steps of the following form:

$$\dfrac{\Gamma \vdash_{S4} \varphi}{\Gamma \vdash_{S4} \square \varphi}$$

Such a rule is not sound, however, for the standard semantics for **S4**, and together with the Deduction Theorem, it leads to unacceptable consequences:

$$\frac{\varphi \vdash_{S4} \varphi}{\frac{\varphi \vdash_{S4} \Box\varphi}{\vdash_{S4} \varphi \supset \Box\varphi}} \quad \begin{array}{l} \text{(Nec)} \\ \text{(Deduction Theorem).} \end{array}$$

In this case, one therefore introduces the notion of consequence from assumptions in another way:

$$\Gamma \vdash_{S4} \varphi \quad \text{iff} \quad \vdash_{S4} \varphi_1 \& \cdots \& \varphi_k \supset \varphi,$$
$$\text{for some } \varphi_1, \ldots, \varphi_k \text{ in } \Gamma.$$

Note that this way of introducing consequences of assumptions has the same drawbacks as **HCPC** had before we introduced proofs from assumptions. The consequence from assumptions in **S4** is defined in terms of *provability* and hence, all the difficulties which adhere to straightforward provability also remain here. The Deduction Theorem holds for the turnstile, though; this is because in **HCPC** one can prove

$$\varphi \& \varphi_1 \& \cdots \& \varphi_k \supset \psi$$

iff one can prove

$$\varphi_1 \& \cdots \& \varphi_k \supset (\varphi \supset \psi).$$

We will give one more example of a system where the Deduction Theorem and its proof are of use, namely an axiomatic system for **CQC=**, classical predicate logic with identity. (Cf. Hodges [Chapter I.1], in particular for the notions of term and free variable.) We use the universal quantifier $\forall$ as a primitive.

The wff ψ is said to be a *generalization* of φ, if for some variables $x_1, \ldots, x_k$, ψ is identical with $\forall x_1 \ldots \forall x_k \varphi$, where the case $k = 0$ is permitted.

The *axiom schemata* are:

(Q1)–(Q3) $=_{\text{def}}$ any generalization of an instance of (A1)–(A3).

Any generalization of the following:

(Q4) $\forall x \varphi \supset \varphi_t^x$, where t is a term *substitutable* for x in φ.

(Q5) $\forall x(\varphi \supset \psi) \supset (\forall x \varphi \supset \forall x \psi)$

(Q6) $\varphi \supset \forall x \varphi$, provided that x does not occur free in φ.

(Q7) $x = x$

I.2: SYSTEMS OF DEDUCTION

(Q8) $x = y \supset (\varphi \supset \varphi')$, where φ is atomic and φ' results from φ by replacing x with y in zero or more (but not necessarily all) places in φ.

MP is the *only rule of proof*.

In (Q4) the notion 'substitutable for x' needs to be explained as well as the substitution notation 'φ_t^x'. The latter stands for the expression which results from φ by replacing the variable x, wherever it occurs free in φ, by the term t. One can define this precisely by an induction:

(0)   For an atomic φ, φ_t^x is the expression obtained by replacing every x in φ by the term t. (The use of 'replacing' can be replaced with another inductive definition.)

(1) $(\neg \varphi)_t^x =_{\text{def}} \neg(\varphi_t^x)$,

(2) $(\varphi \supset \psi)_t^x =_{\text{def}} \varphi_t^x \supset \psi_t^x$,

(3) $(\forall y \varphi)_t^x =_{\text{def}} \forall y \varphi$,   if x and y are the same variable,
               $\forall y(\varphi_t^x)$ otherwise.

Hence, $(x = y)_x^y$ is equal to $(x = x)$ and $\forall x(x = x)_t^x$ is equal to $\forall x(x = x)$.

Consider the wff $\varphi =_{\text{def}} \neg \forall y(x = y)$. Then $\varphi_y^x =_{\text{def}} \neg \forall y(y = y)$, and $(\forall x \varphi \supset \varphi)_y^x =_{\text{def}} (\forall x \neg \forall y(x = y) \supset \neg \forall y(y = y))$. This last sentence is not logically valid, because the antecedent is true whenever the individual domain has got more than one element and the consequence is never true. Thus, there is no lack of counter-models. In this phenomenon, sometimes known as 'clashes between bound variables', lies the reason for the restriction on schema Q4. One says that *t is substitutable for x in φ*, if no free occurrence of x in φ lies within the scope of a quantifier which binds a variable of the term t. This notion can be precisely defined in the following way:

(i) If φ is atomic, then t is substitutable for x in φ.

(ii) If t is substitutable for x in φ, then t is substitutable for x in $(\neg \varphi)$.

(iii) If t is substitutable for x in φ and in ψ, then t is substitutable for x in $(\varphi \supset \psi)$.

(iv) (the crucial clause)
If either, x does not occur free in $\forall y \varphi$, or, y does not occur in t *and t is substitutable for x in φ*, then t is substitutable for x in $\forall y \varphi$.

In place of 'substitutable for x' one sometimes sees the phrase 'free for x'.

We illustrate the use of the present formulation of **HCQC=** in an important

METATHEOREM. *If x does not occur free in any wff of Γ and $\Gamma \vdash \varphi$, then $\Gamma \vdash \forall x \varphi$.*

(The notion of consequence from assumptions is defined via proof trees, just in the same way as before for **HCPC**. The proof of the Deduction Theorem works.)

Proof. Consider a proof tree D for φ from Γ. Then D must, in principle, be of the following form:

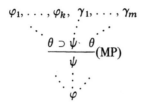

where $\varphi_1, \ldots, \varphi_k$ are members of the set of assumptions Γ, and $\gamma_1, \ldots, \gamma_m$ are axioms. We show that for each wff δ which occurs in D, $\Gamma \vdash \forall x \delta$. Consider first the cases where δ is a top formula of D.

(i) δ is one of the axioms $\gamma_1, \ldots, \gamma_m$, but any generalization of an axiom is an axiom. Therefore, $\forall x \delta$ is an axiom and, hence, $\Gamma \vdash \forall x \delta$.

(ii) δ is an element of Γ. By the hypothesis of the theorem, x does not occur free in Γ and, hence, *a fortiori*, also not in δ. But then $\delta \supset \forall x \delta$ is an instance of (Q6) and an application of MP proves $\forall x \delta$ from assumptions in Γ only. This case provides us with the reason for the inclusion of axiom schema (Q6), just as the previous case gives the explanation for why every generalization of an axiom is an axiom.

What remains to be considered is the case of *modus ponens*. So assume that we have already established that $\Gamma \vdash \forall x(\theta \supset \psi)$ and $\Gamma \vdash \forall x \theta$. We must show that $\Gamma \vdash \forall x \psi$. This is accomplished by the use of a suitable instance of (Q5), viz. $\forall x(\theta \supset \psi) \supset (\forall x \theta \supset \forall x \psi)$ and two applications of MP. Here, then, we see the reason for the inclusion of axiom schema (Q5).

Thus, the proof of our theorem is completed and this so-called *Generalization rule* holds as a derivable rule of inference.

This sort of system for **CQC=**, which uses only MP as a rule of proof, is treated in great detail in a series of papers by Tarski [1965], Kalish and Montague [1965], and Monk [1965]. An elegant exposition is in Enderton [1972], on which we have relied. This sort of system is sometimes of importance in quantified modal logic if one wishes to avoid the so called *Barcan formula*, cf. Kripke [1963].

We just want to remark that another common axiomatization is obtained by using (A1)–(A3) plus (Q4) with the same restriction and

(Q$'$5) $\forall x(\varphi \supset \psi) \supset (\varphi \supset \forall x \psi)$ if x does not occur free in φ.

Here, only the instances of the schemata, but *not their generalizations*, are axioms. There are, however, *two* rules of proof:

R$'$1:   MP
R$'$2:   If φ is a theorem, then so is $\forall x \varphi$.

This last rule is also known as '*Generalization*', but note that here it is a rule of *proof* and previously we showed that it was a (derived) rule of *inference* in **HCQC**=.

The equivalence between **HCQC** and the latter, primed, version (call it **H$'$CQC** – for simplicity we leave = out) is readily established. We will not enter into details, but only note that the effect of the **HCQC** condition that every generalization of an axiom is an axiom, is taken care of via the Generalization rule of proof in **H$'$CQC**. Detailed expositions of the **H$'$CQC** type of system can be found in Church [1956] and Mendelson [1964].

The **HCQC** system is a system of *pure* predicate calculus. If we wish to deal with a specific first-order theory T we have to specify a language L_T and to define the 'non-logical axioms' of T, although the deductive machinery and the development of the theory remain, on the whole, unchanged.

In order to facilitate comparisons with the two other essentially different ways of presenting logical deduction, we find it convenient at this point to change the basic syntactic set-up used hitherto. As we hinted at in the *Remark on notation*, we shall use a separate category of *parameters*: $a, b, \ldots$ possibly with subscripts. The definition of *individual terms* then runs:

(i)   Individual constants are terms.
(ii)   Parameters are terms.
(iii) If f^i is an i-place function symbol and $t_1, \ldots, t_i$ are all terms, then $f^i(t_1, \ldots, t_i)$ is a term.
(iv)   Nothing is a term except by a finite number of (i)–(iii).

In future, we shall leave out the 'extremal clause' (iv) from our inductive definitions.

The language for **QC**, which we shall use in the sequel, is based on the full set of connectives: &, ∨, ⊃, ⊥, ∀ and ∃, where, however, we introduce negation by definition $\neg \varphi =_{def} \varphi \supset \bot$.

The *definition of well-formed formula* runs as usual, except that we use a different clause for the quantifiers:

(+)   If φ is a wff, b is a parameter and x is a variable *which does not occur in* φ, then $\forall x \varphi'$ and $\exists x \varphi'$ are both wffs,

where φ' is the result of writing 'x' for 'b' wherever it occurs in φ.

The net effect of the two main changes – the use of parameters in place of free individual variables in the theory and the prohibition of quantifying with a variable over a wff, which already contains this variable – is to ensure that the same variable does not occur as both free and bound in a wff (this is given by the use of parameters in place of free occurrences of variables) and that no variable occurs bound 'twice over' in a wff. The properties are not important *per se* in the development of a Hilbert-Frege style system for **QC**, but they prove indispensible in the case of the sequent calculus. Note that although the restriction on quantification rules out such expressions as

$$\exists x(P(x) \& \forall x Q(x)),$$

where P and Q are predicate variables, from the class of wffs, this is not an impoverishment of the language because

$$\exists y(P(y) \& \forall x Q(x))$$

is a wff and has the same meaning as the forbidden expression.

We will use the same substitution notation as before, e.g., 'φ_t^a' denotes then the result of substituting the expression t everywhere for the expression a in the expression φ. Sometimes we wish to consider expressions which are just like wffs, except that they contain free variables in place of parameters. Such expressions are called *pseudo-wffs*. An example of a pseudo-wff is obtained by removing the quantifier prefix from a wff, e.g. $P(y) \& \forall x Q(x)$ is a pseudo-wff. All the pseudo-wffs we shall have occasion to consider will be of this type.

We now give a version of intuitionistic predicate logic with identity, **HIQC=**, in a form which is particularly suited for establishing connections with the other main types of systems. The system is, essentially, due to Hilbert and Gentzen, cf. Gentzen [1934].

Axiom schemata: propositional part

(A $\supset_1$) $\varphi \supset (\psi \supset \varphi)$

I.2: SYSTEMS OF DEDUCTION 145

(A $\supset_2$) $(\varphi \supset (\psi \supset \theta)) \supset ((\varphi \supset \psi) \supset (\varphi \supset \theta))$

(A & I) $(\varphi \supset (\psi \supset (\varphi \& \psi)))$

(A & E_i) $\varphi_0 \& \varphi_1 \supset \varphi_i$, $i = 0, 1$.

(A $\vee$ I_i) $\varphi_i \supset \varphi_0 \vee \varphi_1$, $i = 0, 1$.

(A $\vee$ E) $(\varphi \supset \theta) \supset ((\psi \supset \theta) \supset (\varphi \vee \psi \supset \theta))$
This is the only complicated axiom so far; it says that given ways to reach θ from φ and ψ, respectively, there is a way to go from $\varphi \vee \psi$ to θ.

These axiom schemata, together with MP, give *Minimal logic*. In intuitionistic logic we have one more axiom schema:

(A$\bot$) $\bot \supset \varphi$.

Modus ponens is the *only rule of proof* in the propositional part.

Hence, the Deduction Theorem holds and we establish two theorem schemata with its use:

$$\vdash (\varphi \& \psi \supset \theta) \supset (\varphi \supset (\psi \supset \theta))$$

and

$$\vdash (\varphi \supset (\psi \supset \theta)) \supset (\varphi \& \psi \supset \theta).$$

We reason informally:
By MP and (A & I): $\varphi, \psi \vdash \varphi \& \psi$
But then, by MP and A & E: $\varphi \& \psi \supset \theta, \varphi, \psi \vdash \theta$
So, by the Deduction Theorem (twice): $\varphi \& \psi \supset \theta \vdash \varphi \supset (\psi \supset \theta)$
So, by the Deduction Theorem: $\vdash (\varphi \& \psi \supset \theta) \supset (\varphi \supset (\psi \supset \theta))$.
The other direction is left as an exercise.

There are further axioms for the quantifiers and the identity symbol.

(A$\forall$E) $\forall x \varphi \supset \varphi_t^x$

(A$\exists$I) $\varphi_t^x \supset \exists x \varphi$.

Note that we have got 'φ_t^x'. Hence, say, $(a = a) \supset \exists x (a = x)$ is an instance of A$\exists$I, because

$$(a = x)_a^x =_{\text{def}} (a = a).$$

There are two rules of proof in the quantificational part:

(R∀I) $\dfrac{\vdash \varphi \supset \psi}{\vdash \varphi \supset \forall x \psi_x^a}$ provided that a does not occur in φ.

(R∃E) $\dfrac{\vdash \varphi \supset \psi}{\vdash \exists x \varphi_x^a \supset \psi}$ provided that a does not occur in ψ.

In these two rules, the parameter a is called the *eigen*-parameter, or the *proper* parameter, of the instance of the rule.

We note that the restrictions are necessary if the rules are to be sound. Clearly, $\vdash P(a) \supset P(a)$, but $P(a) \supset \forall x P(x)$ is not logically valid. (Consider an interpretation with the domain of persons and interpretation of P as the property of holding the world championship of chess. Over this particular interpretation, the assignment (at the moment of writing) of Anatoly Karpov to the parameter a, produces a counter-model. It is certainly not true that if Karpov is world champion, then anyone is. I, for one, am not.) It is also essential to grasp that (R∀I) is a rule of proof. The corresponding rule of inference is not sound without further restrictions. Semantically, and also deductively in the *propositional* part, where $\forall y Q(y)$ can be viewed just as another wff with no further structure,

$$P(a) \vdash \forall y Q(y) \supset P(a), \quad \text{say via } (A \supset_1) \text{ and MP}$$

if we define proof-trees in the usual way, but we cannot allow that

$$P(a) \vdash \forall y Q(y) \supset \forall x P(x)$$

as a similar counter-model to the Karpov one will show.

We therefore need to take particular care in the definition of proof trees from assumptions. The crucial restriction is this: If D is a proof tree for $\varphi \supset \psi$ from certain top formulae, where $\varphi_1, \ldots, \varphi_k$ are all the top formulae in D which are not axioms, and the parameter a does not occur in φ, then

$$\begin{array}{c} D \\ \varphi \supset \psi \\ \varphi \supset \forall x \psi_x^a \end{array}$$

is a proof tree for $\varphi \supset \forall x \psi_x^a$ from assumptions $\varphi_1, \ldots, \varphi_k$, *provided that a does not occur in any of $\varphi_1, \ldots, \varphi_k$*.

A similar *eigen*-parameter condition is imposed on applications of (R∃E) in proofs from assumptions.

With these restrictions, the Deduction Theorem is valid and the '$\varphi \supset$' transformation method of proof works.

I.2: SYSTEMS OF DEDUCTION

Consider a proof tree D

$$\frac{\begin{array}{c}\varphi, \varphi_1, \ldots, \varphi_k \\ \vdots \\ \theta \supset \psi\end{array}}{\theta \supset \forall x \psi_x^a}(\text{R}\forall\text{I}).$$

As (R∀I) is permissible here, we know that (i) the *eigen*-parameter a does not occur in θ, nor (ii) in any of the assumption formulae $\varphi_1, \ldots, \varphi_k$ and φ. The transformation gives a proof tree D' of $\varphi \supset (\theta \supset \psi)$ from assumptions $\varphi_1, \ldots, \varphi_k$, because in the restoration of provability from assumptions, we use only auxiliary proof trees of the form

$$\frac{\varphi_1 \supset (\varphi \supset \varphi_1) \quad \varphi_1}{\varphi \supset \varphi_1}(\text{MP})$$

and this leaves only φ_1 as an assumption formula. So we now continue D' as follows:

$$\frac{(\varphi \& \theta \supset \forall x \psi_x^a) \supset (\varphi \supset (\theta \supset \forall x \psi_x^a)) \quad \dfrac{\dfrac{(\varphi \supset (\theta \supset \psi)) \supset (\varphi \& \theta \supset \psi) \quad \begin{array}{c}D'\\\varphi \supset (\theta \supset \psi)\end{array}}{\varphi \& \theta \supset \psi}(\text{MP})}{\varphi \& \theta \supset \forall x \psi_x^a}(\text{R}\forall\text{I})}{(\varphi \supset (\theta \supset \forall x \psi_x^a))}(\text{MP})$$

The application of (R∀I) is permissible because we know that the parameter a does not occur in $\varphi \& \theta$, nor in any of the assumption formulae of D'. The rest is just dotting the i's and crossing the t's using the two derivable schemata from the propositional part. The treatment of (R∃E) is similar.

The details of the entire development for rules of inference with parameters and the resulting Deduction Theorem are given meticulous treatment in Hilbert and Bernays [1934] and Kleene [1952] Sections 21–24.

We can now write (R∀I) as one condition of the consequence relation:

If $\Gamma \vdash \theta \supset \psi$, then $\Gamma \vdash \theta \supset \forall x \psi_x^a$

provided that the *eigen*-parameter a does not occur in Γ, θ. Note the similarity with the rule of inference which we showed was derivable in **HCQC**.

We still have to give the identity axioms:

(A=I) $(a = a)$

(A=E) $(a = b) \supset (\varphi_a^x \supset \varphi_b^x)$, where φ is *atomic*.

As these are *axioms*, the presence of parameters does not further complicate the proof trees.

Let us finally conclude our treatment of the Hilbert-Frege style systems by remarking that for all three systems just presented, propositional, quantificational and identity logic, the corresponding classical system results simply by adding either of the two axiom schemata

(DN) $\neg\neg\varphi \supset \varphi$ ('DN' for 'double negation')

(Excluded middle) $\varphi \vee \neg\varphi$

and that if we wish to use negation as a primitive the relevant intuitionistic rules are

(A¬I) $(\varphi \supset \psi) \supset ((\varphi \supset \neg\psi) \supset \neg\varphi)$

(A¬E) $\varphi \supset (\neg\varphi \supset \psi)$

The Hilbert-Frege style systems are particularly well suited for arithmetization of metamathematics, because the inductively defined objects have a very simple structure. Detailed treatment can be found in Smoryński [1977] and Feferman [1960].

2. NATURAL DEDUCTION

The Hilbert-Frege style systems have, as we have seen in Section 1, a reasonably smooth theory, but they suffer from one essential drawback: if one is interested in *actually carrying out* derivational work, they are hopelessly cumbersome, because even the simplest inferences have to be brought back to the fixed and settled axioms. The use of proofs from assumptions, and the ensuing Deduction Theorem, is an attempt to ease the derivational burden which is at least partially successful, particularly for the last of the formulations given above.

In Natural Deduction, on the other hand, one of the two main features is that all rules are *rules of inference* rather than rules of proof and, consequently, theoremhood is defined as the limiting case of derivability from the empty collection of assumptions. The other main feature of Natural Deduction is that the derivational use of each operator $ – connective, quantifier, inductively defined predicate etc. – is regulated by two rules: one, the *introduction rule* for $, ($I), which tells us how a sentence with $ as its main operator may be inferred as a conclusion, how $ may be introduced, and another rule, the *elimination rule* for $, ($E), which tells us how further

I.2: SYSTEMS OF DEDUCTION

consequences may be drawn from a premise with $\$$ as its main operator, how $\$$ can be eliminated.

We now proceed directly to presenting these rules for a Natural Deduction version of **IQC**, which we henceforth call **NIQC**.

(A) ASSUMPTION. For any wff φ, the tree which consists of φ only is a derivation of φ which depends on the assumption φ.

If

(&I) $\quad \begin{array}{c} D_0 \\ \varphi_0 \end{array}$ and $\begin{array}{c} D_1 \\ \varphi_1 \end{array}$

are derivations of φ_0 and φ_1, respectively, which depend on assumptions $\varphi_1, \ldots, \varphi_k$ and $\psi_1, \ldots, \psi_m$, respectively, then

$$\frac{\begin{array}{cc} D_0 & D_1 \\ \varphi_0 & \varphi_1 \end{array}}{\varphi_0 \,\&\, \varphi_1}(\&I)$$

is a derivation of $\varphi_0 \,\&\, \varphi_1$, which depends on all of $\varphi_1, \ldots, \varphi_k$, $\psi_1, \ldots, \psi_m$.

(&E$_i$) $\quad i = 0, 1$. The elimination rule for conjunction is, properly speaking, not one, but two rules.

If

$$\begin{array}{c} D \\ \varphi_0 \,\&\, \varphi_1 \end{array}$$

is a derivation of $\varphi_0 \,\&\, \varphi_1$, depending on $\psi_1, \ldots, \psi_m$, then

$$\frac{\begin{array}{c} D \\ \varphi_0 \,\&\, \varphi_1 \end{array}}{\varphi_i}(\&E_i)$$

is a derivation of φ_i, depending on the same assumptions.

($\supset$I) This is the most characteristic of Natural Deduction rules. It is also difficult to state precisely.

If

$$\begin{array}{c} D \\ \psi \end{array}$$

is a derivation of ψ, depending on assumptions $\varphi_1, \ldots, \varphi_m$, then

$$\frac{\begin{array}{c}D\\ \psi\end{array}}{\varphi \supset \psi}(\supset I)$$

is a derivation of $\varphi \supset \psi$ depending on assumptions $\varphi'_1, \ldots, \varphi'_m$, where this list results from $\varphi_1, \ldots, \varphi_m$ by removing some (all or no) occurrences of φ. We say that the removed occurrences have been *discharged* or *closed*.

($\supset$E) If

$$\begin{array}{ccc} D_0 & \text{and} & D_1 \\ \varphi \supset \psi & & \varphi \end{array}$$

are derivations of $\varphi \supset \psi$ and φ, respectively, depending on $\varphi_1, \ldots, \varphi_k$ and $\psi_1, \ldots, \psi_m$, respectively, then

$$\frac{\begin{array}{cc} D_0 & D_1 \\ \varphi \supset \psi & \varphi \end{array}}{\psi}(\supset E)$$

is a derivation of ψ depending on all the assumptions $\varphi_1, \ldots, \varphi_k$, $\psi_1, \ldots, \psi_m$. The elimination-rule for $\supset$ is nothing but MP construed as a *rule of inference*.

We now have enough rules to give a simple example:

$$\frac{\frac{\frac{\cancel{\varphi}^1 \quad \cancel{\psi}^2}{(\varphi \& \psi)}(\&I)}{\psi \supset (\varphi \& \psi)^2}(\supset I)}{\varphi \supset (\psi \supset (\varphi \& \psi))^1}(\supset I)$$

This derivation tree is a *proof*, i.e. a derivation in which the assumptions have all been closed (there are no *open* assumptions left). It also illustrates how one sets out the derivations in practice. The assumptions are indexed with a numeral and, at the inference where an assumption is discharged, the numeral is written again to indicate closure. In practice, when the system is familiar, one does not always give the names of the rules, nor does one always indicate where the assumptions are discharged, but confines oneself just to crossing them out to indicate closure. The mechanism of the above derivation is thus: assume φ and ψ. By (&I) we get $(\varphi \& \psi)$ depending on the assumptions φ and ψ. Therefore, by the use of ($\supset$I): $\psi \supset (\varphi \& \psi)$, now depending only on φ, and finally by one more use of ($\supset$I): $\varphi \supset (\psi \supset (\varphi \& \psi))$. The use of ($\supset$I) corresponds to the Deduction Theorem in **HIQC**. Note that we have here given a proof in **NIQC** of the **HIQC**-axiom (A&I). The reader may wish to try his hand at (A&E) as an exercise.

The above elementary example illustrates a point of principal importance. It is given as a derivation *schema* and we naturally wish that each instance thereof shall be a derivation. Consider then the, albeit somewhat extreme, choice of the propositional variable p *both* for φ and ψ. The result is

$$\frac{\dfrac{p^1 \quad p^2}{p \& p}(\&\text{I})}{\dfrac{p \supset (p \& p)^2}{p \supset (p \supset (p \& p))^1}(\supset\text{I})}(\supset\text{I})$$

We see that both assumptions p are struck out, but the discharge takes place at *different inferences*. The moral of this example is that not *all* assumptions of the form φ have to be discharged at an application of ($\supset$I) giving $\varphi \supset \psi$ as a conclusion. In fact, no assumption needs to be discharged:

$$\frac{\dfrac{\varphi^1}{\psi \supset \varphi}(\supset\text{I})}{\varphi \supset (\psi \supset \varphi)^1}(\supset\text{I})$$

Here at the first application of ($\supset$I) *no* discharge takes place. We are given a derivation of φ, depending on certain assumptions – in fact only on φ – and we go on to a derivation of ($\psi \supset \varphi$) as we have the right to do by ($\supset$I). As we search for an occurrence to discharge, we see that there is none. Thus, with this permissiveness and the resulting liberal use of ($\supset$I) Natural Deduction is not suitable for *Relevant logic* (but c.f. Prawitz [1965, Chapter VII]).

To sum up: *Discharge is a right, but not an obligation.*

As not all assumptions of the same form need to be discharged at the same place, one speaks of '*assumption classes*', where those assumptions of the same form which are discharged by the same inference belong to the same assumption class. Leivant [1979] contains an exhaustive discussion of the need for assumption classes.

The rules given above were formulated in a rather elaborate way in order to be precise. In practice, one often sees the following sort of definition for the rule ($\supset$I): If

$$\begin{array}{c}\varphi\\D\\\psi\end{array}$$

is a derivation of ψ, depending on, among others, the assumption φ, then

$$\frac{\begin{array}{c}\cancel{\varphi}\\D\\\psi\end{array}}{\varphi\supset\psi}(\supset\text{I})$$

is a derivation of $\varphi\supset\psi$, where the indicated assumptions of the form φ have been *closed* (have been *discharged* or *cancelled*). This is a somewhat loose way of formulating the rule as the discussion of assumption-classes etc. shows. It has, however, a great intuitive appeal, and the rest of our rules will be set out in this fashion on the understanding that the precise versions of the rules have to be formulated in analogy with the exact statement of ($\supset$I).

We now give the rest of the rules:

(∨I) $i = 0, 1$. (This is not one but two rules, just as in the case of (&E).) If

$$\begin{array}{c}D\\\varphi_i\end{array}$$

is a derivation of φ_i, depending on certain assumptions, then

$$\begin{array}{c}D\\\dfrac{\varphi_i}{\varphi_0\vee\varphi_1}(\vee\text{I})_i\end{array}$$

is a derivation of $\varphi_0\vee\varphi_1$, depending on the same assumptions.

(∨E)   If

$$\begin{array}{c}D\\\varphi\vee\psi\end{array}$$

is a derivation of $\varphi\vee\psi$, and

$$\begin{array}{cc}\varphi&\psi\\D_1\text{ and }&D_2\\\theta&\theta\end{array}$$

are derivations of θ, depending respectively on, among others, the assumptions φ and ψ, then

I.2: SYSTEMS OF DEDUCTION

$$\frac{D \quad \overset{\cancel{\varphi}}{D_1} \quad \overset{\cancel{\psi}}{D_2}}{\varphi \vee \psi \quad \theta \quad \theta}(\vee E)$$

is a derivation of θ, where the indicated assumptions have been closed. (This could, and for exactness, should, be reformulated in a manner corresponding to the exact statement of $(\supset I)$.)

The rule ($\vee E$) is a formal version of the type of reasoning known as 'constructive dilemma': we know that A or B is true. First case: A is true. Then C is also true. Second case: B is true. Again, C is also true. Therefore, C is true.

Using ($\vee E$) we can, of course, give a proof, i.e. a derivation without open assumptions, of the axiom ($A \vee E$):

$$\frac{\cancel{\varphi \vee \psi}^1 \quad \dfrac{\cancel{\varphi \supset \theta}^2 \quad \cancel{\varphi}^3}{\theta} \quad \dfrac{\cancel{\psi \supset \theta}^4 \quad \cancel{\psi}^5}{\theta}}{\dfrac{\dfrac{\theta}{\varphi \vee \psi \supset \theta}(1)}{\dfrac{(\psi \supset \theta) \supset (\varphi \vee \psi \supset \theta)}{(\varphi \supset \theta) \supset ((\psi \supset \theta) \supset (\varphi \vee \psi \supset \theta))}(2)}(4)}\,{}_{3,5}(\vee E)$$

($\forall I$)   If

$$D$$
$$\varphi$$

is a derivation of φ, depending on $\psi_1, \ldots, \psi_k$, then

$$D$$
$$\frac{\varphi}{\forall x \varphi_x^a}$$

is a derivation of $\forall x \varphi_x^a$, depending on the same assumptions, *provided* that the parameter a, the *eigen*-parameter of the inference, does not occur in any of the assumptions $\psi_1, \ldots, \psi_k$.

This rule is a codification of a type of reasoning very well-known to calculus students: "Pick an $\epsilon > 0$. Then there is for this ϵ such and such a $\delta > 0$. But ϵ was chosen arbitrariiy. Therefore, for *every* $\epsilon > 0$ there is such a δ." The restriction on the *eigen*-parameter a is, of course, necessary. The Karpov counter-example given in Section 1 in a similar situation works here as well, and for the same reason.

($\forall$E) If
$$D \\ \forall x \varphi$$
is a derivation of $\forall x\varphi$, depending on certain assumptions, then

$$\frac{\begin{array}{c} D \\ \forall x\varphi \end{array}}{\varphi_t^x}(\forall E)$$

is a derivation of φ_t^x, depending on the same assumptions.

($\exists$I) If
$$D \\ \varphi_t^x$$
is a derivation of φ_t^x, depending on certain assumptions, then

$$\frac{\begin{array}{c} D \\ \varphi_t^x \end{array}}{\exists x\varphi}$$

is a derivation of $\exists x\varphi$, depending on the same assumptions.

We now come to the most complicated of the rules, viz.

($\exists$E) If
$$D \\ \exists x\varphi$$
is a derivation of $\exists x\varphi$, and

$$\varphi_a^x \\ D_1 \\ \theta$$

is a derivation of θ depending on, among others, the assumption φ_a^x, then

$$\frac{\begin{array}{cc} & \cancel{\varphi_a^x} \\ D & D_1 \\ \exists x\varphi & \theta \end{array}}{\theta}(\exists E)$$

is a derivation of θ depending on all the assumptions used in D and D_1, except those of the indicated form φ_a^x, *provided* that the *eigen*-parameter a does not occur in $\exists x\varphi$, nor in θ or any of the assumptions in D_1, except those of the form φ_a^x. For examples illustrating the need for the restrictions, we refer to Tennant [1978] Chapter 4.8.

(=I)   For any term t, $t = t$ (=I) is a derivation of $t = t$, depending on no assumptions.

(=E)   If

$$\begin{array}{cc} D & D_1, \\ t_0 = t_1 & \varphi_{t_0}^x \end{array}$$

are derivations of $t_0 = t_1$ and $\varphi_{t_0}^x$, depending on certain assumptions, then

$$\frac{\begin{array}{cc} D & D_1 \\ t_0 = t_1 & \varphi_{t_0}^x \end{array}}{\varphi_{t_1}^x}(=E)$$

is a derivation of $\varphi_{t_1}^x$ depending on all the assumptions used in D and D_1.

Examples:

(1) $$\frac{(a = a)}{\exists x(a = x)}(\exists I)^{(=I)}$$

and

(2) $$\frac{(a = a)}{\exists x(x = x)}(\exists I)^{(=I)}$$

are correct proofs of their respective conclusions; in (1) $(a = a) =_{\text{def}} (a = x)_a^x$ and in (2) $(a = a) =_{\text{def}} (x = x)_a^x$.

In order to complete the description of **HIQC=** we must add one more rule to the above set of rules for *Minimal* predicate logic with identity. The extra rule is of course

($\bot$)   If

$$\begin{array}{c} D \\ \bot \end{array}$$

is a derivation of $\bot$, depending on certain assumptions, then

$$\begin{array}{c} D \\ \bot \\ \hline \varphi \end{array} (\bot)$$

is a derivation of φ, depending on the same assumptions.

This ends the presentation of the rules. In order to summarize the above description of the system of Natural Deduction we now give the rules as inference figures:

(&I) $\quad \dfrac{\varphi \quad \psi}{\varphi \,\&\, \psi}$ $\qquad$ (&E) $\quad \dfrac{\varphi \,\&\, \psi}{\varphi} \quad \dfrac{\varphi \,\&\, \psi}{\psi}$

(∨I) $\quad \dfrac{\varphi}{\varphi \vee \psi} \quad \dfrac{\psi}{\varphi \vee \psi}$ $\qquad$ (∨E) $\quad \dfrac{\varphi \vee \psi \quad \begin{array}{c}[\varphi] \\ \vdots \\ \theta\end{array} \quad \begin{array}{c}[\psi] \\ \vdots \\ \theta\end{array}}{\theta}$

(⊃I) $\quad \dfrac{\begin{array}{c}[\varphi] \\ \vdots \\ \psi\end{array}}{\varphi \supset \psi}$ $\qquad$ (⊃E) $\quad \dfrac{\varphi \supset \psi \quad \varphi}{\psi}$

(∀I) $\quad \dfrac{\varphi}{\forall x \varphi_x^a}$ $\qquad$ (∀E) $\quad \dfrac{\forall x \varphi}{\varphi_t^x}$

provided that the *eigen-parameter* a does not occur in any assumptions on which φ depends

(∃I) $\quad \dfrac{\varphi_t^x}{\exists x \varphi}$ $\qquad$ (∃E) $\quad \dfrac{\exists x \varphi \quad \begin{array}{c}[\varphi_a^x] \\ \vdots \\ \theta\end{array}}{\theta}$

provided that the *eigen-parameter* a does not occur in any assumptions on which θ depends (except φ_a^x), nor in $\exists x \varphi$ or θ.

I.2: SYSTEMS OF DEDUCTION

$(=I) \quad t = t$ $\qquad\qquad (=E) \quad \dfrac{s = t \quad \varphi_s^x}{\varphi_t^x}$

$$(\bot) \quad \dfrac{\bot}{\varphi}$$

The system thus given was introduced, essentially, by Gentzen [1934].

We write

$$\varphi_1, \ldots, \varphi_k \vdash_N \varphi$$

if there is a derivation of φ where all open assumptions are among $\varphi_1, \ldots, \varphi_k$, and similarly for sets Γ.

Above we gave a proof of (A&I), using the homonymous rule and the rules for $\supset$. Later we did the same for (A∨E). The pattern thus presented is perfectly general and it should be clear to the reader how to prove every **HIQC=** axiom by using the corresponding Natural Deduction rule together with the implication rules. The Hilbert-Frege style axioms were chosen just because they are the linearizations of the Natural Deduction rules. When discussing the discharge of assumptions, we gave a proof of $(A\supset_1)$. We now give a proof of $(A\supset_2)$: in order to do this it is clearly enough to show that

$$\varphi \supset (\psi \supset \theta), (\varphi \supset \psi), \varphi \vdash_N \theta$$

because then a series of three ($\supset$I) will establish the axiom. This is readily done by means of the following derivation tree:

$$\dfrac{\dfrac{\varphi \supset (\psi \supset \theta) \quad \varphi}{\psi \supset \theta} \quad \dfrac{\varphi \supset \psi \quad \varphi}{\psi}}{\theta}$$

Here we have used a more relaxed style for setting out the derivation; the assumptions are not indexed and the names of the rules have not been indicated. 'Officially' this must be done, but in practice they are often left out.

We have seen that the Hilbert-Frege style axioms are all provable from our Natural Deduction rules. What about the rules of inference in the Hilbert-Frege style system? We consider (R∀I); then one is given a derivation D of $\varphi \supset \psi$, from assumptions $\psi_1, \ldots, \psi_k$ where a does not occur in any of the assumptions nor in φ, and is allowed to proceed to a derivation of $\varphi \supset \forall x \psi_x^a$ from the same assumptions. Can this be effected using also Natural Deduction? One would expect the answer to be yes, and this is indeed found to be the case by means of the following derivation tree:

$$\frac{\varphi \supset \psi \qquad \cancel{\varphi}^{\times}}{\dfrac{\dfrac{\psi}{\forall x \psi_x^a}(\forall I)}{\varphi \supset \forall x \psi_x^{a(1)}}}(\supset E)$$

with D above $\varphi \supset \psi$.

Here the use of ($\forall I$) is permitted because, by hypothesis, the *eigen-*parameter a does not occur in any of the assumptions of D nor in the assumption φ. Hence, the Natural Deduction system 'is closed' under the Hilbert-Frege style rule (R$\forall I$). The other quantifier rule (R$\exists E$) can be given an analogous treatment. We have therefore established the following

THEOREM (Gentzen [1934]). *If* $\Gamma \vdash_H \varphi$, *then* $\Gamma \vdash_N \varphi$.

For the other direction, the main work has already been carried out in the form of the Deduction Theorem. This theorem enables us to capture the effect of the discharge of assumptions within a Hilbert-Frege style system.

Consider the case of (&I). We must show that the system **HIQC=** is closed under this rule. To this effect assume that $\varphi_1, \ldots, \varphi_k \vdash_H \varphi$ and $\psi_1, \ldots, \psi_m \vdash_H \psi$. We must show that $\varphi_1, \ldots, \varphi_k, \psi_1, \ldots, \psi_m \vdash_H \varphi \& \psi$. But by use of (A&I) and MP we obtain first $\varphi_1, \ldots, \varphi_k \vdash_H \psi \supset \varphi \& \psi$; and then by a use of the second given consequence relation and MP we reach our desired conclusion. The other rules are established in the same simple way by the use of the homonymous axiom and MP. The two eigen-parameter rules demand a separate treatment, however. We consider ($\exists E$). So let there be given H-derivations D and D_1 which establish that $\Gamma \vdash_H \exists x \varphi$ and $\varphi_a^x, \Delta \vdash_H \theta$. Assume further that the parameter a does not occur in $\exists x \varphi$, θ, or Δ. We have to show that $\Gamma, \Delta \vdash_H \theta$, i.e. the desired conclusion of ($\exists D$) from the given premises. This is readily done, however:

$$\frac{\Gamma \qquad \dfrac{\dfrac{\cancel{\varphi_a^{x\times}}, \Delta}{D_1}}{\dfrac{\theta}{\varphi_a^x \supset \theta^{(1)}}(\supset I)}}{\dfrac{\exists x \varphi \supset \theta}{\theta}(R\exists E)}(MP) \qquad (= \text{The Deduction Theorem})$$

with D above $\exists x \varphi$.

The use of the rule (R$\exists E$) is permitted, because by hypothesis the parameter a does not occur in Δ (this justifies use of the Deduction Theorem as well),

I.2: SYSTEMS OF DEDUCTION

nor in $\exists x\varphi$ or θ. Hence, the Hilbert-Frege style system is closed under the quantifier rule ($\exists$E); the treatment of ($\forall$I) is similar. Thus, the second half of Gentzen's

THEOREM. *If* $\Gamma \vdash_N \varphi$, *then* $\Gamma \vdash_H \varphi$

is established.

The above system used intuitionistic logic; in order to obtain smooth systems for classical logic (whether, propositional or predicate, with or without identity) one can add any one of the following three rules:

(i) The axiom *tertium non datur*: $\varphi \vee \neg\varphi$ (TND)
(ii) The rule of indirect proof:

$$\dfrac{\begin{array}{c}\neg\varphi\\ \vdots\\ \bot\end{array}}{\varphi}(C) \quad \text{(C for classical!)}$$

(iii) The rule of Non-constructive Dilemma:

$$\dfrac{\begin{array}{cc}\varphi & \neg\varphi\\ \vdots & \vdots\\ \theta & \theta\end{array}}{\theta}(NCD)$$

These all give classical logic when added to the intuitionistic rules for propositional calculus. We prove this by establishing a chain of implications:

If (ii), then (i). This shall be understood in such a way that given the schematic rule C, one can derive all instances of TND:

$$\dfrac{\neg(\varphi \vee \neg\varphi)^2 \quad \dfrac{\dfrac{\varphi^1}{\varphi \vee \neg\varphi}(\vee I)}{\bot}(\supset E)}{\dfrac{\dfrac{\neg(\varphi \vee \neg\varphi)^2 \quad \dfrac{\neg\varphi^{(1)}}{\varphi \vee \neg\varphi}(\vee I)}{\bot}}{\varphi \vee \neg\varphi^{(2)}}(C)}$$

If (i), then (iii). This is immediate: given TND, the rule NCD simply becomes an instance of ($\vee$E).

If (iii), then (ii). Here we assume that we are given a derivation

$$\begin{array}{c} \neg\varphi \\ D \\ \bot \end{array}$$

of $\bot$ from the assumption $\neg\varphi$. Using only intuitionistic logic plus NCD we have to find a derivation of φ, *not depending on* $\neg\varphi$. First we note that

$$\begin{array}{c} \neg\varphi \\ D \\ \dfrac{\bot}{\varphi}(\bot) \end{array}$$

is an *intuitionistic* derivation of φ *from the assumption* $\neg\varphi$. A use of NCD does the rest:

$$\dfrac{\varphi^1 \qquad \dfrac{\dfrac{\neg\varphi^2}{D}}{\dfrac{\bot}{\varphi}}}{\varphi}{}_{(1,2)}(\text{NCD})$$

As φ is trivially derivable from the assumption φ, we have obtained a derivation of φ from the assumption φ and one of φ from the assumption $\neg\varphi$. The rule NCD gives us the right to discharge the assumptions φ and $\neg\varphi$.

With the given definition of negation in terms of the absurdity $\bot$, one is able to keep the idea that each symbol is governed by its introduction and elimination rules; the absurdity $\bot$ has no *introduction* rules and only one elimination rule, viz. ($\bot$). If one wishes to retain $\neg$ as a primitive, the appropriate intuitionistic rules become:

$$(\neg\text{I}) \qquad \dfrac{\begin{array}{cc} \varphi & \varphi \\ \vdots & \vdots \\ \psi & \neg\psi \end{array}}{\neg\varphi}$$

and

$$(\neg\text{E}) \qquad \dfrac{\varphi \qquad \neg\varphi}{\theta}$$

I.2: SYSTEMS OF DEDUCTION

In the classical case the elimination rule becomes:

$$(\neg E_C) \quad \frac{\neg\neg\varphi}{\varphi}$$

The ($\neg$I) and ($\neg$E) rules are less satisfactory than the other rules because the sign to be introduced occurs already in the premises of the rule.

The intuitionistic rules have a pleasing symmetry; each connective has its use regulated by two rules, one of which is, in a certain sense, the inverse of the other. To give this remark a bit more substance, consider a *maximum* formula, i.e. a formula occurrence in a derivation which is the conclusion of an I rule and the major premise of the corresponding E rule, e.g. & in:

$$\frac{\dfrac{\varphi^1 \quad \psi^2}{\varphi \,\&\, \psi}(\&I)}{\varphi}(\&E)$$

This maximum can be removed, i.e. we can find a derivation of the same conclusion from (at most) the same assumptions, in a maximally simple way: φ^1.

This maximum-removing operation is called a *reduction* and similar reductions can also be given for other types of maxima, e.g. the $\supset$ reduction is given by:

$$(\supset I) \quad \frac{\dfrac{\varphi^1}{\vdots}}{\dfrac{\psi}{\varphi \supset \psi}^{(1)}} \quad \varphi \atop \psi \quad (\supset E) \quad \text{reduces to} \quad \begin{array}{c}\varphi\\ \vdots\\ \psi\end{array}$$

and the $\forall$-reduction is given by:

$$(\forall I) \quad \dfrac{\dfrac{\vdots}{\varphi}}{\dfrac{\forall x \varphi_x^a}{(\varphi_x^a)_t^x}} \quad \text{reduces to} \quad \varphi_t^a \qquad \begin{array}{l} a \\ \vdots \\ t \end{array} \text{(i.e. everywhere in the derivation of } \varphi \text{ replace } a \text{ with } t)$$

($\forall E$)

For the other reductions, cf. Prawitz [1965] pp. 36–38 (where they were first introduced) and Prawitz [1971] pp. 252–253.

Although each individual maximum can be removed via a reduction, the situation is not altogether as simple as one would wish because *the removal of one maximum can create a new maximum*, e.g. in:

$$
\begin{array}{c}
\quad\quad\quad\quad \varphi^1 \\
\quad\quad\quad\quad \vdots \\
(\supset I) \quad \dfrac{\psi}{\varphi \supset \psi}^{(1)} \quad\quad \vdots \\
(\&I) \quad \dfrac{(\varphi \supset \psi) \,\&\, \theta}{\varphi \supset \psi} \quad \vdots \\
(\&E) \quad\quad\quad \dfrac{\varphi \supset \psi \quad\quad \varphi}{\psi}
\end{array}
$$

Here $(\varphi \supset \psi) \,\&\, \theta$ is a maximum and an & reduction gives:

$$
\begin{array}{c}
\quad\quad \varphi^1 \\
\quad\quad \vdots \\
\supset I \quad \dfrac{\psi}{\varphi \supset \psi}^{(1)} \quad \varphi \\
\quad\quad \dfrac{}{\psi} \supset E
\end{array}
$$

Now the wff $\varphi \supset \psi$, which previously was not a maximum, has been turned into a maximum. A $\supset$ reduction will, of course, suffice to remove *this* new maximum, but what about the general case? If we consider the complexities of the old and the new maximum, respectively, we find that the new one has a lower complexity than the old one, and this phenomenon holds in general. Hence, one has something to use induction on in a proof that every maximum can be removed by successive reductions (even though new ones may arise along the way). In fact Prawitz has proved the

NORMALIZATION THEOREM. *Every derivation can be brought to maximum-free, or normal, form by means of successive reductions.*

For a proof and precise statements of this and related results, we refer to the works by Prawitz and Tennant just cited. Dummett [1977] also contains an exhaustive discussion of normalization.

The reduction and the normalization procedures have formed a basis for various attempts to give a 'theory of meaning' not using the Tarski truth definition as a key concept. Cf. Prawitz [1977] and Sundholm [Chapter III.8] for a description of this use of Natural Deduction and its metatheoretical properties.

I.2: SYSTEMS OF DEDUCTION

Natural Deduction formulations can be given not only for pure propositional and predicate logic, as above, but also for, say, modal logic. Define:

(0) $\bot$ is essentially modal,
(i) $\Box\varphi$ is essentially modal,
(ii) If φ and ψ are both essentially modal, then so are $\varphi \& \psi$ and $\varphi \lor \psi$.

Using this concept of an essentially modal formula, one then gives an attractive version of classical **S4** by adding the following two rules to **CPC**.

($\Box$E) $\quad \dfrac{\Box\varphi}{\varphi}$ (This rule is, of course, a rule version of the **T** axiom: $\Box\varphi \supset \varphi$)

and

($\Box$I) $\quad \dfrac{\varphi}{\Box\varphi}$ *provided* that all assumption on which φ depends are essentially modal.

This formulation of **S4** is given in Prawitz [1965] Chapter VI. We derive the S4 axioms. For these axioms, cf. Bull and Segerberg [Chapter II.1]:

K: We have to prove $\Box(\varphi \supset \psi) \supset (\Box\varphi \supset \Box\psi)$

($\Box$E)
$$\dfrac{\dfrac{\dfrac{\dfrac{\dfrac{\Box(\varphi \supset \psi)^1}{\varphi \supset \psi}(\Box E) \quad \dfrac{\Box\varphi^2}{\varphi}(\Box E)}{\psi}(\supset E)}{\Box\psi}(\Box I)}{\Box\varphi \supset \Box\psi}(2)(\supset I)}{\Box(\varphi \supset \psi) \supset (\Box\varphi \supset \Box\psi)}(1)$$

(All assumptions are essentially modal.)

T: We have to prove $\Box\varphi \supset \varphi$.

$$\dfrac{\dfrac{\dfrac{\Box\varphi^1}{\varphi}(\Box E)}{\Box\varphi \supset \varphi}(1)(\supset I)}{}$$

4. We have to prove $\Box\varphi \supset \Box\Box\varphi$.

$$\dfrac{\dfrac{\dfrac{\Box\varphi^1}{\Box\Box\varphi}(\Box I)}{\Box\varphi \supset \Box\Box\varphi}(1)(\supset I)}{}$$

The rule *Nec* is just a special case of ($\Box$I), with the class of assumptions empty.

Thus we have proved that **NS4** includes the previously given **HS4**. For the other direction one uses the two readily verified facts:

(i) If $\varphi_1, \ldots, \varphi_k \vdash_{\mathbf{HS4}} \varphi$, then $\Box\varphi_1, \ldots, \Box\varphi_k \vdash_{\mathbf{HS4}} \Box\varphi$.
(ii) For every essentially modal φ, $\vdash_{\mathbf{HS4}} \varphi \supset \Box\varphi$.

The details are left as an exercise. Finally we remark that if we change (ii) to (ii′) in the definition of essentially modal wffs, we obtain a formulation of **S5**, where (ii′) says: If φ and ψ are essentially modal, then so are $\varphi \& \psi$, $\varphi \vee \psi$ *and* $\varphi \supset \psi$.

For another, final, example of a Natural Deduction version of a wider system, we consider a smooth version of Heyting Arithmetic, **HA**, cf. Van Dalen [Chapter III.4]. We use a new predicate $N(x)$ – 'x is a natural number', a constant 0 and a 'successor function' $s(x)$ (plus some further function constants which we need not bother about now). The introduction rule for N gives an inductive definition of the natural numbers:

(N I): $N(0)$ $\dfrac{N(a)}{N(s(a))}(N\text{ I})$

By this rule, every numeral 0, $s(0)$, $s(s(0))$, ... is in N. The elimination rule, on the other hand, says that nothing else is in N:

(NE)

$$\dfrac{N(t) \quad \varphi_0^x \quad \varphi_{s(a)}^{x}\overset{\varphi_a^x}{\vdots}}{\varphi_t^x}\text{(1)}(N\text{E}) \quad (\textit{eigen-parameter condition on } a)$$

This version of the induction axiom thus says that if t is in N and the property φ is closed under successor and contains 0, then t has the property φ. This form of spelling out inductive definitions can be applied not just to the inductively-generated natural numbers, but to a wide range of inductive definitions, cf. Martin-Löf [1971].

On the positive side of Natural Deduction formulations there are, as we have seen, several advantages, the foremost of which is the great ease with which derivations can actually be *carried out*. On the minus score however, one must note that Natural Deduction is not suitable for all sorts of systems. In several modal systems, say between S4 and S5, it is not at all clear that one can isolate the behaviour of $\Box$ in the form of introduction and elimination

I.2: SYSTEMS OF DEDUCTION

rules. It was already noted that the conventions on assumptions made it difficult to find suitable systems for relevance logic. For some efforts in this area, cf. Prawitz [1965], Chapter VII.

The arithmetization of a Natural Deduction formulation is, *prima facie* also a little more cumbersome than in the case of an Hilbert-Frege style formulation. The Gödel number of a derivation has to contain the following information: (i) the end formula, (ii) the rule used at the last inference, (iii) Gödel numbers for derivation of the premisses, (iv) the assumption class possibly discharged at the inference, (v) the still open assumption classes. Such Gödel numbers are best constructed as sequence numbers containing the above items, cf. Kleene [1952], Section 51-52 and Troelstra [1973] Chapter IV. (The latter reference contains much valuable material on various aspects of the meta-theory of Natural Deduction.)

The Natural Deduction rules obviously impose the following properties on the derivability relation (for intuitionistic predicate logic):

(0) $\varphi \vdash \varphi$ – to make an assumption.
(i) If $\Gamma \vdash \varphi$, then $\Gamma, \Delta \vdash \varphi$ – we may add further assumptions.
(ii) If $\Gamma \vdash \varphi$ and $\Delta \vdash \psi$, then $\Gamma, \Delta \vdash \varphi \& \psi$ – by (& I)
(iii) If $\Gamma \vdash \varphi_0 \& \psi_1$, then $\Gamma \vdash \varphi_i$ – by (&E) for $i = 0, 1$.
(iv) If $\varphi, \Gamma \vdash \psi$, then $\Gamma \vdash \varphi \supset \psi$ – by ($\supset$I)
(v) If $\Gamma \vdash \varphi \supset \psi$ and $\Delta \vdash \varphi$, then $\Gamma, \Delta \vdash \psi$ – by ($\supset$E)
(vi)–(vii) The corresponding clauses for ($\vee$ I) and ($\vee$ E) are left as exercises.
(viii) If $\Gamma \vdash \varphi$ and a does not occur in Γ, then $\Gamma \vdash \forall x \varphi_x^a$ – by ($\forall$ I)
(ix) If $\Gamma \vdash \forall x \varphi$, then $\Gamma \vdash \varphi_t^x$ – by ($\forall$ E)
(x) If $\Gamma \vdash \varphi_t^x$, then $\Gamma \vdash \exists x \varphi$ – by ($\exists$ I)
(xi) If $\Gamma \vdash \exists x \varphi$ and $\varphi_a^x, \Delta \vdash \theta$ and a does not occur in any of Δ, θ and $\exists x \varphi$, then $\Gamma, \Delta \vdash \theta$ – by ($\exists$ E)
(xii) If $\Gamma \vdash \bot$, then $\Gamma \vdash \varphi$ – by ($\bot$).

Define a *sequent* to be an expression of the form '$\Gamma \vdash \varphi$' where Γ is a finite set of wffs – possibly empty – and φ a wff. Then the above clauses (0)–(xii) give the axioms and rules of a certain *Hilbert-Frege style system for deriving sequents*; the clause (0) gives the only axiom, which says that the sequent $\varphi \vdash \varphi$ is derivable. (Here I adhere to the same conventions about notation as before; strictly speaking '$\{\varphi\} \vdash \varphi$' is the axiom.) The other clauses are *rules of proof.* The present system is just a notational variant of the tree arrangement version of Natural Deduction which we have used in the present section; in particular it is not to be confused with Gentzen's Sequent Calculi.

The main feature of the sequential formulation of Natural Deduction is that there are both elimination rules and introduction rules. In the Sequent Calculus, on the other hand, there are *only introduction rules*. The sequential version was first given by Gentzen [1936] in his 'first consistency proof'. There is a certain formal difference between Gentzen's system and ours, because for him the sequent part Γ is not a set of wffs but of a *list* of wffs, and he therefore has rules to the effect that one may permute assumption formulae in the lists and contract two identical assumption formulae to one. The latter rule neglects the problem of assumption classes. The best way to treat that in the sequential version seems to be to let Γ be a set of *labelled* formulae, where a labelled formula is an order pair $\langle \varphi, k \rangle$ where k is a natural number (and φ is a wff, of course). The details can be found in Leivant [1979].

One has several options as to the choice of rules for a sequential version – for an indication of the various possibilities, cf. Dummett [1977] Chapter IV who treats the sequential formulations in considerable detail – and in particular the *'thinning'* rule (ii) allows us to restrict the multi-premise rules to the special cases of always using the same set of assumptions. Thus, one can formulate the rule for conjunction-introduction as follows:

$$\frac{\Gamma \vdash \varphi \qquad \Gamma \vdash \psi}{\Gamma \vdash \varphi \& \psi}$$

The effect of the old rule (ii) can be simulated. From $\Gamma \vdash \varphi$, by thinning, one obtains $\Gamma, \Delta \vdash \varphi$ and from $\Delta \vdash \psi$ one obtains $\Gamma, \Delta \vdash \psi$, also by thinning. Finally, the (new) conjunction introduction gives the desired conclusion: $\Gamma, \Delta \vdash \varphi \& \psi$.

Another variation is to formulate the axioms in a more general way: $\Gamma \vdash \varphi$ is an axiom where φ is an element of Γ, or equivalently: $\varphi, \Gamma \vdash \varphi$ is an axiom.

If we carry out both of these modifications the resulting calculus for deriving sequents will look like:

Axiom: $\varphi, \Gamma \vdash \varphi$

Thinning: $\dfrac{\Gamma \vdash \varphi}{\Gamma, \Delta \vdash \varphi}$

(& I): $\dfrac{\Gamma \vdash \varphi \qquad \Gamma \vdash \psi}{\Gamma \vdash \varphi \& \psi}$

$(\& \text{E})$: $\dfrac{\Gamma \vdash \varphi_0 \& \varphi_1}{\Gamma \vdash \varphi_i}$ $\quad i = 0, 1$

$(\vee \text{I})$: $\dfrac{\Gamma \vdash \varphi_i}{\Gamma \vdash \varphi_0 \vee \varphi_1}$ $\quad i = 0, 1$

$(\vee \text{E})$: $\dfrac{\Gamma \vdash \varphi \vee \psi \quad \varphi, \Gamma \vdash \theta \quad \psi, \Gamma \vdash \theta}{\Gamma \vdash \theta}$

$(\supset \text{I})$: $\dfrac{\varphi, \Gamma \vdash \psi}{\Gamma \vdash \varphi \supset \psi}$

$(\supset \text{E})$: $\dfrac{\Gamma \vdash \varphi \supset \psi \quad \Gamma \vdash \varphi}{\Gamma \vdash \psi}$

$(\forall \text{I})$:* $\dfrac{\Gamma \vdash \varphi}{\Gamma \vdash \forall x \varphi_x^a}$

$(\forall \text{E})$: $\dfrac{\Gamma \vdash \forall x \varphi}{\Gamma \vdash \varphi_t^x}$

$(\exists \text{I})$: $\dfrac{\Gamma \vdash \varphi_t^x}{\Gamma \vdash \exists x \varphi}$

$(\exists \text{E})$:* $\dfrac{\Gamma \vdash \exists x \varphi \quad \varphi_a^x, \Gamma \vdash \theta}{\Gamma \vdash \theta}$

$(\bot)$: $\dfrac{\Gamma \vdash \bot}{\Gamma \vdash \varphi}$

We conclude our treatment of Natural Deduction by mentioning Tennant [1978], which treats the traditional metamathematics of first-order logic and arithmetic on the basis of Natural Deduction formulations of the systems concerned, and Van Dalen [1980], which treats model theory in a similar way. Gentzen's treatment of Natural Deduction, which still seems the most natural to me, inspired many other attempts to keep the derivations from assumptions and the discharge of certain of these as a basic feature. In particular, many authors have experimented with *linear* arrangements of the derivations as opposed to Gentzen's tree formulations. A list of variant treatments of Natural Deduction is found in Hodges [Chapter I.1], Section 7.

* On these rules there is an eigen-parameter condition; a must not occur in Γ, $\exists x \varphi$ or θ.

3. SEQUENT CALCULI

The (reformulated) sequential version of Natural Deduction, which we considered at the end of Section 2, was nothing but a direct linearization of the introduction and elimination rules, where the latter were thought of as imposing conditions on the derivability relation $\vdash$. The pattern thus obtained is, however, not the only way of axiomatizing the derivability relation on the basis of the Natural Deduction rules. Crucial in the above treatment was that the elimination rules were formulated as such and operated on the right-hand side of the turnstile $\vdash$. A completely different approach to the elimination rules is possible, though, and leads to Gentzen's [1934] *Sequent Calculi*.

Consider a derivable sequent

$$\varphi, \varphi_1, \ldots, \varphi_k \vdash \theta;$$

hence if we read the derivation in the sequential system as a description of how to build a derivation tree for θ from assumptions $\varphi, \varphi_1, \ldots, \varphi_k$, we can find a derivation tree D such that

$$\begin{array}{c} \varphi^1, \varphi_1, \ldots, \varphi_k \\ D \\ \theta \end{array}$$

where, in particular, φ is an undischarged assumption. Therefore, using (one of) (&E) we obtain a derivation tree D' of θ from assumptions $\varphi_1, \ldots, \varphi_k$ and $\varphi \& \psi$. In this tree D' the assumption φ has 'disappeared'.

$$D' =_{\text{def}} \quad (\&E)\dfrac{\varphi \& \psi^1, \quad \varphi_1, \ldots, \varphi_k}{\begin{array}{c}\varphi \\ D \\ \theta\end{array}}$$

So the sequent $\varphi \& \psi, \varphi_1, \ldots, \varphi_k \vdash \theta$ is derivable, and on the given interpretation the rule

$$\dfrac{\varphi, \Gamma \vdash \theta}{\varphi \& \psi, \Gamma \vdash \theta}$$

is a sound *rule of proof*.

Likewise, the rule

$$\dfrac{\Gamma \vdash \varphi \qquad \psi, \Gamma \vdash \theta}{\varphi \supset \psi, \Gamma \vdash \theta}$$

I.2: SYSTEMS OF DEDUCTION

is a sound rule of proof, because given derivation trees

$$\begin{array}{cc} \Gamma & \psi^1, \Gamma \\ \vdots & \vdots \\ \varphi & \theta \end{array}$$
and

one readily finds a derivation tree in which the place of the assumption ψ is taken by $\varphi \supset \psi$:

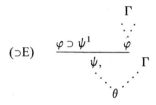

($\supset$E)

The use of ($\vee$ E) immediately justifies the step:

$$\frac{\varphi, \Gamma \vdash \theta \qquad \psi, \Gamma \vdash \theta}{\varphi \vee \psi, \Gamma \vdash \theta}$$

and ($\forall$ E) justifies:

$$\frac{\varphi_t^x, \Gamma \vdash \theta}{\forall x \varphi, \Gamma \vdash \theta}$$

Finally, if a does not occur in the conclusion, then trivially ($\exists$ E) justifies the step:

$$\frac{\varphi_a^x, \Gamma \vdash \theta}{\exists x \varphi, \Gamma \vdash \theta}$$

From the present point of view, the elimination rules are viewed as *left-hand side introduction rules*. Define a *sequent* to be an expression of the form

$$\Gamma \to \theta$$

where Γ is a finite set of wffs. This (re-) definition is only for historical reasons; by custom, one uses an arrow in the sequents.

The axioms and rules of Gentzen's Sequent Calculus are given by:

(Axiom) $\quad \varphi, \Gamma \to \varphi$

(Thinning) $\quad \dfrac{\Gamma \vdash \varphi}{\Delta, \Gamma \vdash \varphi}$

$(\rightarrow \&)$ $\quad \dfrac{\Gamma \rightarrow \varphi \quad \Gamma \rightarrow \psi}{\Gamma \rightarrow \varphi \& \psi}$

$(\& \rightarrow)$ $\quad \dfrac{\varphi, \Gamma \rightarrow \theta}{\varphi \& \psi, \Gamma \rightarrow \theta} \qquad \dfrac{\psi, \Gamma \rightarrow \theta}{\varphi \& \psi, \Gamma \rightarrow \theta}$

$(\rightarrow \vee)$ $\quad \dfrac{\Gamma \rightarrow \varphi}{\Gamma \rightarrow \varphi \vee \psi} \qquad \dfrac{\Gamma \rightarrow \psi}{\Gamma \rightarrow \varphi \vee \psi}$

$(\vee \rightarrow)$ $\quad \dfrac{\varphi, \Gamma \rightarrow \theta \quad \psi, \Gamma \rightarrow \theta}{\varphi \vee \psi, \Gamma \rightarrow \theta}$

$(\rightarrow \supset)$ $\quad \dfrac{\varphi, \Gamma \rightarrow \psi}{\Gamma \rightarrow \varphi \supset \psi}$

$(\supset \rightarrow)$ $\quad \dfrac{\Gamma \rightarrow \varphi \quad \psi, \Gamma \rightarrow \theta}{\varphi \supset \psi, \Gamma \rightarrow \theta}$

$(\rightarrow \forall)$ $\quad \dfrac{\Gamma \rightarrow \varphi}{\Gamma \rightarrow \forall x \varphi_x^a}$ provided that a does not occur in Γ

$(\forall \rightarrow)$ $\quad \dfrac{\varphi_t^x, \Gamma \rightarrow \theta}{\forall x \varphi, \Gamma \rightarrow \theta}$

$(\rightarrow \exists)$ $\quad \dfrac{\Gamma \rightarrow \varphi_t^x}{\Gamma \rightarrow \exists x \varphi}$

$(\exists \rightarrow)$ $\quad \dfrac{\varphi_a^x, \Gamma \rightarrow \theta}{\exists x \varphi, \Gamma \rightarrow \theta}$ provided that a does not occur in the conclusion

$(\bot)$ $\quad \dfrac{\Gamma \rightarrow \bot}{\Gamma \rightarrow \theta}$

A sequent $\Gamma \rightarrow \varphi$ is provable in this 'cut-free' intuitionistic sequent calculus if it has got a proof tree regulated by the above rules and where all the top nodes are axioms. In symbols we write:

$$\vdash_{IS} \Gamma \rightarrow \varphi$$

(**IS** for intuitionistic sequent calculus).

The introductory discussion implicitly proved the following inclusion:

If $\vdash_{IS} \Gamma \rightarrow \varphi$, then the sequent $\Gamma \vdash \varphi$ is derivable in the sequential formulation of Natural Deduction.

A formal proof by induction on the length of the **IS** proof of $\Gamma \to \varphi$ can safely be left to the patient reader.

Of course, one would like to establish the converse relation between the two axiomatizations of the Natural Deduction turnstile, i.e. that the idea of treating the elimination rules as left-hand side *introductions* is as strong as the original formulations where the eliminations operate to the right. This can, in fact, be done but a direct proof would be quite unwieldy. Instead, Gentzen introduced a rule:

$$\text{(CUT)} \quad \frac{\Gamma \to \varphi \qquad \varphi, \Gamma \to \theta}{\Gamma \to \theta}$$

such that in $\mathbf{IS^+} = \mathbf{IS} + \text{CUT}$ one readily shows that the effect of the full Natural Deduction eliminations can be simulated. Then, by means of his famous *Hauptsatz*, he established that if $\vdash_{\mathbf{IS^+}} \Gamma \to \varphi$, then $\vdash_{\mathbf{IS}} \Gamma \to \varphi$, and the equivalence is established.

We treat the case of (&E). Hence, we are given an $\mathbf{IS^+}$ proof of $\Gamma \to \varphi \,\&\, \psi$ and we wish to find one for $\Gamma \to \psi$. This we do as follows:

$$\cfrac{\Gamma \to \varphi \,\&\, \psi \qquad \cfrac{\cfrac{\psi, \Gamma \to \psi}{\varphi \,\&\, \psi, \Gamma \to \psi}(\&\to)}{}(\text{Axiom})}{\Gamma \to \psi}(\text{CUT})$$

The same technique using the CUT rule works uniformly for all the Natural Deduction elimination rules. To treat one more case, we choose the most complex, viz. ($\supset$E): Here we are given $\mathbf{IS^+}$ proofs of $\Gamma \to \varphi$ and $\Gamma \to \varphi \supset \psi$, and have to find one for $\Gamma \to \psi$. This is done as follows:

$$\cfrac{\Gamma \to \varphi \supset \psi \qquad \cfrac{\Gamma \to \varphi \qquad \psi, \Gamma \to \psi}{\varphi \supset \psi, \Gamma \to \psi}(\supset\to)}{\Gamma \to \psi}(\text{CUT})$$

The same method proves the other cases and, hence, it is established that

If $\Gamma \vdash \varphi$ is a provable Natural Deduction sequent, then $\vdash_{\mathbf{IS^+}} \Gamma \to \varphi$.

In order to get the inclusion in **IS** we have to discuss Gentzen's Hauptsatz that CUT can be eliminated. What is the significance of CUT for the Natural Deduction derivations? Here it licenses the step from provable sequents

$\Gamma \vdash \varphi$ and $\varphi, \Gamma \vdash \psi$ to $\Gamma \vdash \psi$. It is simply a derivable rule of proof by means of ($\supset$ I) and ($\supset$ E), viz:

$$(\supset \text{I}) \quad \dfrac{\dfrac{\varphi, \Gamma \vdash \psi}{\Gamma \vdash \varphi \supset \psi} \quad \Gamma \vdash \varphi}{\Gamma \vdash \psi}$$
$(\supset \text{E})$

Here we created a new maximum, viz. $\varphi \supset \psi$ which is the conclusion of an introduction and the major premise of an elimination. Another way to view CUT in Natural Deduction contexts is that it expresses the closure under substituting a derivation of an assumption for that assumption, e.g. consider:

The result of putting the first derivation of on top of the other looks like:

$$\begin{array}{c} \Gamma \\ \vdots \\ \varphi, \quad \Gamma \\ \vdots \\ \psi \end{array}$$

Here the *assumption* φ has disappeared, but a new *maximum* may have arisen, viz. φ. If the last rule of the derivation of φ from Γ is an introduction – say that $\varphi = \varphi_0 \& \varphi_1$, inferred by an (&I) – and the first rule applied to φ in the derivation of ψ from assumptions φ and Γ is an elimination – in this case, then, an (&E) – then φ has been turned into a maximum, although the derivations are closed under the rule of CUT. Hence, cut-free derivations in the Sequent Calculus correspond to normal derivations in the Natural Deduction systems, and if we may assume that the Natural Deduction derivation we have of a sequent is *normal*, one can directly find a cut-free **IS** proof of the corresponding sequent, cf. Prawitz [1965] Appendix A, pp. 88–93. Hence, one needs to prove the Normalization Theorem, or the Hauptsatz, to have an easy proof of the inclusion of the Natural Deduction in the cut-free system **IS**. In view of the above discussion, this is hardly surprising, since there is a correspondence between cuts and maxima.

Gentzen's proof of his Hauptsatz consists of a series of permutations of cuts in such a fashion that the complexity of the 'cut formula' is lowered. We can illustrate this by considering the simplest of the cases:

$$(\to\&)\quad \frac{\dfrac{\vdots}{\Gamma\to\varphi}\quad \dfrac{\vdots}{\Gamma\to\psi}}{\Gamma\to\varphi\&\psi}\quad \frac{\dfrac{\vdots}{\varphi,\Gamma\to\theta}}{\varphi\&\psi,\Gamma\to\theta}(\&\to)$$
$$\overline{\Gamma\to\theta}\text{(CUT)}$$

Here we replace the cut on the wff $\varphi\ \&\ \psi$ by a cut on the wff φ in this way:

$$\frac{\dfrac{\vdots}{\Gamma\to\varphi}\quad \dfrac{\vdots}{\varphi,\Gamma\to\theta}}{\Gamma\to\theta}\text{(CUT)}$$

The situation is not always as simple as this but it gives the idea. The proof of the Hauptsatz can be found in many places. Apart from Gentzen's own [1934], Kleene [1952] and Takeuti [1975], as well as Dummett [1977], contain good expositions of the proof.

Gentzen's own Sequent Calculi used finite *lists* of formulae (in place of finite sets) and, as already remarked apropos his formulation of the sequential Natural Deduction system, he needs further 'structural' rules to permute and contract formulae in the finite lists. The presence of a contraction rule:

$$\frac{\varphi,\varphi,\Gamma\to\theta}{\varphi,\Gamma\to\theta}$$

again raises the question of assumption classes and mandatory discharge of assumptions (via the correspondence with Natural Deduction). The interested reader is referred to Zucker [1974] and Pottinger [1977] for a thorough treatment of such matters.

The Sequent Calculus just presented was set up so as to make the transition from Natural Deduction particularly easy and, hence, it was expedient to keep $\bot$ as a primitive (with negation defined). In the Sequent Calculus, however, another more congenial treatment consists of dropping $\bot$ and adding $\neg$ as a primitive. The concept of sequent is widened to include sequents '$\Gamma\to\Delta$', where both the *'antecedent'* Γ and the *'succedent'* Δ are *finite* sets of formulae, with the important restriction that Δ *has got at most one element*. The sequent '$\Gamma\to\{\varphi\}$' will be written '$\Gamma\to\varphi$' and a sequent $\Gamma\to\phi$, with empty succedent, is written '$\Gamma\to$ '.

In the calculus one drops the ($\bot$) rule, of course, but in order to secure its effect, one also adds a *Thinning* for the succedent:

$$\frac{\Gamma\to}{\Gamma\to\theta}$$

This makes the empty succedent behave like absurdity, and the following two rules for negation suggest themselves:

$(\rightarrow \neg)$ $\quad \dfrac{\varphi, \Gamma \rightarrow}{\Gamma \rightarrow \neg \varphi}$

and

$(\neg \rightarrow)$ $\quad \dfrac{\Gamma \rightarrow \varphi}{\neg \varphi, \Gamma \rightarrow}$

The system thus modified we call **IS'**.

The underlying idea for our way of looking at the Sequent Calculus, up till this point, has been to regard the Sequent Calculus as a sort of 'meta-descriptive' system for how Natural Deduction derivation are put together. There is also another approach to the sequent apart from this consequence relation interpretation, which other approach is particularly well suited to effect the transition to the fully symmetric, classical Sequent Calculus.

On a naive semantical level, we may call a sequent $\Gamma \rightarrow \Delta$ *valid*, if whenever all the members of Γ are true, then at least some member of Δ is true. (Hence, a sequent $\Gamma \rightarrow$ is valid iff Γ is inconsistent.) The rules of **IS'** then give rules of passage between valid sequents; in particular, every axiom is valid. On the other hand, let us call a sequent *falsifiable* if one can make all the members of Γ true and all the members of Δ false. We note that these explanations of validity and falsifiability also work for sequents with more than one element in the succedent.

Now the other approach suggests itself; instead of reading the rules of **IS** as expressing validity conditions of the form, say, "if S_1 and S_2 are both valid sequents, then so is S" we may read them as *expressing falsifiability conditions* of the form, say, "if the sequent S if falsifiable, then S_1 is falsifiable or S_2 is falsifiable". Let us work this out for the conjunction rules: Assume that the sequent '$\Gamma \rightarrow \varphi \& \psi$' is falsifiable. Then one can make all of Γ true and yet make $\varphi \& \psi$ false. Hence, one can make at least one of φ and ψ false, and therefore at least one of the premises $\Gamma \rightarrow \varphi$ and $\Gamma \rightarrow \psi$ must be a falsifiable sequent. Likewise, for the antecedent conjunction rule: Assume that $\varphi \& \psi, \Gamma \rightarrow \Delta$ is a falsifiable sequent. Then one can make all of $\varphi \& \psi, \Gamma$ true and Δ false. But then both of $\varphi, \Gamma \rightarrow \Delta$ and $\psi, \Gamma \rightarrow \Delta$ are falsifiable sequents, so $(\& \rightarrow)$ expresses two falsifiability conditions.

The same sort of reasoning applies to the other rules of **IS'**. It is worthwhile considering the negation rules a bit more carefully. Take $(\neg \rightarrow)$ first.

If $\neg\varphi, \Gamma \to$ is falsifiable, this means one can make all of $\neg\varphi, \Gamma$ true, so one can make φ false. Hence, $\Gamma \to \varphi$ is falsifiable. As for $(\to\neg)$, assume that $\Gamma \to \neg\varphi$ is falsifiable. Then all of Γ can be made true while $\neg\varphi$ is false. Hence, all of φ, Γ can be made true.

Both of these rules thus express good falsifiability conditions, but they are too 'weak' to make use of full classical reasoning. The same argument would also justify the rules with *extra* members in the succedents:

$$\frac{\varphi, \Gamma \to \Delta}{\Gamma \to \Delta, \neg\varphi} \qquad \frac{\Gamma \to \Delta, \varphi}{\neg\varphi, \Gamma \to \Delta}$$

The calculus which results is particularly nice because the rules express necessary and *sufficient* conditions for the falsifiability of the conclusion, i.e. in order that the conclusion be falsifiable it is necessary and sufficient that at least one of the premises be falsifiable. The axioms indicate unfalsifiability, because it is impossible to give φ both the value true *and* the value false. In general, we see that the complexity of the premises is lower – the quantifier rules form exceptions – than the complexity of the conclusion and, hence, the falsifiability condition for the conclusion is – in general – expressed in terms of falsifiability conditions of lower complexity. This leads to a practical way of systematically searching for a proof in the Sequent Calculus.

Axiom $\quad \varphi, \Gamma \to \Delta, \varphi$

$(\to \&) \quad \dfrac{\Gamma \to \Delta, \varphi \qquad \Gamma \to \Delta, \psi}{\Gamma \to \Delta, \varphi \& \psi}$

$(\& \to) \quad \dfrac{\varphi, \psi, \Gamma \to \Delta}{\varphi \& \psi, \Gamma \to \Delta}$

$(\to \vee) \quad \dfrac{\Gamma \to \Delta, \varphi, \psi}{\Gamma \to \Delta, \varphi \vee \psi}$

$(\vee \to) \quad \dfrac{\varphi, \Gamma \to \Delta \qquad \psi, \Gamma \to \Delta}{\varphi \vee \psi, \Gamma \to \Delta}$

$(\to \supset) \quad \dfrac{\varphi, \Gamma \to \Delta, \psi}{\Gamma \to \Delta, \varphi \supset \psi}$

$(\supset \to) \quad \dfrac{\Gamma \to \Delta, \varphi \qquad \psi, \Gamma \to \Delta}{\varphi \supset \psi, \Gamma \to \Delta}$

$(\to \neg)$ $\quad \dfrac{\varphi, \Gamma \to \Delta}{\Gamma \to \Delta, \neg\varphi}$

$(\neg \to)$ $\quad \dfrac{\Gamma \to \Delta, \varphi}{\neg\varphi, \Gamma \to \Delta}$

$(\to \forall)$ $\quad \dfrac{\Gamma \to \Delta, \varphi}{\Gamma \to \Delta, \forall x \varphi_x^a}$ provided that a does not occur in Γ, Δ

$(\forall \to)$ $\quad \dfrac{\varphi_t^x, \forall x\varphi, \Gamma \to \Delta}{\forall x\varphi, \Gamma \to \Delta}$

$(\to \exists)$ $\quad \dfrac{\Gamma \to \Delta, \exists x\varphi, \varphi_t^x}{\Gamma \to \Delta, \exists x\varphi}$

$(\exists \to)$ $\quad \dfrac{\varphi, \Gamma \to \Delta}{\exists x\varphi_x^a, \Gamma \to \Delta}$ provided that a does not occur in Γ, Δ.

This system we call **CS** – '*C*' for classical.

The reader should convince himself that **CS** is sound for the standard truth-value semantics for classical logic, i.e. that if $\vdash_{\mathbf{CS}} \Gamma \to \Delta$, then $\models \Gamma \to \Delta$, where $\models \Gamma \to \Delta$ is defined to mean that if all the members of Γ are true in an interpretation, then at least one member of Δ is true in the same interpretation.

Hence, **CS** does not give us more than classical logic, and as a matter of fact **CS** is *complete* and gives us *all* of classical logic. First one notes that **CS** does prove all instances of TND:

$$\dfrac{\dfrac{\dfrac{\varphi \to \varphi}{\to \varphi, \neg\varphi}(\to\neg)}{\to \varphi \vee \neg\varphi}(\to\vee)}{}\text{(Axiom)}$$

Secondly, one notes that **CS** is closed under *Thinning*. Let D be a **CS** proof of the sequent $\Gamma \to \Delta$. One has to show that $\Gamma, \Gamma' \to \Delta, \Delta'$ is **CS** provable for any finite Γ' and Δ'. Inspect the parameters which occur in Γ', Δ'; these are finitely many. Therefore, every *eigen*-parameter of D which also occurs in Γ', Δ', can be exchanged for a new *eigen*-parameter which does not occur in D or in Γ', Δ' (as we have infinitely many parameters at our disposal). The result is still a derivation D' of the sequent $\Gamma \to \Delta$. If we now add Γ' and Δ' as side formulae everywhere, the result is still a derivation D'' in **CS** (as axioms go into axioms and applications of rules become applications of the same rules), *without violating any eigenparameter conditions*.

I.2: SYSTEMS OF DEDUCTION

Hence, we have found the desired derivation of the sequent $\Gamma, \Gamma' \to \Delta, \Delta'$ which is, thus, **CS** provable.

Thirdly, observe that from the premise of an **IS** rule, the premise of the corresponding **CS** rule can be inferred by Thinning, and then the conclusion of the **IS** rule follows by the **CS** rule from the **CS** premise, e.g. from the **IS** premise $\varphi, \Gamma \to \Delta$ by thinning one obtains the **CS** premise $\varphi, \psi, \Gamma \to \Delta$, and the **CS** rule ($\&\to$) yields the **IS** conclusion $\varphi \& \psi, \Gamma \to \Delta$. Hence, **CS** includes **IS**, and proves TND and is thus complete for classical logic.

The system **CS** is treated in Kleene [1967] Chapter VI, where, in particular, an elegant completeness-proof is given, using the technique of treating the sequent as expressing a falsifiability condition and applying the sequent rules backwards one searches for a counter-model effecting the falsifying interpretation. The whole treatment is very systematic and leads to a canonical search method. We will not enter into the details of the proof for the present calculus but postpone the matter until we discuss the next system to be considered, viz. a 'Tableaux system' for *signed formulae*. Before we discuss this, however, we wish to consider one more, fairly complicated, derivation in the system **CS**. For this we choose to derive in **CS** the schema CD – 'constant domains', so called because it is valid in Kripke models with constant domains of individuals only, cf. Van Dalen [Chapter III.5] – which is not **IS**-provable.

$$CD =_{def} \forall x(\varphi \lor \psi) \supset \varphi \lor \forall x\psi, \text{ where } x \text{ does not occur in } \varphi.$$

We derive CD as follows: Pick a parameter a which does not occur in φ.

$$
\begin{array}{ll}
(\lor\to) & \dfrac{\varphi, \forall x(\varphi \lor \psi) \to \varphi, \psi_a^x \qquad \psi_a^x, \forall x(\varphi \lor \psi) \to \varphi, \psi_a^x}{\varphi \lor \psi_a^x, \forall x(\varphi \lor \psi) \to \varphi, \psi_a^x} \\
(\forall\to) & \\
(\to\forall) & \forall x(\varphi \lor \psi) \to \varphi, \psi_a^x \\
(\to\lor) & \forall x(\varphi \lor \psi) \to \varphi, \forall x\psi \\
(\to\supset) & \forall x(\varphi \lor \psi) \to \varphi \lor \forall x\psi \\
 & \to \forall x(\varphi \lor \psi) \supset \varphi \lor \forall x\psi
\end{array}
$$

The cut-elimination theorem holds for **CS** as well by the same type of syntactic manipulations as in the case of **IS** and **IS**$^+$. The fact that the cut-free calculus is complete is very useful, because cut-free derivations have the *sub-formula property*, i.e. that every formula which occurs in the derivation is a sub-formula of some formula in the conclusion (here one has to count φ as a sub-formula of φ). This simple fact that the conclusion gives a bound on the complexity of the derivation is the source of a host of metamathematical information. We refer to Takeuti [1975] for a comprehensive survey of how to use the cut-free derivations for metamathematical purposes.

Before we leave the system **CS**, we wish to mention what is basically a notational variant due to Schütte [1951]. The semantic interpretation of a sequent $\varphi_1, \ldots, \varphi_k \to \psi_1, \ldots, \psi_m$ is given by the formula $\varphi_1 \& \cdots \& \varphi_k \supset \psi_1 \vee \cdots \vee \psi_m$, or equivalently, by classical logic, $\neg\varphi_1 \vee \cdots \vee \neg\varphi_k \vee \psi_1 \vee \cdots \vee \psi_m$. The sequent-rules of **CS** can therefore by thought of as giving rules of passage between such disjunctions. The axioms, for instance, become $\varphi \vee \psi \vee \neg\varphi$ (both sets of side formulae are joined into ψ). The rule ($\vee\to$) becomes:

$$(\neg\vee) \quad \frac{\neg\varphi \vee \theta \qquad \neg\psi \vee \theta}{\neg(\varphi \vee \psi) \vee \theta}$$

whereas ($\to\neg$) disappears: the premise has the form $\varphi, \Gamma \to \Delta$ and the conclusion $\Gamma \to \Delta, \neg\varphi$, but their 'Schütte-translations' are the same. We leave the other reformulations to the reader and just consider cut:

$$\frac{\psi \vee \varphi \qquad \neg\varphi \vee \theta}{\psi \vee \theta}$$

NB. It is customary to restrict oneself to the connectives &, $\vee$ and $\neg$ in the Schütte-style formulation.

A readily available description of a Schütte-type system can be found in Mendelson [1964] Appendix, where an exposition of Schütte's consistency proof using the infinitary ω-rule also can be found. (Unfortunately, the appendix was dropped from later editions of the book.)

Another variant of the same idea is found in Tait [1968]. We now explain the mechanism for the propositional case. With each propositional letter p we associate another propositional letter $\bar{p}$, called the *complement* of p. As we work in classical logic, we confine ourselves to using &, $\vee$ and $\neg$ only. One then also defines the complements for complex formulae:

$$\bar{\bar{p}} =_{\text{def}} p$$

$$\overline{(p \vee q)} =_{\text{def}} \bar{p} \& \bar{q}$$

$$\overline{(p \& q)} =_{\text{def}} \bar{p} \vee \bar{q}$$

The negation of a formula is identified with its complement. This has the advantage that all formulae are in negation-normal form, i.e. negation can be thought of as applying only to propositional letters.

I.2: SYSTEMS OF DEDUCTION

The sequents are finite sets of formulae; on the intended interpretation they are read as disjunctions (so one has taken one step beyond Schütte; the disjunction signs are not written out). The system of axioms and rules is particularly easy to give:

Axiom $\quad \Gamma, \varphi, \overline{\varphi}$

$(\vee)$ $\quad \dfrac{\Gamma, \varphi}{\Gamma, \varphi \vee \psi}$ and $\dfrac{\Gamma, \psi}{\Gamma, \varphi \vee \psi}$

$(\&)$ $\quad \dfrac{\Gamma, \varphi \quad \Gamma, \psi}{\Gamma, \varphi \& \psi}$

(CUT) $\quad \dfrac{\Gamma, \psi \quad \Gamma, \neg \varphi}{\Gamma}$

The rules $(\vee)$ and $(\&)$ correspond, in the first place, to $(\rightarrow \vee)$ and $(\rightarrow \&)$, respectively. But the original $(\& \rightarrow)$ rule corresponds to the inference of $\neg(\varphi \& \psi), \Gamma$ from the premise $\neg \varphi, \Gamma$, say. As one now identifies the negation of a formula with the complement of that formula, this corresponds to the inference of $\overline{\varphi \& \psi}, \Gamma$ from $\overline{\varphi}, \Gamma$, but by definition, the complement of a conjunction is the disjunction of the complements. Hence, we have an instance of the rule $(\vee)$.

The present approach thus manages to keep the number of rules at a minimum, and this can be put to meta-mathematical use. The present sort of Tait-like system is used by Schwichtenberg [1977] (for full predicate logic and various systems of arithmetic) to present Gentzen's cut elimination theorem in a very compact way. We give a derivation of CD within the Schütte-type formulation so as to bring out the essential equivalence with the original formulation:

$(\neg \vee)$
$(\neg \forall)$
$(\forall)$
$\dfrac{\dfrac{\dfrac{\neg \varphi \vee \varphi \vee \psi_a^x \quad \neg \psi_a^x \vee \varphi \vee \psi_a^x}{\neg(\varphi \vee \psi_a^x) \vee \varphi \vee \psi_a^x}}{\neg \forall x(\varphi \vee \psi) \vee \varphi \vee \psi_a^x}}{\neg \forall x(\varphi \vee \psi) \vee \varphi \vee \forall x \psi}$

The rest of the steps fall away since implication is no longer a connective.

We conclude our treatment of the systems **IS** and **CS** (and their notational variants) by drawing the reader's attention to the survey article Bernays [1965] in which a wealth of information as to various options in formulating the rules and systems is systematically presented. In particular, the connections between general consequence relations and conditions on sequents are examined. The reader interested in this area should also consult the series of papers by Scott, cf. Scott [1974] and other references given therein.

There now remains only the last of the main variants of the sequent calculus, namely systems of *semantic tableaux*. These systems arose out of the completeness proofs for the cut-free Sequent Calculus which were independently discovered in the mid-fifties by Beth, Hintikka, Kanger and Schütte. (Cf. Prawitz [1975] for historical references and a lucid exposition of the completeness proof in question. Another good reference is, as already remarked, Kleene [1967] Chapter VI, p. 285.) The method of tableaux can be applied both in the intuitionistic and classical settings, although the former strikes the author as being a bit artificial. Here we confine ourselves to the classical case, for which the *locus classicus* is Smullyan [1968] and whose presentation we follow. For the intuitionistic case refer, e.g., to Fitting [1969], or Bell and Machover [1977] (another good source for information about tableaux).

We first need a

DEFINITION. If φ is a wff, then $T\varphi$ and $F\varphi$ are both *signed formulae*.

Note that signed formulae cannot be joined together by connectives etc. to form other signed formulae; their intended interpretations are, of course, 'φ is true' and 'φ is false', respectively.

The main idea is now to take the falsifiability condition interpretation of **CS** seriously. Thus, the *sequent* $\{\varphi_1, \ldots, \varphi_k\} \to \{\psi_1, \ldots, \psi_m\}$ is now transformed into a *finite set of signed formulae* $T\varphi_1, \ldots, T\varphi_k, F\psi_1, \ldots, F\psi_m$ and a further change is that the **CS** rules are *turned up-side-down*. One therefore treats a derivation as a systematic search for a falsifying interpretation, and the branching rules, which in **CS** have got more than one premise, now indicate that a falsifying interpretation has to be sought along different possibilities.

We begin by presenting the derivation of CD (before we give the rules): Consider CD; we wish to prove it classically so we wish to show the absence of a falsifying interpretation. Therefore, we begin by assuming that

$$F(\forall x(\varphi \vee \psi) \supset \varphi \vee \forall x\psi)$$

A necessary (and sufficient) condition for this to be the case, is:

$$T\forall x(\varphi \vee \psi), \quad F(\varphi \vee \forall x\psi)$$

I.2: SYSTEMS OF DEDUCTION

because an implication is false precisely when its antecedent is true and the consequent false. It is a necessary and sufficient condition for this to hold that

$$T\forall x(\varphi \vee \psi), \quad F\varphi, \quad F\forall x\psi$$

because a disjunction is false precisely when both disjuncts are false.

But in order to falsify $\forall x\psi$, we must falsify ψ_a^x for some a, *about which we have assumed nothing in particular* up till now, so:

$$T\forall x(\varphi \vee \psi), \quad F\varphi, \quad F\psi_a^x$$

is a falsifiability condition for CD. Now, in order to make $\forall x(\varphi \vee \psi)$ true we also need to make, among other instances, $\varphi \vee \psi_a^x$ true. Thus,

$$T\forall x(\varphi \vee \psi), \quad T(\varphi \vee \psi_a^x), \quad F\varphi, \quad F\psi_a^x \quad \text{(Here we use that } x \text{ does not occur in } \varphi.)$$

is a falsifiability condition for CD. Now, the condition splits into two possibilities for falsifying CD, as in order to make a disjunction true it is sufficient to make only one of the disjuncts true:

(1st possibility): $\quad T\forall x(\varphi \vee \psi), \quad T\varphi, \quad F\varphi, \quad F\psi_a^x$.

This possibility will not yield a falsifying interpretation though, because we cannot assign both the value true and the value false to the same wff. Hence, the search along this possibility may safely be abandoned or *closed*.

(2nd possibility): $\quad T\forall x(\varphi \vee \psi), \quad T\psi_a^x, \quad F\varphi, \quad F\psi_a^x$.

Also this possibility has to be closed, because ψ_a^x has to be made true and false. (It will not have escaped the attentive reader that the closure conditions correspond exactly to the *axioms* of **CS**.) But we have now exhausted *all* the possibilities for finding a falsifying interpretation; thus CD must be classically valid as none of the possible routes along which we could hope for a falsifying interpretation is able to yield one.

The above search can be set out more compactly:

$$\frac{\frac{\frac{\frac{\frac{F(\forall x(\varphi \vee \psi) \supset \varphi \vee \forall x\psi)}{T\forall x(\varphi \vee \psi), F\varphi \vee \forall x\psi}}{T\forall x(\varphi \vee \psi), F\varphi, F\forall x\psi}}{T\forall x(\varphi \vee \psi), F\varphi, F\psi_a^x}}{T\forall x(\varphi \vee \psi), T\varphi \vee \psi_a^x, F\varphi, F\psi_a^x}}{T\forall x(\varphi \vee \psi), T\varphi, F\varphi, F\psi_a^x \mid T\forall x(\varphi \vee \psi), T\psi_a^x, F\varphi, F\psi_a^x}$$

This search tree, however, is nothing but the **CS** derivation of CD turned up-side-down and rewritten using other notation.

We now give the rules for the tableaux system **T**. We will use 'S' as a notation for finite sets of signed formulae.

$F\&$ $\quad \dfrac{S, F\varphi \& \psi}{S, F\varphi \mid S, F\psi}$

$T\&$ $\quad \dfrac{S, T\varphi \& \psi}{S, T\varphi, T\psi}$

$F\vee$ $\quad \dfrac{S, F\varphi \vee \psi}{S, F\varphi, F\psi}$

$T\vee$ $\quad \dfrac{S, T\varphi \vee \psi}{S, T\varphi \mid S, T\psi}$

$F\supset$ $\quad \dfrac{S, F\varphi \supset \psi}{S, T\varphi, F\psi}$

$T\supset$ $\quad \dfrac{S, T\varphi \supset \psi}{S, T\psi \mid S, F\varphi}$

$F\neg$ $\quad \dfrac{S, F\neg\varphi}{S, T\varphi}$

$T\neg$ $\quad \dfrac{S, T\neg\varphi}{S, F\varphi}$

$F\forall$ $\quad \dfrac{S, F\forall x\varphi}{S, F\varphi_a^x}$ provided that a does not occur in S.

$T\forall$ $\quad \dfrac{S, T\forall x\varphi}{S, T\forall x\varphi, T\varphi_t^x}$

$F\exists$ $\quad \dfrac{S, F\exists x\varphi}{S, F\exists x\varphi, F\varphi_t^x}$

$T\exists$ $\quad \dfrac{S, T\exists x\varphi}{S, T\varphi_a^x}$ provided that a does not occur in S.

Some of the rules give a *branching*, viz $T\vee$, $F\&$ and $T\supset$. This is because of the fact that in these cases there is more than just one way of fulfilling the falsifiability condition above the line and all of these must be investigated.

A tableaux proof for φ is a finite tree of finite sets of signed formulae, regulated by the above rules such that (i) the top node of the tree is the set $\{F\varphi\}$ and (ii) every branch in the tree is *closed*, i.e. ends with a set of the form S, $T\varphi$, $F\varphi$. Hence, the tableaux proofs should best be regarded as failed attempts at the construction of a counter-model.

We also show how one can define a counter-model to the top node set from an open branch in the tableaux, where one cannot apply the rules any longer.

Consider the propositional wff $(p \supset q) \supset (p \& q)$. This is not a tautology, and, hence, it is not provable. An attempt at a tableaux proof looks like:

$$\frac{F(p \supset q) \supset (p \& q)}{Tp \supset q, Fp \& q}$$

$$\frac{Tp \supset q, Fp \quad | \quad Tp \supset q, Fq}{Fp, Fp \mid Tq, Fp \quad Fp, Fq \mid Tq, Fq}$$

The last of the branches in the search tree is closed, but from the third, say, we can define a valuation v by

$$v(p) = F \quad \text{and} \quad v(q) = F.$$

Then the value of the formula $(p \supset q) \supset (p \& q)$ under v is F, so we have found our counter-model. This simple example contains the germ of the completeness proof for tableaux proofs; one describes a systematic procedure for generating a search tree (in particular, for how to choose the *eigen*-parameters) such that either the procedure breaks off in a tableaux proof or it provides a counter-model. For the details we refer to Smullyan [1968].

The above formulation, using finite sets of signed formulae, was used mainly to show the equivalence with the classical Sequent Calculus **CS**. It is less convenient in actual practice, as one must duplicate the side formulae in S all the time. A more convenient arrangement of the tableaux consists of using trees where the nodes are not sets of signed formulae but just one signed formula. To transform a 'set' tableaux by a more convenient tableaux one takes the top node which has the form, say, $\{T\varphi_1, \ldots, T\varphi_k, F\psi_1, \ldots, F\psi_m\}$ and places it upright instead:

$T\varphi_1$
$\vdots$
$T\varphi_k$
$F\psi_1$
$\vdots$
$F\psi_m$.

Such an arrangement can then be continued by use of the rules:

$$\begin{array}{c} F\varphi \& \psi \\ \diagup \quad \diagdown \\ F\varphi \quad F\psi \end{array} \qquad \begin{array}{c} T\varphi \& \psi \\ \hline T\varphi \\ T\psi \end{array}$$

$$\begin{array}{c} F\varphi \vee \psi \\ \hline F\varphi \\ F\psi \end{array} \qquad \begin{array}{c} T\varphi \vee \psi \\ \diagup \quad \diagdown \\ T\varphi \quad T\psi \end{array}$$

$$\begin{array}{c} F\varphi \supset \psi \\ \hline T\varphi \\ F\psi \end{array} \qquad \begin{array}{c} T\varphi \supset \psi \\ \diagup \quad \diagdown \\ T\psi \quad F\varphi \end{array}$$

$$\begin{array}{c} F\neg\varphi \\ \hline T\varphi \end{array} \qquad \begin{array}{c} T\neg\varphi \\ \hline F\varphi \end{array}$$

$\dfrac{F\forall x\varphi}{F\varphi^x_a}$ provided that a is new to the branch $\qquad \dfrac{T\forall x\varphi}{T\varphi^x_t}$

$\dfrac{F\exists x\varphi}{F\varphi^x_t} \qquad \dfrac{T\exists x\varphi}{T\varphi^x_a}$ provided that a is new to the branch

We do not wish to trouble the reader with a more detailed description of the modified system, and so we confine ourselves to showing how the rules are used in practice for proving CD:

(1) $\quad F\forall x(\varphi \vee \psi) \supset \varphi \vee \forall x\varphi$

(2) $\quad T\forall x(\varphi \vee \psi)$

(3) $\quad F\varphi \vee \forall x\psi$

(4) $\quad F\varphi$

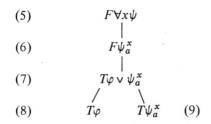

(5) $F\forall x\psi$

(6) $F\psi_a^x$

(7) $T\varphi \vee \psi_a^x$

(8) $T\varphi$ $T\psi_a^x$ (9)

The tableaux proof begins with the signed formula FCD, as we try to show that there is no counter-model to CD. Lines (2) and (3) result from (1) by breaking down the implication. Lines (4) and (5) likewise result from (3) and, as the parameter a is new to the (only) branch the step from (5) to (6) is permitted. Line (7) results from (2) as we have the right to choose any term here. Finally, (8) and (9) come from (7) by breaking up a true disjunction. Both branches are closed at once; (8) closes off against (4) and (9) against (6).

Although this is very simple, there is yet another simplification which can be performed: *drop the signed formulae altogether* and in place of '$F\varphi$' write '$\neg\varphi$' and in place of '$T\varphi$' write 'φ'. The resulting rules look like:

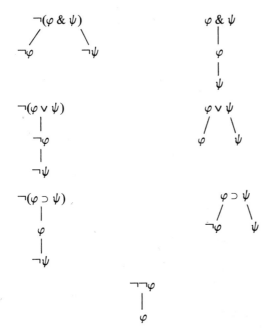

$$\neg \forall x \varphi$$
$$|$$
$$\neg \varphi_a^x$$
provided that a is new

$$\forall x \varphi$$
$$|$$
$$\varphi_t^x$$

$$\neg \exists x \varphi$$
$$|$$
$$\neg \varphi_t^x$$

$$\exists x \varphi$$
$$|$$
$$\varphi_a^x$$
provided that a is new.

We consider our stock example once again:

$$\neg (\forall x(\varphi \vee \psi) \supset (\varphi \vee \forall x\psi))$$
$$|$$
$$\forall x(\varphi \vee \psi)$$
$$|$$
$$\neg(\varphi \vee \forall x\psi)$$
$$|$$
$$\neg \varphi$$
$$|$$
$$\neg \forall x\psi$$
$$|$$
$$\neg \psi_a^x$$
$$|$$
$$\varphi \vee \psi_a^x$$
$$/ \quad \backslash$$
$$\varphi \qquad \psi_a^x$$

The elucidating remarks after the previous example apply almost word for word also here. There remains only to point out that this last style of 'unsigned' tableaux is used in readily available elementary texts, e.g. Jeffrey [1967] and Hodges [1977].

The order of presentation used in the present chapter was also used by the author in those sections of Scott *et al.* [1981], Volume II which were drafted by him. Those notes, Volumes I and II, were intended as a supplement to Hodges [1977] and, in particular, Volume I contains a very detailed development of a tableaux system.

ACKNOWLEDGEMENT

I wish to thank my colleague Ton Weijters who read the manuscript of the present chapter and offered helpful comments.

Katholieke Universiteit, Nijmegen

REFERENCES

Bell, J. and Machover, M.: 1977, *A Course in Mathematical Logic*, North-Holland, Amsterdam.
Bernays, P.: 1965, 'Betrachtungen zum Sequenzen-Kalkül', in A.-T. Tymieniecka (ed.), *Logic and Methodology*, North-Holland, Amsterdam, pp. 1-44.
Church, A.: 1956, *Introduction to Mathematical Logic*, Vol. 1, Princeton U.P., Princeton.
Van Dalen, D.: 1980, *Logic and Structure*, Springer, Berlin.
Dummett, M.: 1977, *Elements of Intuitionism*, Oxford U.P., Oxford.
Enderton, H. B.: 1972, *A Mathematical Introduction to Logic*, Academic Press, New York.
Feferman, S.: 1960, 'Arithmetization of metamathematics in a general setting', *Fundamenta Math.* 49, 35-92.
Fitting, M.: 1969, *Intuitionistic Logic, Model Theory and Forcing*, North-Holland, Amsterdam.
Frege, G.: 1879, *Begriffsschrift*, Nebert, Halle. Complete English translation in Van Heijenoort [1967].
Gentzen, G.: 1934, 'Untersuchungen über das logische Schliessen', *Math. Zeitschrift* 39, 176-210, 405-431. Complete English translation in Szabo [1969].
Gentzen, G.: 1936, 'Die Widerspruchsfreiheit der reinen Zahlentheorie', *Math. Annalen* 112, 493-565. Complete English translation in Szabo [1969].
Van Heijenoort, J.: 1967, *From Frege to Gödel*, Harvard U.P., Cambridge, Mass.
Hilbert, D. and Bernays, P.: 1934, *Grundlagen der Mathematik*, Vol. 1, Springer, Berlin.
Hodges, W.: 1977, *Logic*, Penguin, Harmondsworth.
Jeffrey, R. C.: 1967, *Formal Logic: Its Scope and Limits*, McGraw-Hill, New York.
Kalish, D. and Montague, R.: 1965, 'On Tarski's formalization of predicate logic with identity', *Archiv für mathematische Logik und Grundlagenforschung* 7, 81-101.
Kleene, S. C.: 1952, *Introduction to Metamathematics*, North-Holland, Amsterdam.
Kleene, S. C.: 1967, *Mathematical Logic*, John Wiley, New York.
Kripke, S.: 1963, 'Semantical considerations on modal logic', *Acta Philosophica Fennica* 16, 83-94.
Leivant, D.: 1979, 'Assumption classes in natural deduction', *Zeitschrift für mathematische Logik und Grundlagen der Mathematik* 25, 1-4.
Martin-Löf, P.: 1971, 'Hauptsatz for the intuitionistic theory of iterated inductive definitions' in J. E. Fenstad (ed.), *Proceedings of the Second Scandinavian Logic Symposium*, North-Holland, Amsterdam, pp. 179-216.
Mendelson, E.: 1964, *Introduction to Mathematical Logic*, Van Nostrand, New York.
Monk, D.: 1965, 'Substitutionless predicate logic with identity', *Archiv für mathematische Logik und Grundlagenforschung* 7, 102-121.
Pottinger, G.: 1979, 'Normalization as a homomorphic image of cut-elimination', *Annals of Math. Logic* 12, 323-357.
Prawitz, D.: 1965, *Natural Deduction*, Almqvist & Wiksell, Uppsala.
Prawitz, D.: 1971, 'Ideas and results in proof theory', in J. E. Fenstad (ed.), *Proceedings of the Second Scandinavian Logic Symposium*, North-Holland, Amsterdam, pp. 235-307.
Prawitz, D.: 1975, 'Comments on Gentzen-style procedures and the classical notion of

truth', in J. Diller and G. H. Müller (eds.), *Proof Theory Symposium*, Lecture Notes in Mathematics 500, Springer, Berlin pp. 290-319.

Prawitz, D.: 1977, 'Meaning of proofs', *Theoria* 43, 2-40.

Scott, D. S.: 1974, 'Rules and derived rules', in S. Stenlund (ed.), *Logical Theory and Semantical Analysis*, Reidel, Dordrecht, pp. 147-161.

Scott, D., Bostock, D., Forbes, G., Isaacson, D., and Sundholm, G.: 1981, *Notes on the Formalization of Logic*, Vols. I, II, Sub-Faculty of Philosophy, Oxford.

Schütte, K.: 1951, 'Schlussweisen-Kalküle der Prädikatenlogik', *Math. Annalen* 122, 47-65.

Schwichtenberg, H.: 1977, 'Proof theory: Some applications of cut-elimination', in J. Barwise (ed.), *Handbook of Mathematical Logic*, North-Holland, Amsterdam, pp. 867-896.

Smoryński, C.: 1977, 'The incompleteness theorems', in J. Barwise (ed.), *Handbook of Mathematical Logic*, North-Holland, Amsterdam, pp. 821-865.

Smullyan, R.: 1968, *First-order Logic*, Springer, Berlin.

Szabo, M.: 1969, *The Collected Papers of Gerhard Gentzen*, North-Holland, Amsterdam.

Tait, W. W.: 1968, 'Normal derivability in classical logic', in J. Barwise (ed.), *The Syntax and Semantics of Infinitary Languages*, Lecture Notes in Mathematics 72, Springer, Berlin, pp. 204-236.

Takeuti, G.: 1975, *Proof Theory*, North-Holland, Amsterdam.

Tarski, A.: 1965, 'A simplified formulation of predicate logic with identity', *Archiv für mathematische Logik und Grundlagenforschung* 7, 61-79.

Tennant, N.: 1978, *Natural Logic*, Edinburgh U.P., Edinburgh.

Troelstra, A. S. (ed.): 1973, *Metamathematical Investigations of Intuitionistic Arithmetic and Analysis*, Lecture Notes in Mathematics 344, Springer, Berlin.

Zucker, J.: 1974, 'The correspondence between cut-elimination and normalization', *Annals of Math. Logic* 7, 1-112.

CHAPTER I.3

ALTERNATIVES TO STANDARD FIRST-ORDER SEMANTICS[1]

by HUGUES LEBLANC

0. Introduction	189
1. First-order syntax and standard semantics	191
2. Substitution in standard semantics and substitutional semantics	198
3. Truth-value semantics	209
4. Probabilistic semantics (I)	225
5. Probabilistic semantics (II)	240
6. In summary and conclusion	258
Appendix	261
Notes	265
References	270

0. INTRODUCTION

Alternatives to standard semantics are legion, some even antedating standard semantics. I shall study several here, among them: *substitutional semantics, truth-value semantics*, and *probabilistic semantics*. All three interpret the quantifiers *substitutionally*, i.e. all three rate a universal (an existential) quantification true if, and only if, every one (at least one) of its substitution instances is true.[2] As a result, the first, which retains models, retains only those which are to be called *Henkin models*. The other two dispense with models entirely, truth-value semantics using instead truth-value assignments (or equivalents thereof to be called *truth-value functions*) and probabilistic semantics using probability functions. So reference, central to standard semantics, is no concern at all of truth-value and probabilistic semantics; and truth, also central to standard semantics, is but a marginal concern of probabilistic semantics.

Each of these alternatives to standard semantics explicates logical entailment – and, hence, logical truth – in its own way.[3] In all three cases, however, it can and will be shown that

(i) *A statement is logically entailed by a set of statements if, and only if, provable from the set,*

and hence that

(ii)   *A statement is logically true if, and only if, provable.*

Statement (i) will legitimize each account of logical entailment, (ii) will legitimize each account of logical truth, and (i) and (ii) together will legitimize each alternative semantics treated here.

I had two options: presenting each of substitutional semantics, truth-value semantics, and probabilistic semantics as a semantics *for* **CQC**, *the first-order quantificational calculus*, or as one *for an arbitrary first-order language* L. I chose the latter option for a simple reason. This essay deals largely with truth and probability, and to me truth and probability (the latter understood as a degree of rational belief) are features of the statements you meet in a language rather than the statement forms you meet in such a language form as **CQC**.

For use in alternatives to standard semantics, though *not* in standard semantics itself, I outfit L with term extensions. One of them, L^∞, will serve to prove various completeness theorems in Sections 2–5. I could, instead, have outfitted each set of statements of L with term rewrites, which was the practice in Leblanc [1976] and earlier writings of mine. However, term extensions, exploited in Dunn and Belnap [1968], are handier and admittedly more natural. So I switched to them.[4]

Several matters are studied below, and in the process are bound together. I define for each semantics the notions of logical truth and logical entailment, plus, of course, such notions as they presuppose; and, as announced, I justify the definitions by showing that of logical entailment *strongly* – and, hence, that of logical truth *weakly* – sound and complete. I further show that (i) substitutional semantics is a by-product of standard semantics, (ii) truth-value semantics is substitutional semantics done without models, and (iii) probabilistic semantics is a generalization of truth-value semantics. While doing truth-value semantics, I also study truth-set semantics, a transcription of truth-value semantics into the idiom of sets, and *model-set semantics*, an intriguing and handy variant by Hintikka of truth-set semantics. And, occasionally in the main text, but more often in the Notes and the Appendix, I supply pertinent names, dates, and references.

However, there is far more to truth-value and probabilistic semantics than the essay conveys. (i) When studying probabilistic semantics, I pay particular attention to *singulary* (i.e. one-argument) probability functions. *Binary* (i.e. two-argument) ones also permit definition of logical truth and logical entailment, as writers from Popper on have shown. They receive some attention here, but deserve far more. (ii) When studying truth-value and probabilistic

I.3: ALTERNATIVES TO FIRST-ORDER SEMANTICS 191

semantics, I largely restrict myself to matters of logical truth, logical entailment, soundness, and completeness. But, as Leblanc [1976] attests, a host of definitions, theorems, and proofs from standard semantics translate into the idiom of truth-values; and many – though, for sure, not all – of them translate as well into the idiom of probabilities. The translations, sometimes easy to come by but sometimes not, should figure in any full-fledged treatment of either semantics. (iii) I restrict myself throughout the essay to *first-order logic without identity*. Yet other logics (first-order logic with identity, higher-order logics, many-valued logics, modal logics, tense logics, intuitionistic logic, conditional logic, etc.) have also been supplied with a truth-value semantics or a probabilistic one. Only study of these extensions of and alternatives to elementary logic would reveal the true scope of either semantics.

The alternatives to standard semantics studied here have, in my opinion, considerable interest and – possibly – merit. As noted above, they are *frugal*, (i) substitutional semantics discarding all but Henkin models, while truth-value semantics and probabilistic semantics discard all models, and (ii) truth-value-semantics – though it retains the notion of truth – discarding that of reference, while probabilistic semantics discards both notions. And they are *innovative*, truth-value semantics assigning truth-values to the atomic 'substatements' of all statements (those with quantifiers as well as those without), while probabilistic semantics assigns to statements degrees of credibility rather than truth-values.

I touch on these matters in Sections 2–5, and devote much of Section 6 to them. My main concern, though, is different: to provide the formal prerequisites to further study of and research in non-standard semantics.

1. FIRST-ORDER SYNTAX AND STANDARD SEMANTICS

The *primitive signs* of L will be (i) one or more predicates, each identified as being of a certain degree d ($d \geq 1$), (ii) $\aleph_0$ individual terms, presumed to come in some order known as their *alphabetic order*, (iii) $\aleph_0$ individual variables, (iv) the three logical operators '¬', '∧', and '∀', (v) the two parentheses '(' and ')', and (vi) the comma ','. The *formulas* of L will be all the finite sequences of the primitive signs of L. I shall refer to the predicates of L by means of 'Q'; to its individual terms in general by means of 'T', and – for each i from 1 on – to its alphabetically ith individual term by means of 't_i'; to its individual variables by means of 'x'; to its individual *signs* (i.e. its individual terms and individual variables) by means of 'I'; to its formulas by means of 'A', 'B', and 'C'; and to sets of its formulas by means of 'S'. And,

(i) A being a formula of L, (ii) $I_1, I_2, \ldots, I_n$ ($n \geq 0$) being distinct individual signs of L, and (iii) $I'_1, I'_2, \ldots, I'_n$ being individual signs of L not necessarily distinct from one another nor from $I_1, I_2, \ldots, I_n$, I shall refer by means of '$((A)(I'_1, I'_2, \ldots, I'_n/I_1, I_2, \ldots, I_n))$' to the result of simultaneously putting I'_1 everywhere in A for I_1, I'_2 for I_2, ..., I'_n for I_n. (When clarity permits I shall omit some of the parentheses in '$((A)(I'_1, I'_2, \ldots, I'_n/I_1, I_2, \ldots, I_n))$', thus writing '$(A)(I'_1, I'_2, \ldots, I'_n/I_1, I_2, \ldots, I_n)$', '$(A(I'_1, I'_2, \ldots, I'_n/I_1, I_2, \ldots, I_n))$', and '$A(I'_1, I'_2, \ldots, I'_n/I_1, I_2, \ldots, I_n)$'.)

The *statements* of L will be all formulas of L of the following sorts: (i) $Q(T_1, T_2, \ldots, T_d)$, where Q is a predicate of L of degree d ($d \geq 1$) and $T_1, T_2, \ldots, T_d$ are (not necessarily distinct) individual terms of L, (ii) $\neg A$, where A is a statement of L, (iii) $(A \wedge B)$, where A and B are (not necessarily distinct) statements of L, and (iv) $(\forall x)A$, where $A(T/x)$ – T here being any term of L you please – is a statement of L.[5] As usual, the statements in (i) will be called *atomic*, and the rest *compound*, those in (ii) being *negations*, (iii) *conjunctions*, and (iv) (*universal*) *quantifications*. Statements that contain no individual terms will be called *termless* and statements that contain no '$\forall$' will be called *quantifierless*. At a few points the statements of L will be presumed to come in some definite order, to be known as their *alphabetic order* (that on pp. 9–10 of Leblanc [1976] would do). For brevity's sake, I shall write '$(A \rightarrow B)$' for '$\neg(A \wedge \neg B)$', '$(A \vee B)$' for '$\neg(\neg A \wedge \neg B)$', '$(A \leftrightarrow B)$' for '$(\neg(A \wedge \neg B) \wedge \neg(B \wedge \neg A))$', '$(\exists x)A$' for '$\neg(\forall x)\neg A$', and '$\Pi_{i=1}^{n} A_i$' for '$((\ldots (A_1 \wedge A_2) \wedge \ldots) \wedge A_n)$'.[6] I shall also drop outer parentheses whenever clarity permits, and I shall talk of terms and variables rather than individual terms and individual variables.

Substatements will behave like Gentzen's subformulas. (i) A statement of L will count as one of its substatements; (ii) A will count as a substatement of a negation $\neg A$ of L, each of A and B as a substatement of a conjunction $A \wedge B$ of L, and $A(T/x)$ for each term T of L as a substatement of a quantification $(\forall x)A$ of L; and (iii) the substatements of any substatement of a statement of L will count as substatements of that statement. The *substatements of a set of statements* of L will be the substatements of the various members of the set. As for the *atomic substatements* of a statement or set of statements of L, they will, of course, be those among its substatements that are atomic. When further precision is needed, I shall refer to the foregoing as the substatements *in* L of a statement or set of statements of L.[7]

The substatements $A(t_1/x)$, $A(t_2/x)$, $A(t_3/x)$, etc., of a universal quantification $(\forall x)A$ – and, by extension, of an existential one $(\exists x)A$ – of L are what I called above the *substitution instances* (*in* L) of the quantification.

I.3: ALTERNATIVES TO FIRST-ORDER SEMANTICS 193

As suggested, they play a critical role in substitutional, truth-value, and probabilistic semantics. (Note that when a quantification $(\forall x)A$ of L is vacuous, it has but one substitution instance, A itself.)

The length $l(A)$ of an atomic statement A of L will be 1; that, $l(\neg A)$, of a negation $\neg A$ of L will be $l(A) + 1$; that, $l(A \wedge B)$, of a conjunction $A \wedge B$ of L will be $l(A) + l(B) + 1$; and that, $l((\forall x)A)$, of a quantification $(\forall x)A$ of L will be $l(A(T/x)) + 1$, where T is any term of L you please.

Lastly, a term will be said to be *foreign* to a statement A of L if it does not occur in A; to *occur* in a set S of statements of L if it occurs in at least one member of S; and to be *foreign* to S if it is foreign to each member of S. And S will be held *infinitely extendible* in L if $\aleph_0$ terms of L are foreign to S.

Borrowing from Quine [1940], Fitch [1948], Rosser [1953], and Leblanc [1979a], I shall take (i) the *axioms* of L to be all the statements of L of the following sorts:

A1. $A \rightarrow (A \wedge A)$
A2. $(A \wedge B) \rightarrow A$
A3. $(A \rightarrow B) \rightarrow (\neg(B \wedge C) \rightarrow \neg(C \wedge A))$
A4. $A \rightarrow (\forall x)A$
A5. $(\forall x)A \rightarrow A(T/x)$
A6. $(\forall x)(A \rightarrow B) \rightarrow ((\forall x)A \rightarrow (\forall x)B)$,

plus all those of the sort $(\forall x)(A(x/T))$, where A is an axiom of L, and (ii) the *ponential* of two statements A and $A \rightarrow B$ of L to be B. (Note as regards **A4** that, with $A \rightarrow (\forall x)A$ presumed here to be a statement, x is sure not to occur in A, and hence $(\forall x)A$ is sure to be a vacuous quantification.)

It follows from this account of an axiom that:

THEOREM 1.1. *Every axiom of L is of the sort*

$$(\forall x_1)(\forall x_2) \ldots (\forall x_n)(A(x_1, x_2, \ldots, x_n/T_1, T_2, \ldots, T_n)),$$

where $n \geq 0$ *and A is one of the six sorts* **A1-A6**;

and

THEOREM 1.2. *If* $(\forall x_1)(\forall x_2) \ldots (\forall x_n)(A(x_1, x_2, \ldots, x_n/T_1, T_2, \ldots, T_n))$ *is an axiom of* L, *so for each i from 1 on is* $((\forall x_2) \ldots (\forall x_n)(A(x_1, x_2, \ldots, x_n/T_1, T_2, \ldots, T_n)))(t_i/x_1)$.

By a *proof* in L of a statement A of L from a set S of statements of L, I shall understand any finite column of statements of L such that (i) each entry in the column is a member of S, an axiom of L, or the ponential of two earlier entries in the column and (ii) the last entry in the column is A. I shall say that a statement A of L is *provable* in L from a set S of statements of L – $S \vdash A$, for short – if there is a proof in L of A from S. I shall say that a statement A of L is *provable* in L – $\vdash A$, for short – if $\emptyset \vdash A$. (When all the axioms that turn up in a proof in L of a statement A of L are of the sorts **A1–A3**, I shall say that A is *provable in L by means of just* **A1–A3**, and write '$\vdash_0 A$' in place of '$\vdash A$'.) And I shall say of a set S of statements of L, (i) when each statement of L provable in L from S belongs to S, that S is *deductively closed* in L or constitutes a *theory* of L (*in Tarski's sense*),[8] (ii) when there is no (there is a) statement A of L such that $S \vdash A \wedge \neg A$, that S is *(in)consistent* in L; (iii) when S is consistent in L and – for any statement A of L not in S – $S \cup \{A\}$ is inconsistent in L, that S is *maximally consistent* in L, and (iv) when – no matter the quantification $(\forall x)A$ of L – $S \vdash (\forall x)A$ if $S \vdash A(T/x)$ for each term T of L, that S is ω-*complete* in L.

Deductively closed sets (i.e. theories) will make one appearance below. Consistent, maximally consistent, and ω-complete sets, on the other hand, will turn up everywhere and, hence, immediately rate a few extra words.

When a set S of statements of L is consistent and infinitely extendible in L, there is a way of extending S to a set – called here the *Henkin extension* $\mathscr{H}(S)$ of S in L – that is maximally consistent and ω-complete in L. The method, due essentially to Henkin, is as follows:

(i) let S_0 be S itself,

(ii) A_n being for each n from 1 on the alphabetically nth statement of L,

let S_n be $\begin{cases} S_{n-1} \text{ if } S_{n-1} \cup \{A_n\} \text{ is inconsistent in L} \\ S_{n-1} \cup \{A_n\} \text{ if } S_{n-1} \cup \{A_n\} \text{ is consistent in L and } A_n \text{ is } \\ \quad \textit{not} \text{ a negated universal quantification of L} \\ S_{n-1} \cup \{A_n, \neg B(T/x)\}, \text{ where } T \text{ is the alphabetically} \\ \quad \text{earliest term of L foreign to } S_{n-1} \cup \{A_n\}, \textit{if } S_{n-1} \cup \\ \quad \{A_n\} \text{ is consistent in L and } A_n \textit{ is } \text{a negated quanti-} \\ \quad \text{fication } \neg(\forall x)B \text{ of L,}[9] \end{cases}$

and

(iii) let $\mathscr{H}(S)$ be $\bigcup_{i=0}^{\infty} S_i$

(Henkin's original instructions for extending a set appeared in Henkin [1949], a seminal paper for substitutional and truth-value semantics. I avail myself here of simplifications to those instructions due to Hasenjaeger and Henkin

I.3: ALTERNATIVES TO FIRST-ORDER SEMANTICS

himself – see Smullyan [1968] pp. 93–97 on this matter. Lindenbaum had already shown in the late Twenties how to extend a consistent set into a maximal one, thus paving the way for Henkin's result – see Tarski [1930] Section 7.)

Proofs of the following theorems are in numerous texts (e.g. Leblanc [1976] pp. 38–40) and will be taken for granted. (To abridge things I write 'iff' for 'if, and only if'.)

THEOREM 1.3. *If a set S of statements of L is maximally consistent in L, then (i) a negation $\neg A$ of L belongs to S iff A does not, and (ii) a conjunction $A \wedge B$ of L belongs to S iff each of A and B does. Further, if S is ω-complete in L as well, then (iii) a quantification $(\forall x)A$ of L belongs to S iff each substitution instance $A(T/x)$ of $(\forall x)A$ in L does.*

THEOREM 1.4. *If a set S of statements of L is consistent and infinitely extendible in L, then the Henkin extension $\mathcal{H}(S)$ of S in L is maximally consistent and ω-complete in L (= Henkin's Extension Lemma).*

With Dunn and Belnap I shall understand by a *term extension* of L any language that is exactly like L except for having *countably* many terms – i.e. finitely many or $\aleph_0$ many terms – besides those of L. Note that by this definition L – having zero and, hence, finitely many terms besides its own – is one of its term extensions. I shall refer to the term extensions of L by means of 'L^+'. And, L^+ being an arbitrary term extension of L, I shall assume that the terms of L^+ come in some *alphabetic order*, refer to the alphabetically ith ($i = 1, 2, 3, \ldots$) of them by means of 't_i^+', and write '$S \vdash^+ A$' for 'A is provable in L^+ from S'.

It will prove convenient, when $(\forall x)A$ is a quantification of L, to talk of the substitution instances of $(\forall x)A$ *in any term extension* L^+ *of L one pleases*, not just in L itself. These will of course be

$$A(t_1^+/x), A(t_2^+/x), A(t_3^+/x), \ldots$$

a list which in the case that L^+ is L boils down to $A(t_1/x), A(t_2/x), A(t_3/x), \ldots$, but otherwise will sport fresh entries – the substitution instances of $(\forall x)A$ *peculiar to* L^+. It will also prove convenient, when S is a set of statements of L, to talk of S as being (or failing to be) infinitely extendible, consistent, maximally consistent, and ω-complete *in any term extension* L^+ *of L one pleases*. S, for example, will be held infinitely extendible in L^+ if $\aleph_0$ terms of L^+ are foreign to S, consistent in L^+ if there is no statement A of L^+ such that $S \vdash^+ A \wedge \neg A$, etc. And it will prove convenient, when S is as above, to talk of the Henkin extension of S *in any term extension of* L *one pleases.*

One term extension of L will play a special role below. It will have $\aleph_0$ terms besides those of L, and for that reason will be known as L^∞.

The $\aleph_0$ terms of L^∞ *peculiar* to L^∞ are foreign of course to any set of statements of L. Hence:

THEOREM 1.5. *Each set of statements of* L *is infinitely extendible in* L^∞.

Note further that:

THEOREM 1.6. *If* $S \vdash^\infty A$, *where* S *is a set of statements of* L *and* A *a statement of* L, *then* $S \vdash A$,

and hence:

THEOREM 1.7. *If a set* S *of statements of* L *is consistent in* L, *then* S *is consistent in* L^∞.

For proof of Theorem 1.6, suppose the column made up of $B_1, B_2, \ldots, B_p$ constitutes a proof *in* L^∞ of A from S, and for each i from 1 through p let C_i be the result of putting t_1 (the alphabetically first term of L) for every term in B_i that is peculiar to L^∞. The column made up of $C_1, C_2, \ldots, C_p$ will constitute a proof *in* L of A from S. For proof of Theorem 1.7, suppose S is inconsistent *in* L^∞. Then by a familiar result $S \vdash^\infty A$ for every statement A of L^∞; hence, by Theorem 1.6, $S \vdash A$ for every statement A of L; and, hence, S is inconsistent *in* L. Hence, Theorem 1.7 by Contraposition.

*

Now for one kind of standard semantics that L might be oufitted with.[10]

Understand by a *domain* any non-empty set. Given a domain D, understand by a *D-interpretation* of (the terms and predicates of) L any result of assigning to each term of L a member of D and to each predicate of L of degree d ($d = 1, 2, 3, \ldots$) a subset of D^d (D^d the Cartesian product of D with itself d times); D being a domain, I_D a D-interpretation of L, and T a term of L, understand by a *T-variant* of I_D any D-interpretation of L that is like I_D except for possibly assigning to T a member of D different from that assigned by I_D; and understand by a *model* for L any pair of the sort $\langle D, I_D \rangle$, where D is a domain and I_D is a D-interpretation of L. A model

I.3: ALTERNATIVES TO FIRST-ORDER SEMANTICS 197

$\langle D, I_D \rangle$ will be termed *finite* if D is, *denumerably infinite* if D is, *countable* if D is, etc. (Many writers, Hodges among them, talk of *structures* where I talk of models.)

This done, let $\langle D, I_D \rangle$ be a model for L, A a statement of L, and S a set of statements of L. I shall say that A is *true in* $\langle D, I_D \rangle$ (for short, *true on* I_D) if (i) in the case that A is an atomic statement $Q(T_1, T_2, \ldots, T_d)$, the d-tuple $\langle I_D(T_1), I_D(T_2), \ldots, I_D(T_d) \rangle$ belongs to $I_D(Q)$, (ii) in the case that A is a negation $\neg B$, B is not true in $\langle D, I_D \rangle$, (iii) in the case that A is a conjunction $B \wedge C$, each of B and C is true in $\langle D, I_D \rangle$, and (iv) in the case that A is a quantification $(\forall x)B$, $B(T/x)$ – T here any term of L *foreign* to A – is true in $\langle D, I'_D \rangle$ for each T-variant I'_D of I_D; and I shall say that S is *true* in $\langle D, I_D \rangle$ if each member of S is. (Note: (ii)–(iv) are known and will be referred to as the *truth-conditions* for negations, conjunctions, and universal quantifications; as indicated in note 2, (iv) embodies the *objectual* interpretation of 'All'.)

This done, I shall declare a statement A of L *logically true in the standard sense* if A is true in every model for L; and, where S is a set of statements of L, I shall declare A *logically entailed by S in the standard sense* if A is true in every model for L in which S is true. Under these definitions, A is logically true in the standard sense iff logically entailed by $\emptyset$ in the standard sense, the point made in note 3.

When a statement A (a set S of statements) of L is true in a model $\langle D, I_D \rangle$, (i) A (S) is often said *to have* $\langle D, I_D \rangle$ *as a model*, and (ii) $\langle D, I_D \rangle$ is said *to be a model of A (S)*. Given the second of these locutions, A is logically entailed by S if every model of S is one of A.

Prominent in Section 2 will be Henkin D-interpretations and Henkin models. I_D will constitute a *Henkin D-interpretation* of L if *each* member of D is assigned by I_D to a term of L (more formally, if for each d in D there is a term T of L such that $I_D(T) = d$); and $\langle D, I_D \rangle$ will constitute a *Henkin model* for L if I_D is a Henkin D-interpretation of L. Since L has only $\aleph_0$ terms, the domain D must be countable in each case. So Henkin models are countable by definition. (The telling use to which Henkin D-interpretations and models were put in Henkin [1949] accounts for their names, bestowed upon them in Leblanc and Wisdom [1972].)

The foregoing definitions of a D-interpretation, a model, a Henkin D-interpretation, and a Henkin model are easily generalized to suit *any* term extension L^+ of L (rather than just L itself): write 'L^+' everywhere for 'L'. Since L^+ – like L itself – has only $\aleph_0$ terms, Henkin models for L^+ are countable by definition.

2. SUBSTITUTION IN STANDARD SEMANTICS AND SUBSTITUTIONAL SEMANTICS

Substitutional semantics was characterized on page 189 as a semantics that interprets 'All' and 'Some' substitutionally and, hence, can own only Henkin models. It might also be characterized as a semantics that owns only Henkin models, and hence can interpret 'All' and 'Some' substitutionally. But, whichever characterization be favored, substitutional semantics is an elaboration of the old, yet resilient, dictum: "A universal quantification is true iff all its substitution instances are (and an existential one true iff at least one of them is)."

The dictum, suitably rephrased, appears below as Theorem 2.6 and will be called the *Substitution Theorem*. Devotees of standard semantics often slight it – *unaccountably so*. Indeed, *their* proof of the completeness theorem

> *A, if logically entailed by S in the standard sense, is provable from S in* L (= Theorem 2.14)

appeals to Theorem 2.6; and, slightly reworded, the half of that proof concerning infinitely extendible S's yields the counterpart of Theorem 2.14 in substitutional semantics. Further, *their* proof of the soundness theorem

> *A, if provable from S in* L, *is logically entailed by S in the standard sense* (= Theorem 2.4)

appeals to half of Theorem 2.6, a half recorded below as Theorem 2.2; and, reworded to concern all Henkin models for all term extensions of L, that proof yields the counterpart of Theorem 2.4 in substitutional semantics. So, Theorem 2.14 and Theorem 2.4, the theorems which together legitimize the standard account of logical entailment, hinge upon a substitution lemma; and editing half the proof of one theorem and all the proof of the other will legitimize the substitutional account of that notion.

Concerned to show substitutional semantics a by-product of standard semantics, I first substantiate the above claims, and stress as I go along the role that Henkin models play in standard semantics. Then I spell out and justify what substitutional semantics understands by logical truth and logical entailment. Last I comment on the substitutional handling of theories.

Demonstrating Theorem 2.4 is easy once you have shown that the axioms of L are true in all models for L. Showing that the axioms of L are true in all models for L is relatively easy once you have shown that those of sorts **A1–A6**

I.3: ALTERNATIVES TO FIRST-ORDER SEMANTICS

are. And showing that the axioms of L of sorts **A1–A4** and **A6** are true in all models for L is quite easy. (In the case of **A4**, refer to Theorem 4.1.2 in Leblanc [1976], which guarantees that if A is true on a D-interpretation I_D of L, then A is true on any T-variant of I_D. T here is an arbitrary term of L foreign to $(\forall x)A$.) But showing that the axioms of L of sort **A5** are true in all models for L, calls for a lemma whose proof is surprisingly difficult. (Indeed, few texts undertake to prove Theorem 2.1. The proof here is borrowed from Leblanc and Wisdom [1972] pp. 313–314 and Leblanc [1976] pp. 86–88, with several simplifications and corrections due to Wisdom.)

THEOREM 2.1. *Let A be a statement of* L, *T and T' be terms of* L, *D be a domain, I_D be a D-interpretation of* L, *and I'_D be the T-variant of I_D such that $I'_D(T) = I_D(T')$. Then $A(T'/T)$ is true on I_D iff A is true on I'_D.*

Proof of Theorem 2.1 is by mathematical induction of the length $l(A)$ of A.

Basis: $l(A) = 1$. Then A is of the sort $Q(T_1, T_2, \ldots, T_d)$, and hence $A(T'/T)$ is of the sort $Q(T'_1, T'_2, \ldots, T'_d)$, where – for each i from 1 through d – T'_i is T_i itself if T_i is distinct from T', otherwise T'_i is T'. But, by the construction of I'_D, $I'_D(T_i) = I_D(T'_i)$ for each i from 1 through d, and $I'_D(Q) = I_D(Q)$. Hence, $\langle I'_D(T_1), I'_D(T_2), \ldots, I'_D(T_d) \rangle$ belongs to $I'_D(Q)$ iff $\langle I_D(T'_1), I_D(T'_2), \ldots, I_D(T'_d) \rangle$ belongs to $I_D(Q)$. Hence $A(T'/T)$ is true on I_D iff A is true on I'_D.

Inductive Step: $l(A) > 1$.

Case 1. A is a negation $\neg B$. By the hypothesis of the induction $B(T'/T)$ is (not) true on I_D iff B is (not) true on I'_D. Hence, $\neg(B(T'/T))$ – i.e. $(\neg B)(T'/T)$ – is true on I_D iff $\neg B$ is true on I'_D.

Case 2. A is a conjunction $B \wedge C$. Proof similar to that of Case 1.

Case 3. A is a quantification $(\forall x)B$. (i) Suppose $((\forall x)B)(T'/T)$ – i.e. $(\forall x)(B(T'/T))$ – is not true on I_D. Then, with T'' an arbitrary term of L foreign to $(\forall x)(B(T'/T))$ and distinct from T, there is a T''-variant I''_D of I_D on which $(B(T'/T))(T''/x)$ is not true. But, given the hypothesis on T'', $(B(T''/x))(T'/T)$ is the same as $(B(T'/T))(T''/x)$. Hence, $(B(T''/x))(T'/T)$ is not true on I''_D. Now let I'''_D be the T-variant of I''_D such that $I'''_D(T) = I''_D(T')$. Then, by the hypothesis of the induction, $B(T''/x)$ is not true on I'''_D. But I'''_D is a T''-variant of I'_D, a point I demonstrate two lines hence. So, there is a T''-variant of I'_D on which $B(T''/x)$ is not true. So, $(\forall x)B$ is not true on I'_D. (For proof that I'''_D is a T''-variant of I'_D: Since I'''_D is a T-variant of I''_D and I''_D is a T''-variant of I_D, I'''_D and I_D can differ only on T. So, I'''_D and I'_D can differ only on T''.) (ii) Suppose $(\forall x)B$ is not true on I'_D. Then, with T'' a term of L foreign to $(\forall x)B$ and distinct from T, there is a T''-variant I''_D of I'_D on which

$B(T''/x)$ is not true. Now let I_D''' be the T''-variant of I_D such that $I_D'''(T'') = I_D''(T'')$. Then I_D'' is the T-variant of I_D''' such that $I_D''(T) = I_D'''(T')$, a point I demonstrate four lines hence. So, by the hypothesis of the induction $(B(T''/x))(T'/T)$ – i.e., $(B(T'/T))(T''/x)$ – is not true on I_D'''. So, there is a T''-variant of I_D on which $(B(T'/T))(T''/x)$ is not true. So, $(\forall x)(B(T'/T))$ – i.e. $((\forall x)B)(T'/T)$ – is not true on I_D. (For proof that I_D'' is the T-variant of I_D''' such that $I_D''(T) = I_D'''(T')$: Since I_D'' is a T''-variant of I_D' and I_D' is a T-variant of I_D, I_D'' and I_D can differ only on T'' and T. But I_D'', being a T''-variant of I_D, can only differ from I_D only on T''. So, I_D'' and I_D''' can differ only on T'' and T. But $I_D''(T'') = I_D'''(T'')$. So, I_D'' and I_D''' can differ only on T. But $I_D''(T) = I_D'''(T')$. So, I_D'' is the T-variant of I_D''' such that $I_D''(T) = I_D'''(T')$.) □

Hence, the following half of Theorem 2.6, which guarantees at once that every axiom of L of the sort $(\forall x)A \to A(T/x)$ (= **A5**) is true in all models for L:

THEOREM 2.2. *Let D be a domain and I_D be a D-interpretation of* L. *If $(\forall x)A$ is true on I_D, so is each substitution instance $A(T/x)$ of $(\forall x)A$ in* L.

Proof. Suppose $(\forall x)A$ is true on I_D, and let T' be an arbitrary term of L foreign to $(\forall x)A$. Then $A(T'/x)$ is true on the T'-variant I_D' of I_D such that $I_D'(T') = I_D(T)$ and, hence, by Theorem 2.1 $(A(T'/x))(T/T')$ is true on I_D. But, with T' foreign to $(\forall x)A$, $A(T/x)$ is the same as $(A(T'/x))(T/T')$. Hence, $A(T/x)$ is true on I_D. □

Proof can now be given that:

THEOREM 2.3. *Every axiom of* L *is true in all models for* L.

Proof. Suppose A is an axiom of L, in which case A is bound by Theorem 1.1 to be of the sort

$$(\forall x_1)(\forall x_2)\ldots(\forall x_n)(B(x_1, x_2, \ldots, x_n/T_1, T_2, \ldots, T_n)),$$

where $n \geq 0$ and B is of one of the sorts **A1–A6**. Proof that A is true in all models for L will be by mathematical induction on n.

Basis: $n = 0$. Then A $(= B)$ is of one of the sorts **A1–A6**, a case treated above.

Inductive Step: $n > 0$. Let $\langle D, I_D \rangle$ be an arbitrary model for L, and T be an arbitrary term of L foreign to A.

$$((\forall x_2)\ldots(\forall x_n)(B(x_1, x_2, \ldots, x_n/T_1, T_2, \ldots, T_n)))(T/x_1)$$

I.3: ALTERNATIVES TO FIRST-ORDER SEMANTICS

is bound by Theorem 1.2 to be an axiom of L, and hence by the hypothesis of the induction to be true in $\langle D, I_D' \rangle$ for every T-variant I_D' of I_D. Hence, A is true in $\langle D, I_D \rangle$. Hence, A is true in all models for L. □

And, with Theorem 2.3 on hand, proof can be given that:

THEOREM 2.4. *Let S be an arbitrary set of statements and A be an arbitrary statement of* L. *If* $S \vdash A$, *then A is true in every model for* L *in which S is true, i.e. A is logically entailed by S in the standard sense* (= *The Strong Soundness Theorem for* L *in Standard Semantics*).

Proof. Suppose the column made up of $B_1, B_2, \ldots, B_p$ constitutes a proof in L of A from S, and let $\langle D, I_D \rangle$ be an arbitrary model for L in which S is true. It is easily shown by mathematical induction on i that, for each i from 1 through p, B_i is true in $\langle D, I_D \rangle$. Indeed, when B_i is a member of S, B_i is true in $\langle D, I_D \rangle$ by the hypothesis on $\langle D, I_D \rangle$; when B_i is an axiom of L, B_i is true in $\langle D, I_D \rangle$ by Theorem 2.3; and, when B_i is the potential of, say, B_g and B_h, B_i is sure to be true in $\langle D, I_D \rangle$ if – as the hypothesis of the induction guarantees – B_g and B_h are. Hence, B_p (= A) is true in $\langle D, I_D \rangle$. Hence, A is true in every model for L in which S is true. Hence, Theorem 2.4. □

Because of Theorem 2.4, A – when provable in L from S – is of course sure to be true in every *countable* model for L in which S is true, and – more particularly – to be true in every *Henkin* model for L in which S is true. With an eye to proving the counterpart of Theorem 2.4 for substitutional semantics, I record the second of these corrollaries as a separate theorem:

THEOREM 2.5. *Let S and A be as in Theorem 2.4. If* $S \vdash A$, *then A is true in every Henkin model for* L *in which S is true.*

*

Theorem 2.4, as previously noted, is one of the two theorems that legitimize the standard account of logical entailment. Now for the other, the converse of Theorem 2.4. Proof of it for the case where S is infinitely extendible in L uses one extra notion, that of a model associate, and two extra lemmas, Theorems 2.6 and 2.7.

Let S be an arbitrary set of statements of L. By the *model associate* of S in L, I shall understand the pair $\langle \mathcal{T}, I_{\mathcal{T}} \rangle$, where (i) $\mathcal{T}$ consists of the $\aleph_0$ terms of L, (ii) for each term T of L, $I_{\mathcal{T}}(T) = T$, and (iii) for each predicate Q of L of degree d ($d = 1, 2, 3, \ldots$) $I_{\mathcal{T}}(Q) = \{\langle T_1, T_2, \ldots, T_d \rangle : Q(T_1, T_2, \ldots, T_d) \in S\}$. $\langle \mathcal{T}, I_{\mathcal{T}} \rangle$ is a Henkin model, and one of particular significance.

Theorem 2.6 is what I called on p. 198 the *Substitution Theorem (for* L). Together with Theorem 1.3, it yields Theorem 2.7, a result guaranteeing that if a set S of statements of L is maximally consistent and ω-complete in L, then a statement of L will be true in the model associate of S in L iff it belongs to S. And Theorem 2.7 yields in turn the converse of Theorem 2.4 for infinitely extendible S, this in a mere seven lines.

THEOREM 2.6. *Let I_D be a Henkin D-interpretation of* L. *Then a quantification $(\forall x)A$ of* L *is true on I_D iff each substitution instance $A(T/x)$ of $(\forall x)A$ in* L *is true on I_D* (= *The Substitution Theorem for* L).

Proof. Suppose $A(T/x)$ is true on I_D for every term T of L; let T' be an arbitrary term of L foreign to $(\forall x)A$; let I'_D be an arbitrary T'-variant of I_D; and let d be the value of I'_D for T'. I_D being a Henkin D-interpretation of L, there is sure to be a term T'' of L such that $I_D(T'') = d$ and, hence, $I_D(T'') = I'_D(T')$. But, since $A(T/x)$ is true on I_D for every term T of L and $(A(T'/x))(T/T')$ is the same as $A(T/x)$, $(A(T'/x))(T''/T)$ is sure to be true on I_D. Hence, by Theorem 2.1 $A(T'/x)$ is true on I'_D. So, if $A(T/x)$ is true on I_D for every term T of L, then $A(T'/x)$ is true on every T'-variant of I_D, and hence $(\forall x)A$ is true on I_D. Hence, Theorem 2.6 by Theorem 2.2. □

THEOREM 2.7. *Let S be a set of statements of* L *that is maximally consistent and ω-complete in* L, *and let $\langle \mathcal{T}, I_{\mathcal{T}} \rangle$ be the model associate of S in* L. *Then a statement A of* L *belongs to S iff A is true in $\langle \mathcal{T}, I_{\mathcal{T}} \rangle$.*

Proof of Theorem 2.7 is by mathematical induction on the length $l(A)$ of A.

Basis: $l(A) = 1$, in which case A is of the sort $Q(T_1, T_2, \ldots, T_d)$. By the construction of $I_{\mathcal{T}}$, (i) $Q(T_1, T_2, \ldots, T_d)$ belongs to S iff $\langle T_1, T_2, \ldots, T_d \rangle$ belongs to $I_{\mathcal{T}}(Q)$ and (ii) $\langle T_1, T_2, \ldots, T_d \rangle$ belongs to $I_{\mathcal{T}}(Q)$ iff $\langle I_{\mathcal{T}}(T_1), I_{\mathcal{T}}(T_2), \ldots, I_{\mathcal{T}}(T_d) \rangle$ does. Hence, Theorem 2.7.

Inductive Step: $l(A) > 1$. Proof of Theorem 2.7 in the case that A is a negation $\neg B$ or a conjunction $B \wedge C$ is by Theorem 1.3(i)–(ii), the hypothesis of the induction, and the truth-conditions for '$\neg$' and '$\wedge$'. Suppose then that A is a universal quantification $(\forall x)B$. By Theorem 1.3(iii) $(\forall x)B$ belongs to S iff each substitution instance $B(T/x)$ of $(\forall x)B$ in L does, hence by the hypothesis of the induction iff each substitution instance $B(T/x)$ of $(\forall x)B$ in L is true in $\langle \mathcal{T}, I_{\mathcal{T}} \rangle$, hence by Theorem 2.6 (and the fact that $\langle \mathcal{T}, I_{\mathcal{T}} \rangle$ is a Henkin model) iff $(\forall x)B$ is true in $\langle \mathcal{T}, I_{\mathcal{T}} \rangle$. □

Suppose now that $S \not\vdash A$, where S is infinitely extendible in L.[11] Then by

I.3: ALTERNATIVES TO FIRST-ORDER SEMANTICS   203

a familiar result $S \cup \{\neg A\}$, a set also infinitely extendible in L, is consistent in L; hence, by Theorem 1.4 the Henkin extension $\mathscr{H}(S \cup \{\neg A\})$ of $S \cup \{\neg A\}$ in L is maximally consistent and ω-complete in L; and, hence, by Theorem 2.7 every member of $\mathscr{H}(S \cup \{\neg A\})$ is true in the model associate of $\mathscr{H}(S \cup \{\neg A\})$ in L. But $S \cup \{\neg A\}$ is a subset of $\mathscr{H}(S \cup \{\neg A\})$. Hence, (every member of) S is true in that model, but A is not. Hence:

THEOREM 2.8. *Let S be a set of statements of L that is infinitely extendible in L, and let A be an arbitrary statement of L. If $S \not\vdash A$, then there is a model for L in which S is true but A is not.*

Hence by Contraposition:

THEOREM 2.9. *Let S and A be as in Theorem 2.8. If A is true in every model for L in which S is true, i.e. if A is logically entailed by S in the standard sense, then $S \vdash A$.*

Since the model associate of $\mathscr{H}(S \cup \{\neg A\})$ in L is a Henkin model for L, Theorems 2.8 and 2.9 can be sharpened to read:

THEOREM 2.10. *Let S and A be as in Theorem 2.8. If $S \not\vdash A$, then there is a Henkin model for L (and, hence, there is a countable model for L) in which S is true but A is not.*

THEOREM 2.11. *Let S and A be as in Theorem 2.8. If A is true in every Henkin model for L (hence, if A is true in every countable model for L) in which S is true, then $S \vdash A$.*

$\emptyset$ is, of course, infinitely extendible in L. So, by Theorems 2.4, 2.5, 2.9, and 2.11:

THEOREM 2.12. *Let A be an arbitrary statement of L.*
   (a) *If $\vdash A$, then A is true in every model for L, i.e. A is logically true in the standard sense (= The Weak Soundness Theorem for L in Standard Semantics).*
   (b) *If A is true in every model for L, i.e. if A is logically true in the standard sense, then $\vdash A$ (= The Weak Completeness Theorem for L in Standard Semantics).*
   (c) *$\vdash A$ iff A is true in every Henkin model for L.*

(d) *A is true in every model for* L *if – and, hence, iff – true in every Henkin model for* L.

Theorem 2.12(a)–(b) legitimizes the standard account of logical truth.

The argument leading to Theorem 2.8, and hence to Theorems 2.9–2.11, fails when S is *not* infinitely extendible in L. $S \cup \{\neg A\}$, if consistent in L, will extend to a set that is maximally consistent in L, but it need not extend to one that is *both* maximally consistent and ω-complete in L. A set of the sort $\{Q(t_1), Q(t_2), Q(t_3), \ldots, \neg(\forall x)Q(x)\}$, for example, will clearly not extend to one maximally consistent and ω-complete in L. In point of fact, for many an S that is not infinitely extendible in L, Theorem 2.10 fails: for many an A such that $S \not\vdash A$, *there is no Henkin model for* L *in which* S *is true but* A *is not*. To exploit the same counterexample as above, though $\{Q(t_1), Q(t_2), Q(t_3), \ldots\} \not\vdash (\forall x)Q(x)$, there is no Henkin model for L in which $\{Q(t_1), Q(t_2), Q(t_3), \ldots\}$ is true but $(\forall x)Q(x)$ is not.[12] However, it can be shown that if $S \not\vdash A$, where S is not infinitely extendible in L, then there is a *non-Henkin* model for L (more specifically, a denumerably infinite non-Henkin model for L) in which S is true but A is not.

Understand (i) by the *double rewrite* A^2 of a statement A of L the result of substituting everywhere in A t_{2i} for t_i ($i = 1, 2, 3, \ldots$), (ii) by the double rewrite S^2 of a set S of statements of L the set $\emptyset$ when S is empty, otherwise that consisting of the double rewrites of the various members of S, (iii) by the double rewrite of a D-interpretation I_D of L the D-interpretation I_D^2 of L such that $I_D^2(Q) = I_D(Q)$ for every predicate Q of L, and $I_D^2(t_i) = I_D(t_{2i})$ for each i from 1 on, and (iv) by the double rewrite of a model $\langle D, I_D \rangle$ for L the model $\langle D, I_D^2 \rangle$, where I_D^2 is the double rewrite of I_D. It is easily verified that (a) whether or not S is infinitely extendible, S^2 is, (b) if S is consistent in L, so is S^2, (c) if the double rewrite of a statement A of L is true in a model $\langle D, I_D \rangle$ for L, then A itself is true in the double rewrite of $\langle D, I_D \rangle$, and (d) although the model associate $\langle \mathcal{T}, I_{\mathcal{T}} \rangle$ in L of a set of statements of L is a Henkin model for L, the double rewrite $\langle \mathcal{T}, I_{\mathcal{T}}^2 \rangle$ of $\langle \mathcal{T}, I_{\mathcal{T}} \rangle$ is not (members $t_1, t_3, t_5, \ldots$, of $\mathcal{T}$ are not assigned in $I_{\mathcal{T}}^2$ to any term of L).

Suppose now that $S \not\vdash A$, where S is not infinitely extendible in L, and hence $S \cup \{\neg A\}$ is consistent in L. By (a)–(b), the double rewrite $S^2 \cup \{\neg A^2\}$ of $S \cup \{\neg A\}$ is both consistent and infinitely extendible in L. Hence, as before, $S^2 \cup \{\neg A^2\}$ is true in the model associate in L of $\mathcal{H}(S^2 \cup \{\neg A^2\})$. Hence, by (c) S is true in the double rewrite of that model but A is not. Hence, there is a model for L in which S is true but A is not.[13] That the model in question is *not* a Henkin one follows from (d).

I.3: ALTERNATIVES TO FIRST-ORDER SEMANTICS 205

Hence:

THEOREM 2.13. *Let S be a set of statements of* L *that is not infinitely extendible in* L, *and let A be an arbitrary statement of* L. *If* $S \not\vdash A$, *then there is a model for* L *(more specifically, a countable model for* L*) in which S is true but A is not.*

Hence, by Contraposition on Theorem 2.13 and (for the case where S is infinitely extendible in L) on Theorem 2.8:

THEOREM 2.14. *Let S be an arbitrary set of statements of* L *and A be an arbitrary statement of* L. *If A is true in every model for* L *in which S is true, i.e. if A is logically entailed by S in the standard sense, then* $S \vdash A$ *(= The Strong Completeness Theorem for* L *in Standard Semantics).*

*

The argument yielding Theorem 2.4 and, hence, Theorem 2.5 can be made to yield:

THEOREM 2.15. *Let S be an arbitrary set of statements of* L *and A be an arbitrary statement of* L. *If* $S \vdash A$, *then – no matter the term extension* L^+ *of* L – *A is true in every Henkin model for* L^+ *in which S is true.*

Substituting 'L^+' for 'L' at places that the very phrasing of Theorem 2.5 dictates will do the trick, as the reader may verify.

Theorem 2.10 fails, we just saw, for S *not* infinitely extendible in L. So the converse of Theorem 2.5 fails. Thanks, however, to the term extension L^∞ of L, the converse of Theorem 2.15 does hold. Recall indeed that a set of statements of L, whether infinitely extendible in L or not, is infinitely extendible in L^∞ (= Theorem 1.5), and is sure to be consistent in L^∞ if consistent in L (= Theorem 1.7). So, suppose $S \not\vdash A$, where S and A are as in Theorem 2.15. Then $S \cup \{\neg A\}$ is consistent in L^∞; hence, by the analogue of Theorem 1.4 for L^∞, the Henkin extension $\mathcal{H}^\infty(S \cup \{\neg A\})$ of $S \cup \{\neg A\}$ in L^∞ is maximally consistent and ω-complete in L^∞; hence, by the analogue of Theorem 2.7 for L^∞ every member of $\mathcal{H}^\infty(S \cup \{\neg A\})$ is true in the model associate of $\mathcal{H}^\infty(S \cup \{\neg A\})$ in L^∞; and, hence, S is true in that model but A is not.[14] But the model associate of $\mathcal{H}^\infty(S \cup \{\neg A\})$ in L^∞ constitutes a Henkin model for L^∞. So, there is a term extension L^+ of L and a Henkin model for L^+ in which S is true but A is not.[15] So, by a slight editing of the proof of Theorem 2.8:

THEOREM 2.16. *Let S and A be as in Theorem 2.15. If – no matter the term extension L^+ of L – A is true in every Henkin model for L^+ in which S is true, then $S \vdash A$.*

As their proofs attest, Theorems 2.15 and 2.16 would hold with L^∞ the only term extension of L (besides L itself), but would thereby lose much of their intuitive appeal.

Now for truth, logical truth, and logical entailment in substitutional semantics. Theorem 2.18(a)–(b), the results of writing 'true$_S$' for 'true' in Theorems 2.15 and 2.16, will legitimize the substitutional account of the third – and, hence, of the second – of these notions.

Let L^+ be an arbitrary term extension of L, A be an arbitrary statement of L^+, and S be an arbitrary set of statements of L^+. I shall say that A is *substitutionally true* – or, for short, *true$_S$* – in a *Henkin* model $\langle D, I_D \rangle$ for L^+ if (i) in the case that A is an atomic statement $Q(T_1, T_2, \ldots, T_d)$, the d-tuple $\langle I_D(T_1), I_D(T_2), \ldots, I_D(T_p)\rangle$ belongs to $I_D(Q)$, (ii) in the case that A is a negation $\neg B$, B is not true$_S$ in $\langle D, I_D \rangle$, (iii) in the case that A is a conjunction $B \wedge C$, each of B and C is true$_S$ in $\langle D, I_D \rangle$, and (iv) in the case that A is a quantification $(\forall x)B$, each substitution instance $B(T/x)$ of $(\forall x)B$ in L^+ is true$_S$ in $\langle D, I_D \rangle$; and I shall say that S is *true$_S$* in $\langle D, I_D \rangle$ if every member of S is.

This done, I shall declare a statement A of L *logically true in the substitutional sense* if – no matter the term extension L^+ of L – A is true$_S$ in every Henkin model for L^+; and, where S is a set of statements of L, I shall declare *A logically entailed by S in the substitutional sense* if – no matter the term extension L^+ of L – A is true$_S$ in every Henkin model for L^+ in which S is true$_S$. Equivalently, but more simply, A may be declared logically true in the substitutional sense if true$_S$ in all Henkin models for L. (By supplying two definitions of logical truth I set here a precedent to be repeatedly heeded in the essay. In each case it is, of course, the infinite extendibility of $\emptyset$ in L which allows for the simpler of the two definitions, the one using just L.)

It follows at once from Theorem 2.6 that:

THEOREM 2.17. *Let L^+ be a term extension of L, A be a statement of L^+, S be a set of statements of L^+, and $\langle D, I_D \rangle$ be a Henkin model for L^+.*
(a) *A is true$_S$ in $\langle D, I_D \rangle$ iff true in $\langle D, I_D \rangle$.*
(b) *S is true$_S$ in $\langle D, I_D \rangle$ iff true in $\langle D, I_D \rangle$.*

Hence by Theorems 2.15 and 2.16:

THEOREM 2.18. *Let S be an arbitrary set of statements and A be an arbitrary statement of* L.
(a) *If* $S \vdash A$, *then – no matter the term extension of* L^+ *of* L *– A is true$_S$ in every Henkin model for* L^+ *in which S is true$_S$, i.e. A is logically entailed by S in the substitutional sense* (= *The Strong Soundness Theorem for* L *in Substitutional Semantics*).
(b) *If – no matter the term extension* L^+ *of* L *– A is true$_S$ in every Henkin model for* L^+ *in which S is true$_S$, i.e. if A is logically entailed by S in the substitutional sense, then* $S \vdash A$ (= *The Strong Completeness Theorem for* L *in Substitutional Semantics*).

And hence:

THEOREM 2.19. *Let A be an arbitrary statement of* L.
(a) *If* $\vdash A$, *then – no matter the term extension* L^+ *of* L *– A is true$_S$ in every Henkin model for* L^+, *i.e. A is logically true in the substitutional sense* (= *The Weak Soundness Theorem for* L *in Substitutional Semantics*).
(b) *If – no matter the term extension* L^+ *of* L *– A is true$_S$ in every Henkin model for* L^+, *i.e. if A is logically true in the substitutional sense, then* $\vdash A$ (= *The Weak Completeness Theorem for* L *in Substitutional Semantics*).

Theorem 2.19 legitimizes the first account of logical truth in the preceding paragraph. Writing 'true$_S$' for 'true' in Theorem 2.12(c) legitimizes the second and simpler one:

THEOREM 2.20. *Let A be as in Theorem 2.19. Then* $\vdash A$ *iff A is true$_S$ in every Henkin model for* L.

*

Suppose there is a model for L in which a set S of statements of L is true. Then there is one in which S is true but, say, $Q(t_1) \wedge \neg Q(t_1)$ is not; hence, by Theorem 2.4 $Q(t_1) \wedge \neg Q(t_1)$ is not provable in L from S; and, hence, by Theorems 2.8 and 2.13 there is a countable model for L in which S is true (and $Q(t_1) \wedge \neg Q(t_1)$ is not). So,

(1) *A set S of statements of* L, *if true in a model for* L, *is sure to be true in a countable model for* L.

The result, known as the *Löwenheim-Skolem Theorem* (or *Skolem's Generalization of Löwenheim's Theorem*) can be edited some (use Theorem 2.18(b) in lieu of Theorems 2.18 and 2.13):

(2) *A set S of statements of* L, *if true in a model for* L, *is sure – for some term extension* L^+ *of* L *– to be true$_S$ in a Henkin model for* L^+.

(1), from which the so-called 'Skolem Paradox' issues, disturbed many when reported in Skolem [1920]. But it comforted others who, uneasy over large infinite cardinals (possibly over all cardinals beyond $\aleph_0$), welcomed word that any consistent (first-order) theory can be trusted to have a countable model. In view of (2) such a theory will even have a Henkin model, the kind of model permitting the quantifiers in the theory to be understood *substitutionally*.

On closer inspection, though, (2) affords only slight comfort to devotees of substitutional semantics. The 'intended model' of a (first-order) theory may very much matter, the theory often owing to its intended model whatever consideration it enjoys. Yet $\langle \mathcal{T}^\infty, I_{\mathcal{T}^\infty} \rangle$, the Henkin model whose existence the above proof of (2) guarantees, is generally no kin of *that* model. Indeed, $t_1^\infty, t_2^\infty, t_3^\infty, \ldots$, rarely figure in the 'intended domain' (= the domain of the 'intended model') of a (first-order) theory. To be sure, some proofs of Theorem 2.16 (the theorem from which Theorem 2.18(b) issues) use $\{1, 2, 3, \ldots\}$ in lieu of $\mathcal{T}^\infty$, but the positive integers – though figuring in more domains than $t_1^\infty, t_2^\infty, t_3^\infty, \ldots$, do – hardly figure in the intended domain of *every* (first-order) theory. So, one might hesitate to turn in the intended model of a (first-order) theory, even if uncountable, for $\langle \mathcal{T}^\infty, I_{\mathcal{T}^\infty} \rangle$.

Well-informed readers will at this point appeal to a sharper version of the Löwenheim-Skolem Theorem. Given a model $\langle D, I_D \rangle$ for L, acknowledge as a *submodel* of $\langle D, I_D \rangle$ any pair $\langle D', I_{D'} \rangle$ such that (i) D' is a subset of D to which belong all of $I_D(t_1), I_D(t_2), I_D(t_3), \ldots$, and (ii) $I_{D'}$ is the restriction of I_D to D'. What indeed Skolem [1920] shows is that

(3) *A set S of statements of* L, *if true in a model* $\langle D, I_D \rangle$ *for* L, *is sure to be true in a countable model for* L *that is a submodel of* $\langle D, I_D \rangle$.

The model in question, call it $\langle D', I_{D'} \rangle$, need not constitute a Henkin model for L. But it easily extends to a Henkin model $\langle D', I'_{D'} \rangle$ for L^∞ in which S is sure to be true$_S$. If the members of D' that are not assigned by $I_{D'}$ to any term of L are finitely many in number, let them be $d_1, d_2, \ldots, d_n$ (Case 1); otherwise, let them be $d_1, d_2, d_3, \ldots$ (Case 2); and let $I'_{D'}$ (i) agree with $I_{D'}$ on all the terms and predicates of L, and (ii) assign in Case 1 d_i to t_i^∞ for each i

I.3: ALTERNATIVES TO FIRST-ORDER SEMANTICS

from 1 through $n-1$ and d_n to the remaining terms of L^∞, and in Case 2 d_i to t_i^∞ for each i from 1 on. It is easily verified that S, if true in $\langle D', I_{D'}\rangle$, will be true$_S$ in $\langle D', I'_{D'}\rangle$.

$\langle D', I'_{D'}\rangle$ escapes the criticism levelled at the Henkin model in (2). It bears to the intended model $\langle D, I_D\rangle$ of a (first-order) theory the closest relationship that logical propriety allows, D' being a subset of D and the interpretation that $I'_{D'}$ places upon the terms and predicates of L being the restriction of I_D to D'. So, some might actually turn in $\langle D, I_D\rangle$ for $\langle D', I'_{D'}\rangle$. Others would not, to be sure. But they might well cite $\langle D', I'_{D'}\rangle$ - a 'fall-back model' of the theory, so to speak - as their reason for retaining $\langle D, I_D\rangle$, and retaining it with a clear conscience.

There are, however, practitioners of substitutional semantics who have no qualms over cardinals beyond $\aleph_0$ and consequently provide room in their texts for Henkin models of *any* size. For example, in Robinson [1951] and Shoenfield [1967] no bound is placed on the number of terms such a language as L might have; and in Hintikka [1955] none is placed on the number of terms the term extensions of L might have.[16] Under these circumstances the intended model of any theory of L, if not a Henkin model for L, extends to one for some extension or other of L; and the quantifiers in the theory can as a result be understood substitutionally, something Robinson and Shoenfield both proceed to do on page 19 of their respective texts, and which Hintikka does (in a beautifully sly way) on page 24 of his.[17]

(The Appendix has further information on the history of substitutional semantics.)

3. TRUTH-VALUE SEMANTICS

What I called on page 192 a *quantifierless* statement of L would normally be counted *logically true* (also *truth-functionally true, tautologous*, etc.) if true on all truth-value assignments *to its atomic components*. Beth in 1959, Schütte in 1962, Dunn and Belnap in 1968, etc., showed in effect that an *arbitrary* statement of L is logically true in the standard sense iff (substitutionally) true on all truth-value assignments *to the atomic statements of* L; and I showed, also in 1968, that such a statement is logically true in the standard sense iff (substitutionally) true on all truth-value assignments *to its atomic substatements*. These may have been the first contributions to what, at Quine's suggestion, I term *truth-value semantics*.

Truth-value semantics is a non-referential kind of semantics: it dispenses with models and with condition (i) on page 206, the one truth condition of a model-theoretic sort that substitutional semantics retains. Instead, atomic statements are assigned a truth-value each, and compound statements are then assigned truth-values via counterparts of conditions (ii)–(iv) on page 206 Though modelless, truth-value semantics is nonetheless a congener of substitutional semantics, a claim I made on page 190 and will substantiate in Theorem 3.1 below.

There are several versions of truth-value semantics. I begin with one sketched in Leblanc [1968], subsequently developed in Leblanc and Wisdom [1972], Leblanc [1976], and various papers in Part 2 of Leblanc [1982b], and recently amended to exploit Dunn and Belnap's term extensions.

Let L^+ be an arbitrary term extension of L. By a *truth-value assignment* to a non-empty set Σ^+ of atomic statements of L^+, I shall understand any function from the members of Σ^+ to $\{T, F\}$. Where A is a statement of L^+, Σ^+ a set of atomic statements of L^+ to which belong all the atomic substatements of A, and α^+ a truth value assignment to Σ^+, I shall say that A is *true* on α^+ if (i) in the case that A is atomic, $\alpha^+(A) = T$, (ii) in the case that A is a negation $\neg B$, B is not true on α^+, (iii) in the case that A is a conjunction $B \wedge C$, each of B and C is true on α^+, and (iv) in the case that A is a quantification $(\forall x)B$, each substitution instance of $(\forall x)B$ in L^+ is true on α^+. Finally, where S is a set of statements of L^+, Σ^+ a set of atomic statements of L^+ to which belong all the atomic substatements of S, and α^+ a truth-value assignment to Σ^+, I shall say that S is *true* on α^+ if each member of S is true on α^+.

This done, I shall declare a statement A of L *logically true in the truth-value sense* if – no matter the term extension L^+ of L and truth-value assignment α^+ to the atomic substatements of A in L^+ – A is true on α^+; and, where S is a set of statements of L, I shall declare A *logically entailed by S in the truth-value sense* if – no matter the term extension L^+ of L and truth-value assignment α^+ to the atomic substatements of $S \cup \{A\}$ in L^+ – A is true on α^+ if S is. Equivalently but more simply, A may be declared logically true in the truth-value sense if (as the matter was put above) A is true on every truth-value assignment to the atomic substatements of A in L.

Proof that, where S is a set of statements and A a statement of L,

(1) $S \vdash A$ *iff A is logically entailed by S in the foregoing sense*

and

(2) $\vdash A$ *iff A is logically true in either of the foregoing senses,*

can be retrieved from Chapter 2 of Leblanc [1976]. So I turn at once to another version of truth-value semantics, a version using *total* rather than *partial* assignments.

Again, let L^+ be an arbitrary term extension of L. By a *truth-value assignment for* L^+ I shall understand any function from *all* the atomic statements of L^+ to {T, F}. Where A is a statement of L^+, S a set of statements of L^+, and α^+ a truth-value assignment for L^+, I shall say that A is *true* on α^+ if conditions (i)–(iv) on page 210 are met, and say that S is *true* on α^+ if each member of S is. And, where A is a statement of L and S a set of statements of L, I shall declare A *logically true in the truth-value sense* if – no matter the term extension L^+ of L – A is true on every truth-value assignment for L^+ (equivalently, but more simply, if A is true on every truth-value assignment for L); and I shall declare A *logically entailed by S in the truth-value sense* if – no matter the term extension L^+ of L – A is true on every truth-value assignment for L^+ on which S is true.

Proof that, where S is a set of statements and A is a statement of L,

(3) $S \vdash A$ *iff A is logically entailed by S in this truth-value sense*

and

(4) $\vdash A$ *iff A is logically true in either of these truth-value senses*,

can be had in two different ways.

Arguing (3) (and, hence, (4)) directly, understand by the *truth-value associate* in L^∞ of a set S^∞ of statements of L^∞ the function α^∞ that assigns T to the atomic statements of L^∞ in S^∞ and F to the rest. It follows from Theorem 1.3 that if S^∞ is maximally consistent and ω-complete in L^∞, then a statement of L^∞ will belong to S^∞ iff true on α^∞. But, if $S \nvdash A$, then by Theorems 1.7 and 1.4 the Henkin extension $\mathscr{H}^\infty (S \cup \{\neg A\})$ of $S \cup \{\neg A\}$ in L^∞ is maximally consistent and ω-complete in L^∞, and hence S is true on α^∞ but A is not. Hence, if – no matter the term extension L^+ of L – A is true on every truth-value assignment for L^+ on which S is true, then $S \vdash A$. But proof of the converse is routine. Hence, (3) and, letting S be $\emptyset$, (4).[18]

You can also argue (3)–(4) by appealing to Theorems 2.18–2.20, the counterparts of (3)–(4) in substitutional semantics, and to equivalence theorems that bind Henkin models and truth-value assignments. (The definitions used are from Leblanc [1976] pp. 92–93.)

With $\langle D, I_D \rangle$ a Henkin model for L^+, understand by the *truth-value counterpart* of $\langle D, I_D \rangle$ the function α^+ that assigns T to the atomic statements of L^+ true in $\langle D, I_D \rangle$ and F to the rest; and, α^+ being a truth-value assignment for L^+, understand by the *model counterpart* of α^+ the pair $\langle \mathscr{T}^+, I_{\mathscr{T}^+} \rangle$, where

(i) $\mathcal{T}^+$ consists as usual of the terms of L^+, (ii) for each term T of L^+, $I_{\mathcal{T}^+}(T) = T$, and (iii) for each predicate Q of L^+ of degree d, $I_{\mathcal{T}^+}(Q) = \{\langle T_1, T_2, \ldots, T_d\rangle : \alpha^+(Q(T_1, T_2, \ldots, T_d)) = T\}$. The truth-value counterpart of $\langle D, I_D\rangle$ constitutes, of course, a truth-value assignment for L^+, and the model counterpart of α^+ a Henkin model for L^+.

Proof that:

THEOREM 3.1. *Let A be an arbitrary statement of L^+.*

(a) *A is true in a Henkin model $\langle D, I_D\rangle$ for L^+ iff A is true on the truth-value counterpart α^+ of the model (= Equivalence Theorem One for Truth-Value Semantics);*

(b) *A is true on a truth-value assignment α^+ for L^+ iff A is true in the model counterpart $\langle\mathcal{T}^+, I_{\mathcal{T}^+}\rangle$ of the assignment (= Equivalence Theorem Two for Truth-Value Semantics);*

is immediate. Suppose indeed that A is atomic. The foregoing definitions see to it that A is true in $\langle D, I_D\rangle$ iff true *on* its counterpart, and that A is true on α^+ iff true *in* its counterpart. Or suppose A is a universal quantification. Since $\langle D, I_D\rangle$ is a Henkin model by hypothesis and $\langle\mathcal{T}^+, I_{\mathcal{T}^+}\rangle$ is one by construction, A will be true in either model iff each of its substitution instances in L^+ is. So Henkin models and truth-value assignments match one-to-one.

Tackling soundness first, suppose that $S \vdash A$ (S a set of statements and A a statement of L); let L^+ be an arbitrary term extension of L; and let α^+ be an arbitrary truth-value assignment for L^+ on which S is true. Then by Theorem 3.1(b) S will be true in the model counterpart of α^+; hence, by Theorem 2.18(a), A will be true in that model; and, hence, by Theorem 3.1(b) again, A will be true on α^+. So:

THEOREM 3.2. *Let S be an arbitrary set of statements and A be an arbitrary statement of L. If $S \vdash A$, then A is logically entailed by S in the truth-value sense of page 211 (= The Strong Soundness Theorem for L in Truth-Value Semantics).*

Tackling completeness next, suppose that – no matter the term extension L^+ of L – A is true on every truth-value assignment for L^+ on which S is true; let L^+ be an arbitrary term extension of L; and let $\langle D, I_D\rangle$ be an arbitrary Henkin model for L^+ in which S is true. Then by Theorem 3.1(a) S will be true on the truth-value counterpart of $\langle D, I_D\rangle$; hence, by hypothesis, A will be true on that assignment; and hence, by Theorem 3.1(a) again, A will be true

I.3: ALTERNATIVES TO FIRST-ORDER SEMANTICS   213

in $\langle D, I_D \rangle$. Hence, A will be true in every Henkin model for L^+ in which S is true. Hence, by Theorem 2.18(b), $S \vdash A$. So:

THEOREM 3.3 *Let S and A be as in Theorem 3.2. If A is logically entailed by S in the truth-value sense of page 211, then $S \vdash A$ (= The Strong Completeness Theorem for L in Truth-Value Semantics).*

The same arguments, but using Theorems 2.5 and 2.11 rather than Theorem 2.18, will guarantee that:

THEOREM 3.4. *Let S be a set of statements of L that is infinitely extendible in L, and let A be an arbitrary statement of L. Then $S \vdash A$ iff A is true on every truth-value assignment for L on which S is true.*

Hence:

THEOREM 3.5. *Let A be an arbitrary statement of L. Then $\vdash A$ iff A is true on every truth-value assignment for L.*

Theorems 3.2 and 3.3 are in Dunn and Belnap [1968]; and variants thereof, using isomorphisms in lieu of term extensions, are in Leblanc [1968]. A special case of Theorem 3.4 – S a set of termless statements of L – is in Beth [1959]; and, as reported on page 209, versions of Theorem 3.5 are in these three texts and in Schütte [1962].

Yet another version of truth-value semantics assigns truth-values to *all* the statements of L^+ (rather than just the atomic statements of L^+ or just the atomic substatements of a statement or set of statements of L^+). I expound it in some detail because of its kinship to truth-set semantics and to probability theory.

Let L^+ be an arbitrary term extension of L. By a *truth-value function* for L^+ I shall understand any function α^+ from the statements of L^+ to $\{T, F\}$ that meets the following three constraints (reminiscent of (i)–(iii) in Theorem 1.3):

B1. $\alpha^+(\neg A) = T$ iff $\alpha^+(A) \neq T$

B2. $\alpha^+(A \wedge B) = T$ iff $\alpha^+(A) = T$ and $\alpha^+(B) = T$

B3. $\alpha^+((\forall x)A) = T$ iff $\alpha^+(A(T/x)) = T$ for each term T of L^+.

I shall then declare a statement A of L *logically true in the truth-value sense* if – no matter the term extension L^+ of L and truth-value function α^+ for L^+ – $\alpha^+(A) = T$ (equivalently, but more simply, if $\alpha(A) = T$ for every truth-value function α for L); and, where S is a set of statements of L^+ – I shall declare *A logically entailed by S in the truth-value sense* if – no matter the term extension L^+ of L and truth-function α^+ for $L^+ - \alpha^+(A) = T$ if $\alpha^+(B) = T$ for each member B of S.

It is clear that *each truth-value function for L^+ is an extension to* all *the statements of L^+ of a truth-value assignment for L^+, and each truth-value assignment for L^+ the restriction to* just *the atomic statements of L^+ of a truth-value function for L^+*. So, truth-value assignments and truth-value functions match one-to-one, the way Henkin models and truth-value assignments did. The present accounts of logical truth are therefore sure in light of Theorems 3.2, 3.3 and 3.5 to be weakly sound and complete, and the present account of logical entailment is sure in light of Theorems 3.2 and 3.3 to be strongly sound and complete.

When showing in Section 5 the truth-value functions for L to issue into probability functions, I shall refer to them by means of 'P' rather than 'α', and I shall use as truth-values the reals 1 and 0 rather than T and F.

(The Appendix has further information on the history of truth-value semantics.)

*

Truth-set semantics, my next concern, is but truth-value semantics in set-theoretic disguise. And like truth-value semantics, which – interestingly enough – it antedates, it comes in several versions. One, due to Carnap[19] but recast here in the Dunn and Belnap manner, matches the semantics of page 211 with its total truth-value assignments.

Where L^+ is an arbitrary term extension of L, understand by a *basic pair* for L^+ any set of the sort $\{A, \neg A\}$, where A is an atomic statement of L^+; understand by a *basic truth set* for L^+ – or, as Carnap would put it, a *state-description* for L^+ – any set consisting of one statement from each basic pair for L^+; take a statement A of L^+ to *hold* in a state-description SD^+ for L^+ if (i) in the case that A is atomic, A belongs to SD^+, (iii) in the case that A is a negation $\neg B$, B does not hold in SD^+, (iii) in the case that A is a conjunction $B \wedge C$, each of B and C holds in SD^+, and (iv) in the case that A is a quantification $(\forall x)B$, each substitution instance of A in L^+ holds in SD^+; and take a set S of statements of L^+ to *hold* in SD^+ if each member of S does.

This done, declare a statement A of L *logically true in the truth-set sense* if – no matter the term extension L^+ of L – A holds in every state-description for L^+ (equivalently, but more simply, if A holds in every state-description for L); and, where S is a set of statements of L, declare *A logically entailed*

I.3: ALTERNATIVES TO FIRST-ORDER SEMANTICS

by S in the truth-set sense if – no matter the term extension L^+ of $L - A$ holds in every state-description for L^+ in which S does.

As truth-value assignments and state-descriptions for L^+ match one-to-one, a statement of L^+ will be true on a truth-value assignment for L^+ iff it holds *in* the matching state-description, and the statement will hold in a state-description for L^+ iff it is true *on* the matching truth-value assignment. So, by Theorems 3.2, 3.3, and 3.5:

THEOREM 3.6. *Let S be an arbitrary set of statements and A be an arbitrary statement of* L.
  (a) $S \vdash A$ *iff A is logically entailed by S in the truth-set sense.*
  (b) $\vdash A$ *iff A is logically true in the truth-set sense.*

That A, if provable in L, holds in every state-description for L was first shown in Carnap [1950]. (See Section 22, which should be consulted for more information on this particular brand of truth-set semantics.)

A second version of truth-set semantics can be retrieved from Quine [1940], Smullyan [1968], and doubtless several other sources. It matches the semantics of pages 213–217 with its truth-value functions, boasts the tersest accounts yet of logical truth and entailment, and can be legitimized in the swiftest manner yet.

With L^+ an arbitrary term extension of L, understand by a *truth set* for L^+ any set S^+ of statements of L^+ that meets the following three conditions (patterned after (i)–(iii) in Theorem 1.3):

  (a) $\neg A$ belongs to S^+ iff A does not,
  (b) $A \wedge B$ belongs to S^+ iff each of A and B does,
  (c) $(\forall x)A$ belongs to S^+ iff $A(T/x)$ does for each term T of L^+.

Declare a statement A of L *logically true in the truth-set sense* if – no matter the term extension L^+ of $L - A$ belongs to every truth set for L^+ (equivalently, but more simply, if A belongs to every truth set for L); and, where S is a set of statements of L, declare A *logically entailed by S in the truth-set sense* if – no matter the term extension L^+ of $L - A$ belongs to every truth set for L^+ of which S is a subset.[20]

Truth-value functions and truth sets for L^+ match one-to-one. Indeed, the set of all statements of L^+ evaluating to T on a truth-value function for L^+ constitutes a truth set for L^+; and the function on which all the statements of L^+ in a truth set for L^+ evaluate to T and the rest evaluate to F constitutes a truth-value function for L^+. So, the accounts of logical truth in the foregoing paragraph are sure to be weakly sound and complete since those of pages 213–214 were, and the account of logical entailment in that paragraph is sure to be strongly sound and complete since the one on page 214 was. However, more direct and compact proofs of these matters can be had.

Suppose that $S \vdash A$, where S is a set of statements and A is a statement of L; let L^+ be an arbitrary term extension of L; and let S^+ be an arbitrary truth set for L^+ of which S is a subset. It is readily verified that each axiom of L belongs to each truth set for L^+, and hence to S^+; and that the ponential of two statements of L belongs to a truth set for L^+, say S^+, if the two statements themselves do. So each entry in any proof of A from S in L will belong to S^+. So:

THEOREM 3.7. *Let S be an arbitrary set of statements and A be an arbitrary statement of* L. *If $S \vdash A$, then – no matter the term extension L^+ of* L *– A belongs to every truth set for L^+ of which S is a subset (= The Strong Soundness Theorem for* L *in Truth-Set Semantics).*

As for the converse of Theorem 3.7, recall that if a set of statements of L is consistent and infinitely extendible in L, then the Henkin extension of the set is sure by Theorem 1.4 to be maximally consistent and ω-complete in L. But by virtue of Theorem 1.3 any set of statements of L that is maximally consistent and ω-complete in L constitutes a truth set for L (and vice-versa), truth sets being sometimes called, as a result, *Henkin sets*. So:

THEOREM 3.8. *Let S be a set of statements of* L *that is infinitely extendible in* L. *If S is consistent in* L, *then there is a truth set for* L *– and, hence, there is for some term extension L^+ of* L *a truth set for L^+ – of which S is a subset.*

But, if $S \nvdash A$, then $S \cup \{\neg A\}$ is consistent in L. So:

THEOREM 3.9. *Let S be as in Theorem 3.7, and let A be an arbitrary statement of* L. *If $S \nvdash A$, then there is a truth set for* L *– and, hence, there is for some term extension L^+ of* L *a truth set for L^+ – of which S is a subset but A is not a member.*

Hence, thanks to L^∞, when S is not infinitely extendible:

THEOREM 3.10. *Let S be an arbitrary set of statements and A be an arbitrary statement of* L. *If – no matter the term extension L^+ of* L *– A*

I.3: ALTERNATIVES TO FIRST-ORDER SEMANTICS 217

belongs to every truth set for L^+ of which S is a subset, then $S \vdash A$ (= The Strong Completeness Theorem for L in Truth-Set Semantics).

Hence also:

THEOREM 3.11. *Let A be an arbitrary statement of L. Then $\vdash A$ iff A belongs to every truth set for L.*

Proof of Theorem 3.11 is in Smullyan [1968], a text to be consulted on this particular brand of truth-set semantics.

Note that $Q(t_1) \wedge \neg Q(t_1)$ cannot belong to any truth set for L^+ since at most one of $Q(t_1)$ and $\neg Q(t_1)$ does. But a set S of statements of L is consistent in L iff $Q(t_1) \wedge \neg Q(t_1)$ is not provable in L from S. Hence, by Theorem 3.7:

THEOREM 3.12. *Let S be an arbitrary set of statements of L. If there is for some term extension L^+ of L a truth set for L^+ of which S is a subset, then S is consistent in L.*

Hence as a transition to our last topic in this section:

THEOREM 3.13. *Let S be as in Theorem 3.12. Then S is consistent in L iff there is for some term extension L^+ of L a truth set for L^+ of which S is a subset (i.e. to which S extends).*

*

Besides extending to truth sets, consistent sets also extend to *model sets*, a kind of set first investigated in Hintikka [1955]. These are intriguing in several respects. Though true on a truth-value assignment, a model set need not comprise – as a truth set would – all the statements true on that assignment, and hence it is not a truth-value function in disguise. A model set may even be finite, in which case it (plus of course all sets extending to it) has a finite model. And routines have been devised which (i) when a finite set S is inconsistent, will invariably apprise us of the fact and (ii) when S is consistent, will frequently – though not invariably – extend it to a finite model set and hand us a model of S.

Model sets also permit definitions of logical truth and logical entailment which, though reminiscent of those on page 215, cunningly differ from them. It is, of course, these definitions which particularly interest us and which we will legitimize by means of suitable soundness and completeness theorems.

But some of the results in the previous paragraph are readily had and will be recorded as well.

L^+ being an arbitrary term extension of L, I shall understand by a *model set* for L^+ any set S of statements of L^+ such that (i) at least one term of L^+ occurs in S and (ii) S meets the following constraints:

(a) where A is an atomic statement of L^+, at most one of A and $\neg A$ belongs to S,
(b) if $\neg\neg A$ belongs to S, then so does A,
(c) if $A \wedge B$ belongs to S, then so does each of A and B,
(d) if $\neg(A \wedge B)$ belongs to S, then at least one of $\neg A$ and $\neg B$ does,
(e) if $(\forall x)A$ belongs to S, then so does $A(T/x)$ for every term T of L^+ *that occurs in S*,
(f) if $\neg(\forall x)A$ belongs to S, then so does $\neg A(T/x)$ for at least one term T of L.

Note: Because of requirement (i) S must, of course, be non-empty and – more importantly – at least one substitution instance of any universal quantification belonging to S must belong to S (see clause (e)). The latter is tantamount, model-theoretically, to requiring that domains be non-empty, a point first made in Hintikka [1955] pp. 34–35. And, because of the qualification '*that occurs in S*' in clause (e), not every substitution instance of $(\forall x)A$ need belong to S (though, as we just saw, at least one must). This is the most distinctive feature of model sets, and the one which most notably allows (some) finite sets of statements of L^+ to qualify as model sets for L^+.

It is easily verified that each truth set for L constitutes a model set for L, a fact which with an eye to later developments I record separately:

THEOREM 3.14. *Each truth set for L constitutes a model set for L.*

(And each model set for L is a subset – though not necessarily more than a subset – of a truth set for L: as noted above and proved on pages 221–222, each model set for L is true on a truth-value assignment for L and, hence, is a subset of the truth-set associate of that assignment.)

This done, I declare a statement A of L *logically true in the model-set sense* if – no matter the term extension L^+ of L – $\neg A$ does not belong to any model set for L^+ (equivalently, but more simply, if $\neg A$ does not belong to any model set for L); and, where S is a set of statements of L, I declare A *logically entailed by S in the model-set sense* if – no matter the term

I.3: ALTERNATIVES TO FIRST-ORDER SEMANTICS 219

extension L^+ of $L - \neg A$ does not belong to any model set for L^+ of which S is a subset. (The definitions are suggested by results in Hintikka [1955].)

Proof that the foregoing definition of logical entailment is strongly sound calls for four lemmas (Theorems 3.15-3.18) and one definition. Theorem 3.15 is crucial to the whole enterprise. As noted above, the substitution instances of a quantification $(\forall x)A$ that belongs to a model set need not all belong to the set. Truth-value assignment α in Theorem 3.15 ensures that the substitution instances not in the set behave exactly like those in it. (Clause (a) of Theorem 3.15 stands to clause (b) and eventually to Theorem 3.17 somewhat as Theorem 2.1 stands to Theorem 2.6.)

THEOREM 3.15. *Let $\mathscr{T}$ be a non-empty set of terms of L; let $\mathscr{T}'$ be the complement of $\mathscr{T}$; T_1 being the alphabetically earliest member of $\mathscr{T}$, let the T_1-rewrite $T_1(A)$ of a statement A of L be the result of putting T_1 for each member of $\mathscr{T}'$ in A;*[21] *and let α be any truth-value assignment for L such that, for each atomic statement A of L, $\alpha(A) = \alpha(T_1(A))$.*

(a) *$T'_1, T'_2, T'_3, \ldots$, being in alphabetic order the various members (if any) of $\mathscr{T}'$, let the $\mathscr{L}$-rewrite $\mathscr{L}(A)$ of a statement A of L be the result of simultaneously putting T_1 for T'_1 in A, T'_1 for T'_2, T'_2 for T'_3, etc.*[22] *Then A is true on α iff $\mathscr{L}(A)$ is.*

(b) *Let $(\forall x)A$ be a quantification of L in which no member of $\mathscr{T}'$ occurs. If $A(T/x)$ is true on α for each term T in $\mathscr{T}$, then $A(T'/x)$ is true on α for each term T' in $\mathscr{T}'$.*

Proof. (a) That A is true on α iff $\mathscr{L}(A)$ is true on α is shown by mathematical induction on the length $l(A)$ of A.

Basis: $l(A) = 1$, in which case A is atomic. By construction A and $\mathscr{L}(A)$ have the same T_1-rewrite. Hence, $\alpha(A) = \alpha(\mathscr{L}(A))$. Hence, (a).

Inductive Step: $l(A) > 1$.

Case 1: A is a negation $\neg B$. By the hypothesis of the induction B is true on α iff $\mathscr{L}(B)$ is. Hence, $\neg B$ is true on α iff $\neg(\mathscr{L}(B))$ is, i.e. iff $\mathscr{L}(\neg B)$ is. Hence, (a).

Case 2: A is a conjunction $B \wedge C$. Proof similar to that of Case 1.

Case 3: A is a quantification $(\forall x)B$. (i) For each term T in $\mathscr{T}$, $(\mathscr{L}(B))(T/x)$ is the same as $\mathscr{L}(B(T/x))$; and, for each i from 1 on, $(\mathscr{L}(B))(T'_i/x)$ is the same as $\mathscr{L}(B(T'_{i+1}/x))$. So, each substitution instance of $(\forall x)(\mathscr{L}(B))$ $(=\mathscr{L}((\forall x)B))$ is the $\mathscr{L}$-rewrite of a substitution instance for $(\forall x)B$. (ii) For each member T of $\mathscr{T}$, $\mathscr{L}(B(T/x))$ is the same as $(\mathscr{L}(B))(T/x)$; $\mathscr{L}(B(T'_1/x))$ is the same as $(\mathscr{L}(B))(T_1/x)$; and, for each i from 2 on, $\mathscr{L}(B(T'_i/x))$ is the same as $(\mathscr{L}(B))(T'_{i-1}/x)$. So, the $\mathscr{L}$-rewrite of each substitution instance of $(\forall x)B$ is a substitution instance of $(\forall x)(\mathscr{L}(B))$. (iii) Suppose first that $(\forall x)B$ is true on α, and hence that each substitution instance of $(\forall x)B$ is true on α. Then, by the hypothesis of the induction, the $\mathscr{L}$-rewrite of each substitution instance of $(\forall x)B$ is true on α, and hence by (i) so is each substitution instance of $\mathscr{L}((\forall x)B)$. Hence, $\mathscr{L}((\forall x)B)$ is true on α. Suppose next that $(\forall x)B$ is not true on α, and hence that at least one substitution instance of $(\forall x)B$ is not true on α. Then, by the hypothesis of the induction, the $\mathscr{L}$-rewrite of at least one substitution instance of $(\forall x)B$ is not true on α, and hence by (ii) at least one substitution instance of $\mathscr{L}((\forall x)B)$ is not either. Hence, $\mathscr{L}((\forall x)B)$ is not true on α. Hence, (a).

(b) (i) Since no member of $\mathscr{T}'$ occurs in A, $\mathscr{L}(A(T_1'/x))$ is the same as $A(T_1/x)$, and for each i from 2 on $\mathscr{L}(A(T_i'/x))$ is the same as $A(T_{i-1}'/x)$. (ii) Suppose $A(T/x)$ is true on α for each member T of $\mathscr{T}$. Then in particular $A(T_1/x)$ is true on α, hence by (i) so is $\mathscr{L}(A(T_1'/x))$, hence by (a) so is $A(T_1'/x)$, hence by (i) so is $\mathscr{L}(A(T_2'/x))$, hence by (a) so is $A(T_2'/x)$, hence by (i) so is $\mathscr{L}(A(T_3'/x))$, hence by (a) so is $A(T_3'/x)$, etc. Hence, if $A(T/x)$ is true on α for each member T of $\mathscr{T}$, so is $A(T'/x)$ for each member T' of $\mathscr{T}'$. □

Now let S be a model set for L, and let T_1 be the alphabetically earliest term of L that occurs in S.[23] By the *truth-value associate* of S in L, I shall understand the function α_S from the atomic statements of L to $\{T, F\}$ such that, for each atomic statement A of L,

$$\alpha_S(A) = \begin{cases} T & \text{if } T_1(A) \text{ belongs to } S \\ F & \text{otherwise.} \end{cases}$$

(The function stems from Leblanc and Wisdom [1972] p. 191, and the proofs of Theorems 3.16–3.18 below are simplifications of proofs in that text, pp. 300–304.)

THEOREM 3.16. *Let S be a model set for* L, *and α_S be the truth-value associate of S in* L. *Then:*
   (a) *α_S constitutes a truth-value assignment for* L, *and*
   (b) *for each atomic statement A of* L, *$\alpha_S(A) = \alpha_S(T_1(A))$.*

As regards (b), note that any term of L that occurs in $T_1(A)$ is sure to occur in S. So, $T_1(A)$ and $T_1(T_1(A))$ are the same. So, $T_1(A)$ belongs to S iff $T_1(T_1(A))$ does. So, by the construction of α_S, $\alpha_S(A) = \alpha_S(T_1(A))$.

I.3: ALTERNATIVES TO FIRST-ORDER SEMANTICS 221

Hence, by Theorem 3.15(b) (with the set of all the terms that occur in S serving as $\mathscr{T}$, and the truth-value associate of S serving as α):

THEOREM 3.17. *Let S be a model set for* L, *and $(\forall x)A$ be a quantification of* L *that belongs to S. If – for each and every term T of* L *that occurs in S – $A(T/x)$ is true on the truth-value associate of S, then $(\forall x)A$ is true on that associate.*

I am now in a position to show that each model set for L is true on its truth-value associate, hence is true in the model associate of that associate, and hence has a (Henkin) model.

THEOREM 3.18. *Each model set for* L *is true on its truth-value associate.*

Proof. Let S be an arbitrary model set for L, A be an arbitrary member of S, and α_S be the truth-value associate of S. That A is sure to be true on α_S is shown by mathematical induction on the length $l(A)$ of A.

Basis: $l(A) = 1$, in which case A is atomic. Since A belongs to S, A is its own T_1-rewrite (T_1 the alphabetically earliest term of L to occur in S); hence, $\alpha_S(A) = \mathsf{T}$; and, hence, A is true on α_S.

Inductive Step: $l(A) > 1$, in which case A is a negation, or a conjunction, or a universal quantification (and, when a negation, one of an atomic statement, or of a negation, or of a conjunction, or of a universal quantification).[24]

Case 1: A is of the sort $\neg B$, where B is an atomic statement. Then by the definition of a model set B does not belong to S. But, since $\neg B$ belongs to S, $\neg B$ is its own T_1-rewrite and, hence, so is B. Hence, $T_1(B)$ does not belong to S. Hence, $\alpha_S(B) = \mathsf{F}$. Hence, B is not true on α_S. Hence, $\neg B$ is.

Case 2: A is of the sort $\neg\neg B$. Then by the definition of a model set B belongs to S; hence, by the hypothesis of the induction, B is true on α_S; and, hence, so is $\neg\neg B$.

Case 3: A is of the sort $\neg(B \wedge C)$. Then by the definition of a model set at least one of $\neg B$ and $\neg C$ belongs to S; hence, by the hypothesis of the induction, at least one of $\neg B$ and $\neg C$ is true on α_S; and, hence, $\neg(B \wedge C)$ is true on α_S.

Case 4: A is of the sort $\neg(\forall x)B$. Then by the definition of a model set $\neg B(T/x)$ belongs to S for some term T of L; hence, by the hypothesis of the induction, $\neg B(T/x)$ is true on α_S for some term T of L; and, hence, $\neg(\forall x)B$ is true on α_S.

Case 5: A is of the sort $B \wedge C$. Then by the definition of a model set each of B and C belongs to S; hence, by the hypothesis of the induction, each of B and C is true on α_S; and, hence, so is $B \wedge C$.

Case 6: A is of the sort $(\forall x)B$. Then by the definition of a model set $B(T/x)$ belongs to S for each term T of L that occurs in S; hence by the hypothesis of the induction, $B(T/x)$ is true on α_S for each term T of L that occurs in S; and, hence, by Theorem 3.17, $(\forall x)B$ is true on α_S. □

Now for the Strong Soundness Theorem for L in model-set semantics.

THEOREM 3.19. *Let S be an arbitrary set of statements and A be an arbitrary statement of* L. *If $S \vdash A$, then $\neg A$ does not belong to any model set for* L *of which S is a subset.*

Proof. Suppose $S \vdash A$, and suppose there is no model set for L of which S is a subset. Then, trivially, $\neg A$ does not belong to any model set for L of which S is a subset. Suppose, on the other hand, there is at least one model set for L, say, S', of which S is a subset, and let $\alpha_{S'}$, be the truth-value associate of S'. Then, by Theorem 3.18, S' is true on $\alpha_{S'}$; hence, so is S; and hence, by Theorem 3.2, so is A. But, if A is true on $\alpha_{S'}$, then $\neg A$ is not; and, if $\neg A$ is not true on $\alpha_{S'}$, then by Theorem 3.18 again $\neg A$ does not belong to S'. Hence, again, $\neg A$ does not belong to any model set for L of which S is a subset. □

But the argument leading to Theorem 3.19 holds with 'L$^+$' in place of 'L'. Hence:

THEOREM 3.20. *Let S and A be as in Theorem 3.19. If $S \vdash A$, then – no matter the term extension* L$^+$ *of* L *– $\neg A$ does not belong to any model set for* L$^+$ *of which S is a subset* (= *The Strong Soundness Theorem for* L *in Model-Set Semantics*).

The converse of Theorem 3.19 for infinitely extendible S and that of Theorem 3.20 for arbitrary S readily issue from Theorems 3.14, 3.9 and 3.10:

THEOREM 3.21. *Let S be a set of statements of* L *that is infinitely extendible in* L, *and A be an arbitrary statement of* L. *If $\neg A$ does not belong to any model set for* L *of which S is a subset, then $S \vdash A$.*

THEOREM 3.22. *Let S be an arbitrary set of statements and A be an*

I.3: ALTERNATIVES TO FIRST-ORDER SEMANTICS

arbitrary statement of L. *If – no matter the term extension* L^+ *of* L *–* $\neg A$ *does not belong to any model set for* L^+ *of which S is a subset, then* $S \vdash A$
(= *The Strong Completeness Theorem for* L *in Model-Set Semantics*).

Hence:

THEOREM 3.23. *Let S and A be as in Theorem 3.22. Then* $S \vdash A$ *iff – no matter the term extension* L^+ *of* L *–* $\neg A$ *does not belong to any model set for* L^+ *of which S is a subset.*

And hence:

THEOREM 3.24. $\vdash A$ *iff – no matter the term extension* L^+ *of* L *–* $\neg A$ *does not belong to any model set for* L^+.

Theorem 3.23 legitimizes the account of logical entailment and Theorem 3.24 the first account of logical truth, on page 218.

Theorems 3.19 and 3.22 combine, of course, to read:

THEOREM 3.25. *Let S be a set of statements of* L *that is infinitely extendible in* L, *and A be an arbitrary statement of* L. *Then* $S \vdash A$ *iff* $\neg A$ *does not belong to any model set for* L *of which S is a subset.*

Hence, the following theorem, which legitimizes the second (and simpler) account of logical truth on page 218:

THEOREM 3.26. $\vdash A$ *iff* $\neg A$ *does not belong to any model set for* L.

Now for some of the results reported on page 217. Note that (i) a set S of statements of L is consistent in L iff $Q(t_1) \wedge \neg Q(t_1)$, for example, is not provable from S in L, and (ii) any model set for L^+ is a subset of one with $\neg(Q(t_1) \wedge \neg Q(t_1))$ as a member. (For proof of (ii) note that any model set for L^+ is a subset of a truth set for L^+ (page 218), $\neg(Q(t_1) \wedge \neg Q(t_1))$ belongs by Theorem 3.6(b) to every truth set for L^+, and any truth set for L^+ is a model set for L^+. So, any model set for L^+ is sure to be a subset of one with $\neg(Q(t_1) \wedge \neg Q(t_1))$ as a member.) Hence, by Theorem 3.22:

THEOREM 3.27. *Let S be an arbitrary set of statements of* L. *Then S is consistent in* L *iff there is, for some term extension* L⁺ *of* L, *a model set for* L⁺ *to which S extends (i.e. of which S is a subset).*

Similarly, but with Theorem 3.25 substituting for Theorem 3.22:

THEOREM 3.28. *Let S be a set of statements of* L *that is infinitely extendible in* L. *Then S is consistent in* L *iff there is a model set for* L *to which S extends (i.e. of which S is a subset).*

The two results stem from Hintikka [1955], where – *thanks to first-order languages having as many terms as the occasion calls for* – sets of first-order statements were shown to be consistent iff they extend to model sets. Since a set of first-order statements is consistent iff it has a model, Hintikka's result read, in effect: "Sets of first-order statements have *models* iff they extend to *model* sets."

As suggested earlier, one model in which a model set for L⁺ is sure to be true can be obtained from pages 211–212 and 220. Indeed, let S be a model set for L⁺; let α_S^+ be the counterpart for L⁺ of the truth-value assignment α_S on page 220; and let $\langle \mathcal{T}^+, I_{\mathcal{T}^+} \rangle$ be the model associate of α_S^+ as per pages 211–212. By the counterpart of Theorem 3.18 for L⁺, S is sure to be true on α_S^+. Hence, S is sure by Theorem 3.1 to have $\langle \mathcal{T}^+, I_{\mathcal{T}^+} \rangle$ as a model. Hence, so are all sets of statements of L extending to S. The model, one will recall, is a Henkin one.

THEOREM 3.29. *Let* L⁺ *be an arbitrary term extension of* L, *and S be an arbitrary model set for* L⁺. *Then there is a (Henkin) model for* L⁺ *in which S (and, hence, each set of statements of* L *that extends to S) is true.*

Since the set $\mathcal{T}^+$ of all the terms of L⁺ is infinite, model $\langle \mathcal{T}^+, I_{\mathcal{T}^+} \rangle$ is infinite as well, a point noted in Smullyan [1968] p. 62. However, when S is a finite model set for L, S is sure to have a finite model as well. Let (i) $\mathcal{T}_S$ consist of the various terms of L that occur in S, (ii) for each term T of L, let $I_{\mathcal{T}_S}(T)$ be T itself if T belongs to $\mathcal{T}_S$, otherwise the alphabetically earliest member of $\mathcal{T}_S$, and (iii) for each predicate Q of L of degree d, let $I_{\mathcal{T}_S}(Q)$ be $\{\langle I_{\mathcal{T}_S}(T_1), I_{\mathcal{T}_S}(T_2), \ldots, I_{\mathcal{T}_S}(T_d)\rangle : Q(T_1, T_2, \ldots, T_d) \in S\}$. Theorems 3.18 and 3.1 are easily edited to show that S has finite model $\langle \mathcal{T}_S, I_{\mathcal{T}_S} \rangle$ as a model. Hence, so do all sets of statements of L extending to S. The model, again, is a Henkin one.

I.3: ALTERNATIVES TO FIRST-ORDER SEMANTICS

THEOREM 3.30. *Let S be a finite model set for* L. *Then there is a finite (Henkin) model for* L *in which S (and, hence, each set of statements of* L *that extends to S) is true.*

It can further be shown of any finite set S of statements of L, that if S has a finite model, then S extends to a finite model set for L. *A finite set of statements of* L *thus has a finite model iff it extends to a finite model set for* L. The result is particularly interesting as regards the routine in Leblanc and Wisdom [1972] (and that in Jeffrey [1967] from which it stems) for making consistency trees. As pointed out on p. 298 of Leblanc and Wisdom [1972], any set of statements declared CONSISTENT by the routine of p. 189 is a subset of a finite model set (as that text puts it, a subset of a finite *Hintikka set*). Any such set thus has a finite model. The routine in question declares CONSISTENT only some of the sets of statements of L that extend to finite model sets. However, I have since found another routine which *does* declare CONSISTENT *all* sets of statements of L that extend to finite model sets.[25]

So much, however, for accounts of logical truth and entailment which – be it overtly or covertly – hinge upon the notion of truth.

4. PROBABILISTIC SEMANTICS (I)

My concern in this section and the next is with probabilities, i.e. with degrees of rational belief. The probability functions I consider first are *singulary* real-valued functions. They will thus take *single* statements of L^+ as their arguments, the way truth-value functions do; but they will take as their values reals from the entire interval [0, 1], unlike truth-value functions, which merely take the end-points 0 and 1. I shall place on the functions seven constraints ensuring that "for any such function P^+ for L^+, a rational agent might *simultaneously* believe the various statements of L^+ – say, A_1, A_2, A_3, etc.[26] – to the respective extents $P^+(A_1)$, $P^+(A_2)$, $P^+(A_3)$, etc."[27] These constraints, adopted in some cases and adapted in others from Kolmogorov [1933], Popper [1955], and Gaifman [1964], are of considerable interest.[28] Not only do they eschew all semantic notions, thus being what one calls *autonomous*; they in fact permit definition of several of them – in particular, logical truth and logical entailment. So, besides freeing probability theory (hence, to some extent, inductive logic) of its past dependence upon deductive logic, they make for a brand-new semantics: *probabilistic semantics*.

Thanks in good part to Popper, the functions studied here thus have a fresh look and a new thrust. They nonetheless are of the most orthodox, and in the present context most welcome, sort. They accord to negations and conjunctions the very values that the functions in Kolmogorov [1933] would to complements and intersections. And they accord to universal quantifications values complying with the substitution interpretation of '∀' (and, quite serviceably, accord such values to *all* universal quantifications).

Popper's interest eventually shifted from the present functions to the binary ones studied in Popper [1959]. It is the latter which figure most prominently in recent contributions to probabilistic semantics. I shall investigate them in the second half of Section 5. For novelty's sake, however, and because of the close relationship they bear to truth-value functions, I shall devote this section and half the next to singulary functions.

Formal details, borrowed from Leblanc [1982c], are as follows.

Let L^+ be an arbitrary term extension of L. By a (*singulary*) *probability function* for L^+ I shall understand any function P^+ from the statements of L^+ to the reals that meets the following constraints:

C1. $\quad 0 \leqslant P^+(A)$

C2. $\quad P^+(\neg(A \wedge \neg A)) = 1$

C3. $\quad P^+(A) = P^+(A \wedge B) + P^+(A \wedge \neg B)$

C4. $\quad P^+(A) \leqslant P^+(A \wedge A)$

C5. $\quad P^+(A \wedge B) \leqslant P^+(B \wedge A)$

C6. $\quad P^+(A \wedge (B \wedge C)) \leqslant P^+((A \wedge B) \wedge C)$

C7. $\quad P^+(A \wedge (\forall x)B) = \text{Limit}_{j \to \infty} P^+(A \wedge \Pi_{i=1}^{j} B(t_i^+/x)).$[29]

I shall declare a statement A of L *logically true in the probabilistic sense* if – no matter the term extension L^+ of L and probability function P^+ for L^+ – $P^+(A) = 1$; and, where S is a set of statements of L, I shall declare A *logically entailed by S in the probabilistic sense* if – no matter the term extension L^+ of L and probability function P^+ for L^+ – $P^+(A) = 1$ if $P^+(B) = 1$ for each member B of S. Equivalently, but more simply, A may be declared logically true in the probabilistic sense if A evaluates to 1 on every probability function for L.

I shall legitimize the foregoing account of logical entailment by showing it strongly sound (= Theorem 4.43) and strongly complete (= Theorem 4.48). That the first account of logical truth is legitimate follows, of course, from

I.3: ALTERNATIVES TO FIRST-ORDER SEMANTICS

Theorems 4.43 and 4.48; and that the second is follows from the theorems leading to Theorems 4.43 and 4.48.

Preparatory to proving Theorem 4.43, the Strong Soundness Theorem for L in probabilistic semantics, I establish that (i) each axiom of L evaluates to 1 on an arbitrary function P for L (= Theorem 4.41) and (ii) the ponential B of two statements A and $A \to B$ of L evaluates to 1 on P if A and $A \to B$ do (= Theorem 4.30). Proof of (i) is in three steps. I first show that each axiom of L of sorts **A1–A3** evaluates to 1 on P (= Theorems 4.18, 4.19, and 4.29). Given this first result and Theorem 4.30, I next show that if $\vdash_0 A$ (i.e. if A is provable in L by means of just **A1–A3**), then A evaluates to 1 on P (= Theorem 4.31). And, given this second result, I then show that each axiom of L of sorts **A4–A6** – and, more generally, each axiom of L – evaluates to 1 on P (= Theorems 4.38 and 4.41).

As the reader will notice, the proofs of Theorems 4.1–4.41 hold with 'P^+' everywhere for 'P'. Hence, so do Theorems 4.1–4.41 themselves. Hence, so do such among Theorems 4.1–4.41 as are needed to prove Theorem 4.43 (and, further on, Theorem 4.48).

THEOREM 4.1. $P(A \wedge B) \leq P(A)$.
Proof. $P(A \wedge \neg B) \geq 0$ by **C1**. Hence, $P(A) \geq P(A \wedge B)$ by **C3**. Hence, Theorem 4.1. □

THEOREM 4.2. $P(A) = P(A \wedge A)$.
Proof by **C4** and Theorem 4.1. □

THEOREM 4.3. $P(A \wedge B) = P(B \wedge A)$.
Proof by **C5**. □

THEOREM 4.4. $P((A \wedge B) \wedge C) = P(A \wedge (B \wedge C))$.
Proof. $P((A \wedge B) \wedge C) = P(C \wedge (A \wedge B))$ (Theorem 4.3)
$\leq P((C \wedge A) \wedge B)$ (**C6**)
$\leq P(B \wedge (C \wedge A))$ (Theorem 4.3)
$\leq P((B \wedge C) \wedge A)$ (**C6**)
$\leq P(A \wedge (B \wedge C))$ (Theorem 4.3).

Hence Theorem 4.4 by **C6**. □

THEOREM 4.5. $P(A) = P(B \wedge A) + P(\neg B \wedge A)$.
Proof by **C3** and Theorem 4.3. □

THEOREM 4.6. If $P(A) \leq 0$, then $P(A) = 0$.
Proof by **C1**. □

THEOREM 4.7. $P(A \wedge B) \leq P(B)$.
Proof by Theorems 4.1 and 4.3. □

THEOREM 4.8. $P(A \wedge \neg A) = 0$.
Proof. $P(A \wedge \neg A) = P(A) - P(A \wedge A)$ (**C3**)

$$= 0 \qquad \text{(Theorem 4.2).} \quad \square$$

THEOREM 4.9. $P((A \wedge \neg A) \wedge B) = 0$.
Proof. $P((A \wedge \neg A) \wedge B) \leq P(A \wedge \neg A)$ (Theorem 4.1)

$$\leq 0 \qquad \text{(Theorem 4.8)}$$

$$= 0 \qquad \text{(Theorem 4.6).} \quad \square$$

THEOREM 4.10. $P(A) = P(\neg(B \wedge \neg B) \wedge A)$.
Proof. $P(A) = P(\neg(B \wedge \neg B) \wedge A) + P((B \wedge \neg B) \wedge A)$ (Theorem 4.5)

$$= P(\neg(B \wedge \neg B) \wedge A) \qquad \text{(Theorem 4.9).} \quad \square$$

THEOREM 4.11. $P(\neg A) = 1 - P(A)$.
Proof. $P(\neg A) = P(\neg(A \wedge \neg A) \wedge \neg A)$ (Theorem 4.10)

$$= P(\neg(A \wedge \neg A)) - P(\neg(A \wedge \neg A) \wedge A) \qquad (\mathbf{C3})$$

$$= 1 - P(\neg(A \wedge \neg A) \wedge A) \qquad (\mathbf{C2})$$

$$= 1 - P(A) \qquad \text{(Theorem 4.10).} \quad \square$$

THEOREM 4.12. $P(A) \leq 1$.
Proof. $P(A) = 1 - P(\neg A)$ by Theorem 4.11. But $P(\neg A) \geq 0$ by **C1**. Hence, Theorem 4.12. □

THEOREM 4.13. If $P(A) \geq 1$, then $P(A) = 1$.
Proof by Theorem 4.12. □

I.3: ALTERNATIVES TO FIRST-ORDER SEMANTICS

THEOREM 4.14. $P(\neg A \wedge (A \wedge B)) = 0$.
Proof.
$$\begin{aligned}P(\neg A \wedge (A \wedge B)) &= P((\neg A \wedge A) \wedge B) &&\text{(Theorem 4.4)}\\ &\leq P(\neg A \wedge A) &&\text{(Theorem 4.1)}\\ &\leq P(A \wedge \neg A) &&\text{(Theorem 4.3)}\\ &\leq 0 &&\text{(Theorem 4.8)}\\ &= 0 &&\text{(Theorem 4.6).} \quad \square\end{aligned}$$

THEOREM 4.15. $P(A \wedge B) = P(A \wedge (A \wedge B))$.
Proof.
$$\begin{aligned}P(A \wedge B) &= P(A \wedge (A \wedge B)) + P(\neg A \wedge (A \wedge B)) &&\text{(Theorem 4.5)}\\ &= P(A \wedge (A \wedge B)) &&\text{(Theorem 4.14).} \quad \square\end{aligned}$$

THEOREM 4.16. $P(A \to B) = 1$ iff $P(A \wedge \neg B) = 0$.
Proof by $D_\to$ and Theorem 4.11. $\quad\square$

THEOREM 4.17. $P(A \to B) = 1$ iff $P(A) = P(A \wedge B)$.
Proof by Theorem 4.16 and **C3**. $\quad\square$

THEOREM 4.18. $P(A \to (A \wedge A)) = 1 \quad (= \textit{Axiom Schema } \mathbf{A1})$.
Proof.
$$\begin{aligned}P(A) &= P(A \wedge A) &&\text{(Theorem 4.2)}\\ &= P(A \wedge (A \wedge A)) &&\text{(Theorem 4.15).}\end{aligned}$$
Hence, Theorem 4.18 by Theorem 4.17. $\quad\square$

THEOREM 4.19. $P((A \wedge B) \to A) = 1 \quad (= \textit{Axiom Schema } \mathbf{A2})$.
Proof.
$$\begin{aligned}P(A \wedge B) &= P(A \wedge (A \wedge B)) &&\text{(Theorem 4.15)}\\ &= P((A \wedge B) \wedge A) &&\text{(Theorem 4.3).}\end{aligned}$$
Hence, Theorem 4.19 by Theorem 4.17. $\quad\square$

THEOREM 4.20. $P((A \wedge B) \to B) = 1$.
Proof.
$$\begin{aligned}P(A \wedge B) &= P((A \wedge B) \wedge B) + P((A \wedge B) \wedge \neg B) &&\textbf{(C3)}\\ &= P((A \wedge B) \wedge B) + P(A \wedge (B \wedge \neg B)) &&\text{(Theorem 4.4)}\\ &= P((A \wedge B) \wedge B) + P((B \wedge \neg B) \wedge A) &&\text{(Theorem 4.3)}\\ &= P((A \wedge B) \wedge B) &&\text{(Theorem 4.9).}\end{aligned}$$
Hence, Theorem 4.20 by Theorem 4.17. $\quad\square$

THEOREM 4.21. *If $P(A \to B) = 1$ and $P(B \to C) = 1$, then $P(A \to C) = 1$.*
Proof. Suppose first that $P(A \to B) = 1$. Then

$$0 = P(A \land \neg B) \qquad \text{(Theorem 4.16)}$$
$$\geqslant P(\neg C \land (A \land \neg B)) \qquad \text{(Theorem 4.7)}$$
$$= P(\neg C \land (A \land \neg B)) \qquad \text{(Theorem 4.6)}$$
$$= P((\neg C \land A) \land \neg B) \qquad \text{(Theorem 4.4)}.$$

Suppose next that $P(B \to C) = 1$. Then

$$0 = P(B \land \neg C) \qquad \text{(Theorem 4.16)}$$
$$\geqslant P((B \land \neg C) \land A) \qquad \text{(Theorem 4.1)}$$
$$= P((B \land \neg C) \land A) \qquad \text{(Theorem 4.6)}$$
$$= P(B \land (\neg C \land A)) \qquad \text{(Theorem 4.4)}$$
$$= P((\neg C \land A) \land B) \qquad \text{(Theorem 4.3)}.$$

Hence, $P(\neg C \land A) = 0$ by **C3**; hence, $P(A \land \neg C) = 0$ by Theorem 4.3; and, hence, $P(A \to C) = 1$ by Theorem 4.16. Hence, Theorem 4.21. □

THEOREM 4.22. *If $P(A \to (B \land C)) = 1$, then $P(A \to B) = 1$.*
Proof. Suppose $P(A \to (B \land C)) = 1$. Then

$$P(A) = P(A \land (B \land C)) \qquad \text{(Theorem 4.17)}$$
$$= P((A \land B) \land C) \qquad \text{(Theorem 4.4)}$$
$$\leqslant P(A \land B) \qquad \text{(Theorem 4.1)}.$$

Hence, $P(A) = P(A \land B)$ by Theorem 4.1; and, hence, $P(A \to B) = 1$ by Theorem 4.17. Hence, Theorem 4.22. □

THEOREM 4.23. $P(A \land \neg\neg B) = P(A \land B)$.
Proof. By **C3**

$$P(A) = P(A \land \neg B) + P(A \land \neg\neg B)$$

and

$$P(A) = P(A \land \neg B) + P(A \land B).$$

Hence, Theorem 4.23. □

I.3: ALTERNATIVES TO FIRST-ORDER SEMANTICS

THEOREM 4.24. *If* $P(A \to (B \to C)) = 1$ *and* $P(A \to B) = 1$, *then* $P(A \to C) = 1$.
Proof. Suppose first that $P(A \to (B \to C)) = 1$. Then

$0 = P(A \land \neg\neg(B \land \neg C))$	(Theorem 4.16 and $D_\to$)
$= P(A \land (B \land \neg C))$	(Theorem 4.23)
$= P((A \land B) \land \neg C)$	(Theorem 4.4)
$= P(\neg C \land (A \land B))$	(Theorem 4.3)
$= P((\neg C \land A) \land B)$	(Theorem 4.4).

Suppose next that $P(A \to B) = 1$. Then

$0 = P(A \land \neg C)$	(Theorem 4.16)
$\geq P(\neg C \land (A \land \neg B))$	(Theorem 4.7)
$= P(\neg C \land (A \land \neg B))$	(Theorem 4.6)
$= P((\neg C \land A) \land \neg B)$	(Theorem 4.4).

Hence, $P(\neg C \land A) = 0$ by **C3**; hence, $P(A \land \neg C) = 0$ by Theorem 4.3; and, hence, $P(A \to C) = 1$ by Theorem 4.16. Hence, Theorem 4.24. □

THEOREM 4.25. *If* $P(A \to B) = 1$ *and* $P(A \to C) = 1$, *then* $P(A \to (B \land C)) = 1$.
Proof. Suppose first that $P(A \to B) = 1$. Then

$P(A) = P(A \land B)$	(Theorem 4.17)
$= P((A \land B) \land C) + P((A \land B) \land \neg C)$	(**C3**)
$= P(A \land (B \land C)) + P((A \land B) \land \neg C)$	(Theorem 4.4)
$= P(A \land (B \land C)) + P(\neg C \land (A \land B))$	(Theorem 4.3).

Suppose next that $P(A \to C) = 1$. Then

$0 = P(A \land \neg C)$	(Theorem 4.16)
$= P(\neg C \land A)$	(Theorem 4.3)
$\geq P((\neg C \land A) \land B)$	(Theorem 4.1)
$= P((\neg C \land A) \land B)$	(Theorem 4.6)
$= P(\neg C \land (A \land B))$	(Theorem 4.4).

Hence, $P(A) = P(A \land (B \land C))$; and, hence, $P(A \to (B \land C)) = 1$ by Theorem 4.17. Hence, Theorem 4.25. □

THEOREM 4.26. *If* $P(A \to (B \wedge C)) = 1$, *then* $P(A \to C) = 1$.
Proof. Suppose that $P(A \to (B \wedge C)) = 1$. Then

$P(A) = P(A \wedge (B \wedge C))$	(Theorem 4.17)
$= P((A \wedge B) \wedge C)$	(Theorem 4.4)
$= P(C \wedge (A \wedge B))$	(Theorem 4.3)
$= P((C \wedge A) \wedge B)$	(Theorem 4.4)
$\leqslant P(C \wedge A)$	(Theorem 4.1)
$\leqslant P(A \wedge C)$	(Theorem 4.3).

Hence, $P(A \to C) = 1$ by the same reasoning as in the proof of Theorem 4.22. Hence, Theorem 4.26. □

THEOREM 4.27. *If* $P(A \to B) = 1$ *and* $P(A \to \neg B) = 1$, *then* $P(A) = 0$.
Proof. Suppose $P(A \to B) = 1$. Then $P(A \wedge \neg B) = 0$ by Theorem 4.16. Suppose further that $P(A \to \neg B) = 1$. Then $P(A \wedge \neg\neg B) = 0$ by Theorem 4.16 again. Hence, $P(A) = 0$ by **C3**. Hence, Theorem 4.27. □

THEOREM 4.28. *If* $P((A \wedge B) \to C) = 1$, *then* $P(A \to (B \to C)) = 1$.
Proof. Suppose $P((A \wedge B) \to C) = 1$. Then

$0 = P((A \wedge B) \wedge \neg C)$	(Theorem 4.16)
$= P(A \wedge (B \wedge \neg C))$	(Theorem 4.4)
$= P(A \wedge \neg\neg(B \wedge \neg C))$	(Theorem 4.23).

Hence, $P(A \to (B \to C)) = 1$ by Theorem 4.17 and $D_\to$. Hence, Theorem 4.28. □

THEOREM 4.29. $P((A \to B) \to (\neg(B \wedge C) \to \neg(C \wedge A))) = 1$ (= *Axiom Schema* **A3**).
Proof. (i) By Theorem 4.20
$$P((((A \to B) \wedge \neg(B \wedge C)) \wedge (C \wedge A)) \to (C \wedge A)) = 1.$$
But by Theorem 4.19
$$P((C \wedge A) \to C) = 1.$$
Hence, by Theorem 4.21
$$P((((A \to B) \wedge \neg(B \wedge C)) \wedge (C \wedge A)) \to C) = 1.$$

I.3: ALTERNATIVES TO FIRST-ORDER SEMANTICS

(ii) Similarly, but with Theorem 4.20 substituting for Theorem 4.19.

$$P((((A \to B) \land \neg(B \land C)) \land (C \land A)) \to A) = 1.$$

But by Theorem 4.19

$$P((((A \to B) \land \neg(B \land C)) \land (C \land A)) \to ((A \to B) \land \neg(B \land C))) = 1,$$

and, hence, by Theorem 4.22

$$P((((A \to B) \land \neg(B \land C)) \land (C \land A)) \to (A \to B)) = 1.$$

Hence, by Theorem 4.24

$$P((((A \to B) \land \neg(B \land C)) \land (C \land A)) \to B) = 1.$$

(iii) By (i)–(ii) and Theorem 4.25

$$P((((A \to B) \land \neg(B \land C)) \land (C \land A)) \to (B \land C)) = 1.$$

But by Theorem 4.19

$$P((((A \to B) \land \neg(B \land C)) \land (C \land A)) \to ((A \to B) \land \neg(B \land C))) = 1,$$

and, hence, by Theorem 4.26

$$P((((A \to B) \land \neg(B \land C)) \land (C \land A)) \to \neg(B \land C)) = 1.$$

Hence, by Theorem 4.27

$$P(((A \to B) \land \neg(B \land C)) \land (C \land A)) = 0,$$

hence, by Theorem 4.23

$$P(((A \to B) \land \neg(B \land C)) \land \neg\neg(C \land A)) = 0,$$

hence, by Theorem 4.16

$$P(((A \to B) \land \neg(B \land C)) \to \neg(C \land A)) = 1,$$

and, hence, by Theorem 4.28

$$P((A \to B) \to (\neg(B \land C) \to \neg(C \land A))) = 1. \qquad \square$$

THEOREM 4.30. *If* $P(A) = 1$ *and* $P(A \to B) = 1$, *then* $P(B) = 1$ (= *Modus Ponens*).

Proof. Suppose $P(A) = 1$ and $P(A \to B) = 1$. Then by Theorem 4.17 $P(A \land B) = 1$; hence, by Theorem 4.7 $P(B) \geq 1$; and hence, by Theorem 4.13 $P(B) = 1$. $\qquad \square$

Hence:

THEOREM 4.31. *If* $\vdash_0 A$, *then* $P(A) = 1$.

Proof. Suppose the column made up of $B_1, B_2, \ldots, B_p$ constitutes a proof of A in L by means of just **A1–A3**. It is easily shown by mathematical induction on i that, for each i from 1 through p, $P(B_i) = 1$. For in the case that B_i is an axiom, $P(B_i) = 1$ by Theorems 4.18, 4.19, and 4.29; and in the case that B_i is the ponential of two previous entries in the column, $P(B_i) = 1$ by Theorem 4.30 and the hypothesis of the induction. Hence, $P(B_p) = 1$, i.e. $P(A) = 1$. Hence, Theorem 4.31. □

And hence:

THEOREM 4.32. *If* $\vdash_0 A \leftrightarrow B$, *then* $P(A) = P(B)$.

Proof. If $\vdash_0 A \leftrightarrow B$, then $\vdash_0 A \to B$ and $\vdash_0 B \to A$; hence, by Theorem 4.17,

$$P(A) = P(A \wedge B)$$

and

$$P(B) = P(B \wedge A),$$

and, hence, by Theorem 4.3

$$P(A) = P(B). \qquad \square$$

With Theorem 4.32 on hand, I am ready to show that all axioms of L of sorts **A4–A6** – and, more generally, all axioms of L – evaluate to 1 on any probability function for L.

THEOREM 4.33. $P((\forall x)A) = \text{Limit}_{j \to \infty} P(\Pi_{i=1}^{j} A(t_i/x))$.

Proof. By C7

$$P(\neg(A \wedge \neg A) \wedge (\forall x)A) = \text{Limit}_{j \to \infty} P(\neg(A \wedge \neg A) \wedge \Pi_{i=1}^{j} A(t_i/x)).$$

Hence, Theorem 4.33 by Theorem 4.10 and the definition of a limit. □

THEOREM 4.34. *If* $P(A) = 1$ *and* $P(B) = 1$, *then* $P(A \wedge B) = 1$.

Proof. Suppose $P(A) = 1$. Then $P(A \wedge B) + P(A \wedge \neg B) = 1$ by C3. Suppose also that $P(B) = 1$. Then $P(\neg B) = 0$ by Theorem 4.11; hence, $P(A \wedge \neg B) \leq 0$ by Theorem 4.7; and, hence, $P(A \wedge \neg B) = 0$ by Theorem 4.6. Hence, Theorem 4.34. □

I.3: ALTERNATIVES TO FIRST-ORDER SEMANTICS   235

THEOREM 4.35. *If $P(A_i) = 1$ for each i from 1 through j, then $P(\Pi_{i=1}^{j} A_i) = 1$.*
Proof by Theorem 4.34 and mathematical induction on j.   □

THEOREM 4.36. *If $P(A(t_i/x)) = 1$ for each i from 1 on, then $P((\forall x)A) = 1$.*
Proof by Theorems 4.35 and 4.33, and the definition of a limit.   □

THEOREM 4.37. *If $P(A \to B(t_i/x)) = 1$ for each i from 1 on, then $P(A \to (\forall x)B) = 1$.*
Proof. Suppose

$$P(A \to B(t_i/x)) = 1$$

for each i from 1 on. Then by Theorem 4.25 and mathematical induction on j

$$P(A \to \Pi_{i=1}^{j} B(t_i/x)) = 1$$

for each j from 1 on, hence, by Theorem 4.17

$$P(A) = P(A \wedge \Pi_{i=1}^{j} B(t_i/x)),$$

hence, by the definition of a limit

$$P(A) = \text{Limit}_{j \to \infty} P(A \wedge \Pi_{i=1}^{j} B(t_i/x)),$$

hence, by **C7**

$$P(A) = P(A \wedge (\forall x)B),$$

and, hence, by Theorem 4.17

$$P(A \to (\forall x)B) = 1.$$

Hence, Theorem 4.37.   □

THEOREM 4.38. $P(A \to (\forall x)A) = 1$ (= *Axiom Schema* **A4**).
Proof. Since x here is sure to be foreign to A, $A(t_i/x)$ is sure to be the same as A. Hence,

$$\vdash_0 (A \wedge \Pi_{i=1}^{j} A(t_i/x)) \leftrightarrow A,$$

hence, by Theorem 4.32 and the definition of a limit

$$\text{Limit}_{j \to \infty} P(A \wedge \Pi_{i=1}^{j} A(t_i/x)) = P(A),$$

hence, by **C7**

$$P(A \wedge (\forall x)A) = P(A),$$

and, hence, by Theorem 4.17

$$P(A \to (\forall x)A) = 1.$$ □

THEOREM 4.39. $P((\forall x)A \to A(T/x)) = 1$ (= *Axiom Schema* **A5**).
Proof. Let T be the kth term of L. So long as $j \geqslant k$,

$$\vdash_0 (A(T/x) \wedge \Pi_{i=1}^{j} A(t_i/x)) \leftrightarrow \Pi_{i=1}^{j} A(t_i/x).$$

Hence, by Theorem 4.32 and the definition of a limit

$$\text{Limit}_{j \to \infty} P(A(T/x) \wedge \Pi_{i=1}^{j} A(t_i/x)) = \text{Limit}_{j \to \infty} P(\Pi_{i=1}^{j} A(t_i/x)),$$

hence, by **C7** and Theorem 4.33

$$P(A(T/x) \wedge (\forall x)A) = P((\forall x)A),$$

hence, by Theorem 4.3

$$P((\forall x)A \wedge A(T/x)) = P((\forall x)A),$$

and, hence, by Theorem 4.17

$$P((\forall x)A \to A(T/x)) = 1.$$ □

THEOREM 4.40. $P((\forall x)(A \to B) \to ((\forall x)A \to (\forall x)B)) = 1$ (*Axiom Schema* **A6**).
Proof. Let T be an arbitrary term of L. By Theorem 4.19

$$P(((\forall x)(A \to B) \wedge (\forall x)A) \to (\forall x)(A \to B)) = 1.$$

But by Theorem 4.39

$$P((\forall x)(A \to B) \to (A \to B)(T/x)) = 1.$$

Hence, by Theorem 4.21

$$P(((\forall x)(A \to B) \wedge (\forall x)A) \to (A \to B)(T/x)) = 1.$$

Similarly, but using Theorem 4.20 in place of Theorem 4.19,

$$P(((\forall x)(A \to B) \wedge (\forall x)A) \to A(T/x)) = 1.$$

Hence, by Theorem 4.24

$$P(((\forall x)(A \to B) \wedge (\forall x)A) \to B(T/x)) = 1,$$

hence, by Theorem 4.37 and the hypothesis on T

$$P(((\forall x)(A \to B) \wedge (\forall x)A) \to (\forall x)B) = 1,$$

I.3: ALTERNATIVES TO FIRST-ORDER SEMANTICS

and, hence, by Theorem 4.28

$$P((\forall x)(A \to B) \to ((\forall x)A \to (\forall x)B)) = 1. \qquad \square$$

THEOREM 4.41. *If A is an axiom of L, then $P(A) = 1$.*

Proof. Suppose A is an axiom of L, in which case A is bound by Theorem 1.1 to be of the sort

$$(\forall x_1)(\forall x_2) \ldots (\forall x_n)(B(x_1, x_2, \ldots, x_n/T_1, T_2, \ldots, T_n)),$$

where $n \geqslant 0$ and B is of one of the sorts **A1–A6**. Proof that $P(A) = 1$ will be by mathematical induction on n.

Basis: $n = 0$. Then A $(= B)$ is of one of the sorts **A1–A6** and, hence, $P(A) = 1$ by Theorems 4.18, 4.19, 4.29, and 4.38–4.40.

Inductive Step: $n > 0$. By Theorem 1.2

$$((\forall x_2) \ldots (\forall x_n)(B(x_1, x_2, \ldots, x_n/T_1, T_2, \ldots, T_n)))(t_i/x_1)$$

constitutes an axiom of L for each i from 1 on. Hence, by the hypothesis of the induction

$$P(((\forall x_2) \ldots (\forall x_n)(B(x_1, x_2, \ldots, x_n/T_1, T_2, \ldots, T_n)))(t_i/x_i)) = 1$$

for each i from 1 on. Hence, by Theorem 4.37,

$$P((\forall x_1)(\forall x_2) \ldots (\forall x_n)(B(x_1, x_2, \ldots, x_n/T_1, T_2, \ldots, T_n))) = 1.$$

Hence, $P(A) = 1$. $\qquad \square$

With Theorems 4.30 and 4.41 on hand, roughly the same induction as on page 234 delivers Theorem 4.42, which in turn delivers the Strong Soundness Theorem for L in probabilistic semantics. Suppose indeed that the column made up of $B_1, B_2, \ldots, B_p$ constitutes a proof of A from S in L, and suppose each member of S evaluates to 1 on P. In the case that B_i belongs to S, $P(B_i) = 1$ by the hypothesis on P; in the case that B_i is an axiom of L, $P(B_i) = 1$ by Theorem 4.41; and in the case that B_i is the ponential of two earlier entries in the column, $P(B_i) = 1$ by Theorem 4.30 and the hypothesis of the induction. Hence, $P(B_i) = 1$ for each i from 1 through p. Hence, $P(B_p) = 1$. Hence, $P(A) = 1$.

THEOREM 4.42. *If $S \vdash A$, then A evaluates to 1 on P if all the members of S do.*

But Theorem 4.30 and 4.34 hold with 'P^+' for 'P'. Hence:

THEOREM 4.43. *If $S \vdash A$, then – no matter the term extension L^+ of L and probability function P^+ for L^+ – A evaluates to 1 on P^+ if all the members of S do, i.e. A is logically entailed by S in the probabilistic sense (= The Strong Soundness Theorem for L in Probabilistic Semantics).*

*

Proof of the converse of Theorem 4.43 calls for an additional definition and two additional lemmas.

Let S be an arbitrary set of statements of L. By the *probability associate* of S in L, I shall understand the function P_S such that for any statement A of L:

$$P_S(A) = \begin{cases} 1 & \text{if } S \vdash A \\ 0 & \text{otherwise.} \end{cases}$$

I first establish that the probability associate of S in L meets Constraints **C1–C2** and **C4–C6** under all circumstances, meets Constraint **C3** as well when S is maximally consistent in L, and meets Constraint **C7** as well when S is ω-complete in L. I then establish that when S is maximally consistent in L, a statement A of L belongs to S iff it evaluates to 1 on the probability associate of S.

THEOREM 4.44. *Let S be a set of statements of L, and P_S be the probability associate of S in L.*
  (a) *P_S meets Constraints **C1–C2** and **C4–C6**.*
  (b) *If S is maximally consistent in L, then P_S meets Constraint **C3** as well.*
  (c) *If S is ω-complete in L, then P_S meets Constraint **C7** as well.*
  (d) *If S is maximally consistent and ω-complete in L, then P_S constitutes a probability function for L.*[30]

Proof. (a) follows from the definition of P_S and elementary facts about provability in L.

(b) Suppose S is maximally consistent in L, and suppose first that $P_S(A) = 1$. Then $S \vdash A$ (by the definition of P_S). Hence $S \vdash A \wedge B$ or $S \vdash A \wedge \neg B$, but not both. Hence, one of $P_S(A \wedge B)$ and $P_S(A \wedge \neg B)$ equals 1 and the other 0. Hence, $P_S(A) = P_S(A \wedge B) + P_S(A \wedge \neg B)$. Suppose next that $P_S(A) = 0$. Then $S \nvdash A$ (by the definition of P_S). Hence neither $S \vdash A \wedge B$ nor $S \vdash A \wedge \neg B$. Hence, both $P_S(A \wedge B)$ and $P_S(A \wedge \neg B)$ equal 0. Hence, again, $P_S(A) = P_S(A \wedge B) + P_S(A \wedge \neg B)$.

(c) Suppose S is ω-complete in L, and suppose first that $P_S(A \wedge (\forall x)B) = 1$. Then $S \vdash A \wedge (\forall x)B$ (by the definition of P_S). Hence, $S \vdash A \wedge \Pi_{i=1}^{j} B(t_i/x)$

I.3: ALTERNATIVES TO FIRST-ORDER SEMANTICS   239

for every j from 1 on. Hence, $P_S(A \wedge \Pi_{i=1}^j B(t_i/x)) = 1$ for every j from 1 on. Hence, $\text{Limit}_{j \to \infty} P_S(A \wedge \Pi_{i=1}^j B(t_i/x)) = 1$. Suppose next that $P_S(A \wedge (\forall x)B) = 0$. Then $S \not\vdash A \wedge (\forall x)B$ (by the definition of P_S). Hence, either $S \not\vdash A$ or $S \not\vdash (\forall x)B$ (or both). Now, if $S \not\vdash A$, then $S \not\vdash A \wedge \Pi_{i=1}^j B(t_i/x)$ for any j, hence $P_S(A \wedge \Pi_{i=1}^j B(t_i/x)) = 0$ for every j, and hence $\text{Limit}_{j \to \infty} P_S(A \wedge \Pi_{i=1}^j B(t_i/x)) = 0$. If, on the other hand, $S \not\vdash (\forall x)B$, then $S \not\vdash B(T/x)$ for at least one term T of L, this because S is ω-complete in L. Hence, there is a k such that, for every j from k on, $S \not\vdash A \wedge \Pi_{i=1}^j B(t_i/x)$. Hence, there is a k such that, for every j from k on, $P_S(A \wedge \Pi_{i=1}^j B(t_i/x)) = 0$; and, hence, $\text{Limit}_{j \to \infty} P_S(A \wedge \Pi_{i=1}^j B(t_i/x)) = 0$. Hence, P_S meets Constraint **C7**. □

THEOREM 4.45. *Let S be a set of statements of L that is maximally consistent and ω-complete in L, and let P_S be the probability associate of S in L. Then a statement A of L belongs to S iff $P_S(A) = 1$.*

Proof. If A belongs to S, then $S \vdash A$ and, hence, $P_S(A) = 1$ by the definition of P_S. If A does not belong to S, then by Theorem 1.3 $\neg A$ belongs to S, hence, $P_S(\neg A) = 1$ by the definition of P_S, hence, $P_S(A) = 0$ by Theorems 4.44(d) and 4.11, and, hence, $P_S(A) \neq 1$. Hence, Theorem 4.45. □

Now suppose, as on earlier occasions, that $S \not\vdash A$ where S is infinitely extendible in L. Then by Theorem 1.4 the Henkin extension $\mathscr{H}(S \cup \{\neg A\})$ of $S \cup \{\neg A\}$ in L is sure, as usual, to be maximally consistent and ω-complete in L; hence, by Theorem 4.45, every member of $\mathscr{H}(S \cup \{\neg A\})$ is sure to evaluate to 1 on the probability associate of $\mathscr{H}(S \cup \{\neg A\})$ in L. But by Theorem 4.44 the probability associate in question constitutes a probability function for L. Hence, by Theorem 4.11:

THEOREM 4.46. *Let S be a set of statements of L that is infinitely extendible in L, and let A be an arbitrary statement of L. If $S \not\vdash A$, then there is a probability function for L on which all members of S evaluate to 1 but A does not.*

But all of Theorems 1.4, 4.45, 4.44, and 4.11 hold with 'L$^+$' in place of 'L' and 'P^+' in place of 'P'. Hence:

THEOREM 4.47. *Let S and A be as in Theorem 4.46. If – no matter the term extension L$^+$ of L and probability function P^+ for L$^+$ – A evaluates to 1 on P^+ if all members of S do, then $S \vdash A$.*

When S is *not* infinitely extendible in L, $S \cup \{\neg A\}$ is sure — as usual — to be infinitely extendible in L^∞. Hence, if $S \not\vdash A$, then by the analogues of Theorems 1.4, 4.44, 4.45 and 4.11 for L^+ there is sure to be a term extension L^+ of L (L^∞) and a probability function P^+ for L^+ (the probability associate in L^∞ of the Henkin extension of $S \cup \{\neg A\}$ in L^∞) such that all members of S evaluate to 1 on P^+ but A does not. Hence, given Theorem 4.47:

THEOREM 4.48. *Let S be an arbitrary set of statements of* L *and A be an arbitrary statement of* L. *If — no matter the term extension L^+ of L and probability function P^+ for L^+ — A evaluates to 1 on P^+ if all members of S do, i.e. if A is logically entailed by S in the probabilistic sense, then $S \vdash A$ (= The Strong Completeness Theorem for* L *in Probabilistic Semantics).*

Hence, appealing to Theorem 4.43 and taking S in each case to be $\emptyset$:

THEOREM 4.49. *Let A be an arbitrary statement of* L. *Then $\vdash A$ iff — no matter the term extension L^+ of L and probability function P^+ for L^+ — A evaluates to 1 on P^+, i.e. iff A is logically true in the probabilistic sense.*

As mentioned earlier, Theorems 4.43 and 4.48 legitimize the account on page 226 of logical entailment; Theorem 4.49 legitimizes the first account there of logical truth; and the following corollary of Theorems 4.42 and 4.46 legitimizes the second:

THEOREM 4.50. *Let A be as in Theorem 4.49. Then $\vdash A$ iff A evaluates to 1 on every probability function for* L.

5. PROBABILISTIC SEMANTICS (II)

Getting on with the agenda of page 226 I establish here that (singulary) probability theory is but a generalization of truth-value theory, and then turn to *binary* probability functions. These too, the reader will recall, permit definition of logical truth, logical entailment, and such.

As documented in Section 3, truth-value semantics can be couched in either of two idioms: that of truth-value assignments and that of truth-value functions. The former idiom makes for a smooth and easy transition from truth-functional truth to logical truth, truth-functional entailment to logical entailment, etc. (write 'substatements' for 'components'). However, when it comes to investigating the relationship between truth-value theory and

I.3: ALTERNATIVES TO FRIRST-ORDER SEMANTICS

probability theory, the latter idiom is the handier one, and I accordingly switch to it.

I indicated on page 214 that I would eventually refer to both the truth-value functions for L and the probability ones by means of a single letter, 'P', and use 0 and 1 as truth values. That time has come. From here on a *truth-value function* for L will thus be any function P from the statements of L to $\{0, 1\}$ that meets the following three constraints:

B1. $P(\neg A) = 1 - P(A)$

B2. $P(A \wedge B) = P(A) \times P(B)$

B3. $P((\forall x)A) = 1$ iff $P(A(T/x)) = 1$ for each term T of L.

Thus understood, a truth-value function for L is obviously a *two-valued* function from the statements of L *to the reals* that meets Constraints **B1–B3**. That, conversely, a two-valued function from the statements of L to the reals is sure, if it meets Constraints **B1–B3**, to have 0 and 1 as its two values and hence be a truth-value function for L, follows from clause (a) in Theorem 5.1 ('$\Pi_{i=1}^{n} P(A_i)$' in the proof of that clause is short of course for '$P(A_1) \times P(A_2) \times \cdots \times P(A_n)$').

THEOREM 5.1. *Let P be a two-valued function from the statements of L to the reals.*
 (a) *If P meets Constraint **B2**, then P has 0 and 1 as its values.*
 (b) *If P meets Constraints **C1–C4**, then P again has 0 and 1 as its values.*
Proof. (a) Suppose P meets Constraint **B2**, in which case

$$P(\Pi_{i=1}^{n} A_i) = \Pi_{i=1}^{n} P(A_i)$$

for each n from 1 on, and suppose at least one value of P were some real r other than 0 and 1. Then P would number all of r, r^2, r^3, etc. among its values and, hence, would not – as supposed – be two-valued. Hence, (a).

 (b) $P(\neg(A \wedge \neg A))$ equals 1 by virtue of Constraint **C2**, and – as the proof of Theorem 4.8 attests – $P(A \wedge \neg A)$ equals 0 by virtue of Constraints **C1, C3**, and **C4**. Hence, (b). □

So:

THEOREM 5.2. *P constitutes a truth-value function for L iff P is a two-valued function from the statements of L to the reals that meets Constraints **B1–B3**.*

Theorem 5.2 is the first, and more trite, step in my proof of Theorem 5.8.
I next show that

THEOREM 5.3. *Let P be a two-valued function from the statements of* L *to the reals. If P meets Constraints* **B1–B3**, *then P meets Constraints* **C1–C7** *as well.*

Proof. Because of the hypothesis on P and Theorem 5.1(a), P has but 0 and 1 as values. (i) Proof that under these circumstances P meets Constraints **C1–C2** and **C4–C6** is immediate. (ii) Proof that under these circumstances P meets Constraint **C3** is as follows:

$$P(A) = (P(A) \times P(B)) + P(A) - (P(A) \times P(B))$$
$$= P(A \wedge B) + P(A) - (P(A) \times P(B)) \quad \text{(B2)}$$
$$= P(A \wedge B) + P(A)(1 - P(B))$$
$$= P(A \wedge B) + (P(A) \times P(\neg B)) \quad \text{(B1)}$$
$$= P(A \wedge B) + P(A \wedge \neg B) \quad \text{(B2)}$$

(iii) Proof that under these circumstances P meets Constraint **C7** is as follows. Suppose first that $P(A \wedge (\forall x)B) = 1$. Then by Constraint **B2** $P(A) \times P((\forall x)B) = 1$; hence, $P(A) = P((\forall x)B) = 1$; hence, by **B3** $P(B(T/x)) = 1$ for every term T of L; hence, by Constraint **B2** $P(A \wedge \Pi_{i=1}^{j} B(t_i/x)) = 1$ for each j from 1 on; and, hence, $\text{Limit}_{j \to \infty} P(A \wedge \Pi_{i=1}^{j} B(t_i/x)) = 1$. Suppose next that $P(A \wedge (\forall x)B) = 0$. Then by Constraint **B2** $P(A) \times P((\forall x)B) = 0$ and, hence, $P(A) = 0$ or $P((\forall x)B) = 0$. Now, if $P(A) = 0$, then by Constraint **B2** $P(A \wedge \Pi_{i=1}^{j} B(t_i/x)) = 0$ for each j from 1 on, and hence $\text{Limit}_{j \to \infty} P(A \wedge \Pi_{i=1}^{j} B(t_i/x)) = 0$. If, on the other hand, $P((\forall x)B) = 0$, then by Constraint **B3** $P(B(T/x)) = 0$ for at least one term T of L, hence by Constraint **B2** there is a k such that for every j from k on $P(A \wedge \Pi_{i=1}^{j} B(t_i/x)) = 0$, and hence $\text{Limit}_{j \to \infty} P(A \wedge \Pi_{i=1}^{j} B(t_i/x)) = 0$. □

Proof of the converse of Theorem 5.3 calls for two lemmas. Note that in the first P may have anywhere from two to $\aleph_0$ values, but in the second it is understood to have but two.

THEOREM 5.4. *Let P be a function from the statements of* L *to the reals that meets Constraints* **C1–C5**.
  (a) $P(\neg A) = 1 - P(A) \quad (= \textbf{B1})$.
  (b) *If* $P(A \wedge B) = 1$, *then* $P(A) = P(B) = 1$.

I.3: ALTERNATIVES TO FRIRST-ORDER SEMANTICS   243

(c) *If $P(A \wedge B) = 1$, then $P(A \wedge B) = P(A) \times P(B)$.*
Proof. (a) is Theorem 4.11. (b) Suppose $P(A \wedge B) = 1$. Then $P(A) = 1$ by Theorems 4.1 and 4.13, and $P(B) = 1$ by Theorems 4.7 and 4.13. (c) By (b). □

THEOREM 5.5. *Let P be a two-valued function from the statements of* L *to the reals that meets Constraints* **C1–C5** *and* **C7**.
  (a) *If $P(A \wedge B) = 0$, then $P(A) = 0$ or $P(B) = 0$.*
  (b) *If $P(A \wedge B) \neq 1$, then $P(A \wedge B) = P(A) \times P(B)$.*
  (c) $P(A \wedge B) = P(A) \times P(B)$ (= **B2**).
  (d) $P((\forall x)A) = 1$ *iff $P(A(T/x)) = 1$ for each term T of* L (= **B3**).
Proof. (a) Suppose $P(A \wedge B) = 0$, in which case $P(A) = P(A \wedge \neg B)$ by **C3**. Suppose further that $P(A) \neq 0$, and hence $P(A) = 1$ by Theorem 5.1(b). Then $P(A \wedge \neg B) = 1$; hence, $P(\neg B) = 1$ by Theorems 4.7 and 4.13; and, hence, $P(B) = 0$ by Theorem 4.11. Hence, $P(A) = 0$ or $P(B) = 0$. Hence, (a).
  (b) By Theorem 5.1(b) and clause (a) of the present theorem.
  (c) By (b) and Theorem 5.4(c).
  (d) Suppose first that $P((\forall x)A) = 1$. Then by Theorem 4.33 $\text{Limit}_{j \to \infty} P(\Pi_{i=1}^{j} A(t_i/x)) = 1$. But by Theorem 5.1(b) each of $P(\Pi_{i=1}^{1} A(t_i/x))$, $P(\Pi_{i=1}^{2} A(t_i/x))$, etc., equals 0 or 1, and by Theorem 4.1 each one of them is equal to or smaller than the preceding one. Hence, $P(\Pi_{i=1}^{j} A(t_i/x)) = 1$ for each j from 1 on and, hence, by Theorem 5.4(a) $P(A(T/x)) = 1$ for each term T of L. Suppose next that $P((\forall x)A) \neq 1$. Then by Theorem 5.1(b) $P((\forall x)A) = 0$; hence by Theorem 4.33 $\text{Limit}_{j \to \infty} P(\Pi_{i=1}^{j} A(t_i/x)) = 0$; hence, by Theorem 5.1(b) there is a k such that, for every j from k on, $P(\Pi_{i=1}^{j} A(t_i/x)) = 0$; hence; by (a) $P(A(T/x)) = 0$ for at least one term T of L; and, hence, $P(A(T/x)) \neq 1$ for at least one such term. Hence, (d). □

Hence:

THEOREM 5.6. *Let P be a two-valued function from the statements of* L *to the reals. If P meets Constraints* **C1–C7**, *then P meets Constraints* **B1–B3** *as well.*[31]

Hence, the second step in my proof of Theorem 5.8:

THEOREM 5.7. *Let P be a two-valued function from the statements of* L *to the reals. Then P meets Constraints* **B1–B3** *iff P meets Constraints* **C1–C7**.

Hence by Theorem 5.2:

THEOREM 5.8. *P constitutes a truth-value function for* L *iff P constitutes a two-valued probability function for* L.

Hence, as claimed earlier, probability theory is but a generalization of truth-value theory: allow truth-value functions – understood as meeting Constraints **C1–C7** – to have anywhere from two to $\aleph_0$ values, and truth-value theory expands into probability theory.[32] Or, to view it the other way round, truth-value theory is but a restriction of probability theory: require probability functions to have but two values, and probability theory reduces to truth-value theory. (To press the point further, probability functions are *measure functions* with reals from the interval [0, 1] *as their only values*, and truth-value functions are *probability functions* with the end-points in this interval *as their only values*.)

However, two-valued probability functions (i.e. truth-value functions) differ from the rest in one fundamental respect: the former are extensional, but the latter are not. Indeed, call a probability function P for L *extensional* if the value of P for a statement of L depends exclusively upon the value(s) of P for the immediate substatement(s) of that statement – more formally, if

(i) given that $P(A) = P(A')$, $P(\neg A) = P(\neg A')$,
(ii) given that $P(A) = P(A')$ and $P(B) = P(B')$, $P(A \wedge B) = P(A' \wedge B')$,
(iii) given that $P(A(T/x)) = P(A'(T/x))$ for each term T of L, $P((\forall x)A) = P((\forall x)A')$.

Since $P(\neg A) = 1 - P(A)$ by Theorem 4.11, (i) holds however many values P may have. When P is two-valued, (ii) and (iii) also hold: (ii) because $P(A \wedge B)$ is then equal by Theorem 5.5(c) to $P(A) \times P(B)$, and (iii) because $P((\forall x)A)$ is equal by Theorem 4.33 to $\text{Limit}_{j \to \infty} P(\Pi_{i=1}^{j} A(t_i/x))$ and, hence, behaves as $P(A(t_1/x))$, $P(A(t_1/x) \wedge A(t_2/x))$, $P((A(t_1/x) \wedge A(t_2/x)) \wedge A(t_3/x))$, etc., do. However, when P has more than two values, (ii) and hence (iii) fail. Suppose, for example, that, for some statement A of L, $P(A)$ – and, hence, $P(\neg A)$ – equals $\frac{1}{2}$. Then by Theorem 4.2 $P(A \wedge A)$ equals $P(A)$ and hence $\frac{1}{2}$, whereas by Theorem 4.8 $P(A \wedge \neg A)$ equals 0. So those among the probability functions for L that have just two values are extensional, but the rest are not.[33] The two results were to be expected. P is understood here as a measure of rational belief, and by all accounts rational belief is an intensional matter. When P has just two values, though, P is but a truth-value function, and truth-value functions *are* extensional.

*

Binary probability functions, also known as *conditional probability functions*, are often defined in terms of singular ones. For example, in Kolmogorov [1933], Carnap [1950], etc., $P(A/B)$ is set at $P(A \wedge B)/P(B)$ when $P(B) \neq 0$, but otherwise is left without a value (recall that, when dealing with Kolmogorov, I think of his intersections as conjunctions). Partial functions, however, are unwieldy and – perforce – of limited service.

An alternative approach, favored by Keynes as early as 1921 and, hence, antedating Kolmogorov's, has now gained wide currency, thanks to such diverse writers as Reichenbach, Jeffreys, Von Wright, Rényi, Carnap (in post-1950 publications), Popper, etc. Handling binary probability functions as you would singular ones, you adopt constraints suiting your understanding of $P(A/B)$ and then own as your binary probability functions all functions meeting these constraints.

Heeding the precedent of Section 4, I shall think of $P^+(A_1/B)$, $P^+(A_2/B)$, $P^+(A_3/B)$, etc., *as degrees to which a rational agent might – in light of (or relative to) statement B of* L^+ *– simultaneously believe the various statements* A_1, A_2, A_3, etc., of L^+.[34] Constraints particularly suiting this understanding of $P^+(A/B)$ can be found in Von Wright [1957]. Slightly edited for the occasion, they run:

(1) $0 \leq P^+(A/B)$
(2) $P^+(A/A) = 1$
(3)   If B is not logically false in the standard sense, then $P^+(\neg A/B) = 1 - P^+(A/B)$
(4) $P^+(A \wedge B/C) = P^+(A/B \wedge C) \times P^+(B/C)$
(5)   If A and B are logically equivalent in the standard sense, then $P^+(A/C) = P^+(B/C)$
(6)   If B and C are logically equivalent in the standard sense, then $P^+(A/B) = P^+(A/C)$.[35]

Note: A statement is held *logically false* in the standard sense if its negation is logically true in that sense; and, as in note 29, two statements are held *logically equivalent* in the standard sense if their biconditional is logically true in that sense.

Substitutes for (3), (5), and (6) must of course be found, as these constraints are not autonomous. A number are available in the literature, among them:

(3′)   If $P^+(C/B) \neq 1$ for at least one statement C of L^+, then $P^+(\neg A/B) = 1 - P^+(A/B)$,
(5′) $P^+(A \wedge B/C) = P^+(B \wedge A/C)$,

and

(6') $\quad P^+(A/B \wedge C) = P^+(A/C \wedge B)$.

The last two of these come from Leblanc [1981]. (Like **C5** they could sport '$\leqslant$' rather than '=', but the present formulations are more commonly employed.) (3') stems from Popper [1957]. It is weaker than (3) (in that certain functions meeting (3') will not meet (3)), but welcomely so. To compare the two constraints, declare a statement B of L^+ P^+-*absurd* if $P^+(C/B) = 1$ for every statement C of L^+, i.e. if in light of B a rational agent might believe any such C. Given either of (3) and (3') you can prove that any logical falsehood is P^+-absurd. And, *given* (3), you can go on and prove that, conversely, any statement P^+-absurd is logically false. *Not so, however, given* (3'). Several writers, Carnap one of them, favor (3) as a result. Equally many, Popper one of them, do not – agreeing with Wellington that there are statements which are not logically false, to be sure, but in light of which one might believe anything.[36] I side with Popper on this and adopt Constraint (3').[37]

Popper placed on his binary probability functions an extra constraint, requiring each of them to have at least two values. I adopt a simplified but equivalent version of it, requiring that – for each binary probability function P^+ for L^+ – at least one statement is not P^+-absurd. The constraint guarantees, among other things, that logical truths are not P^+-absurd, a matter of some comfort, and that singulary and binary probability functions bear to each other the relationship reported in my last theorem, Theorem 5.34.

Formal details are as follows, with the more commonly used 'P^+-abnormal' substituting for 'P^+-absurd'.

Let L^+ be an arbitrary term extension of L. By a *binary probability function* for L^+ I shall understand any function P^+ that takes each pair of statements for L^+ into a real and meets the following constraints:

D0. There is a statement A and a statement B of L^+ such that $P^+(A/B) \neq 1$

D1. $0 \leqslant P^+(A/B)$

D2. $P^+(A/A) = 1$

D3. If there is a statement C of L^+ such that $P^+(C/B) \neq 1$, then $P^+(\neg A/B) = 1 - P^+(A/B)$

D4. $P^+(A \wedge B/C) = P^+(A/B \wedge C) \times P^+(B/C)$

D5. $P^+(A \wedge B/C) = P^+(B \wedge A/C)$

D6. $P^+(A/B \wedge C) = P^+(A/C \wedge B)$

D7. $P^+((\forall x)A/B) = \text{Limit}_{j \to \infty} P^+(\Pi_{i=1}^{j} A(t_i^+/x)/B)$.

Where L^+ is a term extension of L and P^+ a binary probability function for L^+, I shall declare a statement B of L^+ P^+-*normal* (P^+-*abnormal*) if there is a (there is no) statement A of L^+ such that $P^+(A/B) \neq 1$. I shall declare a statement A of L *logically true in the binary probabilistic sense* if – no matter the term extension L^+ of L, binary probability function P^+ for L^+, and statement B of L^+ – $P^+(A/B) = 1$; and, where S is a set of statements of L, I shall declare A *logically entailed by S in the binary probabilistic sense* if – no matter the term extension L^+ of L, binary probability function P^+ for L^+, and statement B of L^+ – $P^+(A/B) = 1$ if $P^+(C/B) = 1$ for each member C of S. Equivalently, but more simply, A may be declared logically true in the binary probabilistic sense if $P(A/B) = 1$ for every binary probability function P for L and every statement B of L. (In the idiom of rational belief A is thus logically true if in light of *any* statement B of L a rational agent could not but believe A.)

I first argue for the foregoing definitions by showing that of logical entailment strongly sound and complete, and hence that of logical truth weakly sound and complete. I then study the relationship between singular and binary probability functions.

Thanks to Harper, proof of my last Strong Soundness Theorem can be breath-takingly simple:

THEOREM 5.9. $P(A/B) \leq 1$.

Proof. When B is P-abnormal, $P(A/B) = 1$ by definition and, hence, $P(A/B) \leq 1$. Suppose then that B is P-normal. By **D1** $P(\neg A/B) \geq 0$ and, hence, by **D3** $P(A/B) \leq 1$. Hence, Theorem 5.9. □

THEOREM 5.10. $P(A/B \wedge A) = 1$.

Proof. By **D2** $P(B \wedge A/B \wedge A) = 1$. Hence, by **D4**, $P(B/A \wedge (B \wedge A)) \times P(A/B \wedge A) = 1$. But by **D1** and Theorem 5.9 each of $P(B/A \wedge (B \wedge A))$ and $P(A/B \wedge A)$ lies in the interval $[0, 1]$ and, hence, each of them here must equal 1. Hence, Theorem 5.10. □

THEOREM 5.11. $P(A/A \wedge B) = 1$.

Proof by Theorem 5.10 and **D6**. □

THEOREM 5.12. *Let B be P-normal. Then:*
  (a) $P(\neg B/B) = 0$.
  (b) $P(\neg B \wedge A/B) = 0$.

Proof. (a) by **D2**, **D3**, and the hypothesis on B. (b) By **D5** $P(\neg B \wedge A/B)$ equals $P(A \wedge \neg B/B)$, which by **D4** equals $P(A/\neg B \wedge B) \times P(\neg B/B)$. Hence (b) by (a). □

THEOREM 5.13. *If $P(A/B) = 0$ for no statement B of L, then $P(A/B) = 1$ for every such B (= Harper's Lemma).*

Proof. Suppose $P(A/B) \neq 1$ for some statement B of L. Then B is P-normal; hence, $P(\neg B \wedge \neg A/B) = 0$ by Theorem 5.12(b); and, hence, $P(\neg B/\neg A \wedge B) \times P(\neg A/B) = 0$ by **D4**. But, since $P(A/B) \neq 1$ and B is P-normal, $P(\neg A/B) \neq 0$. Hence, $P(\neg B/\neg A \wedge B) = 0$; and, hence, $\neg A \wedge B$ is P-normal. But $P(\neg A/\neg A \wedge B) = 1$ by Theorem 5.11. Hence, $P(A/\neg A \wedge B) = 0$ by **D3**; and, hence, there is a statement B of L such that $P(A/B) = 0$. Hence, Theorem 5.13. □

THEOREM 5.14. $P(A \wedge B/C) \leqslant P(B/C)$.

Proof. By **D4** $P(A \wedge B/C) = P(A/B \wedge C) \times P(B/C)$. But by **D1** and Theorem 5.9, each of $P(A/B \wedge C)$ and $P(B/C)$ lies in the interval $[0, 1]$ and, hence, neither can be less than $P(A \wedge B/C)$. Hence, Theorem 5.14. □

THEOREM 5.15. $P(A \wedge B/C) \leqslant P(A/C)$.
Proof by Theorem 5.14 and **D5**. □

THEOREM 5.16. *If $P(A \wedge B/C) = 1$, then $P(A/C) = P(B/C) = 1$.*

Proof. Suppose $P(A \wedge B/C) = 1$. Then $P(A/C) = 1$ by Theorems 5.15 and 5.9, and $P(B/C) = 1$ by Theorems 5.14 and 5.9. Hence, Theorem 5.16. □

THEOREM 5.17. *If $P(A \to B/C) = 0$, then $P(A/C) = 1$ and $P(B/C) = 0$.*

Proof. Suppose $P(A \to B/C) = 0$, in which case C is P-normal. Then $P(A \wedge \neg B/C) = 1$ by $D_\to$ and **D3**; hence, $P(A/C) = P(\neg B/C) = 1$ by Theorem 5.16; and, hence, $P(A/C) = 1$ and $P(B/C) = 0$ by **D3**. □

THEOREM 5.18. $P(A \wedge A/B) = P(A/B)$.

Proof. By **D4** $P(A \wedge A/B) = P(A/A \wedge B) \times P(A/B)$. Hence, Theorem 5.18 by Theorem 5.11. □

THEOREM 5.19. *If C is P-normal, then $P(A/C) = P(A \wedge B/C) + P(A \wedge \neg B/C)$.*

I.3: ALTERNATIVES TO FIRST-ORDER SEMANTICS 249

Proof. Suppose C is P-normal. By definition when $A \wedge C$ is P-abnormal but otherwise by **D3**,

$$P(B/A \wedge C) + P(\neg B/A \wedge C) = P(C/A \wedge C) + P(\neg C/A \wedge C),$$

hence

$$P(B/A \wedge C) \times P(A/C) + P(\neg B/A \wedge C) \times P(A/C)$$
$$= P(C/A \wedge C) \times P(A/C) + P(\neg C/A \wedge C) \times P(A/C),$$

hence by **D4**

$$P(B \wedge A/C) + P(\neg B \wedge A/C)$$
$$= P(C/A \wedge C) \times P(A/C) + P(\neg C \wedge A/C),$$

and, hence, by Theorems 5.10 and 5.12(b)

$$P(B \wedge A/C) + P(\neg B \wedge A/C) = P(A/C).$$

Hence, Theorem 5.19 by **D5**. □

THEOREM 5.20. *If* $P(A \to B/C) = 1$ *and* $P(A/C) = 1$, *then* $P(B/C) = 1$ *(= Modus Ponens)*.

Proof. When C is P-abnormal, $P(B/C) = 1$ by definition. So suppose that C is P-normal and $P(A \to B/C) = 1$. Then by **D→** and **D3**, $P(A \wedge \neg B/C) = 0$; and, hence, by Theorem 5.19, $P(A \wedge B/C) = P(A/C)$. Suppose further that $P(A/C) = 1$. Then $P(A \wedge B/C) = 1$; and, hence, $P(B/C) = 1$ by Theorems 5.14 and 5.9. Hence, Theorem 5.20. □

THEOREM 5.21. *If* $P(A/C) = P(B/C) = 1$, *then* $P(A \wedge B/C) = 1$.

Proof. When C is P-abnormal, $P(A \wedge B/C) = 1$ by definition. So suppose that C is P-normal and $P(A/C) = 1$. Then by Theorem 5.19 $P(A \wedge B/C) + P(A \wedge \neg B/C) = 1$. Suppose further that $P(B/C) = 1$. Then by **D3** $P(\neg B/C) = 0$; hence, by Theorem 5.14 and **D1** $P(A \wedge \neg B/C) = 0$; and, hence, by Theorem 5.19 $P(A \wedge B/C) = 1$. Hence, Theorem 5.21. □

That, no matter the statement B of L, $P(A/B) = 1$ for any axiom A of L of sorts **A1–A6** follows by six applications of Harper's Lemma:

THEOREM 5.22. *If A is an axiom of L of sorts* **A1–A6**, *then* $P(A/B) = 1$ *for every statement B of* L.

Proof. Let A be an axiom of L of sorts **A1–A6**. I show that if there were a statement B of L such that $P(A/B) = 0$, a contradiction would ensue. So $P(A/B) = 0$ for no statement B of L and, hence, by Theorem 5.13 $P(A/B) = 1$ for every such B.

Case 1: A is of the sort $A' \to (A' \wedge A')$. Suppose $P(A/B) = 0$ for some statement B of L. Then by Theorem 5.17

(1) $\quad P(A'/B) = 1$

and

(2) $\quad P(A' \wedge A'/B) = 0$.

But by (2) and Theorem 5.18

(3) $\quad P(A'/B) = 0$.

So a contradiction ensues. So $P(A/B) = 1$ for every B.

Case 2: A is of the sort $(A' \wedge B') \to A'$. Suppose $P(A/B) = 0$ for some statement B of L. Then by Theorem 5.17

(1) $\quad P(A' \wedge B'/B) = 1$

and

(2) $\quad P(A'/B) = 0$.

But by (1) and Theorem 5.16

(3) $\quad P(A'/B) = 1$.

So a contradiction ensues. So $P(A/B) = 1$ for every B.

Case 3: A is of the sort $(A' \to B') \to (\neg(B' \wedge C') \to \neg(C' \wedge A'))$. Suppose $P(A/B) = 0$ for some statement B of L. Then by Theorem 5.17

(1) $\quad P(A' \to B'/B) = 1$

and

(2) $\quad P(\neg(B' \wedge C') \to \neg(C' \wedge A')/B) = 0$.

But by (2) and Theorem 5.17

(3) $\quad P(\neg(B' \wedge C')/B) = 1$

and

(4) $\quad P(\neg(C' \wedge A')/B) = 0$.

Hence by **D3** (and the fact that, since $P(A/B) = 0$, B is P-normal)

I.3: ALTERNATIVES TO FIRST-ORDER SEMANTICS 251

(5) $\quad P(C' \wedge A'/B) = 1$,

and, hence, by Theorem 5.16

(6) $\quad P(A'/B) = 1$.

But by (1), $D_\rightarrow$, and **D3**

(7) $\quad P(A' \wedge \neg B'/B) = 0$.

Hence, by (6) and Theorem 5.19

(8) $\quad P(A' \wedge B'/B) = 1$,

and, hence, by Theorem 5.16

(9) $\quad P(B'/B) = 1$.

But by (3) and **D3**

(10) $\quad P(B' \wedge C'/B) = 0$,

and, hence, by **D5**

(11) $\quad P(C' \wedge B'/B) = 0$,

whereas by (5) and Theorem 5.16

(12) $\quad P(C'/B) = 1$.

Hence, by (11) and Theorem 5.19

(13) $\quad P(C' \wedge \neg B'/B) = 1$,

hence, by Theorem 5.16

(14) $\quad P(\neg B'/B) = 1$,

and, hence, by **D3**

(15) $\quad P(B'/B) = 0$.

So a contradiction (= (9) and (15)) ensues. So $P(A/B) = 1$ for every B.

Case 4: A is of the sort $A' \rightarrow (\forall x)A'$. Suppose $P(A/B) = 0$ for some statement B of L. Then by Theorem 5.17

(1) $\quad P(A'/B) = 1$

and

(2) $\quad P((\forall x)A'/B) = 0$.

But by (2) and **D7**

(3) $\text{Limit}_{j \to \infty} P((\underbrace{\cdots (A' \wedge A') \wedge \cdots}_{j \text{ times}}) \wedge A'/B) = 0$

and, hence, by Theorem 5.18 and the definition of a limit

(4) $P(A'/B) = 0$.

So a contradiction ensues. So $P(A/B) = 1$ for every B.

Case 5: A is of the sort $(\forall x)A' \to A'(T/x)$. Suppose $P(A/B) = 0$ for some statement B of L. Then by Theorem 5.17

(1) $P((\forall x)A'/B) = 1$

and

(2) $P(A'(T/x)/B) = 0$ for some term T of L.

But by (1) and **D7**

(3) $\text{Limit}_{j \to \infty} P(\Pi_{i=1}^{j} A(t_i/x)/B) = 1$,

hence, by Theorem 5.15 and familiar considerations

(4) $P(\Pi_{i=1}^{j} A(t_i/x)/B) = 1$ for each j from 1 on,

and, hence, by Theorem 5.16

(5) $P(A(T/x)/B) = 1$ for each term T of L.

So a contradiction ensues. So $P(A/B) = 1$ for every B.

Case 6: A is of the sort $(\forall x)(A' \to B') \to ((\forall x)A' \to (\forall x)B')$. Suppose $P(A/B) = 0$ for some statement B of L. Then by Theorem 5.17

(1) $P((\forall x)(A' \to B')/B) = 1$

and

(2) $P((\forall x)A' \to (\forall x)B'/B) = 0$.

But by (2) and Theorem 5.17

(3) $P((\forall x)A') = 1$

and

(4) $P((\forall x)B') = 0$.

I.3: ALTERNATIVES TO FIRST-ORDER SEMANTICS

Hence by (1), (3), and **D7**

(5) $\quad \text{Limit}_{j\to\infty} P(\Pi_{i=1}^{j}(A' \to B')(t_i/x)/B) = 1$

and

(6) $\quad \text{Limit}_{j\to\infty} P(\Pi_{i=1}^{j} A'(t_i/x)/B) = 1$,

and, hence, by Theorem 5.15 and familiar considerations

(7) $\quad P((A' \to B')(T/x)/B) = 1$ for each term T of L

and

(8) $\quad P(A'(T/x)/B) = 1$ for each term T of L.

Hence, by Theorem 5.20

(9) $\quad P(B'(T/x)/B) = 1$ for each term T of L,

hence, by Theorem 5.21 and the definition of a limit

(10) $\quad \text{Limit}_{j\to\infty} P(\Pi_{i=1}^{j} B'(t_i/x)/B) = 1$,

and, hence, by **D7**

(11) $\quad P((\forall x)B'/B) = 1$.

So a contradiction ($= (4)$ and (11)) ensues. So $P(A/B) = 1$ for every B. □

So by an induction like the first on page 237 (with Theorem 5.21 substituting for Theorem 4.37):

THEOREM 5.23. *If A is an axiom of* L, *then* $P(A/B) = 1$ *for every statement B of* L.

So by an induction like the second on page 237 (with Theorem 5.22 substituting for Theorem 4.41, and Theorem 5.20 for Theorem 4.30):

THEOREM 5.24. *If* $S \vdash A$, *then – no matter the statement B of* L *–* $P(A/B) = 1$ *if* $P(C/B) = 1$ *for every member C of S.*

But Theorems 5.20 and 5.23 hold with 'P^+' for 'P'. Hence, the following counterpart of Theorem 4.42 for binary probability functions:

THEOREM 5.25. *If* $S \vdash A$, *then – no matter the term extension* L^+ *of* L, binary probability function P^+ for L^+, and statement B of $L^+ - P^+(A/B) = 1$ if $P^+(C/B) = 1$ for every member C of S, i.e. A is logically implied by S in the binary probabilistic sense.

Slight editing of the material on pages 238–240 yields proof of the counterpart of Theorem 4.46 for binary probability functions.

Where S is an arbitrary set of statements of L, understand by the *binary probability associate* of S in L the function P_S such that, for any statement A and any statement B of L:

$$P_S(A/B) = \begin{cases} 1 & \text{if } S \vdash B \to A \\ 0 & \text{otherwise.} \end{cases}$$

As the reader may wish to verify:

THEOREM 5.26. *Let S be a set of statements of L, and P_S be the binary probability associate of S in L.*
  (a) *P_S meets Constraints* **D0–D2** *and* **D4–D6**.
  (b) *If S is maximally consistent in L, then P_S meets Constraint* **D3** *as well.*
  (c) *If S is ω-complete in L, then P_S meets Constraint* **D7** *as well.*
  (d) *If S is maximally consistent and ω-complete in L, then P_S constitutes a binary probability function for L.*

THEOREM 5.27. *Let S be a set of statements of L that is maximally consistent in L, and let P_S be the binary probability associate of S in L.*
  (a) *A conditional $B \to A$ of L belongs to S iff $P_S(A/B) = 1$.*
  (b) *$P_S(A/\neg(B \land \neg B))$ equals 1 if A belongs to S, but 0 if $\neg A$ does.*

Note for proof of (b) that, with S maximally consistent in L, (i) $\neg(B \land \neg B) \to A$ belongs to S if A does, but (ii) $\neg(B \land \neg B) \to A$ does not if $\neg A$ does (and, hence, A does not). So, (b) by (a).

Now suppose, for the last time, that $S \not\vdash A$, where S is infinitely extendible in L; let $\mathcal{H}(S \cup \{\neg A\})$ be the Henkin extension of $S \cup \{\neg A\}$ in L; let $P_{\mathcal{H}(S \cup \{\neg A\})}$ be the binary probability associate of $\mathcal{H}(S \cup \{\neg A\})$ in L; and let B' be an arbitrary statement of L. Since by Theorem 1.4 $\mathcal{H}(S \cup \{\neg A\})$ is maximally consistent in L, we know by Theorem 5.27(b) that relative to $\neg(B' \land \neg B')$ each member C of S evaluates to 1 on $P_{\mathcal{H}(S \cup \{\neg A\})}$, whereas A evaluates to 0. But, since by Theorem $\mathcal{H}(S \cup \{\neg A\})$ is ω-complete as well as maximally consistent in L, we know by Theorem 5.27(b) that $P_{\mathcal{H}(S \cup \{\neg A\})}$ constitutes a binary probability function for L. Hence:

I.3: ALTERNATIVES TO FIRST-ORDER SEMANTICS

THEOREM 5.28. *Let S be a set of statements of* L *that is infinitely extendible in* L, *and A be an arbitrary statement of* L. *If* $S \not\vdash A$, *then there is a binary probability function P for* L *and a statement B of* L *such that* $P(C/B) = 1$ *for each member C of S but* $P(A/B) \neq 1$.

But all of Theorems 1.4, 5.26, and 5.27 hold with 'P^+' in place of 'P'. Hence:

THEOREM 5.29. *Let S and A be as in Theorem 5.28. If – no matter the binary probability function P of* L *and statement B of* L – $P(A/B) = 1$ *if* $P(C/B) = 1$ *for each member C of S, then* $S \vdash A$.

When S is not infinitely extendible in L, resort to L^∞ will do the trick as usual. Hence, given Theorem 5.29, the following counterpart of Theorem 4.46 for binary probability functions:

THEOREM 5.30. *Let S be an arbitrary set of statements of* L, *and A be an arbitrary statement of* L. *If – no matter the term extension* L^+ *of* L, *binary probability function* P^+ *for* L^+, *and statement B of* L^+ – $P^+(A/B) = 1$ *if* $P^+(C/B) = 1$ *for each member C of S, i.e. if A is logically entailed by S in the binary probabilistic sense, then* $S \vdash A$.

Hence, appealing to Theorem 5.25 and taking S in each case to be $\emptyset$:

THEOREM 5.31. *Let A be an arbitrary statement of* L. *Then* $\vdash A$ *iff – no matter the term extension* L^+ *of* L, *binary probability function* P^+ *for* L^+, *and statement B of* L^+ – $P^+(A/B) = 1$, *i.e. iff A is logically true in the binary probabilistic sense.*

And, with S again set at $\emptyset$, Theorems 5.24 and 5.29 yield:

THEOREM 5.32. *Let A be as in Theorem 5.31. Then* $\vdash A$ *iff – no matter the probability function P for* L *and statement B of* L – $P(A/B) = 1$.

The binary account of logical entailment on page 247 is thus strongly sound and complete, and the two binary accounts of logical truth on that page weakly sound and complete.

*

No appeal has been made so far to **D0**: the soundness and completeness theorems I have just proved thus hold with P^+ presumed to meet only

Constraints **D1–D7**. However, **D0** will now do some work, namely: ensure that $\neg(Q(t_1) \wedge \neg Q(t_1))$ (Q a predicate of L of degree 1 and t_1 the alphabetically first term of L) is P-normal, and thereby allow for an important connection between singular and binary probability functions. To abridge matters I refer to $\neg(Q(t_1) \wedge \neg Q(t_1))$ by means of 'T'.

D0, as stressed earlier, demands that at least one statement of L be P-normal. After proving one ancillary result, I establish by means of just **D1–D7** (indeed, just **D1–D6**) that T is P-normal *if* any statement of L is, and hence by dint of **D0** that T *is* P-normal.

THEOREM 5.33. *Let P be an arbitrary binary probability function for* L.
(a) $P(A/B \wedge T) = P(A/B)$.
(b) *If any statement of* L *is P-normal, then* T *is*.
(c) T *is P-normal*.

Proof. (a) By **D4** $P(T \wedge A/B) = P(T/A \wedge B) \times P(A/B)$. But by Theorem 5.24 $P(T/A \wedge B) = 1$. Hence, $P(T \wedge A/B) = P(A/B)$. But by the same reasoning $P(A \wedge T/B) = P(A/T \wedge B)$. Hence, by **D5**, $P(A/T \wedge B) = P(A/B)$ and, hence, by **D6**, $P(A/B \wedge T) = P(A/B)$.

(b) Suppose $P(A/T) = 1$ for every statement A of L, and let B be an arbitrary statement of L. Then $P(A \wedge B/T) = 1$; hence, by **D4** $P(A/B \wedge T) \times P(B/T) = 1$; hence, by **D1** and Theorem 5.9 $P(A/B \wedge T) = 1$; and, hence, by (a) $P(A/B) = 1$. Hence, $P(A/B) = 1$ for every statement A and every statement B of L. Hence, by Contraposition, if $P(A/B) \neq 1$ for any statement A and any statement B of L, then $P(A/T) = 1$ for some statement A of L. Hence, (b),

(c) by **D0** and (b). □

The argument is easily edited to show that *any* logical truth of L, not just T, is P-normal. (It follows, incidentally, from Theorem 5.33(c) and **D2** that $P(\neg T/T) = 0$ and, hence, that each binary probability function P for L has at least two values. So **D0** does the same work as Popper's constraint A1.)

The reader will note that by virtue of Theorem 5.33(c) and 5.19

$$P(A/T) = P(A \wedge B/T) + P(A \wedge \neg B/T),$$

a T-version – so to speak – of **C3**. But similar T-versions of **C1–C2** and **C4–C7** can be had as well. Indeed, the T-versions of **C1, C5,** and **C7**:

$$0 \leq P(A/T), \qquad P(A \wedge B/T) \leq P(B \wedge A/T),$$

and

I.3: ALTERNATIVES TO FIRST-ORDER SEMANTICS 257

$$P((\forall x)A/T) = \text{Limit}_{j\to\infty} P(\Pi_{i=1}^{j} A(t_i/x)/T),$$

are special cases of **D1**, **D5**, and **D7**. The T-versions of **C2** and **C4**:

$$P(\neg(A \land \neg A)/T) = 1$$

and

$$P(A/T) \leq P(A \land A/T),$$

hold by Theorem 5.24 and Theorem 5.18. And the T-version of **C6**:

$$P(A \land (B \land C)/T) \leq P((A \land B) \land C/T),$$

can be had by repeated applications of **D4** and Theorem 5.33(a). What is known as the *restriction of P to* T, i.e. the function P_T such that – for each statement A of L – $P_T(A) = P(A/T)$, thus constitutes a singulary probability function for L.

On the other hand, let P be a singulary probability function for L; let P' be the function such that – for any statement A and any statement B of L

$$P'(A/B) = \begin{cases} P(A \land B)/P(B) & \text{if } P(B) \neq 0 \\ 1 & \text{otherwise;} \end{cases}$$

and let P'_T be the restriction of P' to T. It is easily verified that (i) P'_T constitutes a binary probability function for L and (ii) $P'_T(A) = P(A)$ for any statement A of L. For proof of (ii) note that by definition $P'_T(A)$ equals $P'_T(A/T)$. But by **C2** $P(T) \neq 0$. Hence by definition $P'_T(A)$ equals $P(A \land T)/P(T)$, which by Theorem 4.10 and **C2** equals $P(A)$. So there is a binary probability function for L, to wit: P'_T, of which P is the restriction to T.

So:

THEOREM 5.34. (a) *Each binary probability function for L has as its restriction to* T *a singulary probability function for* L.
(b) *Each singulary probability function for L is the restriction to* T *of a binary probability function for* L.

Like results hold of course with 'L^+' for 'L', thus binding together all the probability functions treated in Sections 4 and 5.[38]

(The Appendix has further information on the history of probabilistic semantics.)

6. IN SUMMARY AND CONCLUSION

The non-standard accounts of logical entailment (and, hence, of logical truth) in this essay were justified thusly – with truth sets and model ones treated first to vary the perspective slightly:

Supposing $S \nvdash A$, I formed the Henkin extension $\mathcal{H}^\infty(S \cup \{\neg A\})$ of $S \cup \{\neg A\}$ in L^∞ (in L itself when S is infinitely extendible). That set proved to be a truth set for L^∞ (for L), which ensured that if $S \nvdash A$, then there is – for some term extension L^+ of L – a truth set for L^+ of which S is a subset but A is not a member. Hence, by dint of a suitable soundness theorem, $S \vdash A$ iff – *no matter the term extension L^+ of L – A belongs to every truth set for L^+ of which S is a subset* (Theorems 3.7 and 3.10). But any truth set for L^+ is a model set for L^+, which ensured that if $S \nvdash A$, then there is – for some term extension L^+ of L – a model set for L^+ of which S is a subset and $\neg A$ is a member. Hence, by dint again of a suitable soundness theorem, $S \vdash A$ iff – *no matter the term extension L^+ of L – $\neg A$ does not belong to any model set for L^+ of which S is a subset* (Theorem 3.20 and 3.23). Two alternatives to the standard account of logical entailment, the truth-set account of page 215 and the model-set account on pages 218–219 were thereby legitimized. So, consequently, were two alternatives to the standard account of logical truth.

But any truth set for L^∞ generates a Henkin model for L^∞, the model associate of that set defined on page 201. The fact ensured that if $S \nvdash A$, then there is – for some extension L^+ of L – a Henkin model for L^+ in which S is true$_S$ but A is not. Hence, by dint of a suitable soundness theorem, $S \vdash A$ iff – *no matter the term extension L^+ of L – A is true$_S$ in every Henkin model for L^+ in which S is true$_S$* (Theorem 2.18). But any truth set for L^∞ also generates a truth-value assignment (equivalently, a truth-value function) for L^∞, the truth-value associate of that set defined on page 211. The fact ensured that if $S \nvdash A$, then there is – for some extension L^+ of L – a truth-value assignment (a truth-value function) for L^+ on which each member of S is true (evaluates to T) but A is not (does not). Hence, by dint again of a suitable soundness theorem, $S \vdash A$ iff – *no matter the term extension L^+ of L – A is true on every truth-value assignment (A evaluates to T on every truth-value function) for L^+ on which all members of S are (do)* (Theorems 3.2 and 3.3).[39] Two extra alternatives to the standard account of logical entailment, the substitutional account of page 206 and the truth-value one of page 210, were thereby legitimized. So, consequently, were two extra alternatives to the standard account of logical truth.

I.3: ALTERNATIVES TO FIRST-ORDER SEMANTICS 259

But any truth set for L^∞ also generates a singular probability function for L^∞, the probability associate of that set defined on page 238. The fact ensured that if $S \vdash A$, then there is – for some term extension L^+ of L – a singular probability function for L^+ on which all members of S evalute to 1 but A does not. Hence, by dint of a suitable soundness theorem, $S \vdash A$ iff – *no matter the term extension L^+ of L – A evaluates to 1 on every singular probability function for L^+ on which all members of S do* (Theorems 4.43 and 4.48). But any truth set for L^∞ also generates a binary probability function for L^∞, the probability associate of that set defined on page 254. The fact ensured that if $S \vdash A$, then there is – for some term extension L^+ of L and some statement B of L^+ – a binary probability function for L^+ on which all members of S evaluate to 1 relative to B but A does not. Hence, by dint of a suitable soundness theorem, $S \vdash A$ iff – *no matter the term extension L^+ of L and statement B of L^+ – A evaluates to 1 relative to B on every binary probability function for L^+ on which all members of S do* (Theorems 5.25 and 5.50). Two extra alternatives to the standard account of logical entailment, the probability account of page 226 and that of page 247, were thereby legitimized. So, consequently, were two extra alternatives to the standard account of logical truth.[40]

The definitions of pages 206, 210, 215, 218–219, 226 and 247, thus known to capture exactly the same logical truths and exactly the same logical entailments as the standard ones do, merit attention on further counts. I largely limit myself in what follows to logical truths. Like remarks apply *mutatis mutandis* to logical entailments.

Logical truths are defined quite grandly in standard semantics as statements true in *all* models or – to spell it out once more – true in $\langle D, I_D \rangle$, *no matter the domain D and D-interpretation I_D* of the terms and predicates in those statements. Unfortunately, domains are sets and since 1895, the year Cantor discovered the Burali-Forti paradox, sets have been very much a concern, one indeed that may never be alleviated.[41] So the standard account of logical truth, with its mention – be it overt or covert – of *all* domains and hence of *all* sets, may be unjudicious: it rests logic on very shaky foundations.[42]

That the models figuring in the above account can be presumed to be countable was therefore major news (though, as noted on page 208, not welcome news to all). Since finite cardinals and $\aleph_0$ are unproblematic (to put it more judiciously, are less problematic than larger cardinals), logic studies could be pursued with little apprehension. As conceded on page 209, one might still deal in uncountable models when attending to certain theories (though countable submodels of those models would always do the trick, we saw). But attending to theories whose intended models are uncountable is one thing, explicating logical truth (and logical entailment) is quite another; and discharging the latter assignment without courting disaster or at the very least embarrassment seems the wiser course.

But, as later developments showed, one can do far better than the Löwenheim Theorem and Skolem's generalization thereof guaranteed. Since countable models will do and first-order languages have $\aleph_0$ terms, Henkin models should do as well, with all items in their domains provided with names. What, given just $\aleph_0$ names, uncountable models ruled out, should now be feasible. And indeed statements true in all countable models (hence, in all models) proved to be, as Henkin's completeness proof intimated, those very statements true *in all Henkin models*. The result buoyed those who like things to bear names and quantifiers to be construed substitutionally, a natural inclination as Wittgenstein admitted.

True, logical entailment does pose a problem. As sets are not all infinitely extendible, one may occasionally run out of terms, even though one has $\aleph_0$ of them and only $\aleph_0$ items to name. But, following Henkin's precedent, one can always send out for extra terms by extending one's original language to such a language as L^∞. The Henkin models for L^∞ will serve all of one's needs, yielding a definition of logical entailment to the same effect as the standard one. The resulting semantics is, of course, that on page 206: *substitutional semantics*.

However, even such minimal use of sets as substitutional semantics makes is unnecessary, as Beth, Schütte, and others showed. With the quantifiers substitutionally construed, sets (*qua* domains) turn up at only one point: the truth condition for *atomic statements* on page 206. That condition, incidentally, is not to everyone's taste: some (Frege, Church, and possibly Whitehead and Russell, among them) prefer one that uses propositional functions and, hence, has an intensional rather than extensional flavor. But, whatever one's preference in these matters, is it really logic's charge to spell out the circumstances under which atomic statements are true? The recounting plays no role in substitutional semantics. So, instead of engaging in such controversial and (in this context) inconsequential business, why not assume with Beth, Schütte, and others that atomic statements have truth-values, however they come by them, and proceed with matters of truly logical import?[43] Thus was *truth-value semantics* born, a semantics that dispenses with domains and, hence, with reference (crucial though that notion may be elsewhere). And, dispensing with reference, truth-value

I.3: ALTERNATIVES TO FIRST-ORDER SEMANTICS

semantics can focus on a single notion: truth. In one version of it, truth-value assignments (to *atomic* statements) and a recounting of when *compound* statements are true on them share the work; in another and even sparer version truth-functions do it all. And if future studies prove as rewarding as past ones, most (if not all) of standard semantics might soon admit of translation into truth-value idiom.

As it turns out, though, even truth can be dispensed with, the truth sets introduced in Quine [1940] permitting definition (we saw) of logical truth and logical entailment, as do the model sets introduced in Hintikka [1955]. Some have objected that for A to belong to *every* truth set (for $\neg A$ to belong to *no* model set) is a purely syntactical feature of A, and consequently that truth-set semantics (model set semantics) is but mislabelled syntax.

I shall not meet the objection *here*, except for recalling that truth sets, for example, automatically convert into truth-value functions, the prerequisites for belonging to a truth set ((a)–(c) on page 205) being those for evaluating to T on a truth-value function (**B1–B3** on page 213). Rather, I will draw attention to *probabilistic semantics*, where the notion of truth yields to that of rational (more specifically, coherent) belief, and accounts of logical truth and logical entailment matching the standard ones are also available. Probabilistic semantics is as spare as truth-value semantics is – probability functions (be they singular ones or binary ones) doing *all* the work. And it offers a whole new perspective on logical truth (and logical entailment): traditionally viewed as statements true in all models (more judiciously, in all countable models) and more recently as statements evaluating to T on all truth-value functions, logical truths can now be seen as statements than which none are more (coherently) believable.[44]

Probabilistic semantics, the one intensional semantics studied here, has yet to match standard semantics in breadth. But it already boasts significant results, some presented in Sections 4–5; and, feeding as it does on probability theory, belief theory, game theory, etc., it should soon boast others.

In any event there are accounts of logical truth and logical entailment other than the standard one and no less legitimate. *This much* the essay *has* shown.[45]

APPENDIX

I supplement here historical information provided in the main text.

(1) *The Substitution Interpretation and Substitutional Semantics*

The substitution interpretation of the quantifiers and, more particularly, substitutional semantics have a long and still unchronicled history. As remarked in note 2, Frege understood the quantifiers substitutionally in 1879, but switched later on to the objectual interpretation. Wittgenstein is quoted in Moore [1959] p. 297, as saying in the course of a 1932 lecture that "there was a temptation, to which he had yielded in the *Tractatus*, to say that $(x) \cdot f(x)$ is identical with the logical product '$fa \cdot fb \cdot fc \cdot \ldots$', and $(\exists x) \cdot fx$ identical with the logical sum '$fa \vee fb \vee fc \vee \ldots$'; but that this was in both cases a mistake." In papers written in 1925 and 1926 F. P. Ramsey reported on the interpretation as Wittgenstein's, defended it against an objection of Hilbert's, and stated that, with the quantifiers interpreted Wittgenstein's way, all axioms of Whitehead and Russell [1910–13] other than the Axiom of Reducibility hold true (see Ramsey [1926a] and [1926b]). Ramsey's remark on Whitehead and Russell's work may well be the first formal contribution to substitutional semantics.

To my knowledge little was heard (in any event, little was made) of the substitution interpretation of the quantifiers for the next fifteen years. It then turned up in a succession of books and papers by Carnap, beginning with Carnap [1942] and culminating in Carnap [1950]. The latter book is quite instructive. The account of truth you find there is essentially the substitutional one I gave on page 206, but the account of logical truth, logical entailment, etc., is the truth-set one that I gave on pages 214–215.

Henkin's completeness paper, to which reference has frequently been made, appeared in 1949; the Robinson book mentioned on page 209 appeared in 1951; the first of several papers in which Ruth Marcus champions the substitution interpretation of the quantifiers, and particularly urges its adoption in modal logic, appeared in 1963; the Shoenfield book mentioned on page 209, appeared in 1967; etc. By then, of course, what I call *truth-value semantics* had come into its own, and it eventually captured some of the interest earlier accorded to substitutional semantics. The beginnings of truth-value semantics were reported in Section 3. Since substitutional and truth-value semantics are close relatives, some of the information supplied there belongs here as well.

As the substitution interpretation gained greater currency, it was subjected – expectedly enough – to intense and often critical scrutiny. Davidson and some of his students wondered, for example, whether the substitutional account of truth satisfies Tarski's celebrated Convention *T*, a matter of considerable importance reviewed in Kripke [1976]. Other concerns were voiced by other writers (see Quine [1969], for instance), but in my opinion have been or could be met. Admittedly, the pros and cons of the substitution interpretation demand further study. As suggested at the outset, though, my attention in this essay goes to the novel accounts of logical truth, logical entailment, etc., that the interpretation allows.

(2) *Truth-value Semantics*

Truth-value semantics, the reader may recall, dates back to 1959, the year that saw the publication of Beth's *Foundations of Mathematics*. Several of the contributions to truth-value semantics between 1959 and 1975 are recapitulated in Leblanc [1976]; they concern first-order logic without and with identity, second-order logic, a variety of modal logics, three-valued logic, and presupposition-free variants of most of these. Further contributions will be found in Leblanc [1973] (the proceedings of a Temple University conference on alternative semantics) and in Part 2 of Leblanc [1982b] (a set of papers on truth-value semantics I authored or co-authored from 1968 onwards). The papers in the bilbiography of Kripke [1976] deal primarily with the substitution interpretation of the quantifiers, but many of them touch on truth-value semantics as well. A few additional titles will be found in the bibliography of this essay: Kearns [1978], which distinguishes three brands of substitution interpretation and – hence – of truth-value semantics; Garson [1979], which investigates which logics are susceptible of a substitutional semantics and – hence – of a truth-value one; Parsons [1971] and Gottlieb and McCarthy [1979], which attend to substitutional and truth-value semantics in the particular context of set theory; McArthur [1976], McArthur and Leblanc [1976], and Barnes and Gumb [1979], which provide a truth-value semantics for tense logic; Leblanc and Gumb [1983] and Leblanc [1982a] which provide one for intuitionistic logic; and Gumb [1978, 1979], which sport an important variant of truth-value semantics and also deal with model sets.

Some of these texts are explicitly labelled contributions to truth-value semantics; others are not. A few, anxious to show logical truth and entailment outgrowths of truth-functional truth and entailment, employ the truth-value assignments of page 210; others employ the truth-value assignments of page 213, or – following the precedent of Schütte [1960] – truth-value functions.

(3) *Probabilistic Semantics*

Popper devised autonomous constraints for singulary probability functions as early as 1938 (see Appendix *ii to Popper [1959]), and attended to binary ones only eighteen years later. Yet contributions to singulary probabilistic semantics are comparatively recent and few in number: Stalnaker [1970], Bendall [1979], Leblanc [1982c], portions of this essay, etc. The first two study the relationship between singulary probability functions and what I call truth-value functions (Stalnaker talks instead of *truth valuation functions* and limits himself to quantifierless statements); the third defines logical truth and logical entailment as on page 226, proves some of the theorems in Sections 4 and 5 (referring the reader to this text for proof of Theorem 4.43), and shows Constraints **C1–C6** to pick out the same functions as Kolmogorov's constraints and Popper's do.

Binary probabilistic semantics has a more eventful history, and one spreading over more than two decades.

The earliest account of *truth-functional truth* to employ binary probability functions may be that in Leblanc [1960]. As pointed out in Stalnaker [1970], it is unfortunately too broad, suiting only the binary probability functions attributed to Carnap in Harper *et al.* [1983]. A correct account of the notion can, incidentally, be retrieved from Stalnaker's paper.

The earliest *published* accounts of *logical truth* and *logical entailment* to employ binary probability functions may be those in Field [1977]. Dissatisfied with Field's resort to limits, I proposed in Leblanc [1979b] substitute – but equivalent – accounts of the two notions, those used in this essay (page 247). However, while my paper was in press, I learned that Harper had used the very same accounts in his unpublished doctoral dissertation of 1974. A summary of the dissertation is to appear in Leblanc *et al.* [1983]. Further, but again equivalent, accounts will be found in Adams [1981], Van Fraassen [1981], Morgan and Leblanc [1983a], and doubtless many other texts. Incidentally, Van Fraassen [1981] uses in lieu of **D7** constraints which render the term extensions of L unnecessary but rule out some of the functions acknowledged here. Like substitutes for **C7** have yet to be found.

The earliest theorem that probabilistic semantics boasts of is in Popper [1959] Appendix *v. A soundness theorem, it is roughly to the effect that if a Boolean identity $A = B$ (A and B here either sets or statements) is provable by means of the 'fourth set' of postulates in Huntington [1933], then $P(A/C) = P(B/C)$ for any set or statement C and any binary probability function P meeting Popper's constraints. I extended the result in Leblanc

[1960], showing in effect that if $\vdash_0 A$, where A is a quantifierless statement of L, then $P(A/B) = 1$ for any quantifierless statement B of L and any binary probability function P meeting Popper's constraints. Soundness and completeness theorems essentially like Theorem 5.25 and Theorem 5.30 in this essay were proved in Harper [1974], Field [1977], Leblanc [1979b], and most probably other texts as well. The strategy used on pages 249–253 to prove Theorem 5.23 is due to Harper and already figures in Harper *et al.* [1983]; details, however, are simpler here. The strategy used on page 254 to prove Theorem 5.28 is essentially that in Leblanc [1979b], but again details are simpler here.

As announced on page 191, I confined myself in this essay to first-order logic without identity. Gaifman [1964], Gumb [1983], and Seager [1983] attend to first-order logic with identity; Van Fraassen [1981] attends to intuitionistic logic, as do Morgan and Leblanc [1983a, b]; Morgan [1982a, b] attend to modal logic, as does Schotch and Jennings [1981]; several papers, Stalnaker [1970] possibly the earliest of them, attend the conditional logic (see Section 9 of Nute's essay in this Handbook [II, 9] for further information); etc. The majority of these results will be surveyed in *Probabilistic Semantics*.

Temple University

NOTES

[1] The essay was written while I held a research grant (SES 8007179) from the National Science Foundation and was on a partial research leave from Temple University. Some of the material was used in lectures delivered at City College of CUNY in 1980–1981 and to be published by D. Reidel Publishing Company under the title *Probabilistic Semantics*.

[2] The rival, and more generally accepted, interpretation of the quantifiers is of course the *objectual* one (that in (iv) on page 197). Goldfarb [1979] intimates that the first correct account of (first-order) logical truth, an objectual one, is in Bernays [1922]. By then a substitutional account of logical truth had already been sketched in Wittgenstein [1921]. To take the matter further back, Frege [1893–1903] interprets the quantifiers objectually, but Frege [1879] interprets them substitutionally (see Stevenson [1973]).

[3] In standard semantics and each alternative to it studied here, logical truth is but logical entailment by the null set, the way provability (i.e. theoremhood) is but provability from that set. As a result more space will generally be devoted to logical entailment than to logical truth, and by the same token to strong than to weak soundness and completeness.

[4] Term extensions – i.e. languages exactly like L except for having extra individual terms – have been in use for a good many years. They figure in Henkin [1949], in Hintikka [1955] (a passage from which is quoted in note 16), in Gaifman [1964], etc. The Dunn-Belnap extensions I use are the least extensions of L that serve all my purposes here.

[5] It follows from the account that (a) individual variables can occur only *bound* and (b) identical quantifiers cannot overlap in a statement of L. Because of (a) all statements of L are so-called *closed* statements. As regards (iv): When x does not occur in A, $(\forall x)A$ is known as a *vacuous* quantification.

[6] Statements of the sorts $(A \to B)$, $(A \vee B)$, $(A \leftrightarrow B)$, and $(\exists x)A$ are, of course, known as *conditionals*, *disjunctions*, *biconditionals*, and *existential quantifications*. The convention whereby $(A \to B)$ is short for $\neg(A \wedge \neg B)$ will be referred to by means of '$D_\to$', that whereby $(A \vee B)$ is short for $\neg(\neg A \wedge \neg B)$ by means of '$D_\vee$', etc.

[7] The (atomic) substatements of a *quantifierless* statement of L are also known as – and on page 209 will be called – its (atomic) *components*.

[8] Tarski himself talked of *deductive systems* rather than *theories* (see Tarski [1930, 1935-36]); but the appellation 'theory in Tarski's sense' has prevailed.

[9] With S presumed to be infinitely extendible in L, there is sure to be – however large the n – a term T of L foreign to $S_{n-1} \cup \{A_n\}$; not so, otherwise.

[10] Standard (i.e. model-theoretic) semantics comes in various brands. The present brand (an early version of which appeared in Leblanc and Wisdom [1972]) especially suits languages such as L whose statements are all closed; the one propounded in Tarski [1936] and used in quite a number of recent texts, Hodges among them, especially suits languages with both open and closed statements. The beginnings of standard semantics are chronicled in Goldfarb [1979].

[11] '$S \not\vdash A$' is short here for 'A is *not* provable from S in L'. The proof I embark upon is essentially due to Henkin, and for this reason is often called *Henkin's Completeness Proof*.

[12] The counterexample in the text was noted in Thomason [1965]. I learned of it from Thomason and reported it in Leblanc [1968]. Dunn and Belnap, who had known of the counterexample for several years, reported it in Dunn and Belnap [1968].

[13] The result is proved in surprisingly few texts. The proof here stems from Leblanc [1966] pp. 177-78, but avoids that text's recourse to L^∞. To illustrate matters, $\{Q(t_1), Q(t_2), Q(t_3),\ldots\}$ is true – but $(\forall x)Q(x)$ is *not* – in $\langle \mathcal{T}, I_\mathcal{T}^2 \rangle$, the double rewrite of the model associate $\langle \mathcal{T}, I_\mathcal{T} \rangle$ of $\mathcal{H}(\{Q(t_2), Q(t_4), Q(t_6),\ldots, \neg(\forall x)Q(x)\})$. $I_\mathcal{T}$ assigns t_i to t_i and, hence, $I_\mathcal{T}^2$ assigns t_{2i} to t_i; but both $I_\mathcal{T}$ and $I_\mathcal{T}^2$ assign $\mathcal{T}$ to Q.

[14] I assume in the text that sets of statements of L^∞ (as well as L) have been provided with model associates.

[15] To illustrate matters, $\{Q(t_1), Q(t_2), Q(t_3),\ldots\}$ is true, but $(\forall x)Q(x)$ is not, on the (Henkin) $I_{\mathcal{T}^\infty}$-interpretation of L^∞ ($\mathcal{T}^\infty$ the set of all the terms of L^∞) that assigns each term of L^∞ to itself and $\{t_1, t_2, t_3, \ldots\}$ to Q. Though not true in any Henkin model for L, the set $\{Q(t_1), Q(t_2), Q(t_3), \ldots, \neg(\forall x)Q(x)\}$ is thus true in one for L^∞.

[16] To quote from p. 52 of Hintikka [1955]: "We assume that on any particular occastion we can choose the cardinal number of the individual constants and of the individual variables as large as we wish, by constructing a new, more comprehensive calculus, if necessary." The passage certainly allows what I call the term extensions of L, and it might be construed as allowing L itself, to have *any* number of terms.

I.3: ALTERNATIVES TO FIRST-ORDER SEMANTICS 267

[17] Hintikka deals principally with *model sets*, a topic I cover in the next section. Robinson and Shoenfield, on the other hand, deal with (*Henkin*) *models*.

[18] Because it uses the truth-value associate of $\mathscr{H}^\infty(S \cup \{\neg A\})$, this proof of (3) will be recalled in Section 6.

[19] To be more exact, sketched in Wittgenstein [1921] and formalized in Carnap [1950]. The definitions in the next paragraph are borrowed from Carnap [1950] Section 18.

[20] Quine's truth sets consist of components of a quantifierless statement, and are accordingly called *the truth set of that statement*. As one would expect, a quantifierless statement is declared tautologous if it belongs to every one of its truth sets.

[21] When ambiguity threatens, I shall enclose '$T_i(A)$' within parentheses.

[22] When ambiguity threatens, I shall enclose '$\mathscr{L}(A)$' within parentheses.

[23] Here as in Theorem 3.15 any other term of L occurring in S could of course substitute for T_1.

[24] An induction of this sort, with each kind of negation treated separately, is called in Leblanc and Wisdom [1972] a *Hintikka induction*.

[25] For further results concerning model (or Hintikka) sets, see Hintikka [1955], Jeffrey [1967], Smullyan [1970], Leblanc and Wisdom [1972], and Leblanc [1976].

[26] A_1 here the alphabetically first statement of L⁺, A_2 the second, A_3 the third, etc.

[27] I.e., ensuring that any such function P^+ for L⁺ is *coherent* in the sense of De Finetti [1937]. The literature on belief functions is considerable. For a survey of early results on coherent belief (particularly, betting) functions, see Carnap and Jeffrey [1971] pp. 105–116; for a recent study of belief systems, see Ellis [1979]. I owe the phrasing in the text to Kent Bendall; hence the quotation marks.

[28] As the reader doubtless knows, Kolmogorov's functions take sets rather than statements as their arguments. However, they convert into statement-theoretic functions once you think of his complements and intersections as negations and conjunctions. Popper's functions take what he calls *elements* as their arguments; these, as pointed out on p. 319 of Popper [1959], may be understood either as sets or as statements. Gaifman's functions do take statements as their arguments.

[29] Information concerning the provenance of **C1–C7** may be welcome.

(i) **C1** and **C2** are statement-theoretic counterparts of two axioms in Kolmogorov [1933]: they are known as *Non-negativity* and *Unit Normalization*. **C3–C6** are borrowed from Popper [1955]; they are known as *Complementation, Idempotence, Commutativity*, and *Associativity*. And **C7** is an adaptation by Bendall of an axiom in Gaifman [1964]. Both Gaifman and Bendall, incidentally, use minima where I use limits, but the difference is immaterial here.

(ii) Kolmogorov had two additional axioms, known as *Additivity* and *Continuity*. The statement-theoretic counterpart of the latter calls for infinite disjunctions and, hence, does not belong here. The statement-theoretic counterpart of the former runs:

C8.   If two statements A and B of L⁺ are logically incompatible in the standard sense (i.e. if $\neg(A \wedge B)$ is logically true in that sense), then $P^+(A \vee B) = P^+(A) + P^+(B)$.

Switching from sets to statements forces adoption of yet another constraint:

C9.   If A and B are logically equivalent in the standard sense (i.e. if $A \equiv B$ is logically true), then $P^+(A) = P^+(B)$. (= *Substitutivity*)

It is readily shown, given Theorem 4.50 below, that **C1–C2** and **C8–C9** pick out exactly the same probability functions as **C1–C6** do.

(iii) Popper uses in place of **C1–C2** the following three constraints:

C10. $P^+(A \wedge B) \leq P^+(A)$ (= *Monotony*)

C11. There are at least two distinct statements A and B of L^+ such that $P^+(A) \neq P^+(B)$ (= *Existence*)

C12. For each statement A of L^+ there is a statement B of L^+ such that $P^+(A) \leq P^+(B)$ and $P^+(A \wedge B) = P^+(A) \times P^+(B)$ (= *Multiplication*).

It can be shown that **C3–C6** and **C10–C12** pick out exactly the same probability functions as **C1–C6** do; see Leblanc [1982c] on the matter.

(iv) That $P^+(\neg A) = 1 - P^+(A)$ (= Theorem 4.11) would not do as a substitute for **C3** is readily shown. Let $P^+(\neg A)$ be $1 - P^+(A)$, and $P^+(A \wedge B)$ be the smaller of $P^+(A)$ and $P^+(B)$. All of **C1–C2**, Theorem 4.11, and **C4–C7** will then 'pan out', but **C3** will not. Whether $P^+((\forall x)A) = \text{Limit}_{j \to \infty} P^+(\Pi_{i=1}^{j} A(t_i^+/x))$ (= Theorem 4.33) would do as a substitute for **C7** is an open question. It does in Gaifman [1964]; but, as Bendall pointed out to me, the non-autonomous constraint that Gaifman would use to pass from Theorem 4.33 to **C7** is not available here.

Note that the two limits $\text{Limit}_{j \to \infty} P^+(A \wedge \Pi_{i=1}^{j} B(t_i^+/x))$ and $\text{Limit}_{j \to \infty} P^+(\Pi_{i=1}^{j} A(t_i^+/x))$ exist for every function P^+ for L^+ and all quantifications $(\forall x)B$ and $(\forall x)A$ of L^+. Carnap would rather construe $P^+((\forall x)A)$ as $\text{Limit}_{j \to \infty} P^+(A(t_j^+/x))$ (see Carnap [1950] p. 302). But, since this third limit does not always exist, the ensuing function P^+ would not accord values to all universal quantifications, a serious drawback this.

[30] Note that if a set of statements of L is infinitely extendible in L, then S is sure to be ω-complete and hence its probability associate is sure to meet **C7**. The result, exploited in Morgan and Leblanc [1982], is of no use here: the probability associate of S cannot meet **C3** unless S is maximally consistent, in which case S sports *every* term of L.

[31] Since Constraint **C6** does not figure in Theorems 5.4 and 5.5, two-valued functions that meet Constraints **C1–C5** are sure by Theorem 5.3 to meet Constraint **C6** as well. I doubt, however, that this holds true of functions with more than two values.

[32] A similar conclusion is reached in Popper [1959], p. 356: "In its logical interpretation, the probability calculus is a genuine generalization of the logic of derivation." It is based, though, on a different result, that reported on page 264.

[33] The point was brought to my attention by Bas C. van Fraassen. The counter-example in the text is a simplification by Michael E. Levin and myself of one originally suggested by Van Fraassen.

[34] I thus construe the present P^+'s as what note 27 called *coherent* belief functions. Since singular probability functions are – in effect – binary ones relative to a logical truth (say, $\neg(Q(t_1) \wedge \neg Q(t_1))$), just one notion of coherence is involved here.

[35] Von Wright weakens (2) to read:

(2*)   If A is not logically false, then $P^+(A/A) = 1$,

and strengthens (3) to read:

(3*) $P^+(\neg A/B) = 1 - P^+(A/B)$,

a course few have adopted. That one or the other of (2) and (3*) must carry a restriction, is obvious. Suppose B is logically false, in which case both $B \to A$ and $B \to \neg A$ are logically true. (2) will then permit proof of $P^+(A/B) = P^+(\neg A/B) = 1$, and (3*) must on pain of contradiction be weakened to read like (3). However, weaken (2) to read like Von Wright's (2*). Then proofs of $P^+(A/B) = 1$ and $P^+(\neg A/B) = 1$ are blocked, and (3) may be strengthened to read like Von Wright's (3*). Carnap, who accords $P^+(A/B)$ a value only when B is not logically false, can make do with (2) and (3*). But, as suggested in the text, partial probability functions are wanting.

[36] "Occasions when (Wellington) exhibited his other peculiarity, the crashing retort, were no doubt multiplied by the wit and inventiveness of his contemporaries; but the celebrated exchange between the Duke and some minor official from a government office is authentic.

'Mr. Jones, I believe,' said the official blandly, accosting the great man in Pall Mall and mistaking him for the secretary of the Royal Academy. The world-famous profile froze.

'If you believe that, you'll believe anything.'" (Elizabeth Longford, *Wellington, Pillar of State*, Harper & Rowe, New York, 1972.)

Because of this anecdote P^+-absurd statements are occasionally called *Jones statements*.

[37] To my knowledge, Popper is the first to have devised autonomous constraints for binary probability functions. The ones he finally settled on appeared in Popper [1957], and are extensively studied in Popper [1959] Appendices *iv-*v. They run (in the symbolism of this essay):

A1. For any two statements A and B of L^+ there are statements A' and B' of L^+ such that $P^+(A/B) \neq P^+(A'/B')$ (*Existence*)

A2. If $P^+(A/C) = P^+(B/C)$ for every statement C of L^+, then $P^+(C/A) = P^+(C/B)$ (*Substitutivity*)

A3. $P^+(A/A) = P^+(B/B)$ (*Reflexivity*)

B1. $P^+(A \wedge B/C) \leq P^+(A/C)$ (*Monotony*)

B2. $P^+(A \wedge B/C) = P^+(A/B \wedge C) \times P^+(B/C)$ (*Multiplication*)

C. If $P^+(C/B) \neq P^+(B/B)$ for at least one statement C of L^+, then $P^+(\neg A/B) = P^+(B/B) - P^+(A/B)$ (*Complementation*).

That **D0–D6** pick out exactly the same binary probability functions as these six constraints do was undoubtedly known to Popper. The result is formally established in Harper *et al.* [1983]. (For a proof of A2 that uses only **D1, D2, D4**, and Theorem 5.9 see Leblanc [1981].) In effect, Popper only dealt with quantifierless statements and hence had no analogue of **D7**.

[38] Binary probability functions, like their singularity brethren, bear a close relationship to truth-value functions. See Gumb [1983] for preliminary results on the matter.

[39] The proof of Theorem 3.3 I just rehearsed is that on page 211 (see note 18).

[40] Schematically:

$S \cup \{\neg A\}$ extends to $\mathcal{H}^\infty (S \cup \{\neg A\})$, a set of L^∞ which constitutes { a truth set for L^∞ generating { a Henkin model for L^∞ / a truth-value assignment (truth-value function) for L^∞ / a singulary and a binary probability function for L^∞ } hence a model set for L^∞ }

[41] The Zermelo-Fraenkel axiomatization of set theory is mathematically serviceable, to be sure, but ontologically and epistemologically how much of a case has ever been made for it?

[42] When teaching standard semantics, how many warn students not to ask whether the set of all domains is itself a domain?

[43] I borrow at this point from a conversation with Alex Orenstein.

[44] Note indeed that $\vdash A$ iff – no matter the probability function P for L – $P(A) = 1$. But $P(B) \leqslant 1$ for every statement B of L. Hence $\vdash A$ iff – no matter the probability function P for L and statement B of L – $P(A) \not< P(B)$.

[45] Thanks are due to Ermanno Bencivenga, Kent Bendall, John Corcoran, Melvin C. Fitting, Raymond D. Gumb, John T. Kearns, Jack Nelson, Muffy E. A. Siegel, George Weaver, William A. Wisdom, and graduate students at Temple University for reading an earlier version of this essay; to Elizabeth Lazeski and Anne D. Ward for assisting in the preparation of the manuscript; to Thomas McGinness and John Serembus for reading the proofs with me; and to my son Stephen for urging me, at a difficult time, to nonetheless write this essay.

REFERENCES

Adams, E. W.: 1981, 'Transmissible improbabilities and marginal essentialness of premises in inferences involving indicative conditionals', *J. Philosophical Logic* **10**, 149–178.

Barnes, R. F., Jr. and Gumb, R. D.: 1979, 'The completeness of presupposition-free tense logics', *Zeitschrift für mathematische Logik und Grundlagen der Mathematik* **25**, 192–208.

Behmann, H.: 1922, 'Beiträge zur Algebra der Logik, in besondere zum Entscheidungsproblem', *Math. Annalen* **86**, 163–229.

Bendall, K.: 1979, 'Belief-theoretic formal semantics for first-order logic and probability', *J. Philosophical Logic* **8**, 375–394.

Bendall, K.: 1982, 'A "definitive" probabilistic semantics for first-order logic', *J. Philosophical Logic* **11**, 255–278.

Bergmann, M., Moore, J., and Nelson, R.: 1980, *The Logic Book*, Random House, New York.

Bernays, P.: 1922, Review of Behmann [1922], *Jahrbuch über die Fortschritte der Mathematik* **48**, 1119.
Beth, E. W.: 1959, *The Foundations of Mathematics*, North-Holland, Amsterdam.
Carnap, R.: 1942, *Introduction to Semantics*, Harvard University Press, Cambridge, Mass.
Carnap, R.: 1950, *Logical Foundations of Probability*, University of Chicago Press, Chicago, Ill.
Carnap, R.: 1952, *The Continuum of Inductive Methods*, University of Chicago Press, Chicago, Ill.
Carnap, R. and Jeffrey, R. C.: 1971, *Studies in Inductive Logic and Probability*, Volume I, University of California Press, Berkeley and Los Angeles, Ca.
De Finetti, B.: 1937, 'La prévision: Ses lois logiques, ses sources subjectives', *Annales de l'Institut Henri Poincaré* **7**, 1–68.
Dunn, J. M. and Belnap, N. D., Jr.: 1968, 'The substitution interpretation of the quantifiers', *Noûs* **2**, 177–185.
Ellis, B.: 1979, *Rational Belief Systems*, APQ Library of Philosophy, Rowman and Littlefield, Totowa, N.J.
Field, H. H.: 1977, 'Logic, meaning, and conceptual role', *J. Philosophy* **74**, 379–409.
Fitch, F. B.: 1948, 'Intuitionistic modal logic with quantifiers', *Portugaliae Mathematica* **7**, 113–118.
Frege, G.: 1879, *Begriffschrift*, Halle.
Frege, G.: 1893–1903, *Grundgesetze der Arithmetik*, Jena.
Gaifman, H.: 1964, 'Concerning measures on first-order calculi', *Israel J. Math.* **2**, 1–18.
Garson, J. W.: 1979, 'The substitution interpretation and the expressive power of intensional logics', *Notre Dame J. Formal Logic* **20**, 858–864.
Gentzen, G.: 1934–35, 'Untersuchungen über das logische Schliessen', *Mathematische Zeitschrift* **39**, 176–210, 405–431.
Goldfarb, W. D.: 1979, 'Logic in the Twenties: the nature of the quantifier', *J. Symbolic Logic* **44**, 351–68.
Gottlieb, D. and McCarthy, T.: 1979, 'Substitutional quantification and set theory', *J. Philosophical Logic* **8**, 315–331.
Gumb, R. D.: 1978, 'Metaphor theory', *Reports on Math. Logic* **10**, 51–60.
Gumb, R. D.: 1979, *Evolving Theories*, Haven Publishing, New York.
Gumb, R. D.: 1983, 'Comments on probabilistic semantics', in Leblanc *et al.* [1983].
Harper, W. L.: 1974, 'Counterfactuals and representations of rational belief', doctoral dissertation, University of Rochester.
Harper, W. L.: 1983, 'A conditional belief semantics for free quantificational logic with identity', in Leblanc *et al.* [1983].
Harper, W. L., Leblanc, H. and Van Fraassen, B. C.: 1983, 'On characterizing Popper and Carnap probability functions', in Leblanc *et al.* [1983].
Hasenjaeger, G.: 1953, 'Eine Bemerkung zu Henkin's Beweis für die Vollständigkeit des Prädikatenkalküls der ersten Stufe', *J. Symbolic Logic* **18**, 42–48.
Henkin, L.: 1949, 'The completeness of the first-order functional calculus', *J. Symbolic Logic* **14**, 159–166.
Hintikka, J.: 1955, *Two Papers on Symbolic Logic*, Acta Philosophica Fennica **8**.
Huntington, E. V.: 1933, 'New sets of independent postulates for the algebra of logic', *Transactions of the American Mathematical Society* **35**, 274–304.

Jeffrey, R. C.: 1967, *Formal Logic: Its Scope and Limits*, McGraw-Hill, New York.
Jeffreys, H.: 1939, *Theory of Probability*, Oxford University Press, London.
Kearns, J. T.: 1978, 'Three substitution-instance interpretations', *Notre Dame J. Formal Logic* 19, 331–354.
Keynes, J. M.: 1921, *A Treatise on Probability*, Macmillan, London.
Kolmogorov, A. N.: 1933, *Grundbegriffe der Wahrscheinlichkeitsrechnung*, Berlin.
Kripke, S.: 1976, 'Is there a problem about substitutional quantification?' in G. Evans and J. McDowell (eds.), *Truth and Meaning*, Clarendon Press, Oxford, 325–419.
Leblanc, H.: 1960, 'On requirements for conditional probability functions', *J. Symbolic Logic* 25, 171–175.
Leblanc, H.: 1966, *Techniques of Deductive Inference*, Prentice-Hall, Englewood Cliffs, N.J.
Leblanc, H.: 1968, 'A simplified account of validity and implication for quantificational logic', *J. Symbolic Logic* 33, 231–235.
Leblanc, H.: 1973, *Truth, Syntax and Modality*, North-Holland, Amsterdam.
Leblanc, H.: 1976, *Truth-Value Semantics*, North-Holland, Amsterdam.
Leblanc, H.: 1979a, 'Generalization in first-order logic', *Notre Dame J. of Formal Logic* 20, 835–857.
Leblanc, H.: 1979b, 'Probabilistic semantics for first-order logic', *Zeitschrift für mathematische Logik und Grundlagen der Mathematik* 25, 497–509.
Leblanc, H.: 1981, 'What price substitutivity? A note on probability theory', *Philosophy of Science* 48, 317–322.
Leblanc, H.: 1982a, 'Free intuitionistic logic: A formal sketch', in J. Agassi and R. Cohen (eds), *Scientific Philosophy Today: Essays in Honor of Mario Bunge*, D. Reidel, Dordrecht, pp. 133–145.
Leblanc, H.: 1982b, *Existence, Truth, and Provability*, SUNY Press, Albany, N.Y.
Leblanc, H.: 1982c, 'Popper's 1955 axiomatization of absolute probability', *Pacific Philosophical Quarterly* 63, 133–145.
Leblanc, H.: 1983, 'Probability functions and their assumption sets: The singulary case', *J. Philosophical Logic* 12.
Leblanc, H.: forthcoming, 'Of consistency trees and model sets'.
Leblanc, H. and Gumb, R. D.: 1983, 'Soundness and completeness proofs for three brands of intuitionistic logic', in Leblanc *et al.* [1983].
Leblanc, H., Gumb, R. D. and Stern, R. (eds.): 1983, *Essays in Epistemology and Semantics*, Haven Publishing, New York.
Leblanc, H. and Morgan, C. G.: forthcoming, 'Probability functions and their assumption sets: The binary case', *Synthèse*.
Leblanc, H. and Wisdom, W. A.: 1972, *Deductive Logic*, Allyn & Bacon, Boston, Mass.
Löwenheim, L.: 1915, 'Über Möglichkeiten im Relativkalkul', *Mathematischen Annalen* 76, 447–470.
Marcus, R. B.: 1963, 'Modal logics I: Modalities and international languages', in M. W. Wartofsky (ed.), *Proceedings of the Boston Colloquium for the Philosophy of Science, 1961–1962*, D. Reidel, Dordrecht.
McArthur, R. P.: 1976, *Tense Logic*, D. Reidel, Dordrecht.
McArthur, R. P. and Leblanc, H.: 1976, 'A completeness result for quantificational tense logic', *Zeitschrift für mathematische Logik und Grundlagen der Mathematik* 22, 89–96.

Moore, G. E.: 1959, *Philosophical Papers*, Allen and Unwin, London.
Morgan, C. G.: 1982a, 'There is a probabilistic semantics for every extension of classical sentence logic', *J. Philosophical Logic* **11**, 431–442.
Morgan, C. G.: 1982b, 'Simple probabilistic semantics for propositional K, T, B, S4, and S5', *J. Philosophical Logic* **11**, 443–458.
Morgan, C. G.: 1983, 'Probabilistic semantics for propositional modal logics', in Leblanc *et al.* [1983].
Morgan, C. G. and Leblanc, H.: 1983a, 'Probabilistic semantics for intuitionistic logic', *Notre Dame J. Formal Logic* **23**, 161–180.
Morgan, C. G. and Leblanc, H.: 1983b, 'Probability theory, intuitionism, semantics and the Dutch Book argument', *Notre Dame J. Formal Logic* **24**, 289–304.
Morgan, C. G. and Leblanc, H.: 1983c, 'Satisfiability in probabilistic semantics', in Leblanc *et al.* [1983].
Orenstein, A.: 1979, *Existence and the Particular Quantifier*, Temple University Press, Philadelphia, Pa.
Parsons, C.: 1971, 'A plea for substitutional quantification', *J. Philosophy* **68**, 231–237.
Popper, K. R.: 1955, 'Two autonomous axiom systems for the calculus of probabilities', *British J. Philosophy of Science* **6**, 51–57, 176, 351.
Popper, K. R.: 1957, 'Philosophy of science: a personal report', in A. C. Mace (ed.), *British Philosophy in Mid-Century*, Allen and Unwin, London, pp. 155–191.
Popper, K. R.: 1959, *The Logic of Scientific Discovery*, Basic Books, New York.
Quine, W. V.: 1940, *Mathematical Logic*, Norton, New York.
Quine, W. V.: 1969, *Ontological Relativity and Other Essays*, Columbia University Press, New York and London.
Ramsey, F. P.: 1926a, 'The foundations of mathematics', *Proceedings of the London Mathematical Society, Series* **2**, 25, 338–384.
Ramsey, F. P.: 1926b, 'Mathematical Logic', *The Mathematical Gazette* **13**, 185–194.
Reichenbach, R.: 1935, *Wahrscheinlichkeitslehre*, Leiden.
Rényi, A.: 1955, 'On a new axiomatic theory of probability', *Acta Mathematica Acad. Scient. Hungaricae* **6**, 285–335.
Robinson, A.: 1951, *On the Mathematics of Algebra*, North-Holland, Amsterdam.
Rosser, J. B.: 1953, *Logic for Mathematicians*, McGraw Hill, New York.
Schotch, P. K. and Jennings, R. E.: 1981, 'Probabilistic considerations on modal semantics', *Notre Dame J. Formal Logic* **22**, 227–238.
Schütte, K.: 1960, 'Syntactical and semantical properties of simple type theory', *J. Symbolic Logic* **25**, 305–326.
Schütte, K.: 1962, *Lecture Notes in Mathematical Logic*, Volume I, Pennsylvania State University.
Seager, W.: 1983, 'Probabilistic semantics, identity and belief', *Canadian J. Philosophy*, **12**.
Shoenfield, J. R.: 1967, *Mathematical Logic*, Addison-Wesley, Reading, Mass.
Skolem, T. A.: 1920, 'Logisch-Kombinatorische Untersuchungen über die Erfüllbarkeit und Beweisbarkeit mathematischen Sätze nebst einem Theoreme über dichte Mengen', *Skrifter utgit av Videnskapsselkapet i Kristiania, I. Mathematisknaturvidenskabelig klasse 1920*, no. 4, 1–36.
Smullyan, R. M.: 1968, *First-Order Logic*, Springer-Verlag, New York.

Stalnaker, R.: 1970, 'Probability and conditionals', *Philosophy of Science* **37**, 64–80.
Stevenson, L.: 1973, 'Frege's two definitions of quantification', *Philosophical Quarterly* **23**, 207–223.
Tarski, A.: 1930, 'Fundamentale Begriffe der Methodologie der deductiven Wissenschaften. I', *Monatshefte für Mathematik und Physik* **37**, 361–404.
Tarski, A.: 1935–36, 'Grundzüge des Systemkalkül', *Fundamenta Math.* **25**, 503–526, **26**, 283–301.
Tarski, A.: 1936, 'Der Wahrheitsbegriff in den formalisierten Sprachen', *Studia Philosophica* **1**, 261–405.
Thomason, R. H.: 1965, 'Studies in the formal logic of quantification', doctoral dissertation, Yale University.
Van Fraassen, B. C.: 1981, 'Probabilistic semantics objectified', *J. Philosophical Logic* **10**, 371–394, 495–510.
Van Fraassen, B. C.: 1982, 'Quantification as an act of mind', *J. Philosophical Logic* **11**, 343–369.
Von Wright, G. R.: 1957, *The Logical Problem of Induction*, 2nd revised edition, Macmillan, New York.
Whitehead, A. N. and Russell, B.: 1910–13, *Principia Mathematica*, Cambridge University Press, Cambridge.
Wittgenstein, L.: 1921, *Tractatus Logico-Philosphicus (= Logisch-philosophische Abhandlung)*, *Annalen der Naturphilosophie* **14**, 185–262.

CHAPTER I.4

HIGHER-ORDER LOGIC

by *JOHAN VAN BENTHEM and KEES DOETS*

Introduction 275
1. First-order logic and its extensions 276
   1.1. Limits of expressive power 276
   1.2. Extensions 277
   1.3. Abstract model theory 278
   1.4. Characterization results: Lindström, Keisler and Fraissé 281
2. Second-order logic 288
   2.1. Language 288
   2.2. Expressive power 289
   2.3. Non-axiomatizability 293
   2.4. Set-theoretic aspects 294
   2.5. Special classes: Σ_1^1 and Π_1^1 297
3. Higher-order logic 305
   3.1. Syntax and semantics 306
   3.2. The prenex hierarchy of complexity 308
   3.3. Two faces of type theory 311
4. Reduction to first-order logic 315
   4.1. General models 316
   4.2. General completeness 318
   4.3. Second-order reduction 322
   4.4. Type theory and the lambda calculus 324
5. Reflections 326
References 327

INTRODUCTION

What is nowadays the central part of any introduction to logic, and indeed to some the logical theory par excellence, used to be a modest fragment of the more ambitious language employed in the logicist program of Frege and Russell. 'Elementary' or 'first-order', or 'predicate logic' only became a recognized stable base for logical theory by 1930, when its interesting and fruitful meta-properties had become clear, such as completeness, compactness and Löwenheim–Skolem. Richer higher-order and type theories receded into the background, to such an extent that the (re-)discovery of useful and interesting extensions and variations upon first-order logic came as a surprise to many logicians in the sixties.

In this chapter, we shall first take a general look at first-order logic, its properties, limitations, and possible extensions, in the perspective of so-called 'abstract model theory'. Some characterizations of this basic system are found in the process, due to Lindström, Keisler–Shelah and Fraïssé. Then, we go on to consider the original mother theory, of which first-order logic was the elementary part, starting from second-order logic and arriving at Russell's theory of finite types. As will be observed repeatedly, a border has been crossed here with the domain of set theory; and we proceed, as Quine has warned us again and again, at our own peril. Nevertheless, first-order logic has a vengeance. In the end, it turns out that higher-order logic can be viewed from an elementary perspective again, and we shall derive various insights from the resulting semantics.

Before pushing off, however, we have a final remark about possible pretensions of what is to follow. Unlike first-order logic and some of its less baroque extensions, second- and higher-order logic have no coherent well-established theory; the existent material consisting merely of scattered remarks quite diverse with respect to character and origin. As the time available for the present enterprise was rather limited (to say the least) the authors do not therefore make any claims as to complete coverage of the relevant literature.

1. FIRST-ORDER LOGIC AND ITS EXTENSIONS

The starting point of the present story lies somewhere within chapter I.1. We will review some of the peculiarities of first-order logic, in order to set the stage for higher-order logics.

1.1. *Limits of Expressive Power*

In addition to its primitives *all* and *some*, a first-order predicate language with identity can also express such quantifiers as *precisely one, all but two, at most three*, etcetera, referring to specific finite quantities. What is lacking, however, is the general mathematical concept of *finiteness*.

EXAMPLE. The notion 'finiteness of the domain' is not definable by means of any first-order sentence, or set of such sentences.

It will be recalled that the relevant refutation turned on the *compactness theorem* for first-order logic, which implies that sentences with arbitrarily large finite models will also have infinite ones.

Another striking omission, this time from the perspective of natural language, is that of common quantifiers, such as *most*, *least*, not to speak of *many* or *few*.

EXAMPLE. The notion 'most A are B' is not definable in a first-order logic with identity having, at least, unary predicate constants A, B. This time, a refutation involves both compactness and the (downward) *Löwenheim–Skolem theorem*: Consider any proposed definition $\mu(A, B)$ together with the infinite set of assertions 'at least $n\,A$ are B', 'at least $n\,A$ are not B' ($n = 1, 2, 3, \ldots$). Any finite subset of this collection is satisfiable in some finite domain with $A - B$ large enough and $A \cap B$ a little larger. By compactness then, the whole collection has a model with infinite $A \cap B$, $A - B$. But now, the Löwenheim–Skolem theorem gives a countably infinite such model, which makes the latter two sets equinumerous – and 'most' A are no longer B: in spite of $\mu(A, B)$. □

One peculiarity of this argument is its lifting the meaning of colloquial 'most' to the infinite case. The use of infinite models is indeed vital in the coming sections. Only in Section 1.4.3 shall we consider the purely *finite* case: little regarded in mathematically-oriented model theory, but rather interesting for the semantics of natural language.

In a sense, these expressive limits of first-order logic show up more dramatically in a slightly different perspective. A given *theory* in a first-order language may possess various 'non-standard models', not originally intended. For instance, by compactness, Peano Arithmetic has non-Archimedean models featuring infinite natural numbers. And by Löwenheim–Skolem, Zermelo–Fraenkel set theory has countable models (if consistent), a phenomenon known as 'Skolem's Paradox'. Conversely, a given model may not be defined categorically by its complete first-order theory, as is in fact known for all (infinite) mathematical standard structures such as integers, rationals or reals. (These two observations are sides of the same coin, of course.) Weakness or strength carry no moral connotations in logic, however, as one may turn into the other. Non-standard models for analysis have turned out quite useful for their own sake, and countable models of set theory are at the base of the independence proofs: first-order logic's loss thus can often be the mathematician's or philosopher's gain.

1.2. *Extensions*

When some reasonable notion falls outside the scope of first-order logic, one rather natural strategy is to add it to the latter base and consider the resulting stronger logic instead. Thus, for instance, the above two examples inspire what is called 'weak second-order logic', adding the quantifier 'there exist finitely many', as well as first-order logic with the added 'generalized quantifier' 'most'. But, there is a price to be paid here. Inevitably, these logics lose some of the meta-properties of first-order logic employed in the earlier refutations of definability. Here is a telling little table:

	Compactness	Löwenheim–Skolem
First-order logic	yes	yes
Plus 'there exist finitely many'	no	yes
Plus 'there exist uncountably many'	yes	no
Plus 'most'	no	no

For the second and third rows, cf. Monk [1976] chapter 30. For the fourth row, here is an argument.

EXAMPLE. Let the most-sentence $\varphi(R)$ express that R is a discrete linear order with end points, possessing a greatest point with more successors than non-successors (i.e. most points in the order are its successors). Such orders can only be finite, though of arbitrarily large size: which contradicts compactness. Next, consider the statement that R is a dense linear order without end points, possessing a point with more successors than predecessors. There are uncountable models of this kind, but no countable ones: and hence Löwenheim–Skolem fails. □

As it happens, no proposed proper extension of first-order logic ever managed to retain both the compactness and Löwenheim–Skolem properties. And indeed, in 1969 Lindström proved his famous theorem (Lindström [1969]) that, given some suitable explication of a 'logic', first-order logic is indeed characterizable as the strongest logic to possess these two meta-properties.

1.3. Abstract Model Theory

Over the past two decades, many types of extension of first-order logic have been considered. Again, the earlier two examples illustrate general patterns. First, there are so-called *finitary* extensions, retaining the (effective) finite syntax of first-order logic. The 'most' example inspires two general directions of this kind.

First, one may add *generalized quantifiers* Q, allowing patterns

$$Qx \cdot \varphi(x) \quad \text{or} \quad Qxy \cdot \varphi(x), \psi(y).$$

E.g., 'the φ's fill the universe' (*all*), 'the φ's form the majority in the universe' (most), 'the φ's form the majority of the ψ's' (most φ are ψ). But also, one may stick with the old types of quantifier, while employing them with new ranges. For instance, 'most A are B' may be read as an ordinary quantification over functions: 'there exists a 1-1 correspondence between A–B and some subset of $A \cap B$, but not vice versa'. Thus, one enters the domain of *higher-order logic*, to be discussed in later sections.

The earlier example of 'finiteness' may lead to finitary extensions of the above two kinds, but also to an *infinitary* one, where the syntax now allows infinite conjunctions and disjunctions, or even quantifications. For instance, finiteness may be expressed as 'either one, or two, or three, or ...' in $L_{\omega_1 \omega}$: a first-order logic allowing countable conjunctions and disjunctions of formulas (provided that they have only finitely many free variables together) and finite quantifier sequences. Alternatively, it may be expressed as 'there are no $x_1, x_2, \ldots$: all distinct', which would belong to $L_{\omega_1 \omega_1}$, having a countably infinite quantifier string. In general, logicians have studied a whole family of languages $L_{\alpha \beta}$; but $L_{\omega_1 \omega}$ remains the favourite (cf. Keisler [1971]).

Following Lindström's result, a research area of 'abstract model theory' has arisen where these various logics are developed and compared. Here is one example of a basic theme. Every logic L 'casts its net' over the sea of all structures, so to speak, identifying models verifying the same L sentences (L-equivalence). On the other hand, there is the finest sieve of *isomorphism* between models. One of Lindström's basic requirements on a logic was that the latter imply the former. One measure of strength of the logic is now to which extent the converse obtains. For instance, where L is first-order logic, we know that elementary equivalence implies isomorphism for *finite* models, but not for countable ones. (Cf. the earlier phenomenon of non-categorical definability of the integers.) A famous result concerning $L_{\omega_1 \omega}$ is Scott's theorem to the effect that, for *countable* models, $L_{\omega_1 \omega}$-equivalence and isomorphism coincide. (Cf. Keisler [1971] Chapter 2 or Barwise [1975] Chapter VII.6.) That such matches cannot last in the long run follows from a simple set-theoretic consideration, however, first made by Hanf. As long as the L sentences form a set, they can distinguish at best $2^{|L|}$ models, up to L-equivalence – whereas the number of models, even up to isomorphism, is unbounded.

A more abstract line of research is concerned with the earlier meta-properties. In addition to compactness and Löwenheim–Skolem, one also considers such properties as recursive axiomatizability of universally valid sentences (*'completeness'*) or *interpolation* (cf. Hodges [I.1] Section 14). Such notions may lead to new characterization results. For instance, Lindström himself proved that elementary logic is also the strongest logic with an effective finitary syntax to possess the Löwenheim–Skolem property and be complete. (The infinitary language $L_{\omega_1\omega}$ has both, without collapsing into elementary logic, however; its countable *admissible fragments* even possess compactness in the sense of Barwise [1975].) Similar characterizations for stronger logics have proven rather elusive up till now.

But then, there are many more possible themes in this area which are of a general interest. For instance, instead of haphazardly selecting some particular feature of first-order, or any other suggestive logic, one might proceed to a systematic description of meta-properties.

EXAMPLE. A folklore prejudice has it that interpolation was 'the final elementary property of first-order logic to be discovered'. Recall the statement of this meta-property: if one formula implies another, then (modulo some trivial cases) there exists an *interpolant* in their common vocabulary, implied by the first, itself implying the second. Now, this assertion may be viewed as a (first-order) fact about the two-sorted 'meta-structure' consisting of all first-order formulas, their vocabulary types (i.e. all finite sets of non-logical constants), the relations of implication and type-inclusion, as well as the type-assigning relation. Now, the complete first-order theories of the separate components are easily determined. The pre-order ⟨formulas, implication⟩ carries a definable Boolean structure, as one may define the connectives (∧ as greatest lower bound, ¬ as some suitable complement). Moreover, this Boolean algebra is countable, and atomless (the latter by the assumption of an infinite vocabulary). Thus, the given principles are complete, thanks to the well-known categoricity and, hence, completeness of the latter theory. The complete logic of the partial order ⟨finite types, inclusion⟩ may be determined in a slightly more complex way. The vindication of the above conviction concerning the above meta-structure would then consist in showing that interpolation provides the essential link between these two separate theories, in order to obtain a complete axiomatization for the whole.

Even more revolutionary about abstract model theory is the gradual reversal in methodological perspective. Instead of starting from a given logic and proving some meta-properties, one also considers these properties as such, establishes connections between them, and asks for (the ranges of) logics exemplifying certain desirable combinations of features.

Finally, a warning. The above study by no means exhausts the range of logical questions that can be asked about extensions of first-order logic. Indeed, the perspective of meta-properties is very global and abstract. One more concrete new development is the interest in, e.g., generalized quantifiers from the perspective of linguistic semantics (cf. Barwise and Cooper [1981], Van Benthem [1982]), which leads to proposals for reasonable constraints on new quantifiers, and indeed to a semantically-motivated classification of all reasonable additions to elementary logic. □

1.4. Characterization Results

A good understanding of first-order logic is essential to any study of its extensions. To this end, various characterizations of first-order definability will be reviewed here in a little more detail than in Chapter I.1.

1.4.1. Lindström's theorem.
Lindström's result itself gives a definition of first-order logic, in terms of its global properties. Nevertheless, in practice, it is of little help in establishing or refuting first-order definability. To see if some property Φ of models is elementary, one would have to consider the first-order language with Φ added (say, as a propositional constant), close under the operations that Lindström requires of a 'logic' (notably, the Boolean operations and relativization to unary predicates), and then find out if the resulting logic possesses the compactness and Löwenheim–Skolem properties. Moreover, the predicate logic is to have an *infinite* vocabulary (cf. the proof to be sketched below): otherwise, we are in for surprises.

EXAMPLE. Lindström's theorem fails for the pure identity language. First, it is a routine observation that sentences in this language can only express (negations of) disjunctions 'there are precisely n_1 or ... or precisely n_k objects in the universe'. Now, add a propositional constant C expressing *countable infinity* of the universe.

This logic retains compactness. For, consider any finitely satisfiable set Σ of its sentences. It is not difficult to see that either $\Sigma \cup \{C\}$ or $\Sigma \cup \{\neg C\}$ must also be finitely satisfiable. In the first case, replace occurrences of C in Σ by some tautology: a set of first-order sentences remains, each of whose finite subsets has a (countably) infinite model. Therefore, it has an infinite model itself and, hence, a countably infinite one (satisfying C) – by ordinary compactness and Löwenheim-Skolem. This model satisfies the original Σ as well. In the second case, replace C in Σ by some contradiction. The resulting set either has a finite model, or an infinite one, and hence an uncountably infinite one: either way, $\neg C$ is satisfied – and again, the original Σ is too.

The logic also retains Löwenheim-Skolem. Suppose that φ has no countably infinite models. Then $\varphi \wedge \neg C$ has a model, if φ has one. Again, replace occurrences of C inside φ by some contradiction: a pure identity sentence remains. But such sentences can always be verified on some finite universe (witness the above description) where $\neg C$ is satisfied too. □

1.4.2. *Keisler's theorem.* A more local description of first-order definability was given by Keisler, in terms of preservation under certain basic operations on models.

THEOREM. *A property Φ of models is definable by means of some first-order sentence iff both Φ and its complement are closed under the formation of isomorphs and ultraproducts.*

The second operation has not been introduced yet. As it will occur at several other places in this Handbook, a short introduction is given at this point. For convenience, henceforth, our standard example will be that of binary relational models $F = \langle A, R \rangle$ (or $F_i = \langle A_i, R_i \rangle$).

A logical fable. A family of models $\{F_i \mid i \in I\}$ once got together and decided to join into a common state. As everyone wanted to be fully represented, it was decided to create new composite individuals as functions f with domain I, picking at each $i \in I$ some individual $f(i) \in A_i$. But now, how were relations to be established between these new individuals? Many models were in favour of consensus democracy:

Rfg iff $R_i f(i) g(i)$ *for all* $i \in I$.

But, this led to indeterminacies as soon as models started voting about whether or *not Rfg*. More often than not, no decision was reached. Therefore, it was decided to ask the gods for an 'election manual' U, saying which sets of votes were to be 'decisive' for a given atomic statement. Thus, votes now were to go as follows:

Rfg iff $\{i \in I \mid R_i f(i) g(i)\} \in U$.

Moreover, although one should not presume in these matters, the gods were asked to incorporate certain requirements of consistency

I.4: HIGHER-ORDER LOGIC

if $X \in U$, then $I-X \notin U$

as well as democracy

if $X \in U$ and $Y \supseteq X$, then $Y \in U$.

Finally, there was also the matter of expediency: the voting procedure for atomic statements should extend to complex decisions:

$$\varphi(f_1,\ldots,f_n) \quad \text{iff} \quad \{i \in I \mid F_i \models \varphi[f_1(i),\ldots,f_n(i)]\} \in U$$

for all predicate-logical issues φ.

After having pondered these wishes, the gods sent them an *ultrafilter* U over I, proclaiming the Łoś Equivalence:

THEOREM. *For any ultrafilter U over I, the stipulation* $(*)$ *creates a structure $F = \langle \Pi_{i \in I} A_i, R \rangle$ such that*

$$F \models \varphi[f_1,\ldots,f_n] \quad \text{iff} \quad \{i \in I \mid F_i \models \varphi[f_1(i),\ldots,f_n(i)]\} \in U.$$

Proof. The basic case is just $(*)$. The negation and conjunction cases correspond to precisely the defining conditions on ultrafilters, viz. (i) $X \notin U$ iff $I-X \in U$; (ii) $X, Y \in U$ iff $X \cap Y \in U$ (or, alternatively, besides consistency and democracy above: if $X, Y \in U$ then also $X \cap Y \in U$; and: if $I-X \notin U$ then $X \in U$). And finally, the gods gave them the existential quantifier step for free:

- if $\exists x \varphi(x, f_1, \ldots, f_n)$ holds then so does $\varphi(f, f_1, \ldots, f_n)$ for some function f. Hence, by the inductive hypothesis for φ, $\{i \in I \mid F_i \models \varphi[f(i), f_1(i), \ldots, f_n(i)]\} \in U$, which set is contained in $\{i \in I \mid F_i \models \exists x \varphi[f_1(i), \ldots, f_n(i)]\} \in U$.
- if $\{i \in I \mid F_i \models \exists x \varphi[f_1(i), \ldots, f_n(i)]\} \in U$, then choose $f(i) \in A_i$ verifying φ for each of these i (and arbitrary elsewhere): this f verifies $\varphi(x, f_1, \ldots, f_n)$ in the whole product, whence $\exists x \varphi(f_1, \ldots, f_n)$ holds. □

After a while, an unexpected difficulty occurred. Two functions f, g who did not agree among themselves asked for a public vote, and the outcome was . . .

$$\{i \in I \mid f(i) = g(i)\} \in U.$$

Thus it came to light how the gift of the gods had introduced an invisible equality $\sim$. By its definition, and the Łoś Equivalence, it even turned out to partition the individuals into equivalence classes, whose members were indistinguishable as to R behaviour:

$Rfg, f \sim f', g \sim g'$ imply $Rf'g'$

But then, such classes themselves could be regarded as the building bricks of society, and in the end they were:

DEFINITION. For any family of models $\{F_i \mid i \in I\}$ with an ultrafilter U on I, the *ultraproduct* $\Pi_U F_i$ is the model $\langle A, R \rangle$ with
  (i) A is the set of classes $f_\sim$ for all functions $f \in \Pi_{i \in I} A_i$, where $f_\sim$ is the equivalence class of f in the above relation,
  (ii) R is the set of couples $\langle f_\sim, g_\sim \rangle$ for which $\{i \in I \mid R_i f(i) g(i)\} \in U$.
By the above observations, the latter clause is well-defined – and indeed the whole Łoś Equivalence remained valid.

Whatever their merits as regards democracy, ultraproducts play an important role in the following fundamental question of model theory:
What structural behaviour makes a class of models *elementary*, i.e. definable by means of some first-order sentence?
First, the Łoś Equivalence implies that first-order sentences φ are preserved under ultraproducts in the following sense:

if $F_i \vDash \varphi$ (all $i \in I$), then $\Pi_U F_i \vDash \varphi$.

(The reason is that I itself must belong to U.)
But conversely, Keisler's theorem told us that this is also enough. *End of fable.*

The proof of Keisler's theorem (subsequently improved by Shelah) is rather formidable: cf. Chang and Keisler [1973] Chapter 6. A more accessible variant will be proved below, however. First, one relaxes the notion of isomorphism to the following partial variant.

DEFINITION. A *partial isomorphism* between $\langle A, R \rangle$ and $\langle B, S \rangle$ is a set I of coupled finite sequences (s, t) from A resp. B, of equal length, satisfying

$(s)_i = (s)_j$ iff $(t)_i = (t)_j$

$(s)_i R (s)_j$ iff $(t)_i S (t)_j$,

which possesses the *back-and-forth property*, i.e. for every $(s, t) \in I$ and every $a \in A$ there exists some $b \in B$ with $(s \frown a, t \frown b) \in I$; and vice versa.

Cantor's zig-zag argument shows that partial isomorphism coincides with total isomorphism on the countable models. Higher up, matters change; e.g. $\langle \mathbb{Q}, < \rangle$ and $\langle \mathbb{R}, < \rangle$ are partially isomorphic by the obvious I without being isomorphic.

First-order formulas φ are preserved under partial isomorphism in the following sense:

$$\text{if } (s, t) \in I, \text{ then } \langle A, R \rangle \vDash \varphi[s] \quad \text{iff} \quad \langle B, S \rangle \vDash \varphi[t].$$

Indeed, this equivalence extends to formulas from arbitrary infinitary languages $L_{\alpha\omega}$: cf. Barwise [1977] Chapter A.2.9 for further explanation.

THEOREM. *A property Φ of models is first-order definable iff both Φ and its complement are closed under the formation of partial isomorphs and countable ultraproducts.*

1.4.3. *Fraïssé's theorem.* Even the Keisler characterization may be difficult to apply in practice, as ultraproducts are such abstract entities. In many cases, a more combinatorial method may be preferable; in some, it's even necessary.

EXAMPLE. As was remarked earlier, colloquial 'most' only seems to have natural meaning on the *finite* models. But, as to first-order definability on this restricted class, both previous methods fail us completely, all relevant notions being tied up with infinite models. Nevertheless, *most A are B* is not definable on the finite models in the first-order language with A, B and identity. But this time, we need a closer combinatorial look at definability.

First, a natural measure of the 'pattern complexity' of a first-order formula φ is its *quantifier depth* $d(\varphi)$, which is the maximum length of quantifier nestings inside φ. (Inductively, $d(\varphi) = 0$ for atomic φ, $d(\neg\varphi) = d(\varphi)$, $d(\varphi \wedge \psi) = \text{maximum}(d(\varphi), d(\psi))$, etcetera, $d(\exists x \varphi) = d(\forall x \varphi) = d(\varphi) + 1$.) Intuitively, structural complexity beyond this level will escape φ's notice. We make this precise.

Call two sets X, Y n-equivalent if either $|X| = |Y| < n$ or $|X|, |Y| \geq n$. By extension, call two models $\langle D, A, B \rangle$, $\langle D', A', B' \rangle$ n-equivalent if all four 'state descriptions' $A \cap B$, $A \setminus B$, $B \setminus A$, $D \setminus (A \cup B)$ are n-equivalent to their primed counterparts.

LEMMA. *If $\langle D, A, B \rangle$, $\langle D', A', B' \rangle$ are n-equivalent then all sequences d, d' with corresponding points in corresponding states verify the same first-order formulas with quantifier depth not exceeding n.*

COROLLARY. *'Most A are B' is not first-order definable on the finite models.*

Proof. For no finite number n, 'most A are B' exhibits the required n-insensitivity. □

This idea of insensitivity to structural complexity beyond a certain level forms the core of our third and final characterization, due to Fraïssé. Again, only the case of a binary relation R will be considered, for ease of demonstration.

First, on the linguistic side, two models are *n-elementary equivalent* if they verify the same first-order sentences of quantifier depth not exceeding n. Next, on the structural side, a matching notion of *n-partial isomorphism* may be defined, by postulating the existence of a chain $I_n \supseteq \cdots \supseteq I_0$ of sets of matching couples (s, t), as in the earlier definition of partial isomorphism. This time, the back-and-forth condition is index-relative, however:

if $(s, t) \in I_{i+1}$ and $a \in A$, then for some $b \in B$, $(s\frown a, t\frown b) \in I_i$, and vice versa.

PROPOSITION. *Two models are n-elementary equivalent iff they are n-partially isomorphic.*

The straightforward proof uses the following auxiliary result, for first-order languages with a finite non-logical vocabulary of relations and individual constants.

LEMMA. *For each depth n, for each fixed number of free variables $x_1, \ldots, x_m$ there exist only finitely many formulas $\varphi(x_1, \ldots, x_m)$, up to logical equivalence.*

This lemma allows us to describe all possible n-patterns in a single first-order formula, a purpose for which one sometimes uses explicit 'Hintikka normal forms'.

THEOREM. *A property Φ of models is first-order definable iff it is preserved under n-partial isomorphism for some natural number n.*

Proof. The invariance condition is obvious for first-order definable properties. Conversely, for n-invariant properties, the disjunction of all complete n-structure descriptions for models satisfying Φ defines the latter property. □

I.4: HIGHER-ORDER LOGIC

Applications. Now, from the Fraïssé theorem, both the weak Keisler and the Lindström characterization may be derived in a perspicuous way. Here is an indication of the proofs.

EXAMPLE. Weak Keisler from Fraïssé.

First-order definable properties are obviously preserved under partial isomorphism and (countable) ultraproducts. As for the converse, suppose that Φ is not thus definable. By Fraïssé, this implies the existence of a sequence of n-partially isomorphic model pairs $\mathfrak{A}_n, \mathfrak{B}_n$ of which only the first verify Φ.

The key observation is now simply this. Any *free* ultrafilter U on $\mathbb{N}$ (containing all tails of the form $[n, \infty)$) will make the countable ultraproducts $\Pi_U \mathfrak{A}_n, \Pi_U \mathfrak{B}_n$ partially isomorphic. The trick here is to find a suitable set I of partial isomorphisms, and this is accomplished by setting, for sequences of functions s, t of length m

$$((s)_U, (t)_U) \in I \quad \text{iff} \quad \{n \geqslant m \mid (s(n), t(n)) \in I^n_{n-m}\} \in U;$$

where '$I^n_n, \ldots, I^n_0$' is the sequence establishing the n-partial isomorphism of $\mathfrak{A}_n, \mathfrak{B}_n$.

So, by the assumed preservation properties, Φ would hold for $\Pi_U \mathfrak{A}_n$ and hence for $\Pi_U \mathfrak{B}_n$. But, so would not-Φ: a contradiction. □

EXAMPLE. Lindström from Fraïssé.

Let L be a logic whose non-logical vocabulary consists of infinitely many predicate constants of all arities. L is completely specified by its *sentences S*, each provided with a finite 'type' (i.e. set of predicate constants), its *models M* (this time: ordinary first-order models) and its *truth relation T* between sentences and models. We assume four basic conditions on L: the truth relation is invariant for *isomorphs*, the sentence set S is closed under *negations* and *conjunctions* (in the obvious semantic sense), and each sentence φ can be relativized by arbitrary unary predicates A, such that a model verifies φ^A iff its A-submodel verifies φ. Finally, we say that L 'contains elementary logic' if each first-order sentence is represented by some sentence in S having the same models. 'Compactness' and 'Löwenheim–Skolem' are already definable in this austere framework. (By the latter we'll merely mean: 'sentences with any model at all have countable models'.)

THEOREM. *Any logic containing elementary logic has compactness and Löwenheim–Skolem iff it coincides with elementary logic.*

The non-evident half of this assertion again starts from Fraïssé's result. Suppose that $\Phi \in S$ is not first-order. Again, there is a sequence $\mathfrak{A}_n, \mathfrak{B}_n$ as above. Now, by suitable diligence, a sentence Π in S may be written down recording the following facts about a putative limit $\mathfrak{A}, \mathfrak{B}$ of this sequence.

(1) A and B are disjoint sets such that Φ^A, not-Φ^B,
(2) $\langle N, < \rangle$ is an index set like the negative integers, but without the condition that there be no initial point,
(3) P is a pairing predicate (needed in order to discuss finite sequences),
(4) I is an indexing predicate between members of N and sequences from A, B, satisfying the partial isomorphism condition w.r.t. the type of Φ as well as the downward index-relative back-and-forth properties mentioned earlier.

Next, add an infinite set Σ of assertions stating that N exceeds every finite length. Each finite subset of $\Sigma \cup \{\Pi\}$ has a model, using a suitably large couple $\mathfrak{A}_n, \mathfrak{B}_n$ in the above sequence. Hence, by compactness there exists a model for the whole set. In this model, N is infinite – and hence we have a model for the single sentence Π^+ consisting of Π plus the assertion that R is an injection from N to N which fails to be surjective (the hall-mark of infinity). By the Löwenheim–Skolem property then, Π^+ has a countable model as well. From this model, structures $\mathfrak{A}, \mathfrak{B}$ may be extracted (using A, B) satisfying Φ, non-Φ respectively, which are partially isomorphic (using the first countable sequence in N). By Cantor's theorem, it follows that $\mathfrak{A}, \mathfrak{B}$ must be truly isomorphic (being countable) and, hence, we have a contradiction with Φ's invariance for isomorphism. □

2. SECOND-ORDER LOGIC

2.1. *Language*

Quantification over properties and predicates, rather than just objects, has a philosophical pedigree. For instance, Leibniz's celebrated principle of Identity of Indiscernibles has the natural form

$$\forall xy(\forall X(X(x) \leftrightarrow X(y)) \rightarrow x = y).$$

There also seems to be good evidence for this phenomenon from natural language, witness Russell's example 'Napoleon had all the properties of a great general'

$$\forall X(\forall y(gg(y) \rightarrow X(y)) \rightarrow X(n)).$$

Moreover, of course, mathematics abounds with this type of discourse, with its explicit quantification over relations and functions. And indeed, logic itself seems to call for this move. For, there is a curious asymmetry in ordinary predicate logic between individuals: occurring both in constant and variable contexts, and predicates: where we are denied the power of quantification. This distinction seems arbitrary: significantly, Frege's 'Begriffsschrift' still lacks it. We now pass on to an account of second-order logic, with its virtues and vices.

The language of second-order logic distinguishes itself from that of first-order logic by the addition of variables for subsets, relations and functions of the universe and the possibility of quantification over these. The result is extremely strong in expressive power; we list a couple of examples in Section 2.2. As a consequence, important theorems valid for first-order languages fail here; we mention the compactness theorem, the Löwenheim-Skolem theorems (Section 2.2) and the completeness theorem (Section 2.3). With second-order logic, one really enters the realm of set theory. This state of affairs will be illustrated in Section 2.4 with a few examples. What little viable logic can be snatched in the teeth of these limitations usually concerns special fragments of the language, of which some are considered in Section 2.5.

2.2. *Expressive Power*

2.2.1. An obvious example of a second-order statement is Peano's induction axiom according to which every set of natural numbers containing 0 and closed under immediate successors contains all natural numbers. Using S for successor, this might be written down as

(1) $\forall Y[Y(0) \wedge \forall x(Y(x) \to Y(S(x))) \to \forall x Y(x)]$

(The intention here is that x stands for numbers, Y for sets of numbers, and $Y(x)$ says, as usual, that x is an element of Y.)

Dedekind already observed that the axiom system consisting of the induction axiom and the two first-order sentences

(2) $\forall x \forall y(S(x) = S(y) \to x = y)$

and

(3) $\forall x(S(x) \neq 0)$

is categorical. Indeed suppose that $\langle A, f, a\rangle$ models (1)–(3). Let $A' = \{a, f(a), f(f(a)), \ldots\}$. Axioms (2) and (3) alone imply that the submodel $\langle A', f\restriction A', a\rangle$ is isomorphic with $\langle \mathbb{N}, S, 0\rangle$ (the isomorphism is clear). But, (1) implies that $A' = A$ (just let X be A'). □

This result should be contrasted with the first-order case. *No set of first-order sentences true of $\langle \mathbb{N}, S, 0\rangle$ is categorical.* This can be proved using either the upward Löwenheim–Skolem theorem or the compactness theorem. As a result, neither of these two extend to second-order logic. The nearest one can come to (1) in first-order terms is the 'schema'

(4) $\quad \varphi(0) \wedge \forall x(\varphi(x) \to \varphi(S(x))) \to \forall x \varphi(x)$

where φ is any first-order formula in the vocabulary under consideration. It follows that in models $\langle A, f, a\rangle$ of (4) the set A' above cannot be defined in first-order terms: otherwise one could apply (4) showing $A' = A$ just as we applied (1) to show this before. (This weakness of first-order logic becomes its strength in so-called 'overspill arguments', also mentioned in Chapter I.1 [Hodges].) We will use the categoricity of (1)–(3) again in Section 2.3 to show non-axiomatizability of second-order logic.

2.2.2. The next prominent example of a second-order statement is the one expressing 'Dedekind completeness' of the order of the reals: every set of reals with an upper bound has a least upper bound. Formally

(5) $\quad \forall X[\exists x \forall y (X(y) \to y \leqslant x) \to$
$\quad\quad \to \exists x (\forall y (X(y) \to y \leqslant x) \wedge$
$\quad\quad \wedge \forall x'[\forall y(X(y) \to y \leqslant x') \to x \leqslant x'])]$

It is an old theorem of Cantor's that (5) together with the first-order statements expressing that $\leqslant$ is a dense linear order without endpoints plus the statement 'there is a countable dense subset', is categorical. The latter statement of so-called 'separability' is also second-order definable: cf. Section 2.2.5. Without it a system is obtained whose models all *embed* $\langle \mathbb{R}, \leqslant\rangle$. (For, these models must embed $\langle \mathbb{Q}, \leqslant\rangle$ for first-order reasons; and such an embedding induces one for $\langle \mathbb{R}, \leqslant\rangle$ by (5).) Thus, the downward Löwenheim–Skolem theorem fails for second-order logic.

2.2.3. A relation $R \subseteq A^2$ is *well-founded* if every non-empty subset of A has an R-minimal element. In second-order terms w.r.t. models $\langle A, R, \ldots \rangle$

(6) $\forall X[\exists x X(x) \to \exists x(X(x) \wedge \forall y[X(y) \to \neg R(y,x)])]$

This cannot be expressed in first-order terms. For instance, every first-order theory about R which admits models with R-chains of arbitrary large but finite length must, by compactness, admit models with infinite R-chains which decrease, and such a chain has no minimal element.

2.2.4. Every first-order theory admitting arbitrarily large, finite models has infinite models as well: this is one of the standard applications of compactness. On the other hand, higher-order terms enable one to define finiteness of the universe. Probably the most natural way to do this uses *third-order* means: a set is finite iff it is in every *collection of* sets containing the empty set and closed under the addition of one element. Nevertheless, we can define finiteness in second-order terms as well: A is finite iff every relation $R \subseteq A^2$ is well-founded; hence, a second-order definition results from (6) by putting a universal quantifier over R in front. Yet another second-order definition of finiteness uses Dedekind's criterion: every injective function on A is surjective. Evidently, such a quantification over functions on A may be simulated using suitable predicates. By the way, to see that these second-order sentences do indeed define finiteness one needs the axiom of choice.

2.2.5. Generalized quantifiers. Using Section 2.2.4, it is easy to define the quantifier $\exists_{<\aleph_0}$ (where $\exists_{<\aleph_0} x \varphi(x)$ means: there are only finitely many x s.t. $\varphi(x)$) in second-order terms; $\exists_{\geq \aleph_0}$ simply is its negation. (In earlier terminology, weak second-order logic is part of second-order logic.) What about higher cardinalities? Well, e.g. $|X| \geq \aleph_1$ iff X has an infinite subset Y which cannot be mapped one–one onto X. This can obviously be expressed using function quantifiers. And then of course one can go on to $\aleph_2, \aleph_3, \ldots$.

Other generalized quantifiers are definable by second-order means as well. For instance, the standard example of Section 1 has the following form. *Most A are B* becomes 'there is no injective function from $A \cap B$ into $A-B$'.

A highly successful generalized quantifier occurs in *stationary logic*, cf. Barwise et al. [1978]. Its language is second-order in that it contains monadic second-order variables; but the only quantification over these allowed is by means of the *almost all* quantifier *aa*. A sentence $aaX\varphi(X)$ is read as: there is a collection C of countable sets X for which $\varphi(X)$, which is closed under the formation of countable unions and has the property that every countable subset of the universe is subset of a member of C. (We'll not take the trouble explaining what 'stationary' means here.) The obvious definition of *aa* in higher-order terms employs third-order means. Stationary logic can define the quantifier $\exists_{\geqslant \aleph_1}$. It has a complete axiomatization and, as a consequence, obeys compactness and downward Löwenheim–Skolem (in the form: if a sentence has an uncountable model, it has one of power $\aleph_1$).

Other compact logics defining $\exists_{\geqslant \aleph_i}$ have been studied by Magidor and Malitz [1977].

2.2.6. The immense strength of second-order logic shows quite clearly when set theory itself is considered.

Zermelo's *separation axiom* says that the elements of a given set sharing a given property form again a set. Knowing of problematic properties occurring in the paradoxes, he required 'definiteness' of properties to be used. In later times, Skolem replaced this by 'first-order definability', and the axiom became a first-order schema. Nevertheless, the intended axiom quite simply is the second-order statement

(8) $\quad \forall X \forall x \exists y \forall z (z \in y \leftrightarrow z \in x \land X(z))$

Later on, Fraenkel and Skolem completed Zermelo's set theory with the substitution axiom: the complete image of a set under an operation is again a set, resulting from the first by 'substituting' for its elements the corresponding images. Again, this became a first-order schema, but the original intention was the second-order principle

(9) $\quad \forall F \forall a \exists b \forall y (y \in b \leftrightarrow \exists x [x \in a \land y = F(x)])$

Here F is used as a variable for arbitrary operations from the universe to itself; $F(x)$ denotes application. The resources of set theory allow an equivalent formulation of (9) with a set (i.e. class) variable, of course. Together with the usual axioms, (9) implies (8).

It must be considered quite a remarkable fact that the first-order versions of (8) and (9) have turned out to be sufficient for every mathematical purpose. (By the way, in ordinary mathematical practice, (9) is seldom used; the proof that Borel-games are determined is a notable exception. Cf. also Section 2.4.)

The Zermelo–Fraenkel axioms intend to describe the cumulative hierarchy with its membership structure $\langle V, \in \rangle$, where $V = \cup_\alpha V_\alpha$ (α ranging over all ordinals) and $V_\alpha = \cup_{\beta < \alpha} \mathscr{P} V_\beta$. For the reasons mentioned in Section 1, the first-order version ZF^1 of these axioms does not come close to this goal, as it has many non-standard models as well. The second-order counterpart ZF^2 using (9) has a much better score in this respect:

THEOREM. $\langle A, E\rangle$ satisfies ZF^2 iff for some (strongly) inaccessible κ: $\langle A, E\rangle \cong \langle V_\kappa, \in\rangle$.

It is generally agreed that the models $\langle V_\kappa, \in\rangle$ are 'standard' to a high degree.

If we add an axiom to ZF^2 saying there are no inaccessibles, the system even becomes categorical, defining $\langle V_\kappa, \in\rangle$ for the first inaccessible κ.

2.3. Non-axiomatizability

First-order logic has an effective notion of proof which is complete w.r.t. the intended interpretation. This is the content of Gödel's completeness theorem. As a result, the set of (Gödel numbers of) universally valid first-order formulas is recursively enumerable. Using example 2.2.1, it is not hard to show that the set of second-order validities is not arithmetically definable, let alone recursively enumerable, and hence that an effective and complete axiomatization of second-order validity is impossible.

Let P^2 be Peano arithmetic in its second-order form, i.e. the theory in the language of $\mathfrak{N} = \langle \mathbb{N}, S, 0, +, \times\rangle$ consisting of (1)–(3) above plus the (first-order) recursion equations for $+$ and $\times$. P^2 is a categorical description of $\mathfrak{N}$, just as (1)–(3) categorically describe $\langle \mathbb{N}, S, 0\rangle$. Now, let φ be any first-order sentence in the language of $\mathfrak{N}$. Then clearly

$$\mathfrak{N} \models \varphi \quad \text{iff} \quad P^2 \to \varphi \text{ is valid.}$$

(Notice that P^2 may be regarded as a single second-order sentence.)

Now the left-hand side of this equivalence expresses a condition on (the Gödel number of) φ which is not arithmetically definable by Tarski's theorem on non-definability of truth (cf. Section 3.2 or, for a slightly different setting, Hodges [I.1] Section 20). Thus, second-order validity cannot be arithmetical either. □

Actually, this is still a very weak result. We may take φ second-order and show that second-order truth doesn't fit in the analytic hierarchy (again, see Section 3.4). Finally, using Section 2.2.6, we can replace in the above argument $\mathfrak{N}$ by $\langle V_\kappa, \in\rangle$, where κ is the smallest inaccessible, and P^2 by $ZF^2 +$ 'there are no inaccessibles', and find that second-order truth cannot be (first-order) defined in $\langle V_\kappa, \in\rangle$, etc. This clearly shows how frightfully complex this notion is.

Not to end on too pessimistic a note, let it be remarked that the logic may improve considerably for certain fragments of the second-order language, possibly with restricted classes of models. An early example is the decidability of second-order monadic predicate logic (cf. Ackermann [1968]). A more recent example is Rabin's theorem (cf. Rabin [1969]) stating that the monadic second-order theory (employing only second-order quantification over subsets) of the structure $\langle 2^\omega, P_0, P_1 \rangle$ is still decidable. Here, 2^ω is the set of all finite sequences of zeros and ones, and P_i is the unary operation 'post-fix i' ($i = 0, 1$).

Many decidability results for monadic second-order theories have been derived from this one by showing their models to be definable parts of the Rabin structure. For instance, the monadic second-order theory of the natural numbers $\langle \mathbb{N}, < \rangle$ is decidable by this method.

The limits of Rabin's theorem show up again as follows. The *dyadic* second-order theory of $\langle \mathbb{N}, < \rangle$ is already non-arithmetical, by the previous type of consideration. (Briefly, $\mathfrak{N} \models \varphi$ iff $\langle \mathbb{N}, < \rangle$ verifies $P \to \varphi$ for all those choices of $0, S, +, \times$ whose defined relation 'smaller than' coincides with the actual $<$. Here, P is first-order Peano Arithmetic minus induction. In this formulation ternary predicates are employed (for $+, \times$), but this can be coded down to the binary case.)

2.4. Set-theoretic Aspects

Even the simplest questions about the model theory of second-order logic turn out to raise problems of set theory, rather than logic. Our first example of this phenomenon was a basic theme in Section 1.3.

If two models are first-order (elementary) equivalent and one of them is finite, they must be isomorphic. What, if we use second-order equivalence and relax finiteness to, say, countability? Ajtai [1979] contains a proof that this question is undecidable in ZF (of course, the *first-order* system is intended here). One of his simplest examples shows it is consistent for there to be two countable well-orderings, second-order (or indeed higher-order) equivalent but not isomorphic.

The germ of the proof is in the following observation: if the Continuum Hypothesis holds, there must be second-order (or indeed higher-order) equivalent well orderings of power $\aleph_1$: for, up to isomorphism, there are $\aleph_2$ such well orderings (by the standard representation in terms of ordinals), whereas there are only $2^{\aleph_0} = \aleph_1$ second-order theories. The consistency-proof itself turns on a refined form of this cardinality-argument, using 'cardinal collapsing'. On the other hand, Ajtai mentions the 'folklore' fact that countable second-order equivalent models *are* isomorphic when the

I.4: HIGHER-ORDER LOGIC

axiom of constructibility holds; in fact, this may be derived from the existence of a second-order definable well ordering of the reals (which follows from this axiom).

Another example belongs to the field of second-order cardinal characterization. Whether a sentence without non-logical symbols holds in a model or not depends only on the cardinality of the model. If a sentence has models of one cardinality only, it is said to *characterize* that cardinal. As we have seen in Section 1, first-order sentences can only characterize single finite cardinals. In the meantime, we have seen how to characterize, e.g. $\aleph_0$ in a second-order way: let φ be the conjunct of (1)–(3) of Section 2.2.1 and consider $\exists S \exists 0 \varphi$ – where S and 0 are now being considered as variables. Now, various questions about the simplest second-order definition of a given cardinal, apparently admitting of 'absolute' answers, turn out to be undecidable set theoretic problems; cf. Kunen [1971].

As a third example, we finally mention the question of cardinals characterizing, conversely, a logic L. The oldest one is the notion of *Hanf number* of a logic, alluded to in Section 1.3. This is the least cardinal γ such that, if an L-sentence has a model of power $\geq \gamma$, it has models of arbitrarily large powers. The *Löwenheim number* λ of a language L compares to the *downward* Löwenheim–Skolem property just as the Hanf number does to the *upward* notion: it is the least cardinal with the property that every satisfiable L sentence has a model of power $\leq \lambda$. It exists by a reasoning similar to Hanf's: for satisfiable φ, let $|\varphi|$ be the least cardinal which is the power of some model of φ. Then λ clearly is the sup of these cardinals. (By the way, existence proofs such as these may rely heavily on ZF's substitution-axiom. Cf. Barwise [1972].)

How large are these numbers pertaining to second-order logic? From Section 2.2.6 it follows, that the first inaccessible (if it exists) can be second-order characterized; thus, the Löwenheim and Hanf number are at least bigger still. By similar reasoning, they are not smaller than the second, third,... inaccessible. And we can go on to larger cardinals; for instance, they must be larger than the first *measurable*. The reason is mainly that, like inaccessibility, defining measurability of κ only needs reference to sets of rank not much higher than κ. (In fact, inaccessibility of κ is a first-order property of $\langle V_{\kappa+1}, \in \rangle$; measurability one of $\langle V_{\kappa+2}, \in \rangle$.) Only when large cardinal properties refer in an essential way to the *whole* set theoretic universe (the simplest example being that of *strong compactness*) can matters possibly change. Thus, Magidor [1971] proves that the Löwenheim number of universal second-order sentences (and hence, by

4.3, of higher-order logic in general) is less than the first *supercompact* cardinal.

As these matters do bring us a little far afield (after all, this is a handbook of *philosophical* logic) we stop here.

In this light, the recommendation in the last problem of the famous list 'Open problems in classical model theory' in Chang and Keisler [1973] remains as problematic as ever: 'Develop the model theory of second- and higher-order logic'.

Additional evidence for the view that second-order logic (and, a fortiori, higher-order logic in general) is not so much logic as set theory, is provided by looking directly at existing set-theoretic problems in second-order terms.

Let κ be the first inaccessible cardinal. In Section 2.2.6 we have seen that every ZF^2 model contains (embeds) $\langle V_\kappa, \in \rangle$. As this portion is certainly (first-order) definable in all ZF^2 models in a uniform way, ZF^2 decides every set theoretic problem which mentions sets in V_κ only. This observation has led Kreisel to recommend this theory to our lively attention, so let us continue.

Now, in fact, already far below κ, interesting questions live. Foremost is the *continuum problem*, which asks whether there are sets of reals in cardinality strictly between $\mathbb{N}$ and $\mathbb{R}$. (Cantor's famous *continuum hypothesis* (CH) says there are not.) Thus, ZF^2 decides CH: either it or its negation follows logically from ZF^2. Since ZF^2 is correct, in the former case CH is true, while it is false in the latter. But of course, this reduction of the continuum problem to second-order truth really begs the question and is of no help whatsoever.

It does refute an analogy however, which is often drawn between the continuum hypothesis and the Euclidean postulate of parallels in geometry. For, the latter axiom is not decided by second-order geometry. Its independence is of a different nature; there are different 'correct' geometries, but only one correct set theory (modulo the addition of large cardinal axioms): ZF^2. (In view of Section 2.2.6, a better formal analogy would be that between the parallel postulate and the existence of inaccessibles – though it has shortcomings as well.)

Another example of a set-theoretic question deep down in the universe is whether there are non-constructible reals. This question occurs at a level so low that, using a certain amount of coding, it can be formulated already in the language of P^2.

ZF^2 knows the answers – unfortunately, we're not able to figure out exactly what it knows.

So, what is the practical use of second-order set theory? To be true, there are *some* things we do know ZF^2 proves while ZF^1 does not; for instance, the fact that ZF^1 is consistent. Such metamathematical gains are hardly encouraging, however, and indeed we can reasonably argue that there is no way of knowing something to follow from ZF^2 unless it is provable in the two-sorted set/class theory of Morse–Mostowski, a theory which doesn't have many advantages over its subtheory $ZF = ZF^1$. (In terms of Section 4.2 below, Morse–Mostowski can be described as ZF^2 under the general-models interpretation with full comprehension-axioms added.)

We finally mention that sometimes, higher-order notions find application in the theory of sets. In Myhill and Scott [1971] it is shown that the class of hereditarily ordinal-definable sets can be obtained by iterating second-order (or general higher-order) definability through the ordinals. (The constructible sets are obtained by iterating first-order definability; they satisfy the ZF-axioms only by virtue of their first-order character.) Also, interesting classes of large cardinals can be obtained by their reflecting higher-order properties; cf. for instance, Drake [1974] Chapter 9.

2.5. Special Classes: Σ_1^1 and Π_1^1

In the light of the above considerations, the scarcity of results forming a subject of 'second-order logic' becomes understandable. (A little) more can be said, however, for certain fragments of the second-order language. Thus, in Section 2.3, the monadic quantificational part was considered, to which belong, e.g. second-order Peano arithmetic P^2 and Zermelo–Fraenkel set theory ZF^2. The more fruitful restriction for general model-theoretic purposes employs quantificational pattern complexity, however. We will consider the two simplest cases here, viz. prenex forms with only existential second-order quantifiers (Σ_1^1 formulas) or only universal quantifiers (Π_1^1 formulas). For the full prenex hierarchy, cf. Section 3.2; note however that we restrict the discussion here to formulas all of whose free variables are first-order. One useful shift in perspective, made possible by the present restricted language, is the following.

If $\exists X_1 \ldots \exists X_k \varphi$ is a Σ_1^1 formula in a vocabulary L, we sometimes consider φ as a first-order formula in the vocabulary $L \cup \{X_1, \ldots, X_k\}$ — now suddenly looking upon the $X_1, \ldots, X_k$ not as second-order *variables* but as non-logical *constants* of the extended language. Conversely, if φ is a first-order L formula containing a relational symbol R, we may consider $\exists R \varphi$ as a Σ_1^1 formula of $L - \{R\}$ — viewing R now as a second-order *variable*. As a matter of fact, this way of putting things has been used already (in Section 2.4).

2.5.1. *Showing things to be Σ_1^1 or Π_1^1.* Most examples of second-order formulas given in Section 2.2 were either Σ_1^1 or Π_1^1; in most cases, it was not too hard to translate the given notion into second-order terms.

A simple result is given in Section 3.2 which may be used in showing things to be Σ_1^1 or Π_1^1-expressible: any formula obtained from a Σ_1^1 (Π_1^1) formula by prefixing a series of first-order quantifications still has a Σ_1^1 (Π_1^1) equivalent.

For more intricate results, we refer to Kleene [1952] and Barwise [1975]. The first shows that if ϕ is a recursive set of first-order formulas, the infinitary conjunct $\bigwedge \phi$ has a Σ_1^1 equivalent (on infinite models). Thus, $\exists X_1, \ldots, \exists X_k \bigwedge \phi$ is also Σ_1^1. This fact has some relevance to resplendency, cf. Section 2.5.4. below. Kleene's method of proof uses absoluteness of definitions of recursive sets, coding of satisfaction and the integer structure on arbitrary infinite models. (It is implicit in much of Barwise [1975] Chapter IV 2/3, which shows that we are allowed to refer to integers in certain ways when defining Σ_1^1 and Π_1^1 notions.)

We now consider these concepts one by one.

2.5.2. *Σ_1^1-sentences.* The key quantifier combination in Frege's predicate logic expresses dependencies beyond the resources of traditional logic: $\forall\exists$. This dependency may be made explicit using a Σ_1^1 formula:

$$\forall x \exists y \varphi(x,y) \longleftrightarrow \exists f \forall x \varphi(x, f(x)).$$

This introduction of so-called *Skolem functions* is one prime source of Σ_1^1 statements. The quantification over functions here may be reduced to our predicate format as follows:

$$\exists X(\forall xyz(X(x,y) \wedge X(x,z) \rightarrow y = z) \wedge \forall x \exists y(X(x,y) \wedge \varphi(x,y))).$$

Even at this innocent level, the connection with set theory shows up (Section 2.4): the above equivalence itself amounts to the assumption of the Axiom of Choice (Bernays).

Through the above equivalence, all first-order sentences may be brought into 'Skolem normal form', e.g. $\forall x \exists y \forall z \exists u A(x,y,z,u)$ goes to $\exists f \forall x \forall z \exists u A(x, f(x), z, u)$, and thence to $\exists f \exists g \forall x \forall z A(x, f(x), z, g(x,z))$. For another type of Skolem normal form (using relations instead), cf. Barwise [1975] Chapter V 8.6.

Conversely, Σ_1^1 sentences allow for many other patterns of dependency. For instance, the variant $\exists f \exists g \forall x \forall z A(x, f(x), z, g(z))$, with g only dependent on z, is not equivalent to any first-order formula, but rather to a so-called 'branching' pattern (first studied in Henkin [1961])

$$\begin{pmatrix} \forall x \exists y \\ \forall z \exists u \end{pmatrix} A(x, y, z, u).$$

For a discussion of the linguistic significance of these 'branching quantifiers', cf. Barwise [1979]. One sentence which has been claimed to exhibit the above pattern is 'some relative of each villager and some relative of each townsman hate each other' (Hintikka). The most convincing examples of first-order branching to date, however, rather concern quantifiers such as (precisely) *one* or *no*. Thus, 'one relative of each villager and one relative of each townsman hate each other' seems to lack any linear reading. (The reason is that any linear sequence of *precisely one*'s creates undesired dependencies. In this connection, recall that 'one sailor has discovered one sea' is not equivalent to 'one sea has been discovered by one sailor'.) An even simpler example might be 'no one loves no one', which has a linear reading $\neg \exists x \neg \exists y \, L(x, y)$ (i.e. everyone loves someone), but also a branching reading amounting to $\neg \exists x \exists y \, L(x, y)$. (Curiously, it seems to lack the inverse scope reading $\neg \exists y \neg \exists x \, L(x, y)$ predicted by Montague Grammar.) Actually, this last example also shows that the phenomenon of branching does not lead inevitably to second-order readings.

The preceding digression has illustrated the delicacy of the issue whether second-order quantification actually occurs in natural language. In any case, if branching quantifiers occur, then the logic of natural language would be extremely complex, because of the following two facts. As Enderton [1970] observes, universal validity of Σ_1^1 statements may be effectively reduced to that of branching statements. Thus, the complexity of the latter notion is at least that of the former. And, by inspection of the argument in Section 2.3 above, we see that

THEOREM. *Universal validity of Σ_1^1 sentences is non-arithmetical, etc.*

Proof. The reduction formula was of the form $P^2 \to \varphi$, where P^2 is Π_1^1 and φ is first-order. By the usual prenex operation, the universal second-order quantifier in the antecedent becomes an existential one in front. □

Indeed, as will be shown in Section 4.3, the complexity of Σ_1^1-universal validity is essentially that of universal validity for the whole second-order (or higher-order) language. Nevertheless, one observation is in order here.

These results require the availability of non-logical constants and, e.g., universal validity of $\exists X \varphi(X, R)$ really amounts to universal validity of the Π_2^1 statement $\forall Y \exists X \varphi(X, Y)$. When attention is restricted to 'pure' cases, it may be shown that universal validity of Σ_1^1 statements is much less complex, amounting to truth in all finite models (cf. Van Benthem [1977]). Thus, in the arithmetical hierarchy (cf. Section 3.2) its complexity is only Π_1^0.

When is a Σ_1^1 sentence, say of the form $\exists X_1 \ldots \exists X_k \varphi(X_1, \ldots, X_k, R)$, equivalent to a first-order statement about its parameter R? An answer follows from Keisler's theorem (Section 1.4.2), by the following observation.

THEOREM. *Truth of Σ_1^1 sentences is preserved under the formation of ultraproducts.*

(This is a trivial corollary of the preservation of first-order sentences, cf. Section 1.4.2.)

COROLLARY. *A Σ_1^1 sentence is first-order definable iff its negation is preserved under ultraproducts.*

(That Σ_1^1 sentences, and indeed all higher-order sentences are preserved under isomorphism should be clear.)

Moreover, there is a consequence analogous to Post's theorem in recursion theory:

COROLLARY. *Properties of models which are both Σ_1^1 and Π_1^1 are already elementary.*

(Of course, this is also immediate from the interpolation theorem which, in this terminology, says that disjoint Σ_1^1 classes can be separated by an elementary class.)

Next, we consider a finer subdivision of Σ_1^1 sentences, according to their first-order matrix. The simplest forms are the following (φ quantifier-free):

(1)   ($\exists\exists$) $\exists X_1 \ldots X_k \exists y_1 \ldots y_m \varphi(X_1, \ldots, X_k, y_1, \ldots, y_m, R)$

(2)   ($\exists\forall$) $\exists X_1 \ldots X_k \forall y_1 \ldots y_m \varphi(X_1, \ldots, X_k, y_1, \ldots, y_m, R)$

(3)   ($\exists\forall\exists$) $\exists X_1 \ldots X_k \forall y_1 \ldots y_m \exists z_1 \ldots z_n \varphi(X_1, \ldots, X_k, y_1, \ldots, y_m, z_1, \ldots, z_n, R)$

I.4: HIGHER-ORDER LOGIC

We quote a few observations from Van Benthem [1983]:
- all forms (1) have a first-order equivalent,
- all forms (2) are preserved under elementary (first-order) equivalence, and hence are equivalent to some (infinite) disjunction of (infinite) conjunctions of first-order sentences,
- the forms (3) harbour the full complexity of Σ_1^1.

The first assertion follows from its counterpart for Π_1^1 sentences, to be stated below. A proof sketch of the second assertion is as follows. If (2) holds in a model $\mathfrak{A}$, then so does its first-order matrix (2)* in some expansion $\mathfrak{A}^+$ of $\mathfrak{A}$. Now suppose that $\mathfrak{B}$ is elementarily equivalent to $\mathfrak{A}$. By a standard compactness argument, (2)* is satisfiable together with the elementary diagram of $\mathfrak{B}$, i.e. in some elementary extension of $\mathfrak{B}$. But, restricting $X_1, \ldots, X_k$ to B, a substructure arises giving the same truth values to formulas of the specific form (2)*; and hence we have an expansion of $\mathfrak{B}$ to a model for (2)* – i.e. $\mathfrak{B}$ satisfies (2). □

Finally, the third assertion follows from the earlier Skolem reduction: with proper care, the Skolem normal form of the first-order matrix will add some predicates to $X_1, \ldots, X_k$, while leaving a first-order prefix of the form $\forall \exists$.

Lastly, we mention the Svenonius characterization of Σ_1^1 sentences in terms of quantifiers of infinite length. In Chapter I.1 an interpretation is mentioned of finite formulas in terms of games. This is a particularly good way of explaining infinite sequences of quantifiers like

(1) $\quad \forall x_1 \exists y_1 \forall x_2 \exists y_2 \forall x_3 \exists y_3 \ldots \varphi(x_1, y_1, x_2, y_2, \ldots)$.

Imagine players $\forall$ and $\exists$ alternately picking $x_1, x_2, \ldots$ resp. $y_1, y_2, \ldots$: $\exists$ wins iff $\varphi(x_1, y_1, x_2, \ldots)$. (1) is counted as *true* iff $\exists$ has a *winning strategy*, i.e. a function telling him how to play, given $\forall$'s previous moves, in order to win. Of course, a winning strategy is nothing more than a bunch of Skolem-functions.

Now, Svenonius' theorem says that, on countable models, every Σ_1^1 sentence is equivalent to one of the form (1) where φ is the conjunction of an (infinite) recursive set of first-order formulas. The theorem is in Svenonius [1965]; for a more accessible exposition, cf. Barwise [1975] Chapter VI.6.

2.5.3. Π_1^1 *sentences.* Most examples of second-order sentences in Section 2.2 were Π_1^1: full induction, Dedekind completeness, full substitution. Also, our recurrent example *most* belonged to this category − and so do, e.g. the modal formulas of intensional logic (compare Van Benthem [II.4]).

Results about Π_1^1 sentences closely parallel those for Σ_1^1. (One notable exception is universal validity, however: that notion is recursively axiomatizable here, for the simple reason that $\models_2 \forall X \varphi(X, Y)$ iff $\models_1 \varphi(X, Y)$.) For instance, we have

THEOREM. *A Π_1^1 sentence has a first-order equivalent iff it is preserved under ultraproducts.*

This time, we shall be little more explicit about various possibilities here. The above theorem refers to *elementary* definitions of Π_1^1 sentences, i.e. in terms of *single* first-order sentences. The next two more liberal possibilities are Δ-*elementary* definitions (allowing an infinite conjunction of first-order sentences) and Σ-*elementary* ones (allowing an infinite disjunction). As was noted in Section 1.2, the non-first-order Π_1^1 notion of finiteness is also Σ-elementary: 'precisely one or precisely two or ... '. The other possibility does not occur, however: all Δ-elementary Π_1^1 sentences are already elementary. (If the conjunction $\bigwedge S$ defines $\forall X_1 \ldots X_n \varphi$, then the following first-order implication holds: $S \models \varphi(X_1, \ldots, X_n)$. Hence, by compactness, $S_0 \models \varphi$ for some finite $S_0 \subseteq S$ − and $\bigwedge S_0$ defines $\forall X_1 \ldots X_n \varphi$ as well.) The next levels in this more liberal hierarchy of first-order definability are ΣΔ and ΔΣ. (Unions of intersections and intersections of unions, respectively.) These two, and in fact all putative 'higher' ones collapse, by the following observation from Bell and Slomson [1969].

PROPOSITION. *A property of models is preserved under elementary equivalence iff it is ΣΔ-elementary.*

Thus, essentially, there remains a hierarchy of the following form:

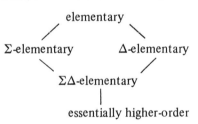

Now, by a reasoning similar to the above, we see that ΣΔ-elementary Π_1^1 sentences are Σ-elementary already. Thus, in the Π_1^1 case, the hierarchy

collapses to 'elementary, Σ-elementary, essentially second-order'. This observation may be connected up with the earlier syntactic classification of Σ_1^1. Using negations, we get for Π_1^1 sentences the types $\forall\forall(1)$, $\forall\exists(2)$ and $\forall\exists\forall(3)$. And these provide precisely instances for each of the above remaining three stages. For instance, that all types (1) are elementary follows from the above characterization theorem, in combination with the observation that type (1) Π_1^1 sentences are preserved under *ultraproducts* (cf. van Benthem [1983]).

We may derive another interesting example of failure of first-order model theory here. One of the classical mother results is the Łoś–Tarski theorem: preservation under submodels is syntactically characterized by definability in universal prenex form. But now, consider well-foundedness (Section 2.2.3). This property of models is preserved in passing to submodels, but it cannot even be defined in the universal form (1), lacking first-order definability.

Our last result shows that even this modest, and basic topic of connections between Π_1^1 sentences and first-order ones is already fraught with complexity.

THEOREM. *The notion of first-order definability for Π_1^1 sentences is not arithmetical.*

Proof. Suppose, for the sake of reductio, that it were. We will then derive the arithmetical definability of arithmetical truth – again contradicting Tarski's theorem. Actually, it is slightly more informative to consider a set-theoretic reformulation (involving only one, binary relation constant): truth in $\langle V_\omega, \in \rangle$ cannot be arithmetical for first-order sentences ψ.

Now, consider any categorical Π_1^1 definition Φ for $\langle V_\omega, \in \rangle$. Truth in $\langle V_\omega, \in \rangle$ of ψ then amounts to the implication $\Phi \models_2 \psi$. It now suffices to show that this statement is effectively equivalent to the following one: '$\Phi \vee \psi$ is first-order definable'. Here, the Π_1^1 statement $\Phi \vee \psi$ is obtained by pulling ψ into the first-order matrix of Φ.

'$\Longrightarrow$': If $\Phi \models_2 \psi$, then $\Phi \vee \psi$ is defined by ψ.

'$\Longleftarrow$': Assume that some first-order sentence α defines $\Phi \vee \psi$.

Consider $\langle V_\omega, \in \rangle$: Φ holds here, and hence so does α. Now let $\mathfrak{A}$ be any proper elementary extension of $\langle V_\omega, \in \rangle$: Φ fails there, while α still holds. Hence ($\Phi \vee \psi$ and so) ψ holds in $\mathfrak{A}$. But then, ψ holds in the elementary submodel $\langle V_\omega, \in \rangle$, i.e., $\Phi \models_2 \psi$. □

2.5.4. *Resplendent models.* One tiny corner of 'higher-order model theory' deserves some special attention. Models on which Σ_1^1 formulas are equivalent with their set of first-order consequences recently acquired special interest in model theory. Formally, $\mathfrak{A}$ is called *resplendent* if for every first-order formula $\varphi = \varphi(x_1, \ldots, x_n)$ in the language of $\mathfrak{A}$ supplemented with some relation-symbol R:

$$\mathfrak{A} \models_2 \forall x_1 \ldots x_n (\bigwedge \psi \to \exists R \varphi),$$

where ψ is the set of all first-order $\psi = \psi(x_1, \ldots, x_n)$ in the language of $\mathfrak{A}$ logically implied by φ. Thus, $\mathfrak{A}$ can be expanded to a model of φ as soon as it satisfies all first-order consequences of φ in its own language.

Resplendency was introduced, in the setting of infinitary admissible languages, by Ressayre [1977] under the name *relation-universality*. A discussion of its importance for the first-order case can be found in Barwise and Schlipf [1976]. The notion is closely related to *recursive saturation* (i.e., saturation w.r.t. recursive types of formulas): every resplendent model is recursively saturated, and, for countable models, the converse obtains as well. In fact, Ressayre was led to (the infinitary version of) this type of saturation by looking at what it takes to prove resplendency.

The importance of resplendent models is derived from the fact that they exist in abundance in all cardinals and can be used to trivialize results in first-order model theory formerly proved by means of saturated and special models of awkward cardinalities; besides, Ressayre took the applicability of the infinitary notions to great depth, deriving results in descriptive set theory as well.

We only mention two easy examples.

Proof of the Craig interpolation theorem: suppose that $\models \varphi(R) \to \psi(S)$; let Φ be the set of R-less consequences of φ. When $\Phi \models \psi$, we are finished by one application of compactness. Thus, let $(\mathfrak{A}, S)$ be a resplendent model of Φ. By definition, we can expand $(\mathfrak{A}, S)$ to a model $(\mathfrak{A}, R, S)$ of φ. Hence, $(\mathfrak{A}, S) \models \psi$. But then, $\Phi \models \psi$, as *every* model has a resplendent equivalent. □

As is the case for saturated and special models, many global definability theorems have local companions for resplendent ones. We illustrate this fact again with the interpolation theorem, which in its local version takes the following form: if the resplendent model $\mathfrak{A}$ satisfies $\forall x (\exists R \varphi(x) \to \forall S \psi(x))$, then there exists a first-order formula $\eta(x)$ in the $\mathfrak{A}$-language such that $\mathfrak{A}$ satisfies both $\forall x (\exists R \varphi \to \eta)$ and $\forall x (\eta \to \forall S \psi)$. To make the proof slightly more perspicuous, we make the statement more symmetrical. Let $\varphi' = \neg \psi$.

The first sentence then is equivalent with $\forall x(\neg \exists R\varphi \vee \neg \exists S\varphi')$ (∗), while the last amounts then to $\forall x(\exists S\varphi' \to \neg \eta)$. Hence, interpolation takes the (local) 'Robinson-consistency' form: disjoint Σ_1^1-definable sets on $\mathfrak{A}$ can be separated by a first-order definable one.

Now for the proof, which is a nice cooperation of both resplendency and recursive saturation. Suppose that our resplendent model $\mathfrak{A} = \langle A, \ldots \rangle$ satisfies (∗). By resplendency, the set of logical consequences of either φ or φ' in the language of $\mathfrak{A}$ is not satisfiable in $\mathfrak{A}$.

Applying recursive saturation (the set concerned is only recursively enumerable according to first-order completeness – but we can use Craig's 'pleonasm' trick to get a recursive equivalent), some finite subset $\Phi \cup \Phi'$ is non-satisfiable already, where we've put the φ consequences in Φ and the φ' consequences in Φ'. We now have $\models \varphi \to \bigwedge \Phi$, $\models \varphi' \to \bigwedge \Phi'$. and, by choice of $\Phi \cup \Phi'$, $\mathfrak{A} \models \forall x \neg \bigwedge(\Phi \cup \Phi')$, which amounts to $\mathfrak{A} \models \forall x(\neg \bigwedge \Phi \vee \neg \bigwedge \Phi')$; hence we may take either $\bigwedge \Phi$ or $\bigwedge \Phi'$ as the 'separating' formula. The local Beth theorem is an immediate consequence: if the disjoint Σ_1^1-definable sets are each other's complement, they obviously coincide with the first-order definable separating set and its complement, respectively. In other words, sets which are both Σ_1^1 and Π_1^1-definable on $\mathfrak{A}$ are in fact first-order definable.

This situation sharply contrasts with the case for (say) $\mathfrak{N}$ discussed in Section 3.2, where we mention that arithmetic truth is both Σ_1^1 and Π_1^1-definable, but not arithmetical.

3. HIGHER-ORDER LOGIC

Once upon the road of second-order quantification, higher predicates come into view. In mathematics, one wants to quantify over functions, but also over functions defined on functions, etcetera. Accordingly, the type theories of the logicist foundational program allowed quantification over objects of all finite orders, as in *Principia Mathematica*. But also natural language offers such as ascending hierarchy, at least in the types of its lexical items. For instance, nouns (such as 'woman') denote properties, but then already adjectives become higher-order phrases ('blonde woman'), taking such properties to other properties. In fact, the latter type of motivation has given type theories a new linguistic lease of life, in so-called 'Montague Grammar', at a time when their mathematical functions had largely been taken over by ordinary set theory (cf. Montague [1974]).

In this section, we will consider a stream-lined relational version of higher-order logic, which leads to the basic logical results with the least amount of effort. Unfortunately for the contemporary semanticist, it does not look very much like the functional Montagovian type theory. In fact, we will not even encounter such modern highlights as lambda-abstraction, because our language can do all this by purely traditional means. Moreover, in Section 4, we shall be able to derive partial completeness results from the standard first-order ones for many-sorted logic in an extremely simple fashion. (In particular, the complicated machinery of Henkin [1950] seems unnecessary.) It's all very elegant, simple, and exasperating. A comparison with the more semantic, categorial grammar-oriented type theories will be given at the end.

3.1. *Syntax and Semantics*

As with first-order languages, higher-order formulas are generated from a given set L of non-logical constants, among which we can distinguish individual constants, function symbols and relation signs. (Often, we will just think of the latter.) Formulas will be interpreted in the same type of models $\mathfrak{A} = \langle A, * \rangle$ as used in the first-order case, i.e. $A \neq \emptyset$, and $*$ assigns something appropriate to every L symbol: ('distinguished') elements of A to individual constants, functions over A to function symbols (with the proper number of arguments) and relations over A to relation signs (again, of the proper arity).

Thus, fix any such set L. Patterns of complexity are now recorded in *types*, defined inductively by

(i) 0 is a type

(ii) a finite sequence of types is again a type.

Here, 0 will be the type of *individuals*, $(\tau_1, \ldots, \tau_n)$ that of *relations* between objects of types $\tau_1, \ldots, \tau_n$. Notice that, if we read clause (ii) as also producing the *empty* sequence, we obtain a type of relation without arguments; i.e. of propositional constants, or *truth values*. Higher up then, we will have propositional functions, etcetera. This possibility will not be employed in the future, as our metatheory would lose some of its elegance. (But see Section 3.3 for a reasonable substitute.)

The *order* of a type is a natural number defined as follows: the order of 0 is 1 (individuals are 'first-order' objects), while the order of $(\tau_1, \ldots, \tau_n)$ equals 1 + max order (τ_i) $(1 \leq i \leq n)$. Thus, the terminology of 'first-order', 'second-order', etcetera, now becomes perfectly general.

For each type τ, the language has a countably infinite set of variables. The order of a variable is the order of its type. Thus, there is only one kind of first-order variable, because the only order 1-type is 0. The second-order variables all have types $(0, \ldots, 0)$. Next, the *terms* of type 0 are generated from the type 0 variables and the individual constants by applying function symbols in the proper fashion. A term of type $\neq 0$ is just a variable of that type. Thus, for convenience, non-logical constants of higher-orders have been omitted: we are really thinking of our former first-order language provided with a quantificational higher-order apparatus. Finally, one might naturally consider a relation symbol with n places as a term of type $(0, \ldots, 0)$ (n times); but the resulting language has no additional expressive power, while it becomes a little more complicated. Hence, we refrain from utilizing this possibility.

Atomic formulas arise as follows:
  (i) $R(t_1, \ldots, t_n)$, where R is an n-place relation symbol, and $t_1, \ldots, t_n$ terms of type 0,
  (ii) $X(t_1, \ldots, t_n)$, where X is a variable of type $(\tau_1, \ldots, \tau_n)$ and t_i a type τ_i-term $(1 \leq i \leq n)$.

We could have added identities $X = Y$ here for all higher types; but these may be thought of as defined through the scheme $\forall X_1 \ldots \forall X_n (X(X_1, \ldots, X_n) \leftrightarrow Y(X_1, \ldots, X_n))$, with appropriate types.

Formulas are defined inductively from the atomic ones using propositional connectives and quantification with respect to variables of all types. The resulting set, based on the vocabulary L is called L_ω. L_n is the set of formulas all of whose variables have order $\leq n$ ($n = 1, 2, \ldots$). Thus, we can identify L_1 with the first-order formulas over L, and L_2 with the second-order ones. (A more sophisticated classification of orders is developed in Section 3.2 below, however.) The reader is requested to formulate the examples of Section 2.2 in this language; especially the L_3-definition of finiteness.

Again, let us notice that we have opted for a rather austere medium: no higher-order constants or identities, no conveniences such as *function quantifiers*, etcetera. One final omission is that of relational *abstracts* taking formulas $\varphi(X_1, \ldots, X_n)$ to terms $\lambda X_1 \ldots X_n \cdot \varphi(X_1, \ldots, X_n)$ denoting the corresponding relation. In practice, these commodities do make life a lot easier; but they are usually dispensable in theory. For instance, the statement $\varphi(\lambda X \cdot \psi(X))$ is equally well expressed by means of $\exists Y(Y = \lambda X \cdot \psi(X) \wedge \varphi(Y))$, and this again by $\exists Y(\forall Z(Y(Z) \leftrightarrow \psi(Z)) \wedge \varphi(Y))$.

From the syntax of our higher-order language, we now pass on to its semantics. Let $\mathfrak{A}$ be an ordinary L model $\langle A, * \rangle$ as described above. To interpret the L_ω formulas in $\mathfrak{A}$ we need the *universes of type τ over A* for all types τ:
  (i) $D_0(A) = A$
  (ii) $D_{(\tau_1, \ldots, \tau_n)}(A) = \mathscr{P}(D_{\tau_1}(A) \times \cdots \times D_{\tau_n}(A))$.

An *A assignment* is a function α defined on all variables such that, if X has type τ, $\alpha(X) \in D_\tau(A)$.

We now lift the ordinary satisfaction relation to L_ω formulas φ in the obvious way. For instance, for an L model $\mathfrak{A}$ and an A assignment α,

$$\mathfrak{A} \vDash_\omega X(t_1, \ldots, t_n)[\alpha] \quad \text{iff} \quad \alpha(X)(t_1^{\mathfrak{A}}[\alpha], \ldots, t_n^{\mathfrak{A}}[\alpha]);$$

where $t^{\mathfrak{A}}[\alpha]$ is the *value* of the term t under α in $\mathfrak{A}$ defined as usual. Also, e.g., $\mathfrak{A} \vDash_\omega \forall X \varphi[\alpha]$ iff for all assignments α' differing from α at most in the value given to X: $\mathfrak{A} \vDash_\omega \varphi[\alpha']$.

The other semantical notions are derived from satisfaction in the usual fashion.

3.2. *The Prenex Hierarchy of Complexity*

The logic and model theory of L_ω exhibit the same phenomena as those of L_2: a fluid border line with set theory, and few systematic results. Indeed, in a sense, higher-order *logic* does not offer anything new. It will be shown in Section 4.2 that there exists an effective reduction from universal validity for L_ω formulas to that for second-order ones, indeed to monadic Σ_1^1 formulas (Hintikka [1955]).

As for the connections between L_ω and set theory, notice that the present logic is essentially that of arbitrary models A provided with a natural set-theoretic superstructure $\bigcup_n V^n(A)$; where $V^0(A) = A$, and $V^{n+1}(A) = \mathcal{P}(A \cup V^n(A))$. As a 'working logic', this is a sufficient setting for many mathematical purposes. (But cf. Barwise [1975] for a smaller, *constructible* hierarchy over models, with a far more elegant metatheory.)

We will not go into the exact relations between the logic of L_ω and ordinary set theory, but for the following remark.

Given a structure $\mathfrak{A} = \langle A, * \rangle$, the structure $\mathfrak{A}^+ = \langle A \cup \bigcup_n V^n(A), \in, * \rangle$ is a model for a set theory with atoms. There is an obvious translation from the L_ω theory of $\mathfrak{A}$ into a fragment of the ordinary first-order theory of $\mathfrak{A}^+$. The reader may care to speculate about a converse (cf. Kemeny [1950]).

What will be considered instead in this section, is one new topic which is typical for a hierarchical language such as the present one. We develop a prenex classification of formulas, according to their patterns of complexity; first in general, then on a specific model, viz. the natural numbers. This is one of the few areas where a coherent body of higher-order theory has so far been developed.

I.4: HIGHER-ORDER LOGIC

There exists a standard classification of first-order formulas in prenex form. $\Sigma_0 = \Pi_0$ is the class of quantifier-free formulas; Σ_{m+1} is the class of formulas $\exists x_1 \ldots \exists x_k \varphi$ where $\varphi \in \Pi_m$; and dually, Π_{m+1} is the class of formulas $\forall x_1 \ldots \forall x_k \varphi$ with $\varphi \in \Sigma_m$. The well-known Prenex Normal Form Theorem now says that every first-order formula is logically equivalent to one in $\bigcup_m (\Sigma_m \cup \Pi_m)$; i.e. to one in prenex form.

The above may be generalized to arbitrary higher-order formulas as follows. We classify quantificational complexity with respect to the $n + 1$st order. $\Sigma_0^n = \Pi_0^n$ is the class of L_ω formulas all of whose quantified variables have order $\leq n$. Thus, Σ_0^0 is the class of quantifier-free L_ω formulas. (Notice that the above Σ_0 is a proper subclass of Σ_0^0, as we allow free variables of higher type in Σ_0^0 formulas. Also, it is not true that $\Sigma_0^1 \subseteq L_1$, or even $\Sigma_0^1 \subseteq L_n$ for some $n > 1$.)

Next, Σ_{m+1}^n is the class of formulas $\exists X_1 \ldots \exists X_k \varphi$, where $\varphi \in \Pi_m^n$ and $X_1, \ldots, X_k$ have order $n + 1$; and dually, Π_{m+1}^n consists of the formulas $\forall X_1 \ldots \forall X_k \varphi$ with $\varphi \in \Sigma_m^n$ and $X_1, \ldots, X_k$ $(n + 1)$st order. (Notice the peculiar, but well-established use of the upper index n: a Σ_2^1 formula thus has quantified *second*-order variables.)

The reader may wonder why we did not just take Σ_0^n to be L_n. The reason is that we do not consider the mere occurrence of, say, second-order variables in a formula a reason to call it (at least) second-order. (Likewise, we do not call first-order formulas 'second-order' ones, because of the occurrence of second-order relational constants.) It is *quantification* that counts: we take a formula to be of order n when its interpretation in a model $\langle A, \ldots \rangle$ presupposes complete knowledge about some nth order universe $D_\tau(A)$ over A. And it is the quantifier over some order n variable which presupposes such knowledge, not the mere presence of free variables of that order. (After all, we want to call, e.g. a property of type ((0)) 'first-order' definable, even if its first-order definition contains a second-order free variable – and it must.) There is an interesting historical analogy here. One way to think of the prenex hierarchy is as one of *definitional complexity*, superimposed upon one of *argument type complexity* (given by the free variable pattern of a formula). This move is reminiscent of Russell's passage from *ordinary* to *ramified* type theory.

THEOREM. *Every L_{n+1} formula has an equivalent in $\bigcup_m (\Sigma_m^n \cup \Pi_m^n)$.*

Proof. Let $\varphi \in L_{n+1}$ be given. First, manipulate it into prenex form, where the order of the quantifiers is immaterial – just as in the first-order case. Now, if we can manage to get quantifiers over $n + 1$st order variables to the front, we are done. But, this follows by repeated use of the valid equivalence below and its dual.

Let x have type τ_0 and order less than $n+1$: the order of the type $(\tau_1, \ldots, \tau_k)$ of the variable X. Let Y be some type $(\tau_0, \ldots, \tau_k)$ variable; its order is then $n+1$ too, and we have the equivalence

$$\forall x \exists X \psi \leftrightarrow \exists Y \forall x \psi'.$$

Here ψ' is obtained from ψ by replacing subformulas $X(t_1, \ldots, t_k)$ by $Y(x, t_1, \ldots, t_k)$; where Y does not occur in ψ. Thanks to the restriction to L_{n+1}, the *only* atomic subformulas of ψ containing X are of the above form and, hence, ψ' does not contain X any longer. (If X could occur in argument positions, it would have to be defined away using suitable Y *abstracts*. But, this addition to the language would bring about a revised account of complexity in any case.)

To show intuitively that the above equivalence is valid, assume that $\forall x \exists X \psi(x, X)$. For every x, choose X_x such that $\psi(x, X_x)$. Define Y by setting $Y(x, y_1, \ldots, y_n) := X_x(y_1, \ldots, y_n)$. Then clearly $\forall x \psi(x, \{\langle y_1, \ldots, y_n \rangle \mid Y(x, y_1, \ldots, y_n)\})$ and, hence, $\exists Y \forall x \psi'$. The converse is immediate. □

We will now pass on to more concrete hierarchies of higher-order definable relations on specific models.

Let $\mathfrak{A} = \langle A, \ldots \rangle$ be some model, $R \in D_\tau(A)$, $\tau = (\tau_1, \ldots, \tau_n)$, and let $\varphi \in L_\omega$ have free variables $X_1, \ldots, X_n$ of types (respectively) $\tau_1, \ldots, \tau_n$. φ is said to *define* R on $\mathfrak{A}$ if, whenever $S_1 \in D_{\tau_1}(A), \ldots, S_n \in D_{\tau_n}(A)$,

$$R(S_1, \ldots, S_n) \text{ iff } \mathfrak{A} \vDash_\omega \varphi[S_1, \ldots, S_n].$$

R is called Σ_m^n (Π_m^n) on $\mathfrak{A}$ if it has a defining formula of this kind. It is Δ_m^n if it is both Σ_m^n and Π_m^n. We denote these classes of definable relations on $\mathfrak{A}$ by $\Sigma_m^n(\mathfrak{A})$, etcetera.

Now, let us restrict attention to $\mathfrak{A} = $ the natural numbers $\mathfrak{N}: \langle \mathbb{N}, +, \times, 0 \rangle$. (In this particular case it is customary to let $\Sigma_0^0(\mathfrak{N}) = \Pi_0^0(\mathfrak{N})$ be the wider class of relations definable using formulas in which *restricted* quantification over first-class variables is allowed.) For any type τ, $\Delta_1^0(\mathfrak{N}) \cap D_\tau(\mathbb{N})$ is the class of *recursive* relations of type τ; the ones in $\Sigma_1^0(\mathfrak{N}) \cap D_\tau(\mathbb{N})$ are called *recursively enumerable*. These are the simplest cases of the *arithmetical hierarchy*, consisting of all Σ_n^0 and Π_n^0-definable relations on $\mathfrak{N}$. Evidently, these are precisely the first-order-definable ones, in any type τ.

At the next level, the *analytic hierarchy* consists of the Σ_n^1 and Π_n^1-definable relations on $\mathfrak{R}$. Those in $\Delta_1^1(\mathfrak{R})$ are called *hyperarithmetical*, and have a (transfinite) hierarchy of their own. One reason for the special interest in this class is the fact that *arithmetic truth* for first-order sentences is hyperarithmetical (though not arithmetical, by Tarski's Theorem).

These hierarchies developed after the notion of recursiveness had been identified by Gödel, Turing and Church, and were studied in the fifties by Kleene, Mostowski and others.

Just to give an impression of the more concrete type of investigation in this area, we mention a few results. Methods of proof are rather uniform: positive results (e.g. '$\varphi \in \Sigma_n^1$') by actual inspection of possible definitions, negative results ('$\varphi \notin \Sigma_n^1$') by diagonal arguments reminiscent of the mother example in Russell's Paradox.

(1) The satisfaction predicate 'the sequence (coded by) s satisfies the first-order formula (coded by) φ in $\mathfrak{R}$' is in $\Delta_1^1(\mathfrak{R}) \cap D_{(0,0)}(\mathbb{N})$.
(2) This predicate is not in $\Sigma_0^1(\mathfrak{R})$.
(3) The *Analytic Hierarchy Theorem* for $D_{(0)}(\mathbb{N})$ relations. All inclusions in the following scheme are proper (for all m):

$$\Delta_m^1(\mathfrak{R}) \begin{array}{c} \subseteq \\ \\ \subseteq \end{array} \begin{array}{c} \Sigma_m^1(\mathfrak{R}) \\ \\ \Pi_m^1(\mathfrak{R}) \end{array} \begin{array}{c} \subseteq \\ \\ \subseteq \end{array} \Delta_{m+1}^1(\mathfrak{R})$$

These results may be generalized to higher orders.
(4) Satisfaction for Σ_0^n formulas (with first-order free variables only) on $\mathfrak{R}$ is in $\Delta_1^n(\mathfrak{R}) - \Sigma_0^n(\mathfrak{R})$.
(5) The Hierarchy Theorem holds in fact for any upper index ≥ 1.

By allowing second-order parameters in the defining formulas, the analytic hierarchy is transformed into the classical hierarchy of *projective* relations. Stifled in set-theoretic difficulties around the twenties, interest in this theory was revived by the set-theoretic revolution of 20 years ago. The reader is referred to the modern exposition Moschovakis [1980].

3.3. *Two Faces of Type Theory*

As was observed earlier, the above language L_ω is one elegant medium of description for one natural type superstructure on models with relations.

Nevertheless, there is another perspective, leading to a more function-oriented-type theory closer to the categorial system of natural language. In a sense, the two are equivalent through codings of functions as special relations, or of relations through characteristic functions. It is this kind of *sous entendu* which would allow an ordinary logic text book to suppress all reference to functional type theories in the spirit of Church [1940], Henkin [1950] or Montague [1974]. (It is this juggling with codings and equivalences also, which makes advanced logic texts so impenetrable to the outsider lacking that frame of mind.)

For this reason, we give the outline of a functional type theory, comparing it with the above. As was observed earlier on, in a first approximation, the extensional part of natural language can be described on the model of a *categorial grammar*, with basic *entity expressions* (e.g. proper names; type e) and *truth value expressions* (sentences; type t), allowing arbitrary binary couplings (a, b): the type of functional expressions taking an a-type expression to a b-type one. Thus, for instance, the intransitive verb 'walk' has type (e, t), the transitive verb 'buy' type $(e, (e, t))$, the sentence negation 'not' has (t, t) while sentence conjunction has $(t, (t, t))$. More complicated examples are quantifier phrases, such as 'no man', with type $((e, t), t)$, or determiners, such as 'no', with type $((e, t), ((e, t), t))$. Again, to a first approximation, there arises the picture of natural language as a huge jigsaw puzzle, in which the interpretable sentences are those for which the types of their component words can be fitted together, step by step, in such a way that the end result for the whole is type t.

Now, the natural matching type theory has the above types, with a generous supply of variables and constants for each of these. Its basic operations will be, at least, *identity* (between expressions of the same type), yielding truth value expressions, and *functional application* combining B with type (a, b) and A with type a to form the expression $B(A)$ of type b. What about the logical constants? In the present light, these are merely constants of specific categories. Thus, binary connectives ('and', 'or') are in $(t, (t, t))$, quantifiers ('all', 'some') in the above determiner type $((e, t), ((e, t), t))$. (Actually, this makes them into binary relations between properties: a point of view often urged in the logical folklore.) Nevertheless, one can single them out for special treatment, as was Montague's own strategy. On the other hand, a truly natural feature of natural language seems to be the phenomenon of *abstraction*: from any expression of type b, we can make a functional one of type (a, b) by varying some occurrence(s) of component a expressions. Formally then, our type theory will have so-called 'lambda abstraction': if B is an expression of type b, and x a variable of type a, then $\lambda x \cdot B$ is an expression of type (a, b).

Semantic structures for this language form a function hierarchy as follows:

(i) D_e is some set (of 'entities' or 'individuals'),
(ii) D_t is the set of truth values $\{0, 1\}$ (or some generalization thereof),
(iii) $D_{(a, b)} = D_b^{D_a}$.

Given a suitable interpretation for constants and assignment for variables, values may be computed for terms of type a in the proper domain D_a through the usual compositional procedure. Thus, in particular, suppressing indices,

$$\text{val}(B(A)) = \text{val}(B)(\text{val}(A))$$

$$\text{val}(\lambda x \cdot B) = \lambda a \in D_a \cdot \text{val}(B)_{x \to a}$$

(Just this once, we have refrained from the usual pedantic formulation.)

In Montague's so-called 'intensional type theory', this picture is considerably complicated by the addition of a realm of possible world-times, accompanied by an auxiliary type s with restricted occurrences. This is a classical example of an unfortunate formalization. Actually, the above set-up remains exactly the same with one additional basic type s (or two, or ten) with corresponding semantic domains D_s (all world-times, in Montague's case). In the terms of Gallin [1975]: once we move up from Ty to $Ty2$, simplicity is restored.

We return to the simplest case, as all relevant points can be made here. What is the connection with the earlier logic L_ω? Here is the obvious translation, simple in content, a little arduous in combinatorial detail.

First, let us embed the Montague hierarchy of domains D_a over a given universe A into our previous hierarchy $D_\tau(A)$. In fact, we shall *identify* the D_a with certain subsets of the $D_\tau(A)$. There seems to be one major problem here, viz. what is to correspond to $D_t = \{0, 1\}$. (Recall that we opted for an L_ω hierarchy without truth-value types.) We choose to define $D_t \subseteq D_{(0)}(A)$: 0 becoming $\emptyset$, and 1 becoming the whole A. Next, of course, $D_e = D_0(A)$. The rule $D_{(a, b)} = D_b^{D_a}$ then generates the other domains. Thus, every Montague universe D_a has been identified with a subset of a certain $D_{\underline{a}}(A)$; where $\underline{a}$ is obviously determined by the rules $\underline{e} := 0$, $\underline{t} := (0)$ and $\underline{(a, b)} := (\underline{a}, \underline{b})$. (Thus, functions have become identified with their graphs; which are binary relations in this case.)

Next, for each Montague type a, one can write down an L_ω formula $T_a(x)$ (with x of type $\underline{a}$) which *defines* D_a in $D_{\underline{a}}(A)$, i.e. for $b \in D_{\underline{a}}(A)$, $\mathfrak{A} \vDash_\omega T_a[b]$ iff $b \in D_a$.

When $E = E(x_1, \ldots, x_n)$ is any type a_0 expression in the Montague system, with the free variables $x_1, \ldots, x_n$ (with types $a_1, \ldots, a_n$, respectively) and $b_i \in D_{a_i}$ ($1 \leq i \leq n$), an object $E^{\mathfrak{A}}[b_1, \ldots, b_n] \in D_{a_0}$ has been defined which is the *value* of E under $b_1, \ldots, b_n$ in $\mathfrak{A}$. We shall indicate now how to write down an L_ω formula $V(x_0, E)$ with free variables $x_0, \ldots, x_n$ (where *now* x_i has type $\underline{a}_i$ ($1 \leq i \leq n$)), which says that x_0 is the value of E under $x_1, \ldots, x_n$. To be completely precise, we will have

(1) $\qquad \mathfrak{A} \vDash_\omega V(x_0, E)[b_0, \ldots, b_n]$ iff $b_0 = E^{\mathfrak{A}}[b_1, \ldots, b_n]$

for objects $b_0, \ldots, b_n$ of the appropriate types.

As a consequence of (1), we obtain

$$\mathfrak{A} \vDash_\omega \exists x(V(x, E_1) \wedge V(x, E_2)) \text{ iff } E_1^{\mathfrak{A}} = E_2^{\mathfrak{A}}$$

for closed expressions E_1, E_2. Thus, the characteristic assertions of Montagovian type theory have been translated into our higher-order logic.

It remains to be indicated how to construct the desired V. For perspicuity, three shorthands will be used in L_ω. First, $x(y)$ stands for the unique z such that $x(y, z)$, if it exists. (Elimination is always possible in the standard fashion.) Furthermore, we will always have $\forall x_1 \ldots x_n \exists ! x_0 V(x_0, E)$ valid when relativized to the proper types. Therefore, instead of $V(x_0, E)$, one may write $x_0 = V(E)$. Third, quantifier relativization to T_a will be expressed by $\forall x \in T_a$ ($\exists x \in T_a$) (where x has type $\underline{a}$). Finally, in agreement with the above definition of the truth values, we abbreviate $\forall y \in T_e x(y)$ and $\forall y \in T_e \neg x(y)$ by $x = \mathsf{T}, x = \bot$, respectively (where x has type (0)).

Here are the essential cases:

(i) E is a two-place relation symbol of the base vocabulary L.

$V(x, E) := x \in T_{(e, (e, t))} \wedge \forall yz \in T_e((x(y))(z) = \mathsf{T} \leftrightarrow E(y, z))$.

(ii) $E = E_1(E_2)$.

$V(x, E) := x = V(E_1)(V(E_2))$.

(iii) $E = \lambda y \cdot F$ (y of type a, F of type b).

$V(x, E) := x \in T_{(a, b)} \wedge \forall y \in T_a(x(y) = V(F))$.

(iv) $E = (E_1 = E_2)$.

$V(x, E) := x \in T_t \wedge (x = \mathsf{T} \leftrightarrow V(E_1) = V(E_2))$.

That these clauses do their job has to be demonstrated by induction, of course; but this is really obvious.

It should be noted that the procedure as it stands does not handle higher-order constants: but, a generalization is straightforward.

For further details, cf. Gallin [1975] Chapter 13. Gallin also has a converse translation from L_ω into functional type theory, not considered here.

The reduction to L_ω makes some prominent features of functional type theory disappear. Notably, lambda abstraction is simulated by means of ordinary quantification. It should be mentioned, however, that this also deprives us of some natural and important questions of functional type theory, such as the search for unique *normal forms*. This topic will be reviewed briefly at the end of the following Section.

4. REDUCTION TO FIRST-ORDER LOGIC

One weak spot in popular justifications for employing higher-order logic lies precisely in the phrase 'all predicates'. When we say that Napoleon has all properties of the great generals, we surely mean to refer to some sort of relevant human properties, probably even definable ones. In other words, the lexical item 'property' refers to some sort of 'things', just like other common nouns. Another, more philosophical illustration of this point is Leibniz' Principle, quoted earlier, of the identity of indiscernibles. Of course, when x, y share *all* properties, they will share that of being identical to x and, hence, they coincide. But this triviality is not what the great German had in mind – witness the charming anecdote about the ladies at court, whom Leibniz made to search for autumn leaves, promising them noticeable differences in colour or shape for any two merely distinct ones.

Thus, there arises the logical idea of re-interpreting second-order, or even higher-order logic as some kind of *many-sorted* first-order logic, with various distinct kinds of objects: a useful, though inessential variation upon first-order logic itself. To be true, properties and predicates are rather abstract kinds of 'things'; but then, so are many other kinds of 'individual' that no one would object to. The semantic net effect of this change in perspective is to allow a greater variety of models for L_ω, with essentially smaller ranges of predicates than the original 'full' ones. Thus, more potential counter-examples become available to universal truths, and the earlier set of L_ω validities decreases; so much so, that we end up with a recursively axiomatizable set. This is the basic content of the celebrated introduction of 'general models' in Henkin [1950]: the remainder is frills and laces.

4.1. General Models

The type structure $\langle D_\tau(A) \mid \tau \in \mathcal{T} \rangle$ ($\mathcal{T}$ the set of types) over a given non-empty set A as defined in Section 3.1 is called the principal or *full* type structure over A; the interpretation of L_ω by means of $\models_\omega$ given there the *standard* interpretation. We can generalize these definitions as follows.

$E = \langle E_\tau \mid \tau \in \mathcal{T} \rangle$ is called a *type structure* over A when

(i) $E_0 = A$ (as before)
(ii) $E_{(\tau_1, \ldots, \tau_n)} \subseteq \mathcal{P}(E_{\tau_1} \times \cdots \times E_{\tau_n})$

Thus, not every relation on $E_{\tau_1} \times \cdots \times E_{\tau_n}$ need be in $E_{(\tau_1, \ldots, \tau_n)}$ any more. Restricting assignments to take values in such more general type structures, satisfaction can be defined as before, leading to a notion of truth with respect to arbitrary type structures. This so-called *general models interpretation* of L_ω admits of a complete axiomatization, as we shall see in due course.

First, we need a certain transformation of higher-order logic into first-order terms. Let L be a given vocabulary. L^+ is the *first-order* language based on the vocabulary

$$L \cup \{\epsilon_\tau \mid 0 \neq \tau \in \mathcal{T}\} \cup \{T_\tau \mid \tau \in \mathcal{T}\};$$

where ϵ_τ is an $n+1$ ary relation symbol when $\tau = (\tau_1, \ldots, \tau_n)$, and the T_τ are unary relation symbols. Now, define the translation $^+: L_\omega \to L^+$ as follows. Let $\varphi \in L_\omega$. First, replace every atom $X(t_1, \ldots, t_n)$ in it by $\epsilon_\tau(X, t_1, \ldots, t_n)$ when X has type τ. Second, relativize quantification with respect to type τ variables to T_τ. Third, consider all variables to be (type 0) variables of L^+. This defines φ^+. (For those familiar with many-sorted thinking (cf. Hodges [Chapter I.1]), the unary predicates T_τ may even be omitted, and φ^+ just becomes φ, in a many-sorted reading.)

On the model-theoretic level, suppose that $(\mathfrak{A}, E)$ is a general model for L; i.e. $\mathfrak{A}$ is an L model with universe A and E is a type structure over A. We indicate how $(\mathfrak{A}, E)$ can be transformed into an ordinary (first-order) model $(\mathfrak{A}, E)^+$ for L^+:

(1) the universe of $(\mathfrak{A}, E)^+$ is $\bigcup_{\tau \in \mathcal{T}} E_\tau$
(2) the interpretation of L symbols is the same as in $\mathfrak{A}$
(3) ϵ_τ is interpreted by ($\tau = (\tau_1, \ldots, \tau_n)$) $\epsilon_\tau^*(R, S_1, \ldots, S_n)$ iff $R \in E_\tau, S_i \in E_{\tau_i} (1 \leq i \leq n)$ and $R(S_1, \ldots, S_n)$.
(4) T_τ is interpreted by E_τ.

There is a slight problem here. When L contains function symbols, the corresponding functions in $\mathfrak{A}$ should be extended on $\cup_{\tau \in \mathcal{T}} E_\tau$. It is irrelevant how this is done, as arguments outside of E_0 will not be used.

The connection between these transformations is the

LEMMA. *Let α be an E assignment, and let $\varphi \in L_\omega$. Then $(\mathfrak{A}, E) \vDash_\omega \varphi[\alpha]$ iff $(\mathfrak{A}, E)^+ \vDash \varphi^+[\alpha]$.*

The proof is a straightforward induction on φ.

There is semantic drama behind the simple change in clause (ii) for $E_{(\tau_1, \ldots, \tau_n)}$ from identity to inclusion. Full type structures are immense; witness their cardinality, which increases exponentially at each level. In stark contrast, a general model may well have an empty type structure, not ascending beyond the original universe. Evidently, the interesting general models lie somewhere in between these two extremes.

At least two points of view suggest themselves for picking out special candidates, starting from either boundary.

'From above to below', the idea is to preserve as much as possible of the global type structure; i.e. to impose various principles valid in the full model, such as Comprehension or Choice (cf. the end of Section 4.2). In the limit, one might consider general models which are L_ω-elementarily equivalent to the full type model. Notice that, by general logic, only Π_1^1 truths are automatically preserved in passing from the full model to its general submodels. Such preservation phenomena were already noticed in Orey [1959], which contains the conjecture that a higher-order sentence is first-order definable if and only if it has the above persistence property, as well as its converse. (A proof of this assertion is in van Benthem [1977].)

Persistence is of some interest for the semantics of natural language, in that some of its 'extensional' fragments translate into persistent fragments of higher-order logic (cf. Gallin [1975] Chapter I.4). Although the main observation (due to Kamp and Montague) is a little beyond the resources of our austere L_ω, it may be stated quite simply. Existential statements $\exists X A(X)$ may be lost in passing from full standard models to their general variants (cf. the example given below). But, *restricted* existential statements $\exists X (P(X, Y) \wedge A(X))$ with all their parameters (i.e. P (!), Y) in the relevant general model, are thus preserved – and the above-mentioned extensional fragments of natural language translate into these restricted forms, which are insensitive, in a sense, to the difference between a general model and its full parent. Therefore, the completeness of L_ω with respect to the general models interpretation (Section 4.2) extends to these fragments of natural language, despite their prima facie higher-order nature.

Conversely, one may also look 'from below to above', considering reasonable constructions for filling the type universes without the above explosive features. For instance, already in the particular case of L_2, a natural idea is to consider predicate ranges consisting of all predicates *first-order definable* in the base vocabulary (possibly with individual parameters). Notice that this choice is stable, in the sense that iteration of the construction (plugging in newly defined predicates into first-order definitions) does not yield anything new. (By the way, the simplest proof that, e.g. von Neumann–Bernays–Gödel set theory is conservative over ZF uses exactly this construction.)

EXAMPLE. The first-order definable sets on the base model $\langle \mathbb{N}, < \rangle$ are precisely all finite and co-finite ones; and a similar characterization may be given for arbitrary predicates. This general model for L_2 is not elementarily equivalent to the standard model, however, as it fails to validate

$$\exists X \forall y ((\exists z (X(z) \wedge y < z) \wedge \exists z (\neg X(z) \wedge y < z)).$$

Second-order general models obtained in this way only satisfy the so-called 'predicative' comprehension axioms. (Referring to the end of Section 4.2, these are the sentences (1) where φ does not *quantify* over second-order variables, but may contain them freely.) We can, however, obtain general models of full ('impredicative') comprehension if we iterate the procedure as follows. For any second-order general model $\langle \mathfrak{A}, E \rangle$, let E^+ consist of all relations on $\mathfrak{A}$ parametrically second-order definable in $\langle \mathfrak{A}, E \rangle$. Thus, the above 'predicative' extension is just $\langle \mathfrak{A}, \emptyset^+ \rangle$. This time, define E_α for ordinals α by $E_\alpha = \cup_{\beta < \alpha} E_\beta^+$. By cardinality considerations, the hierarchy must stop at some γ (by first-order Löwenheim–Skolem, it can in fact be proved that γ has the same cardinal as $\mathfrak{A}$), which obviously means that $\langle \mathfrak{A}, E_\gamma \rangle$ satisfies full comprehension.

For $\mathfrak{A} = \mathfrak{N}$, the above transfinite hierarchy is called *ramified analysis*, γ is Church–Kleene ω_1 and there is an extensive literature on the subject. Barwise [1975] studies related things in a more set-theory oriented setting for arbitrary models.

4.2. General Completeness

As a necessary preliminary to a completeness theorem for L_ω with its new semantics, we may ask which L^+ sentences hold in every model of the form

$(\mathfrak{A}, E)^+$, where $\mathfrak{A}$ is an L model and E a type structure over its universe A. As it happens, these are of six kinds.
   (i) $\exists x T_0 x$. This is because T_0 is interpreted by $E_0 = A$, which is not empty. The other type levels of E might indeed be empty, if E is not full.
   (ii) The next sentences express the fact that the L symbols stand for distinguished elements, functions and relations over the set denoted by T_0:

(1) $T_0(c)$, for each individual constant of L.
(2) $\forall x_1 \ldots \forall x_n(T_0(x_1) \wedge \cdots \wedge T_0(x_n) \to T_0(F(x_1, \ldots, x_n)))$, for all n-place function symbols F of L.
(3) $\forall x_1 \ldots \forall x_n(R(x_1, \ldots, x_n) \to T_0(x_1) \wedge \cdots \wedge T_0(x_n))$, for all n-place relation symbols R of L.

Finally, there are sentences about the type levels.
   (iii) $\forall x(T_\tau(x) \to \neg T_{\tau'}(x))$, whenever $\tau \neq \tau'$.
As a matter of fact, there is a small problem here. If A has elements which are sets, then we might have simultaneously $a \in A$ and $a \subseteq A$. It could happen then that $a \in E_{(0)}$ also, and hence $E_0 \cap E_{(0)} \neq \emptyset$. To avoid inessential sophistries, we shall resolutely ignore these eventualities.
   (iv) $\forall x \bigvee_{\tau \in \mathcal{T}} T_\tau(x)$.

The content of this statement is clear; but unfortunately, it is not a first-order sentence of L^+, having an infinite disjunction. We shall circumvent this problem eventually.
   (v) $\forall x \forall y_1 \ldots \forall y_n(\epsilon_\tau(x, y_1, \ldots, y_n) \to T_\tau(x) \wedge T_{\tau_1}(y_1) \wedge \cdots \wedge T_{\tau_n}(y_n))$, whenever $\tau = (\tau_1, \ldots, \tau_n)$. (Compare the earlier definition of $(\mathfrak{A}, E)^+$: especially the role of ϵ_τ^*.)

The sentences (i)–(v) are all rather trivial constraints on the type framework. The following *extensionality axioms* may be more interesting:
   (vi) $\forall x \forall y(T_\tau(x) \wedge T_\tau(y) \wedge \forall z_1 \ldots \forall z_n(T_{\tau_1}(z_1) \wedge \cdots \wedge T_{\tau_n}(z_n) \to$
       $\to (\epsilon_\tau(x, z_1, \ldots, z_n) \leftrightarrow \epsilon_\tau(y, z_1, \ldots, z_n))) \to x = y)$;
   whenever $\tau = (\tau_1, \ldots, \tau_n)$.

That this holds in $(\mathfrak{A}, E)^+$ when E is full, is due to the extensionality axiom of set theory. But it is also easily checked for general type structures.

This exhausts the obvious validities. Now, we can ask whether, conversely, every L^+ model of (i)–(vi) is of the form $(\mathfrak{A}, E)^+$, at least, up to isomorphism. (Otherwise, trivial counter-examples could be given.) The answer is positive, by an elementary argument. For any L^+ model $\mathfrak{B}$ of our six principles, we may construct a general model $(\mathfrak{A}, E)$ and an isomorphism $h: \mathfrak{B} \to (\mathfrak{A}, E)^+$ as follows.

Writing $h_\tau := h \upharpoonright T_\tau^{\mathfrak{B}}$, we shall construct h_τ and E_τ simultaneously by induction on the order of τ, relying heavily on (vi). (This construction is really a particular case of the Mostowski collapsing lemma in set theory.)

First, let $A = E_0 := T_0^{\mathfrak{B}}$, while h_0 is the identity of $T_0^{\mathfrak{B}}$. (i) says that $A \neq \emptyset$, and (ii) adds that we can define $\mathfrak{A}$ by taking over the interpretations that $\mathfrak{B}$ gave to the L symbols. Trivially then, h_0 preserves L structure. Next, suppose $\tau = (\tau_1, \ldots, \tau_n)$, where $E_{\tau_i}, h_{\tau_i} (1 \leq i \leq n)$ have been constructed already. Define h_τ on $T_\tau^{\mathfrak{B}}$ by setting

$$h_\tau(b) := \{(h_{\tau_1}(a_1), \ldots, h_{\tau_n}(a_n)) \mid \epsilon_\tau^{\mathfrak{B}}(b, a_1, \ldots, a_n)\}$$

(by (v), this stipulation makes sense); putting $E_\tau := h_\tau[T_\tau^{\mathfrak{B}}]$. Clearly, $E_\tau \subseteq \mathscr{P}(E_{\tau_1} \times \cdots \times E_{\tau_n})$. We are finished if it can be shown that h_τ is one–one, while $\epsilon_\tau^{\mathfrak{B}}(b, a_1, \ldots, a_n)$ iff $h_\tau(b)(h_{\tau_1}(a_1), \ldots, h_{\tau_n}(a_n))$. But, the first assertion is immediate from (vi), and it implies the second. Finally, put $h := \bigcup_{\tau \in \mathcal{T}} h_\tau$. (iii) is our licence to do this. That h is defined on all of $\mathfrak{B}$ is implied by (iv). □

The previous observations yield a conclusion:

LEMMA. *An L_ω sentence φ is true in all general models iff its translation φ^+ logically follows from* (i)–(vi) *above.*

Proof. The direction from right to left is immediate from the definition of the translation $^+$, and its semantic behaviour. From left to right, we use the above representation. □

The value of the Lemma is diminished by the fact that (iv) has an infinite disjunction, outside of L^+. But we can do better.

THEOREM. $\varphi \in L_\omega$ *is true in all general models iff φ^+ follows from* (i), (ii), (iii), (v) *and* (vi).

Proof. The first half is as before. Next, assume that φ is true in all general models, and consider any L^+ model $\mathfrak{B}$ satisfying the above five principles. Now, its submodel $\mathfrak{B}^*$ with universe $\bigcup_{\tau \in \mathcal{T}} T_\tau^{\mathfrak{B}}$ satisfies these principles as well, but in addition, it also verifies (iv). Thus, as before, $\mathfrak{B}^* \models \varphi^+$. But then, as all quantifiers in φ^+ occur restricted to the levels T_τ, $\mathfrak{B} \models \varphi^+$, and we are done after all. □

This theorem effectively reduces L_ω truth under the general model interpretation to first-order consequence from a recursive set of axioms: which shows it to be recursively enumerable and, hence, recursively axiomatizable (by Craig's Theorem). This strongly contrasts with the negative result in Section 2.3. We conclude with a few comments on the situation.

Henkin's original general models (defined, by the way, with respect to a richer language) form a proper subclass of ours. This is because one may strengthen the theorem a little (or much-depending on one's philosophy) by adding to (i)–(vi) translations of L_ω sentences obviously true in the *standard model* interpretation, thereby narrowing the class of admissible general models. Of course, Section 2.3 prevents an effective narrowing down to *exactly* the standard models!

Here are two examples of such additional axioms, bringing the general models interpretation closer to the standard one.

1. Comprehension Axioms for type $\tau = (\tau_1, \ldots, \tau_n)$:

$$\forall X_1 \ldots \forall X_m \exists Y \forall Z_1 \ldots \forall Z_n (Y(Z_1, \ldots, Z_n) \leftrightarrow \varphi),$$

where Y has type τ, Z_i type $\tau_i (1 \leq i \leq n)$ and the free variables of φ are among $X_1, \ldots, X_m, Z_1, \ldots, Z_n$. Thus, all definable predicates are to be actually present in the model.

2. Axioms of Choice for type $\tau = (\tau_1, \ldots, \tau_n, \tau_{n+1})$:

$$\forall Z_1 \exists Z_2 \forall X_1 \ldots \forall X_n (\exists Y Z_1(X_1, \ldots, X_n, Y) \rightarrow$$
$$\rightarrow \exists! Y Z_2(X_1, \ldots, X_n, Y));$$

where Z_1, Z_2 have type τ, X_i has $\tau_i (1 \leq i \leq n)$ and Y has type τ_{n+1}. Thus, every relation contains a function: cf. Bernays' Axiom of Choice mentioned in Section 2.5.1.

There is also a more 'deductive' motivation for these axioms. When one ponders which principles of deduction should enter into any reasonable higher-order logic, one immediate candidate is the ordinary complete first-order axiom set, with quantifiers now also of higher orders (cf. Enderton [1972] last chapter, for this line). All usual principles are valid in general models without further ado, except for Universal Instantiation, or equivalently, Existential Generalization:

$$\forall X \varphi(X) \rightarrow \varphi(T) \quad \text{or} \quad \varphi(T) \rightarrow \exists X \varphi(X).$$

These two axioms are valid in all general models when T is any variable or constant of the type of X. But, in actual practice, one wants to substitute further instances in higher-order reasoning. For example, from $\forall X \varphi(X, R)$, with X of type (0), one wants to conclude $\varphi(\psi)$ for any *first-order* definable property ψ in R, = (cf. van Benthem [Chapter II.4]). In terms of Comprehension, this amounts to closure of predicate ranges under first-order definability, mentioned in Section 4.1. A further possibility is to allow *predicative* substitutions, where ψ may be higher-order, but with its quantifiers all ranging over orders lower than that of X. Finally, no holds barred, there is the use of *arbitrary* substitutions, whether predicative or not; as in the above Comprehension Schema.

One consequence of Comprehension is the following Axiom of Descriptiveness:

$$\forall x \exists! y \varphi(x, y) \to \exists f \forall x \varphi(x, f(x)).$$

If we want to strengthen this to the useful existence of Skolem functions (cf. Section 2.5.2), we have to postulate

$$\forall x \exists y \varphi(x, y) \to \exists f \forall x \varphi(x, f(x));$$

and this motivates the above Axioms of Choice.

No further obvious logical desiderata seem to have been discovered in the literature.

By the way, our above formulation of the Axiom of Choice cannot be strengthened when all types are present, assuming the comprehension axioms. If this is not the case, it can be. For instance, in the second-order language, the strongest possible formulation is just the implication $\forall x \exists X \psi \to \exists Y \forall x \psi'$ (where ψ' is obtained from ψ by substituting $Y(x, t_1, \ldots, t_n)$ for $X(t_1, \ldots, t_n)$) used to prove the prenex theorem in Section 3.2.

In a sense, this form gives more than just choice; conceived of set-theoretically, it has the flavour of a 'collection' principle. It plays a crucial role in proving reflectivity of second-order theories containing it, similar to the role the substitution (or collection) axiom has in proving reflection principles in set theory.

The general picture emerging here is that of an ascending range of recursively axiomatized higher-order logics, formalizing most useful fragments of L_ω-validity that one encounters in practice.

4.3. Second-Order Reduction

The general completeness theorem, or rather, the family of theorems in Section 4.2, by no means exhausts the uses of the general model idea of Section 4.1. For instance, once upon this track, we may develop a 'general model theory' which is much closer to the first-order subject of that description. A case in point are the 'general ultraproducts' of van Benthem [1983], which allow for an extension of the fundamental characterization theorems of Section 1.4 to higher-order logic. This area remains largely unexplored.

Here we present a rather more unexpected application, announced in Section 3.2: L_ω-standard validity is effectively reducible to standard validity in monadic L_2, in fact in the monadic Σ_1^1 fragment.

Consider the *first-order* language L^+ (relative to a given base language L) introduced in Section 4.1. Extend it to a *second-order* language L_2^+ by adding second-order variables of all types $(0, \ldots, 0)$, with which we can form atoms $X(t_1, \ldots, t_n)$. Consider the following L_2^+-principles ($\tau = (\tau_1, \ldots, \tau_n)$):

PLENITUDE (τ)

$$\forall X \exists x \forall y_1 \ldots \forall y_n (T_\tau(x) \wedge (T_{\tau_1}(y_1) \wedge \cdots \wedge T_{\tau_n}(y_n) \rightarrow$$
$$\rightarrow (\epsilon_\tau(x, y_1, \ldots, y_n) \leftrightarrow X(y_1, \ldots, y_n)))).$$

Evidently, Plenitude holds in all $^+$ transforms of full standard models of L_ω. Conversely, if the L^+ model $\mathfrak{B}$ satisfies Plenitude (τ) for all types τ, then its submodel $\mathfrak{B}^*$ (cf. the proof of the main theorem in Section 4.2) is isomorphic to a model of the form $(\mathfrak{A}, E)^+$ with a full type structure E.

THEOREM. $\varphi \in L_\omega$ *is true in all standard models iff* φ^+ *follows from* (i), (ii), (iii), (v), (vi) *and the Plenitude axioms.*

As φ can only mention a finite number of types and non-logical constants, the relevant axioms of the above-mentioned kinds can be reduced to a finite number and hence a single sentence ψ.

THEOREM. *With every* $\varphi \in L_\omega$ *a* Π_1^1 *sentence* ψ *of* L_2^+ *can be associated effectively, and uniformly, such that*

$$\vDash_\omega \varphi \quad \text{iff} \quad \vDash_2 \psi \rightarrow \varphi^+.$$

As $\psi \in \Pi_1^1$ and φ^+ is first-order, this implication is equivalent to a Σ_1^1 sentence; and the promised reduction is there.

But, Plenitude has been formulated using second-order variables of an arbitrary type. We finally indicate how this may be improved to the case of only monadic ones. Consider the variant

PLENITUDE*(τ)

$$\forall X \exists x \forall y_1 \ldots \forall y_n (T_\tau(x) \wedge (T_{\tau_1}(y_1) \wedge \cdots \wedge T_{\tau_n}(y_n) \rightarrow$$
$$\rightarrow (\epsilon_\tau(x, y_1, \ldots, y_n) \leftrightarrow \exists y (T_\tau(y) \wedge X(y) \wedge \epsilon_\tau(y, y_1, \ldots, y_n)))).$$

When E is full, this will obviously hold in $(\mathfrak{A}, E)^+$. To make this monadic variant do its job, it has to be helped by the following first-order principle stating the existence of singleton sets of ordered sequences:

SINGLETONS (τ)

$$\forall z_1 \ldots \forall z_n \exists x \forall y_1 \ldots \forall y_n (T_{\tau_1}(z_1) \wedge \cdots \wedge T_{\tau_n}(z_n) \rightarrow (T_\tau(x) \wedge$$
$$\wedge (T_{\tau_1}(y_1) \wedge \cdots \wedge T_{\tau_n}(y_n) \rightarrow (\epsilon_\tau(x, y_1, \ldots, y_n) \leftrightarrow$$
$$\leftrightarrow y_1 = z_1 \wedge \cdots \wedge y_n = z_n)))).$$

Suppose now that $\mathfrak{B}$ satisfies all these axioms and $(\mathfrak{A}, E)^+ \cong \mathfrak{B}^*$. Let $S \subseteq E_{\tau_1} \times \cdots \times E_{\tau_n}$ be arbitrary: we must show that $S \in E_\tau$. Notice that Singletons(τ) implies that, if $s \in E_{\tau_1} \times \cdots \times E_{\tau_n}$ (in particular, if $s \in S$), then $\{s\} \in E_\tau$. Now let $S' := \{\{s\} \mid s \in S\}$. Clearly, $S = \cup S'$ and $S' \subseteq E_\tau$. That $S \in E_\tau$ follows from one application of Plenitude*(τ), taking S' as value for X. □

4.4. Type Theory and Lambda Calculus

Readers of Section 4.2 may have been a little disappointed at finding no preferred *explicit* axiomatized 'first-order' version of L_ω logic. And indeed, an extreme latitude of choices was of the essence of the situation. Indeed, there exist various additional points of view leading to, at least, interesting logics. One of these is provided by the earlier functional type theory of Section 3.3. We will chart the natural road from the perspective of its basic primitives.

Identity and *application* inspire the usual identity axioms, including replacement of identicals. *Lambda abstraction* really contributes only one further principle, viz. the famous 'lambda conversion'

$$\lambda x \cdot B(A) = [A/x]B;$$

for x, B, A of suitable types, and modulo obvious conditions of freedom and bondage. Thus, there arises a simple kind of *lambda calculus*. (Actually, a rule of 'alphabetic bound variants' will have to be added in any case, for domestic purposes.)

I.4: HIGHER-ORDER LOGIC

Lambda conversion is really a kind of simplification rule, often encountered in the semantics of natural, or programming languages. One immediate question then is if this process of simplification ever stops.

THEOREM. *Every lambda reduction sequence stops in a finite number of steps.*

Proof. Introduce a suitable measure of type complexity on terms, so that each reduction lowers complexity. □

This theorem does not hold for the more general *type free* lambda calculi of Barendregt [1980]; where e.g. $\lambda x \, . \, x(x)(\lambda x \, . \, x(x))$ runs into an infinite regress.

Another immediate follow-up question concerns the *unicity* (in addition to the above *existence*) of such irreducible 'normal forms'. This follows in fact from the 'diamond property':

THEOREM (Church–Rosser). *Every two lambda reduction sequences starting from the same term can be continued to meet in a common term (up to alphabetic variance).*

Stronger lambda calculi arise upon the addition of further principles, such as *extensionality*:

$$\lambda x \, . \, A(x) = \lambda x \, . \, B(x) \text{ implies } A = B \quad (\text{for } x \text{ not free in } A, B).$$

This is the lambda analogon of the earlier principle (vi) in Section 4.2.

Still further additions might be made reflecting the constancy of the truth value domain D_t. Up till now, all principles considered would also be valid for arbitrary truth value structures. (In some cases, this will be a virtue, of course.)

Let us now turn to traditional logic. Henkin has observed how all familiar logical constants may be *defined* (under the standard interpretation) in terms of the previous notions. Here is the relevant list (Henkin [1963]):

$$\mathsf{T} \text{ (a tautology)} := \lambda x \, . \, x = \lambda x \, . \, x$$
$$\bot \text{ (a contradiction)} := \lambda x_t . x_t = \lambda x_t . \mathsf{T}$$
$$\neg \text{ (negation)} := \lambda x_t . x_t = \bot$$

The most tricky case is that of conjunction:

$$\wedge := \lambda x_t . \lambda y_t (\lambda f_{(t,t)} . (f_{(t,t)}(x_t) = y_t) =$$
$$= \lambda f_{(t,t)} \, . \, f_{(t,t)}\mathsf{T}))$$

One may then define $\vee$, $\rightarrow$ in various ways. Finally, as for the quantifiers,

$$\forall xA := \lambda x.A = \lambda x.\mathsf{T}.$$

The induced logic has not been determined yet, as far as we know.

With the addition of the axiom of *bivalence*, we are on the road to classical logic:

$$\forall x_t . f_{(t,\,t)} x_t = f_{(t,\,t)}\mathsf{T} \wedge f_{(t,\,t)}\bot.$$

For a fuller account, cf. Gallin [1975] Chapter 1.2.

One may prove a general completeness theorem for the above identity, application, abstraction theory in a not inelegant direct manner, along the lines of Henkin's original completeness proof. (Notably, the familiar 'witnesses' would now be needed in order to provide instances $f(c) \ne g(c)$ when $f \ne g$.) But, the additional technicalities, especially in setting up the correct account of general models for functional-type theory, have motivated exclusion here.

Even so, the differences between the more 'logical' climate of functional-type theory and the more 'set-theoretic' atmosphere of the higher-order L_ω will have become clear.

5. REFLECTIONS

Why should a Handbook of (after all) Philosophical Logic contain a chapter on extensions of first-order logic; in particular, on higher-order logic? There are some very general, but also some more specific answers to this (by now) rather rhetorical question.

One general reason is that the advent of competitors for first-order logic may relativize the intense preoccupation with the latter theory in philosophical circles. No specific theory is sacrosanct in contemporary logic. It is rather a certain logical perspective in setting up theories, weaker or stronger as the needs of some specific application require, that should be cultivated. Of course, this point is equally valid for *alternatives* to, rather than *extensions* of classical first-order logic (such as intuitionistic logic).

More specifically, two themes in Section 1 seem of a wider philosophical interest: the role of limitative results such as the Löwenheim–Skolem, or the Compactness Theorem for scientific theory construction; but also the new systematic perspective upon the nature of logical constants (witness the remarks made about generalized quantifiers). Some authors have even claimed that proper applications of logic, e.g. in the philosophy of science or of language, can only get off the ground now that we have this amazing diversity of logics, allowing for conceptual 'fine tuning' in our formal analyses.

As for the specific case study of higher-order logic, there was at least a convincing prima facie case for this theory, both from the (logicist) foundations of mathematics and the formal semantics of natural language. Especially in the latter area, there have been recurrent disputes about clues from natural language urging higher-order descriptions. (The discussion of branching quantifiers in Section 2.5.1 has been an example; but many others could be cited.) This subject is rather delicate, however, having to do with philosophy as much as with linguistics. (Cf. van Benthem [1981] for a discussion of some issues.) For instance, the choice between a standard model or a general model approach to higher-order quantification is semantically highly significant and will hopefully undercut at present rather dogmatic discussions of the issue. For instance, even on a Montagovian type theoretic semantics, we are not committed to a non-axiomatizable logic, or models of wild cardinalities: contrary to what is usually claimed. (General models on a countable universe may well remain countable throughout, no matter how far the full type structure explodes.)

One might even hazard the conjecture that natural language is partial to restricted predicate ranges which are *constructive* in some sense. For instance, Hintikka [1973] contains the suggestion to read branching quantifier statements on countable domains in terms of the existence of Skolem functions which are recursive in the base predicates. If so, our story might end quite differently: for, the higher-order logic of constructive general models might well lapse into non-axiomatizability again. Thus, our chapter is an open-ended one, as far as the philosophy and semantics of language are concerned. It suggests possibilities for semantic description; but on the other hand, this new area of application may well inspire new directions in logical research.

University of Groningen
University of Amsterdam

REFERENCES

Ackermann, W.: 1968, *Solvable Cases of the Decision Problem*, North-Holland, Amsterdam.
Ajtai, M.: 1979, 'Isomorphism and higher-order equivalence', *Annals of Math. Logic* 16, 181–203.
Barendregt, H.: 1980, *The Lambda Calculus*, North-Holland, Amsterdam.

Barwise, K. J.: 1972, 'The Hanf-number of second-order logic'. *J. Symbolic Logic* **37**, 588-594.
Barwise, K. J.: 1975, *Admissible Sets and Structures*, Springer, Berlin.
Barwise, K. J. (ed.): 1977, *Handbook of Mathematical Logic*, North-Holland, Amsterdam.
Barwise, K. J.: 1979, 'On branching quantifiers in English', *J. Philos. Logic* **8**, 47-80.
Barwise, K. J. and Cooper, R.: 1981, 'Generalized quantifiers and natural language', *Linguistics and Philosophy* **4**, 159-219.
Barwise, K. J. and Schlipf, J.: 1976, 'An introduction to recursively saturated and resplendent models', *J. Symbolic Logic* **41**, 531-536.
Barwise, K. J., Kaufman, M. and Makkai, M.: 1978, 'Stationary logic', *Annals of Math. Logic* **13**, 171-224. A correction appeared in *Annals of Math. Logic* **16**, 231-232.
Bell, J. L. and Slomson, A. B.: 1969, *Models and Ultraproducts*, North-Holland, Amsterdam.
Chang, C. C. and Keisler, H. J.: 1973, *Model Theory*, North-Holland, Amsterdam.
Church, A.: 1940, 'A formulation of the simple theory of types', *J. Symbolic Logic* **5**, 56-68.
Copi, I. M.: 1971, *The Logical Theory of Types*, Routledge & Kegan Paul, London.
Drake, F. R.: 1974, *Set Theory. An Introduction to Large Cardinals*, North-Holland, Amsterdam.
Enderton, H. B.: 1970, 'Finite partially-ordered quantifiers', *Zeitschrift für mathematische Logik und Grundlagen der Mathematik* **16**, 393-397.
Enderton, H. B.: 1972, *A Mathematical Introduction to Logic*, Academic Press, New York.
Gallin, D.: 1975, *Intensional and Higher-Order Modal Logic*, North-Holland, Amsterdam.
Garland, S. J.: 1974, 'Second-order cardinal characterizability', in *Proceedings of Symposia in Pure Mathematics*, A.M.S., vol. 13, part II, 127-146.
Henkin, L. A.: 1950, 'Completeness in the theory of types', *J. Symbolic Logic* **15**, 81-91.
Henkin, L. A.: 1961, 'Some remarks on infinitely long formulas', in *Infinitistic Methods. Proceedings of a Symposium on the Foundations of Mathematics*, Pergamon Press, London, pp. 167-183.
Henkin, L. A.: 1963, 'A theory of propositional types', *Fundamenta Math.* **52**, 323-344.
Hintikka, K. J. J.: 1955, 'Reductions in the theory of types', *Acta Philosophica Fennica* **8**, 61-115.
Hintikka, K. J. J.: 1973, 'Quantifiers versus quantification theory', *Dialectica* **27**, 329-358.
Hodges, W.: 1983, Elementary predicate logic, Chapter I.1, this Handbook.
Keisler, H. J.: 1971, *Model Theory for Infinitary Logic*, North-Holland, Amsterdam.
Kemeny, J.: 1950, 'Type theory vs. set theory', *J. Symbolic Logic* **15**, 78.
Kleene, S. C.: 1952, 'Finite axiomatizability of theories in the predicate calculus using additional predicate symbols', in *Two Papers on the Predicate Calculus*, Memoirs of the Amer. Math. Soc. vol. 10, pp. 27-68.
Kunen, K.: 1971, 'Indescribability and the continuum', in *Proceedings of Symposia in Pure Mathematics*, A.M.S., vol. 13, part I, 199-204.
Lindström, P.: 1969, 'On extensions of elementary logic', *Theoria* **35**, 1-11.

Magidor, M.: 1971, 'On the role of supercompact and extendible cardinals in logic', *Israel J. Math.* **10**, 147–157.

Magidor, M. and Malitz, J.: 1977, 'Compact extensions of L_Q', *Annals of Math. Logic* **11**, 217–261.

Monk, J. D.: 1976, *Mathematical Logic*, Springer, Berlin.

Montague, R.: 1974, in R. H. Thomason (ed.), *Formal Philosophy: Selected Papers of Richard Montague*, Yale University Press, New Haven.

Moschovakis, Y. N.: 1980, *Descriptive Set Theory*, North-Holland, Amsterdam.

Myhill, J. and Scott, D. S.: 1971, 'Ordinal definability', in *Proceedings in Symposia in Pure Mathematics*, A.M.S., vol. 13, part I, 271–278.

Orey, S.: 1959, 'Model theory for the higher-order predicate calculus', *Transactions of the American Mathematical Society* **92**, 72–84.

Rabin, M. O.: 1969, 'Decidability of second-order theories and automata on infinite trees', *Transactions of the American Mathematical Society* **141**, 1–35.

Ressayre, J. P.: 1977, 'Models with compactness properties relative to an admissible language', *Annals of Math. Logic* **11**, 31–55.

Svenonius, L.: 1965, 'On the denumerable models of theories with extra predicates', in *The Theory of Models*, North-Holland, Amsterdam, pp. 376–389.

Väänänen, J.: 1982, 'Abstract logic and set theory: II Large cardinals', *J. Symbolic Logic* **47**, 335–346.

Van Benthem, J. F. A. K.: 1977, 'Modal logic as second-order logic', report 77–04, Mathematisch Instituut, University of Amsterdam.

Van Benthem, J. F. A. K.: 1982, Questions about Quantifiers, to appear in *J. Symbolic Logic*.

Van Benthem, J. F. A. K.: 1983, *Modal Logic and Classical Logic*, Bibliopolis, Naples.

CHAPTER I.5

PREDICATIVE LOGICS
by ALLEN HAZEN

Preface   331
1. Historical background   332
2. Some formal systems   338
3. Substitutional quantification   351
4. Consistency and other examples of deductive weakness   355
5. Reducibility and heterologicality   365
6. Predicative mathematics   375
7. Tarski and such   382
8. Concluding ruminations   391
Notes   396
References   403

PREFACE

The best introduction available to the formalism of ramified type theory (perhaps the best introduction available to any topic in the whole of logic) is provided by Church [1976] together with the relevant sections of Church [1956], after which the reader should go directly to the classic Russell [1908]. (Readers particularly interested in predicative mathematics and its metamathematics should probably start with Feferman [1964].) In what follows I have attempted an overview of ramified type theory and topics connected to it, giving references which I hope will be useful to someone becoming interested in the field, or particular aspects of it, for the first time.

Salving my conscience with the thought that Church's exposition is better than anything I could produce, I have made my account of formal matters concise rather than precise. My emphasis has been on general philosophical issues concerned with the interpretation and application of ramified theory. This has led me, particularly in the last two sections, to make speculative suggestions about extensions of ramified type theory; some of these I hope to pursue in my own future research.

The hope is that this exposition will be useful to first year graduate students and undergraduate philosophy majors. Despite my efforts to keep them under control, I am aware that a fair bit of jargon and some horrendously complicated syntax have crept in, but the actual amount of technical knowledge of logic presupposed is moderate. A familiarity with the notation of predicate logic is a must. Section D presupposes a knowledge of the key notions of Gentzen [1934]; I have attempted to make my formulations in that section neutral as between Gentzen's N and L systems and a variety of other natural deduction formulations. An acquaintance with the rudiments of the theory of ordinal numbers (but no more of an acquaintance than could be gained from a standard introduction to set theory, or from Quine [1963]) is presupposed in Sections 6 and 7, and in sporadic passages elsewhere. Finally, Section 7 is addressed to readers already acquainted with Tarski [1936] (or one of the multitudinous secondary expositions of Tarski's theory) and Kripke [1975] (for these cf. also Visser [Chapter IV.14]).

I am indebted, for questions answered, references to the literature provided, and philosophical discussions which have at the very least made me more aware of some of the areas of my own ignorance, to an immense number of people. Those I remember include Anil Gupta, Saul Kripke, Charles Parsons, Wilfried Sieg, and Hao Wang. (They are, of course, not responsible for, and in general do not yet know, how badly I have used their advice.) Solomon Feferman, Charles Parsons, and Richmond Thomason have shown me typescripts of unpublished papers. Edith Hazen has assisted in the preparation of the bibliography and the preparation of the manuscript. Lots of people, listed above and not, have given encouragement and moral support. The editors of the *Handbook of Philosophical Logic* have given me time and more time. I am in all their debt.

1. HISTORICAL BACKGROUND

In one sense all of the objects of mathematical investigation are abstract entities. In another they come in different grades of abstractness. Some, like the natural numbers or the straightedge and compass constructions of elementary geometry, have standard concrete representations: numerals and diagrams. They are, therefore, intuitive in at least the sense that someone familiar with the system of representation can determine the mathematically important properties of a particular number or construction by examining its numeral or diagram. The most elementary parts of mathematics refer to and generalize about or, when formalized in symbolic logic, quantify over, only such intuitive objects. From the beginning, however, there has been a tendency to go beyond the intuitive, to postulate the existence of and generalize about more abstract objects. Thus, even in elementary geometry, arbitrary lines and figures are considered without regard to their possibility of construction, and the first lesson in secondary school geometry is that the diagram must not be trusted as a guide to the properties of the object. In post-medieval mathematics, the abstract trend has accelerated. Thus, among the functions considered by modern analysis, only an exceptional minority can be given by equations or represented by curves on a graph.

Abstraction leads to mathematical power, but also to the risk of meaninglessness and contradiction. The infinitesimals of 18th-century analysis are a typical example: with their aid, classical mathematical physics was developed, but their definitions were too contradictory to define anything, and, whatever reconstructions our logically more self-conscious age has managed to provide, it was far from clear to contemporaries what consistent *principle* would permit the separation of the useful from the absurd consequences of their postulation. The 19th century reacted against them, and attempted to give a rigorous foundation for analysis. The new foundation, however, again postulated and depended on generalizations about abstract objects as far removed from any sort of geometrical or physical intuition as the supposed infinitesimals had been. *Real numbers* were constructed as limits of arbitrary converging sequences of rationals (or, in Dedekind's version, of arbitrary bounded sets of rationals), without regard to how these sequences (sets) were to be defined, and *functions* as arbitrary sets of ordered pairs of these real numbers (subject to the sole condition that every real be a first element of a pair in the set, and be paired only with one second element), again without regard to how these sets were to be defined. When pursued beyond the immediate needs of the foundations of analysis, the principles postulated for these sets allowed the definition of an infinite hierarchy of ever increasing infinite 'numbers'. By the end of the century another reaction had set in, with one school of mathematicians, led by Poincaré, rejecting the set theoretic foundations as resting on unjustified assumptions. They were delighted when, at around the turn of the century, the fundamental principles of the new theory, implicit in the work of others and formulated to new standards of logical explicitness by Frege, were discovered to lead to explicit contradictions.

Bertrand Russell, who had discovered the simplest of these contradictions, had been impressed by the power and elegance of the new methods. They seemed to allow all of mathematics to be reduced to – its concepts to be defined in terms of, and its theorems to be derived from the principles governing – a small number of basic concepts which he, like Frege, saw as logical in nature. This was the *logicist* program of Russell [1903]. (For a spirited recent summary, cf. Coffa [1982].) The epistemological and ontological problems of the philosophy of mathematics were to be solved at a single stroke, by defining all mathematical notions in terms of logical ones and defining all mathematical theorems from the analytic truths of logic. No special mathematical objects were to be postulated, and no sort of mathematical intuition recognized as a source of synthetic *a priori* knowledge in mathematics. (This is in contrast to Poincaré and the later intuitionists, who recognized the principle of mathematical induction as a primitive given, irreducible to logic or anything else.) He thus spent the first few years of the new century seeking a system which would be better founded than the old, inconsistent, 'logic', but would allow much the same mathematical development. Since the techniques of set theory were essential to this development, the central part of the task was the discovery of something like a set theory whose principles could be seen to be logically evident: it would not be enough to present a set of postulates formally analogous enough to the old to allow the reconstruction of the desired set theoretic proofs without allowing the known derivations of contradictions. In order to fulfill the requirements of the logicist program, the fundamental concepts of the new theory would have to be clearly logical in nature, and the principles governing them clearly logical truths. In practice, in order to establish this, and to give confidence that no new contradictions would be forthcoming, what was needed was an *analysis* of the contradictions, showing how they had resulted from overlooking some subtlety in some fundamental logical concept, and how the new theory embodied a recognition of that subtlety.

At the same time he was conducting a debate with Poincaré, each in turn criticizing the other's views of the foundations of mathematics and responding to the other's criticisms of his own.[1] In the course of the debate, each man seems to have creatively misunderstood the other. Perhaps the key idea arising out of the debate was the notion that the contradictions were attributable to a kind of circularity of definition. This, however, was not the ordinary kind of circularity, in which a *word* is defined, perhaps through a chain of intermediate definitions, in terms of itself. Rather, a typical example would be a definition of a set of, say, natural numbers in which the *definiens* contained quantified variables ranging over all sets of natural numbers – a definition of a set or property, to generalize, in which the set defined is itself supposed to be among the values of the quantified variables appearing in the *definiens*. Now, as has been pointed out repeatedly by adherents of Platonistic and Pragmatic philosophies of mathematics (the former really believing in the existence of the abstract entities of set theory and the latter willing to accept any consistent set of postulates that allows the easy derivation of physically applicable theorems), there is no circularity or other flaw in such a definition if the sets are thought of as ontologically independent of their definitions. If, however, sets could somehow be conceived as owing their existence to their definitions, such a definition would in a very clear sense *presuppose* the thing defined. In order for the *definiens* to have a definite meaning, the quantified variables occurring in it must have definite ranges. In the sort of definition under consideration, therefore, the propriety of the definition will depend on the existence, and presence in the range of the bound variables of the *definiens*, of the very set to be defined, and the set, supposing sets to be ontologically dependent on their definitions, will only exist if the *definiens* has a definite meaning. The definition and the set defined depend on each other, and there seems to be no principled way of deciding whether either is legitimate. In particular cases, however, the circularity shows its viciousness by producing a contradiction. Thus consider Richard's paradox (Richard [1905], discussed in Poincaré [1906]). Suppose that sets owe their existence to their definitions, and that every set of natural numbers[2] can be defined in some one language which can be written in a finite alphabet. Then, since all the possible definitions can be alphabetized,[3] sets of natural numbers can be thought of as being arranged in a linear order isomorphic to the order of the natural numbers themselves. If, therefore, definitions of sets using quantified variables including the defined sets in their ranges were legitimate, we might define a set as containing the natural number N if and only if the Nth set of natural numbers in alphabetical order did *not* contain N. Since this set – call it the *Richard set* – differs in respect of containing or not containing N from the Nth set in the sequence for every natural number N, it cannot itself be one of the sets in the sequence. But this contradicts our initial supposition that the sequence resulted from alphabetizing *all* the sets of natural numbers.[4]

By Russell's own account, the first major breakthrough in his efforts to 'solve' the paradoxes – to analyze them and discover a consistent and well-founded theory – came with the discovery of the theory of descriptions (Russell [1905]). The avowal sounds odd to the modern reader. The contextual definition of definite descriptions, neat trick though it may be, hardly seems like much of a step in the direction of the formulation of the theory of types. What it did for Russell (and for those later philosophical analysts who took it as "a paradigm of philosophy") was to set a precedent, freeing him from the naive and ontologically extravagant semantics he had earlier assumed. If definite descriptions didn't have to stand for real entities in order for the sentences in which they occurred to be meaningful, perhaps other sorts of expressions didn't either . . . . The way was open to the elimination, by contextual definition, of whole categories of entities in which he had previously believed. The technique is perhaps most simply illustrated with a non-Russellian example. Suppose Russell had believed in natural numbers, but found fractions somehow dubious. Then he might have been attracted by a contextual definition of 'fraction talk' which replaced each quantifier binding a variable ranging over fractions with *two* quantifiers over natural numbers, and analyzed each *n*-adic predicate of fractions in terms of a 2*n*-adic relation of natural numbers. (Replacing a relation holding of *n* fractions, in effect, with the equivalent relation holding of their numerators and denominators.) In fact, the category Russell chose to eliminate was that of *sets*, or, as he called them, *classes*. The details varied over the years in which he was wrestling with the problem, but his own name for the approach that finally succeeded was the *no-class theory*, and the common feature of the various versions he experimented with was that they eliminated references to and quantifications over sets by contextual definition in terms of some other category, or categories, of entities.

In the version he ultimately adopted (in Russell [1908]; this formed the basis for *Principia Mathematica*), the reducing entities were what he called *propositional functions*, and what might also be called *concepts* or *properties*.[5] Again, the modern reader may find puzzling the stress Russell lays on what is, after all, merely a reduction of one category of abstract entity to another. For Russell, however, there was an essential difference between sets on the one hand and properties on the other. The notion of a set or a class was, for him, that of an *extensional* entity, owing its existence, as it were, only to its members, and being determined uniquely by them. (Bernays [1935] speaks of this notion of a set as *combinatorial*, the idea being that the notion of an arbitrary set of, say, numbers, is a generalization to the infinite case of the combinatorial notion of an arbitrary combination of m out of n objects.) On such a conception, sets would exist independently of their being defined, and there would be no reason not to admit the sorts of circular, or *impredicative*, definitions considered. Concepts or properties, on the other hand, are plausibly thought of as existing only in so far as they have definitions. A concept, after all, exists only if it is possible for someone to grasp it, and we may (subject to some hedges that would have occurred to Russell, and more that will occur to readers of Putnam [1975]) think of someone who grasps a concept as doing something like formulating, in his own mind, its definition. When the basic entities underlying set theory are conceived in this conceptualistic manner, as existing in virtue of their definability, the restriction to *predicative* definitions becomes plausible and, indeed, inevitable.

The theory of Russell [1908], on the basis of which the developments of *Principia Mathematica* were constructed, met at least some of the requirements. It analyzed the paradoxes in two stages. The first stage transformed the explicitly set theoretic paradoxes into a form in which predicativity considerations would be relevant to them, by explaining apparent references to and quantifications over sets as *façons de parler*, so that only a theory of properties remained. The fundamental notions of the remaining theory, those of a proposition and of a concept or propositional function, were part of the traditional subject matter of logical theory, and in the second stage they were subjected to a more penetrating analysis than Frege, or the Russell of Russell [1903], had given them. Frege had considered what sorts of properties concepts would have to have in order for them to serve as meanings for the expressions of a language for which classical predicate logic is appropriate. With his hierarchy of *Stufen*, he in effect postulated a simple theory of types for concepts. (Ironically, the simple theory of types, which is closely related to the standard modern systems of axiomatic set theory stemming from Zermelo, appears to avoid the set theoretic paradoxes, so that he would have been safe had he, like Russell, explained set theory as a *façon de parler* for talking about concepts, instead of postulating sets, or rather *Werthverläufe*, as a separate category of untyped entities. This would still, however, have left him exposed to the other paradoxes that Ramsey [1926] called *epistemological*.) On the other hand, he neglected to consider the properties concepts needed in order to be graspable. By considering them, Russell was led to the ramified theory of types, which avoided *all* the paradoxes: not just the set theoretic ones, which indeed did not appear when set theoretic notions were contextually defined, but also the paradoxes of definability.

By analyzing the paradoxes, and presenting a paradox-free system justified by his analysis, Russell won the battle for logicism. The war, however, was lost, for it has become increasingly clear that the content of significant mathematical theories can only be captured with the aid of additional assumptions beyond the truths of (ramified type theoretic) logic. Poincaré was right after all, and mathematical induction is a primitive given, not reducible to logic.

2. SOME FORMAL SYSTEMS

In this section I will sketch a variety of formal systems inspired by the predicative conception of property. These systems are related to, though somewhat simpler than, the imperfectly formalized system of *Principia Mathematica*, and in the subsequent literature metamathematical results about precisely defined systems like those to be described have been taken as applying to the system of *Principia Mathematica*: they are, in other words, accepted as formalizations of the same intuitive ideas as those behind *Principia Mathematica*. Their notation, if all the type and level distinctions are marked by explicit subscripts and superscripts, can appear quite daunting, but their actual complexity is comparable to (and formalized proofs in them at least as perspicuous as proofs in) the simpler systems of axiomatic set theory. In practice, of course, everyone from Russell on who has done any extensive work with a formalized system of (simple or ramified) type theory has, with or without explicitly stating abbreviational conventions, abbreviated or left implicit many type specifications.

The systems are all many-sorted predicate calculi, with one sort of variable ranging over individuals and other sorts ranging over properties of various types. Thus a basic part of the axiomatization in each case, which I shall not describe in any detail, is that the full classical logic of connectives and quantifiers is assumed in each of them, with the quantificational principles holding with each sort of variable. Corresponding intuitionistic systems can be defined by adopting instead the intuitionistic propositional and quantificational logics; whereas there is considerable doubt that the intuitionistic analogue of the simple theory of types, the "full theory of species", has a genuine justification from the intuitionistic standpoint, it is generally agreed that intuitionistic ramified systems are intuitionistically acceptable.[6]

In order to start with the simplest predicative system, I shall first describe two second-order logics.[7] A third system has variables of all finite orders; this is the system most directly comparable to that of *Principia Mathematica*. Finally, some simplifications of this system will be considered.

Start by picking a first-order theory, or at least a first-order language. Our first system, or more accurately, our first schema for forming systems, will extend it by quantifying over those properties of its individuals that are *first-order definable*.[8] We add to the language for each natural number n, n-adic predicate variables. (The 0-adic ones, of course, are more familiarly known as *propositional* variables: let us extend to syntactic terminology the practice of letting one word do duty for propositions, monadic properties, and relations.) In axiomatizing the expanded system, we must, in addition to extending the principles of pure quantificational logic to formulas containing the new variables, include something to show that the properties quantified over include those expressed by first-order formulas of the language. Conceptually the simplest of several equivalent ways of doing this is to rely on the quantificational principles that an existential quantification is implied by its instances and a universal implies its, and to count first-order formulas themselves (including those containing free variables) as forming the substitution classes of the predicate variables.[9]

For quantifications in which the quantified variable is *propositional*, this is fairly straightforward: the instances of such a quantification are precisely the formulas obtained from it by lopping off the quantifier and replacing in the matrix each occurrence of the propositional variable bound to it with an occurrence of some one first-order formula. With monadic and relational predicate variables, where in general different occurrences of the bound variable in the matrix of the quantification may take different individual variables or constants as arguments, the precise statement of the substitution rules becomes more complicated. (Of course, though giving a precise statement is difficult, anyone capable of following an exposition of first-order logic that uses schematic letters should be capable of using these rules in practice, for they are what you get if you simply treat the usual schematic letters as quantifiable variables.) Thus the interest in more simply describable equivalents.

One such expedient is the use of an axiom scheme of comprehension. Take as axioms all formulas (or, if your favorite version of quantificational logic allows only sentences and not open formulas as theorems, all universal closures of formulas) of the form

(Comp) $\quad \exists F \forall x_1 \ldots \forall x_n ((Fx_1 \ldots x_n \to A) \land (A \to Fx_1 \ldots x_n))$

where F is an n-adic predicate variable, $x_1, \ldots, x_n$ are distinct individual variables, and A is a first-order formula not containing F. The resulting system is equivalent to the one with the complicated definition of the instances of a second-order quantification. Note, however, that, strictly speaking, axioms of the form Comp only assert the existence of properties holding of the same individuals as satisfy the formula A, and not of the property *expressed* by the formula. This distinction would become important in an extension of the system whose language contained *intensional* contexts, and there the comprehension axioms would have to be supplemented or, better, reformulated. (As an example of the sort of thing that would be called for, note that schema (3) on p. 752 of Church [1976] has much the character of a comprehension schema.)

A third alternative is to rely, as in the first, on the principles that universal quantifications imply and existentials are implied by their instances, but to use the machinery of *abstraction* to simplify the definition of an instance of a second-order quantification. Where $x_1, \ldots, x_n$ are n distinct individual variables and A a formula,

$$\{x_1 \ldots x_n : A\}$$

is an *n-adic abstract* and may be treated as an *n*-adic predicate in constructing further formulas; if A is a first-order formula, it is an *n*-adic abstract of level 1. (Note that a 0-adic, or propositional, abstract is simply the result of writing curly brackets around a formula, and that, just as *n*-adic abstracts for n greater than 0 count as *predicates* grammatically, propositional abstracts are grammatically formulas.) We assume rules or axioms for abstraction providing that, where $a_1, \ldots, a_n$ are arbitrary individual terms (constants or variables, and not necessarily distinct) and $A(a_1/x_1) \ldots (a_n/x_n)$ is obtained from A by simultaneously replacing each free occurrence of x_i with one of a_i, $1 \leq i \leq n$, relettering bound variables where necessary, the atomic formula

$$\{x_1 \ldots x_n : A\} a_1 \ldots a_n$$

is equivalent to $A(a_1/x_1) \ldots (a_n/x_n)$. (Note that in the special case of 0-adic abstracts this simply amounts to the provision that a formula with an outer pair of curly brackets is equivalent to the formula without it.)

By themselves abstracts and the rules governing them are logically trivial, and a first-order system containing them is an inessential extension of the corresponding system without them, for, if not taken to be in the substitution classes of quantifiable variables, the abstracts could be taken as *virtual* in the sense of Quine [1963]. We obtain another formulation of the system quantifying over first-order definable properties by taking as the allowable substituents for a quantified *n*-adic predicate variable, in the definition of the *instances* of a second order quantification, *n*-adic predicates (including free predicate variables as well as the predicates of the underlying first-order language) and *n*-adic abstracts of level 1.

Systems of this sort are both more familiar, and richer in their expressive power, than they may at first seem. On the score of familiarity, note first that, under certain circumstances, simplifications are possible. If the domain of individuals is closed under the formation of finite sequences, so that ordered pairs, triples, etc., of individuals are themselves taken as individuals, and the notions of sequence and *n*th item in a sequence are expressible in the first-order language, monadic properties can stand in for relations, and all but the monadic predicate variables can be dropped from the formal language. Next, in the absence of intensional contexts in our formal language, we can, harmlessly, pretend to accept an extensional criterion of identity for properties: we may write

$$[F = G]$$

as an abbreviation for

$$\forall x((Fx \to Gx) \wedge (Gx \to Fx));$$

since the predicate expressions of our formal language (predicates of the underlying first-order language, predicate variables, and abstracts) occur in atomic formulas *only* as predicates, and since the only modes of construction for non-atomic formulas are those involving the extensional operators of classical logic, this definition licenses an unrestricted rule of substitution of identicals. Finally, a change of notation, on which nothing of any logical importance hangs: we change the order of subject and predicate in atomic formulas (putting the subject *first*) and put some letter of the Greek alphabet (different authors have chosen different letters of that alphabet, with the fifth a particularly popular choice), and – *voilà*: a theory of classes! Now choose as one's domain of individuals the *sets*, as conceived in Zermelo–Fraenkel set theory, and take that set theory itself as the underlying first-order system.[10] Clearly the conditions about finite sequences of individuals are met. When this first-order theory is enriched in the way described with quantification over first-order definable properties (classes), the result is the Von Neumann–Bernays–Gödel system of set theory, whence the description (in Quine [1963]) of NBG as the *predicative enlargement* of ZF.

As to the expressive power of a language with quantification over first-order definable properties or classes, it is well known among mathematical logicians, cf. Wang [1952]; (there is also a good brief exposition in Parsons [1974]) that if the underlying first-order language has a finite set of constants and allows for the formulation of enough[11] of its own syntax, the addition of quantified variables ranging over first-order definable properties allows the notions of *satisfaction* and *truth* for formulas and sentences of the original, first-order, language to be defined. The proof is easy, related to the point that *partial* truth definitions, good for sentences with less than some given number of quantifiers, can be given within the first-order language itself; we sketch it for the special case where the first-order language is that of ZF and the language with quantification over properties is that of NBG. For convenience, we suppose that the same, numbered, sequence of variables serves both as free and as bound variables, and that ZF is formulated with no individual constants, membership as its only predicate, negation, conjunction, and universal quantification as its only logical operations, and we define, relative to some representation of syntax in set theory, satisfaction as a relation between finite sequences of sets, and formulas.

Define a sequence as *long enough* for a formula if and only if it contains at least n sets, where n is the number of the highest-numbered variable occurring in the formula. (This is a purely first-order condition.) Next, we say that F is a *satisfaction class* if and only if

   (1) Every object in F is an ordered pair of a finite sequence of sets and a formula, with the sequence long enough for the formula.
   (2) There is a natural number, n, such that
      (i) F contains no ordered pair of a sequence with a formula containing more than n quantifiers, and
      (ii) For every formula containing n or fewer quantifiers, and every sequence of sets long enough for the formula, F contains either the set and the formula or the pair of the set with the negation of the formula, but not both.
   (3) For every atomic formula, $x_i \in x_j$, and every sequence of at least i or j (whichever is greater) sets, F contains the pair of the sequence and the formula if and only if the ith set of the sequence is a member of the jth.
   (4) If a sequence of sets is long enough for a conjunction, and the total number of quantifiers in the conjunction is small enough (cf. clause (2)), F contains the ordered pair of the sequence and the conjunction if and only if it contains both the ordered pair of the sequence and its left conjunct, and the ordered pair of the sequence with its right conjunct.
   (5) If a sequence of sets is long enough for a universal quantification, and the total number of quantifiers in that formula is small enough, and the variable bound by its initial quantifier is the ith, then F contains the ordered pair of the sequence and the quantification if and only if, for every set x, F contains the ordered pair of the sequence, otherwise just like the given one, having x as its ith item with the formula that is the matrix of the quantification.

Except for containing the property (class) variable F, this condition too is entirely in the first-order language of ZF. Finally, quantifying that class variable, we may say in the language of NBG that a sequence of sets *satisfies* a formula if and only if there is some satisfaction class which contains the ordered pair of the sequence with the formula; a sentence is *true* if and only if some (equivalently, if every) sequence long enough for it satisfies it.[12]

Having enriched the basic first-order theory with quantification over the first-order definable properties of its individuals, we may repeat the process, and allow (using a new lot of variables) quantification over those properties of individuals definable with the aid, in addition to the resources of the first-order language, of quantification over first-order definable properties. (Thus, in the sort of case described above, the property of being a true sentence of the first-order language, though definable with the help of our first lot of second-order variables, was not in their range, but would be one of the properties quantified over with the new lot.) And having done that, we may repeat again, and again .... Thus, we come to what Church [1956] calls the *ramified functional calculus of second order*. For vocabulary we have, in addition to what the first-order language provides, an alphabet of n-adic predicate variables of *level* m, for every natural number n and every *positive* integer m. Counting the predicates of the original first-order language as of level 1, we define the level of a formula as the level of the highest level predicate, abstract, or free variable occurring in it, or one greater than the level of the highest level variable quantified in it, whichever is higher, and the level of an abstract as the level of the formula it contains. The substitution range of an n-adic predicate variable of level m, for the purpose of defining the instances that universal quantifications imply and existentials are implied by, include the predicates of the first-order language, free n-adic predicate variables of level m or lower, and the n-adic abstracts of level m or lower,[13] with the *or lower* clauses marking our cumulative interpretation of the notion of level.

Before we go on to consider higher-order logics, quantifying over properties, it is worth remarking that even *second*-order logic can be extended into the transfinite by admitting transfinite ordinal numbers as levels. As the example of the truth definition suggests, if the domain of individuals is infinite and that of first-level properties – the properties expressable in the underlying first-order language – moderately rich, the addition of quantification over properties of each finite level allows the definition of new properties, differing even in extension from any properties definable by quantification over lower-level entities alone. There will also be properties which are not definable by quantification over entities of any finite level, but which could be defined if we could somehow quantify over the properties of *all* finite levels, and it is natural to think of these as properties of a transfinite level.

(It is technically convenient to let limit-ordinal levels be simply the unions of lower levels, and to let the first new properties, defined by quantification over the union of the properties of levels below the limit, have as their level the successor of the limit. Thus the properties of level ω will be simply the properties of finite level, and properties definable by quantification over properties of all finite levels will be of level $\omega + 1$.) We can specify *formal* systems of ramified second-order logic as far into the transfinite as we can devise effective systems of notation for transfinite ordinal numbers to use in distinguishing variables of different levels. Thus, for example, we may specify a system of ramified second-order logic quantifying over properties of individuals of every level less than, say, $\omega + \omega$ in almost exactly the way we specified one quantifying only over properties of finite levels. The only real change in the description given in the last paragraph would be in the specification of the variables: we will have, for every natural number n and every ordinal α strictly greater than 0 and strictly less than $\omega + \omega$, an alphabet of n-adic predicate variables of level α. Given the new variables, the definitions of the levels of formulas and abstracts in terms of the levels of the expressions occurring in them carry over without change,[14] and once again the substitution range of a variable will consist of the constant predicates, free predicate variables, and abstracts of its own and lower levels.[15]

The systems so far considered have quantified only over properties of individuals. This has made them technically simpler, but the restriction to second-order, does not, from a conceptualist standpoint, have the sort of *conceptual* naturalness it might have on, e.g., a conception of classes as *aggregates*. For a predicativist, the existence of a property consists in the possibility of defining it. If, when some item in a sentence is replaced with a variable, the resulting open formula is accepted as having a fully definite meaning, as being either true or false for every value of the variable, and the variable itself is taken to have a fully definite range, then the formula is being taken to define a property.

If a property is defined, then, in the sense appropriate to a conceptualist theory of properties, it exists, and so should be admitted in the range of some sort of quantified variable. The second-order systems described above admit variables to the places of predicates of individuals and allow the full classical logic of quantification in connection with these variables; in accepting such a system, therefore, we are signifying in the strongest possible way our acceptance of formulas containing free variables for properties of individuals as defining properties of these properties.[16] From the standpoint of a predicativist conception of properties, therefore, there can be no reason not

I.5: PREDICATIVE LOGICS 345

to quantify over these properties as well. This same argument, which we have just used to establish the predicative acceptability of ramified *third*-order logic, can be used again to justify fourth- and higher-order quantification. The system to be described next, therefore, is a ramified theory of types with quantified variables for properties of all finite orders. This is the system, of those described in this section, that is closest to that of *Principia Mathematica*.[17]

As in ramified second-order logic, we have infinitely many alphabets of variables, each, this time, assigned a *type* and a *level*. One type is that of individual variables; they are of level 0. Another is the type of propositional variables; as in ramified second-order logic, there is an alphabet of propositional variables for every level ≥ 1. For n-adic predicate variables with $n \geq 1$, the type is given by an n-tuple of (type, level) pairs, representing the types and the levels of the arguments they take; those in which each argument must be of the type of individuals correspond to the monadic and relational variables of ramified second-order logic, and come at every level ≥ 1. Finally, the level of a monadic or relational variable taking non-individual arguments at some or all of its argument places must be *greater by at least 1* than the highest level permitted among its arguments. Any predicate *constants* of the language must similarly be specified as to type and level. The well-formed *atomic* formulas are those in which a predicate is applied to an appropriate number of terms, which may be variables, constants, or (if we formulate the logic with them) abstracts, each such term having the type and level specified for the corresponding argument place of the predicate by the predicate's type. As with ramified second-order logic, it is convenient to allow levels to be cumulative; with higher-order predicates this cumulativity affects the set of atomic formulas, in that if the level specified for a given argument place by the type of a predicate is not the lowest possible given the type specified for that argument place, that place may in atomic formulas be filled with terms of lower level than specified. Thus, for example, if the type of a certain predicate specifies that for its first argument it takes level 7 dyadic relations of individuals, the first term after it when it is the predicate of an atomic formula may be any dyadic predicate of individuals, be it a constant, a variable, or an abstract, of any of the levels 1 through 7 inclusive.

The level of a formula is defined in terms of those of its components as before: it is the level of the highest level constant, abstract, or variable occurring free in it or one greater than the level of the highest level variable quantified in it, whichever is higher. It is probably still simplest to describe the deductive rules of the system if abstracts are used. (In the actual writing out of formal deductions this formulation would involve a lot of boring abstraction and concretion steps, and if these were omitted the result would resemble a derivation in a system formulated with direct substitution of formulas for variables along the lines of the formulations of second-order logic in Church [1956].) We may use the same notation for abstracts as in second-order logic, but here the sequence of distinct variables to the left of the colon in a (non-propositional) abstract need not all be individual variables but may be drawn from any alphabet. In the conversion (abstraction/concretion) equivalences, each of the terms a_i must be of the same type as, and of the same or (cumulativity strikes again) lower level than the corresponding variable x_i. The type of a (non-propositional) abstract is given by the sequence of types and levels of the variables on the left of its colon, and its level is either that of the formula to the right of its colon or one higher than the level of the highest level variable on the left, whichever is higher. Finally, the substitution class of a variable, for the purposes of the quantifier rules, comprises the constants, free variables, and abstracts of its own type and of its own or lower level.

So far nothing has been said about how *identity* can be made to fit into the system. For simplicity, we might imitate the usual practice in axiomatic set theories with *Urelemente*, and include an identity predicate for individuals as one of our predicate constants[18] (of, say, level 1) with an additional rule or axiom scheme of substitution of identicals, but this would still leave the question of identity for properties up in the air. The most familiar approach in work on the foundation of mathematics is the adoption of an axiom of extensionality in set theory, but the conceptualist notion of property hardly seems to imply extensionality, and it may not at first be obvious that extensionality can consistently be assumed. In fact, if primitive predicates taking non-individual arguments are included in the system, extensionality becomes very dubious, for we might intuitively think of some such predicate as expressing a property that distinguishes between coextensive properties. Plausible axioms for a predicate expressing such a property might well *contradict* extensionality. If the only primitive predicate constants in a ramified system are predicates of individuals, however, this cannot happen, and no axioms for such predicates can contradict extensionality. In this case (which includes the sorts of systems proposed by foundational researchers to formalize parts of classical mathematics) we can adopt holding of the same arguments as a definition of identity for properties (having the same truth value for propositions), and add a new rule or axiom scheme for the substitution of identicals, so defined.[19] Ramified type theory with this sort of extensional treatment of identity is not really a stronger system than ramified type theory without it, for the system with these definitions of identity and extra rules can be interpreted within a system without them, and even within a system with axioms that contradict extensionality, by using quantifiers *restricted* to properties which do not distinguish between coextensive arguments to represent the quantifications of the extensional system. That is, if universal quantification *for every property . . .* of the extensional system are translated as restricted quantifications *for every property that does not distinguish between coextensive arguments . . .* , etc., translations of the theorems of the extensional system are derivable in the non-extensional.

Two different kinds of transfinite extension of this *Principia*-like system may be considered. On the one hand, we may add transfinite *levels* without altering the *type* hierarchy.[20] This is as simple a matter here as it is in the case of ramified second-order logic: indeed, nothing I said above in describing the system requires that only the natural numbers, and not some segment of the ordinals, figure as levels. We might also wish to include variables for properties of transfinite *order*, but since the purely type theoretic structure of the system as so far described is *not* cumulative, this will involve a major reformulation. It will be convenient to introduce certain simplifications in the type structure before we tackle it.

Fitch [1938] presents a system comparable to the extensional version of our type theory,[21] but with its type structure radically simplified in several respects. First, only dyadic relations are considered. This is a harmless restriction, for monadic properties may be represented by dyadic ones (for each monadic property consider a relation an object bears to itself just in case it has the property), and the treatment of identity provided by extensionality allows for the representation of ordered pairs, and so of relations of higher degree. Second, types are construed as cumulative with regard to order, so that, e.g., a third-order relation can hold between individuals as well as between relations between individuals. This too can be shown to be harmless, by the move which is standard in all liberalizations of type theories: if you want a relation relating only objects of some specified order and not of lower ones, just take a relation which is *false*, rather than undefined, of the unwanted objects. Further, it is possible, given extensionality, to represent lower-order objects by higher (e.g. by the relation which holds *only* between the given object and itself), so this cumulativity does not really strengthen either the expressive or the deductive power of the system. Finally, relations are taken to be of the highest *order*, in terms of the arguments they are defined for, compatible with their *levels*. Once again, this is harmless, for if one wants, say, a level 2 relation between individuals, one can simply take a level 2 relation which, though defined for level 1 relations, is false of them.

As long as extensionality allows the representation of ordered pairs anyway, we can simplify yet further by imitating set theory and quantifying only over monadic properties. (Fitch's choice of the dyadic as primitive is justified by the simplications it allows in the formulation of the infinity, choice, and ancestry axioms.) We thus arrive at a system like that presented by Quine in Section 6 of his 'Logic and the Reification of Universals' (Chapter VI of Quine [1953]) as a formalization of the conceptualistic approach to mathematics.[22] Here we have an alphabet of variables corresponding to each natural number, the number correlated with a sort of variable being its level, and at the same time determining its order. The variables of level 0 range over the individuals, those of level 1 over individuals and level 1 properties of individuals, those of level 2 over individuals, level 1 properties of individuals, and level 2 properties which may be had by either individuals or level 1 properties of individuals, and so on. Any individual constants or primitive predicates of individuals included in the system will be of level 0 or 1 respectively (it will simplify things to include a primitive for identity of individuals). Individual constants and variables of any level may occur as terms in atomic formulas with primitive predicates; other atomic formulas consist of two juxtaposed terms, individual constants or variables of any level, the first acting as grammatical predicate and the second as grammatical subject. This freedom of association on the part of terms of different levels may seem contrary to the spirit of type theory, but since orders are cumulative on the intended interpretation, the range of a higher-level variable will include objects for which properties in the range of lower-level predicate variables are defined, so there is no reason not to allow formulas with predicates of level lower than that of their subjects. Intuitively, we justify the use of classical logic by stipulating that an atomic formula is to be false if its predicate is assigned an individual, or its subject an object not in the range of definition of the property assigned its predicate.[23] Note that there will be valid formulas relating objects of different levels, on this interpretation, which will not be derivable from the axioms enumerated below; one might consider adding at least some of them as additional axioms.

By the above stipulations, the identity predicate of the formal language may stand between terms of any level, but we have only assumed it defined for individuals. We must specify its meaning for objects of higher level, and lay down appropriate axioms. We treat identity of properties extensionally,

I.5: PREDICATIVE LOGICS 349

taking the identity predicate to be true of two properties just in case they hold of all the same things. (Even properties of different level can be identical in this sense, for as long as they hold of the same things it doesn't matter whether they are false or undefined of the others.) Thus we adopt an axiom scheme of extensionality,

$$\neg \exists y(s = y) \rightarrow (\forall x((sx \rightarrow tx) \land (tx \rightarrow sx)) \rightarrow (s = t)),$$

where y is of level 0, at least one of s and t is of level > 0, and x is of the level immediately beneath that of whichever of s, t is of higher level. We also adopt the usual principles of the logic of identity for terms of level 0, and an unrestricted rule or axiom scheme of substitution of identicals with identities having terms of any level.

Given extensionality, it is simpler to leave abstracts out of the language, and to get property existence theorems from an axiom scheme of comprehension. The substitution range of a variable, for the purposes of the quantifier rules, will be limited to free variables of its own and lower levels, and individual constants. As usual, we define the level of a formula as either that of the predicate (constant or variable) of highest level occurring free in it, or one greater than the level of the highest level variable quantified in it, whichever is higher. We take as comprehension axioms all formulas of the form

(Comp) $\exists F \forall x((Fx \rightarrow A) \land (A \rightarrow Fx)),$

where F is of the same level as the formula A and does not occur free in it, and x is of the level immediately beneath that of F.

Such a system is easily extended to transfinite levels. Variables of level ω (and other limit levels) are thought of as ranging over the union of the ranges of variables of lower levels; since the substitution range of a variable is taken to include the free variables of lower levels, simple quantifier logic suffices to derive sentences of the form

(Summation) $\forall x \exists y(x = y),$

where y is a variable of limit level and x is of any lower level. Since limit ordinals do not have predecessors, there are no variables of level *immediately* beneath a limit level, so a variable of limit level cannot be the existentially quantified variable of an instance of the comprehension scheme. Comprehension axioms appear again at higher levels, however: in particular, there are instances of Comp with the existentially quantified F of level one greater than a limit level, the x of the limit level, and in which variables of the limit level (as well as of any lower level) may occur bound in the formula A. Thus, the new properties of level $\omega + 1$ whose existence is asserted by the comprehension axioms are those (which may hold of objects of any finite level) definable by quantification over the objects of all finite levels.

The hierarchy of systems Σ_α (for 'constructive' ordinals α from Cantor's second number class) proposed by Wang [1954] for the formalization of those parts of classical mathematics not dependent on impredicative methods are systems of essentially this kind (but with some additional axioms and assuming the natural numbers, or some mathematically equivalent denumerable set of objects, as the individuals), with α the level of the highest level variables of a system Σ_α. The system Σ_ω of Chihara [1973] is the special case of this schema with ω the highest level of variable: its language therefore differs from that of Quine's system by having one additional alphabet of variables. Since Chihara's book, in addition to its stimulating discussion of a variety of issues in the philosophy of mathematics, contains the most detailed published account of one of Wang's systems, a few comments may be appropriate.[24]

First, Chihara has chosen for his domain of individuals, not the natural numbers but (what is mathematically equivalent) the hereditarily finite sets. Since he uses the same (set theoretic) notation for membership among these 'individuals' as he does for having a property, his system could also be described as assuming no individuals and having no variables of finite level, but as having variables of levels ω to $\omega + \omega$ inclusive. Since every hereditarily finite set can be defined explicitly, they may be construed as properties on the conceptualist view, and their existence proved in a predicative system such as Quine's where orders as well as levels are taken as cumulative. Thus, the things Chihara, using the individual variables of his system, says about the hereditarily finite sets, may be reconstrued as part of the theory of finite level properties as expressed in terms of variables of level ω; in general, for ordinals $\alpha \leq \omega$, Chihara's variables of level (he calls it *order*) α can be thought of as of level $\omega + \alpha$. Some of Chihara's metamathematical argument (such as his use of the *nesting* function) might be easier to follow if the system were so redescribed.

Second, though it may not be possible to derive it from our comprehension axioms for higher successor levels, there is a plausible comprehension scheme for limit levels which might be incorporated into Σ_ω. (Chihara was unable simply to adopt Wang's axioms because some of them need higher level variables for their statement.) Where X, Y, and Z are variables of level ω, the formula A may contain free or quantified variables of level ω as well as of finite levels, but all quantifications in it are *restricted* to X, where R is the conjunction of the restricting formulas Xa for all the variables a other than X and Z occurring free in A, and where Y does not occur free in A, we may take as axioms formulas of the form

(ω_{Comp}) $\forall X[R \to \exists Y \forall Z[(YZ \to (XZ \wedge A)) \wedge ((XZ \wedge A) \to YZ)]]$.

Since the variables of level ω are construed as ranging over the union of the ranges of the variables of finite levels, this scheme in essence *summarizes* the comprehension axioms of the finite levels. This is spelled out a bit in Section 3 below. The function of the 'restricting' variable X is to ensure that the objects of which the property whose existence is asserted must hold, or that enter into its definition, are all of level less than some *one* finite level; this may be compared to the restriction on the comprehension scheme in Zermelo's axiom scheme of Separation or *Aussonderung*. The axioms, *Intersection*, etc., mentioned on p. 225 of Chihara [1973], of which only one is stated, are evidently related to this axiom scheme, perhaps in roughly the way Von Neumann's finite axiomatization of the principles of class existence in NBG is related to a (predicative) axiom scheme of class existence.

Finally, systems whose highest level variables are of limit levels, though they may be technically elegant, and allow formalizations of fragments of classical mathematics that will look familiar to those used to the usual impredicative systems, are not, from the standpoint of the conceptualist or predicative notion of property, *conceptually* very natural. Once a sort of variable is admitted, the immediate next step is to use it in the definition of properties which will be the values of a new sort: the limit level cries out for a successor. Thus, though Russell and Whitehead may have believed that any property at all was defined at some finite level, a consistent predicativist is bound to see Σ_ω as merely a fragment of a more comprehensive system. Cf. further discussion in Sections 6 and 8 below.

3. SUBSTITUTIONAL QUANTIFICATION

Even on the predicative approach, the notion of a property is of an abstract character. Since, however, on the predicative notion properties are all explicitly definable, the defining formulas themselves would seem to provide a class of more nearly concrete (or *anschaulich*, in the sense of Gödel [1958]) stand-ins for the abstract properties. Combining a construal of predicative quantification over properties as expressing generalizations about formulas of a formal language with a reasonably simple description of the grammar of the ramified systems,[25] however, we get a *substitutional* interpretation of the quantifiers: a universal (existential) quantification is true just in case each (at least one) of its closed instances[26] is true. Thus even people whose general philosophical orientation inclines them to be leary of substitutional interpretations of quantifiers elsewhere are sometimes attracted to the substitutional interpretation for the higher-order quantifiers of predicative logics. What is essentially a substitution interpretation of a ramified type theory appears in a technical context in Fitch [1938], after which there appears to be a gap – the discussions of substitutional quantification in the philosophical literature of the 1940s through the 1960s make no mention of ramified type theory. Then, in the early 1970s, a number of writers, with a variety of philosophical axes to grind, presented substitutional interpretations of ramified theories: Parsons [1971] and [1974], Chihara [1973], Leblanc and Weaver [1973], and (for ramified propositional quantification), Grover [1973].[27]

There is one slight complication, discussed in Parsons [1971]. If each of the individuals in the intended range of the individual variables has a name, so that the individual variables can also be interpreted substitutionally, all will be well, but if not there will be a problem with formulas that (intuitively) imply the existence of *relational* properties. Thus, for example, if being taller than is a relation among individuals, there will be, for each individual, the property of being taller than *that* individual, and the property of being taller than some unnamed individual will not be expressed by a formula of the system. In this case we will have to generalize our interpretation to a *relatively* substitutional one. We consider the language of the ramified system to contain *both* substitutionally interpreted variables (those of higher order than the first) and objectually interpreted (individual) variables; the fundamental semantic notion is that of the *satisfaction* of a formula by a sequence of individuals. (As usual, the formulation is *simplest* if we assume infinite sequences, simultaneously assigning variables to all individual variables, but finite sequences and slightly more complicated definitions will also work.) A formula universally (existentially) quantified with respect to a substitutionally interpreted variable is *satisfied* by a sequence just in case each (at least one) of its instances is satisfied by every (some) sequence agreeing with the given sequence in its assignments to objectually interpreted variables occurring free in the original formula, and *truth* is defined in terms of satisfaction in the usual way. In effect, properties are not, as on a pure substitution interpretation, being represented by formulas alone, but by ordered pairs of a formula and an assignment of values to the objectually interpreted free variables[28] in it.

A consideration of the substitutional interpretation can show the naturalness of what may seem like arbitrary choices in our formulations of the ramified logics. For example, quantifying, in the definition of a property, over higher-order properties and over higher-level properties of low order, should both raise the level of the property defined, but one may wonder about the justification of our precise numbers. For example, why should a property of individuals defined by quantification over level 2 properties of individuals have the same level (viz. 3) as one defined by quantification over lowest possible level properties of level 1 properties of individuals? One answer is given by the fact that under certain special assumptions – that the domain of individuals contain the formulas of the language, that the various *expression* relations between formulas and properties of various levels all be expressed in the language, and that the ramified logic be supplemented with axioms to the effect that all properties are expressed, for example – the two sorts of definition can do each other's mathematical work. A more general answer is that the two kinds of quantification are the same number of steps removed from a pure first-order language. The quantified variable in each case has as its substitution range formulas containing quantification over only individuals and level 1 properties of individuals.

A more serious example is provided by the comprehension scheme for limit levels, ω_{Comp}, given at the end of Section 2 above, which was claimed to summarize the comprehension axioms at lower levels. Consider a sentence of this form. It will begin with a string of universal quantifiers: the one binding the schematic X and those binding the remaining free variables (other than the schematic Z) of the formula A. On the substitutional interpretation (we ignore here the complications of relative substitutionality), this (multiple) universal quantification is true just in case all of the appropriate instances are true. Since the system under consideration was formulated with comprehension axioms and not with direct substitution of formulas for variables, these instances are a bit complicated to define, but when all the details are worked out, it emerges that, in those instances in which the restrictive antecedent R (or rather, the sentence replacing it) is *true*, the schematic variable X has been replaced by a formula of *higher* finite[29] level than any of the formulas replacing the variables bound by the other initial quantifiers. Further, since the variables quantified in the formula A are restricted by X, the instance is equivalent to the formula formed from it by rewriting A replacing any variable quantified in it of level greater than or equal to the level of the formula replacing X with a variable of the level immediately beneath that level. Now, in the part of the instance beginning with the quantifiers $\exists Y \forall Z$, replace Y with a variable of the level of the formula replacing X and Z each with a variable of the level immediately beneath that. In those instances in which the restrictive antecedent R is (replaced by) a true sentence, the consequent beginning with these quantifiers will have turned into a comprehension axiom of the form Comp for some finite level.

The substitution interpretation of the variables works so nicely, indeed, that one is almost tempted to identify it as the *intended* interpretation of the ramified logics. If one were interested solely in mathematical applications, there would be little objection to such an identification, but in the general case it runs afoul of one very simple objection: the intuitive idea is that ramified logic provides a way of generalizing about properties, but for any given formal language, there is no guarantee that every property is expressed (even relative to an assignment of values to individual variables) by a formula of *that language*. Thus, for example, Russell at about the time of the writing of *Principia Mathematica* was a phenomenalist and a concept empiricist (cf. Russell [1912]). He thought of sense data as typical individuals, and such properties as, perhaps, being a visual sense datum of a particular determinate shade of red as typical level 1 properties of individuals. For him, then, the puzzle of the 'missing shade of blue', which has been worried over for generations in the literature of empiricism, would have provided a counterexample to the identification of the substitutional with the intended interpretation of ramified logic. For suppose there were some determinate hue such that Russell had never been acquainted with any sense datum having it. Then, on his theory, he would never have been acquainted with the hue, and would only have been able to specify it by quantifying over properties – by saying that it was *the* hue related to certain other hues in the spectrum in such and such ways. But then no level 1 formula of a language *he* was in a position to use (in the rarified sense in which one *uses* an 'ideal language') would express the property of being of that particular hue, and quantification over level 1 properties would be narrower on the substitutional than on the intended interpretation. Similar examples could be given from the standpoint of more modern metaphysical and epistemological theories. Thus, for example, a version of scientific realism might take physical objects as individuals and identify as of level 1 those properties whose presence or absence in an object can be experimentally measured. (This account would certainly need some elaboration before it could be plausible, but it seems unlikely that any such elaboration would change the argument against the substitutional interpretation of the quantification over properties in predicative logics.) The physicists, however, are still discovering new 'effects' and explaining them by imputing weird properties to particles, so that it would be rash to assume that our current language had predicates for all level 1 properties. On the other hand, some sort of quantification over properties would seem to be just what is needed in order to discuss intelligibly the possibility of making meaningful *additions* to one's language, so it may be that some of the appeals nominalistically minded philosophers are in the habit of making to substitutional interpretations (and more generally to talk of 'language' or 'languages') should be directed instead to a predicative theory of properties.

Parsons [1974] makes the point that predicative quantification over properties (or rather, in the context of his article, classes) is 'more general' than a more strictly linguistic quantification over predicates for precisely this reason – that it allows for properties that are not expressed in the language we have learned so far.

4. CONSISTENCY AND OTHER EXAMPLES OF DEDUCTIVE WEAKNESS

Although the substitutional interpretation of the ramified theory of types may not be exactly the intended one, it is of undoubted fertility as a source of metamathematical information, and it is close enough to the intended interpretation to inspire confidence that metamathematical results obtained through it are genuinely relevant to the intuitive project of discovering a predicative theory of higher-type entities.

A system of ramified second or higher-order logic may, obviously, be reformulated as a *natural deduction* system, either along the lines of the original **N** systems of Gentzen [1934] or of the Fitch-style 'subordinate proof' systems, as in Fitch [1952], Thomason [1970], and many other textbooks. Similarly, it may be reformulated as a 'sequent' or 'consecution' calculus on the model of the **L** systems of Gentzen [1934]. When this is done, Gentzen's purely finitistic proof of his *Hauptsatz*, or 'cut-elimination' theorem for first-order logic (formulated as a sequent calculus – the analogue for a natural deduction system would be a 'normalization' theorem, as in Prawitz [1965, 1971]) carries over almost without change. As a preliminary to establishing connections with the substitution interpretation, note that a key part of the procedure by which a formal proof is transformed into a cut-free or normalized version of itself is the replacement of axioms $A \to A$ and uses of the cut rule (the analogues in a system of natural deduction might be, in the classical case, excluded middle axioms and uses of the rule *ex falso quodlibet* to infer

B from A and $\neg A$) with logically simpler instances of the axiom scheme and cuts (*quodlibets*) with logically simpler cut (contradictory) formulas. It is further essential that the instances of a quantification be logically simpler, in the relevant sense, than the quantification itself. Thus, to give a natural deduction example, in normalizing a proof in which some arbitrary formula B was inferred from the contradictories $\exists x Ax$, $\neg \exists x Ax$, we might first infer the instance Aa from the unnegated formula (with, of course, a 'flag' on the a or, in Fitch's variant, within a subordinate proof marked as general for a), then, since $\neg \exists x Ax$ implies $\forall x \neg Ax$, infer the corresponding negated instance from the negated quantification, and infer B by the *quodlibet* rules from Aa and $\neg Aa$. (In a natural deduction formulation, classical logic runs more smoothly in practice, and its metatheory would be more elegant, with rules like those in Fitch [1952], taking the Double Negation and De Morgan laws, and the quantificational analogues of the latter, as primitive.) The process can be iterated until a version of the proof is produced in which the axiom scheme is assumed, and the cut (*quodlibet*) rule used, only for *atomic* formulas, at which point it is clear that it can be transformed into a fully normal proof in which the cut rule or its natural deduction analogue is not used at all.[30] The only aspect of the proof that must be changed when ramified higher-order quantification is added is the precise measure of the 'logical complexity' of a formula. In first-order logic, it is possible to use the *degree* of a formula, which may as well be defined simply as the number of logical operators it contains, as the measure. In the ramified systems, however, an instance of a second- or higher-order quantification may contain more logical operators than the quantification itself, and be simpler only in the sense that the 'new' quantifiers in the instance will bind variables of lower level than the one replaced. The measure of logical complexity – the analogue of degree – must therefore be so defined as to ensure that the addition of a higher-level quantifier can increase complexity more than that of any number of lower-level quantifiers and connectives.[31] Where the induction on the degree of formulas in the proof of the *Hauptsatz* for first-order logic is a simple induction on natural numbers, the corresponding induction for, e.g., a system with quantified variables of all finite levels will be a transfinite induction up to ω^2. This will, obviously, make the proof of the *Hauptsatz* a bit more complex, but it will leave it, for systems that aren't ramified *too* far into the transfinite, safely within the bounds of Hilbert's *finitist* standpoint, even on the strict construal of that notion defended by Tait [1981]. In contrast, analogues of the *Hauptsatz* for impredicative higher-order systems have only been established by non-finitist methods.[32] The induction on logical complexity of formulas breaks down for systems like the simple theory of types, because there an instance of a quantification may be on any measure more complex than the quantification – indeed, an example may be given of a quantification which is identical to one of its own instances!

The finitistic provability of the *Hauptsatz* for predicative systems is of both technical and conceptual interest. On the conceptual side, for example, Hacking [1979] takes it as a key part of his argument that the ramified theory of types is part of *logic* in a sense in which the simple theory of types is not. (Note, however, that there are problems with Hacking's 'do-it-yourself' semantics, cf. Sundholm [1981].) Again, the completeness of a ramified theory relative to a certain sort of substitutional interpretation, as in Leblanc and Weaver [1973], can be demonstrated as a corollary of the *Hauptsatz*, just as Gödel's completeness theorem for first-order predicate logic can be proven on the basis of *its Hauptsatz*. On the other side, perhaps the most striking technical use of the *Hauptsatz* is in giving consistency proofs. The consistency of a ramified system pure and simple – that is, without any axiom of infinity or other special axiom – is immediate from the *Hauptsatz*. This, however, is of comparatively little interest, for in the absence of an axiom of infinity, even the simple theory of types can be given a finitistic consistency proof along lines inspired by semantic rather than proof theoretic ideas. (Cf. Gentzen [1936]; the consistency proof in Section 53 of Church [1956] is based on similar ideas.) More significant is the fact that the *Hauptsatz* allows us to prove consistency even in the presence of an axiom of infinity.

Axioms of infinity say that there are infinitely many individuals, or more accurately, since this is not strictly expressible in predicative terms, they say something that implies that there are infinitely many individuals. A typical such axiom, then, will assert the existence of a transitive and irreflexive relation that each individual (or at least each individual in its – assumed nonempty – field) bears to at least one individual. It will be convenient to consider a system with a constant dyadic predicate of individuals, of lowest level, and axioms to the effect that this predicate expresses such a relation. Our axiom of infinity, therefore, may be the sentence

(Inf) $\quad \forall x \forall y \forall z \exists w (G(x,w) \land \neg G(x,x) \land ((G(x,y) \land G(y,z)) \to G(x,z)))$.

The consistency of the *first-order* theory based on this axiom is trivial: the axiom is true in the natural numbers when $G(x,y)$ is interpreted as "y is greater than x", and the set of sentences containing this as the only predicate true in the natural numbers is decidable (For a detailed proof of this (for a somewhat richer language) in the style of an elementary textbook, see

Chapter 7 of Fisher [1982].) The consistency problem for a ramified higher-order system with this axiom of infinity is thus comparable to that for the system of first-order arithmetic without induction whose consistency is proved as a corollary to the Strengthened *Hauptsatz* in Section 4 of Gentzen [1934]. In each case, the theory divides into two strata – on the one hand a basic and unproblematic stratum (truth functions of numerical equations in arithmetic, the first-order theory of the greater than relation in our theory) for which consistency is obvious, and on the other, a quantified stratum of generalizations about the first stratum which are established by purely logical means starting with premises taken from the first stratum. In each case, the Strengthened *Hauptsatz* shows that if a contradiction were derivable in the whole theory, it would already be derivable in the unproblematic first stratum. The key point is that in both theories, *generalizations* about the first stratum are derivable only by logical means: the quantifier rules (first-order in Gentzen's theory, higher-order in ours) and their machinery of free variables. The consistency problem for full first order arithmetic is harder because the rule or axiom of induction allows a second way of establishing generalizations.

It is instructive to recast the consistency proof in a more semantic (but still elementary) form, making use of the possibility of a substitution interpretation of the quantifiers. In order to be able to interpret the individual as well as the higher-order quantifiers substitutionally, we need to add a denumerable set of names to our language. The easiest way to do this is by adding one individual constant, 0, and one function symbol, S. (Note that when these are added to the language with their usual arithmetic interpretations, as denoting Zero and the Successor function, the set of first-order sentences true in the natural numbers remains decidable, though we need not appeal to this fact in our proof). We first define truth and falsity for atomic formulas containing no variables. Since the only variable-free terms are 0, $S(0)$, $S(S(0))$, and so on, this is trivial: an atomic sentence, $G(t, t')$, is true if the term t' contains more occurrences of S than the term t, and is false otherwise. We then define truth and falsity *inductively* – that is, by recursion on the analogue of degree mentioned in connection with the proof of the *Hauptsatz* – at each stage specifying the truth value of a logically-complex sentence in terms of the truth values of simpler sentences. The clauses for the truth functional connectives are the usual ones; for the quantifiers *of every level*, including the individual quantifiers, we say that a universal (existential) quantification is true [false] just in case all the sentences (at least one of the sentences) which are its instances are (is) true [at least one of its instances (all of its instances) is (are) false]. It is immediately evident that there are false and true sentences, that no sentence is both true and false, and that the axiom of infinity is true. It remains to be shown that everything provable is true. For this purpose it is useful to formulate the theory as a sequent calculus; we adopt as initial sequents (in addition to Gentzen's logical axioms $A \to A$) all sequents of the forms $G(s, t) \to$ where s and t are terms, possibly containing variables, either both containing 0 or both containing the same variable, with t containing no more occurrences of S than s, all sequents of the form $\to G(s, t)$ where s and t are again terms, possibly containing variables, either both containing 0 or both containing the same variable, with t containing more occurrences of S than s, and all sequents of the form $G(s, t), G(t, u) \to G(s, u)$ where s, t, and u are any terms, possibly containing variables, at all. Theorems are those sentences A for which the sequent $\to A$ has a (cut-free) derivation; it is easy to see these include all the formulas derivable in ordinary formulations with the axiom of infinity. Call a sequent all of whose formulas are sentences (a *closed* sequent) *true* if and only if it contains one or more true sentences on the right of the arrow or one or more false sentences on the left of the arrow (or both). Call a sequent containing free variables true if and only if all the closed sequents resulting from making grammatically appropriate uniform substitutions for all the free variables are true. We show that all derivable sequents are true, from which it follows immediately that all theorems are true sentences.

Clearly all initial sequents are true, and the structural rules and rules for the truth functional connectives present no problems. We are left with the rules introducing quantifiers. In the rules introducing the universal quantifier on the right of the arrow and the existential quantifier on the left, the introduced quantifier binds a variable occupying exactly the places in the formula occupied, in the corresponding formula of the premise sequent, by a free variable having no other occurrences in the sequent. Since the premise sequent is derivable, so is any sequent formed from it by making a grammatically appropriate substitution for this free variable in that formula; by hypothesis of induction, all such sequents will be true. These sequents, however, will be exactly the sequents obtainable from the conclusion sequent of the inference by replacing the newly introduced quantification with one of its instances. If the truth of the conclusion requires this quantification to have a certain truth value, the truth of these sequents will require its instances to have the same truth value. The truth of the premise, therefore, guarantees that of the conclusion.

In the rules introducing the universal quantifier on the left and the existential quantifier on the right, the newly quantified variable may occupy places occupied, in the corresponding formula of the premise sequent, by occurrences of any grammatically appropriate expression whatsoever, regardless of what other occurrences it may have in the sequent. The truth conditions for existential and the falsity conditions for universal quantifications, however, are correspondingly more lenient, so that once again the derivability and truth of the premise sequent (and of sequents obtained from it by uniform substitutions for free variables) will guarantee the existence of instances of the newly introduced quantification with any truth value required in order to make the conclusion sequent true. All derivable sequents, therefore, are true.[33]

When one attempts to work out the details of the consistency proofs, it soon becomes apparent that they are almost the same whether the system to be shown consistent is of ramified second-order logic (as in Church [1956]) or is more like that of *Principia Mathematica* in allowing higher types.

The ramified theory of types was originally proposed by Russell [1908] as part of the logicistic program of reducing mathematics to logic. The consistency proofs give at best a very weak sort of support to that program, for by themselves they give no support to the claim that the ramified theory is in some important sense *logical*, or that its principles are analytic or knowable *a priori*. An inconsistent theory, to be sure, cannot be a correct logic, but false or merely empirically true theories can be consistent. (The particular way in which consistency is proved, *via* the *Hauptsatz* with its corollary, the possibility of a substitution interpretation, supports Russell's analysis more strongly than the consistency *result*, by itself, does.) Whatever slight encouragement logicists might draw from the consistency results, that with the ramified theory they were at least on the right track and that the disaster of the paradoxes would not be repeated, is more than outweighed by the depressing fact that the mere *possibility* of a consistency *proof* shows the logicistic program of *reduction* to be hopeless. Alas! It follows from the second incompleteness theorem of Gödel [1931] that if the consistency of a system of ramified logic with an axiom of infinity can be proved by finitistic means (or even by means formalizable in arithmetic), then full first-order arithmetic is not derivable in that system. There is no translation of the language of first-order arithmetic into that of the ramified theory of types (even when the latter is enriched with a 'greater than' predicate and the resources for forming infinitely many individual names) which will make the translations of the axioms of first-order arithmetic, including the instances of the axiom scheme of induction, come out as theorems of the ramified system (even with the axiom of infinity).

This is somewhat surprising, for when our axiom of infinity is added to the *simple* theory of types, or even to impredicative second-order logic, full arithmetic *is* derivable. In third- or higher-order impredicative logic, the Frege–Russell definition of a natural number could be used, but, given our formulation of the axiom of infinity, it is far more convenient to take as natural numbers some or all of the individuals ordered by our 'greater than' relation. The axiom of infinity guarantees that there will be enough such individuals to play the role, but we must define the class of natural numbers to ensure that only a series of individuals constituting a standard model of arithmetic is, considered. Thus, in impredicative second-order logic, we might define '*a* is a *natural number*' as

(DefN) $\forall F(F(0) \to (\forall x(F(x) \to F(S(x))) \to F(a))).$[34]

Similar quantifications of second-order variables would allow us to convert the usual recursion equations for addition and multiplication into explicit definitions. Given the definitions, all the usual axioms of arithmetic become derivable, instances of the induction scheme by little more than an instance of second-order Universal Instantiation.

Since the consistency of the ramified theories is provable in arithmetic, this development cannot be carried out in them. The problem, not surprisingly, concerns the induction scheme. We can formulate DefN in a predicative logic, but the bound predicate variable F will have to be of some definite level; whatever level is chosen, the defined predicate 'is a natural number' will, since F is bound, be of a higher level. If, now, we attempt to formulate an instance of the induction scheme containing quantified number variables in the condition, these will have to be represented by individual quantification *restricted* to natural numbers (as defined by DefN). The condition of the induction, therefore, will be of a higher level than F, and so not in its substitution-range, with the result that the instantiation leading from DefN to this particular induction axiom will not be a legitimate inference. It is, in fact, not hard to imagine models of ramified logics with axioms of infinity in which, when natural numbers are defined by DefN or some analogue, the translations of some instances of the induction scheme are false. Suppose the individuals ordered by our 'greater than' relation form a *non-standard* model of arithmetic, so that there are, in addition to a sequence of individuals of the same order type as the natural numbers, further individuals, 'greater than' any of them, in the field of the 'greater than' relation. We may, for simplicity, assume that the language contains primitive predicates of individuals which could serve as 'sum of' and 'product of', and that, when arithmetic is formulated with these predicates, the individuals in the field of the 'greater than' relation form a model for, say, the axioms of the inductionless arithmetic proven consistent in Gentzen [1934], but that they *falsify* some instance of the induction scheme that don't follow logically from these axioms. Depending on just what properties are in the range of the bound predicate variable, DefN may or may not exclude the non-standard numbers that come after the standard ones in the series. In particular, the ramified logic admits of a substitution interpretation, so one *possible* range for the predicate variable (of, e.g., lowest level) is simply the class of properties first-order definable in terms of the primitive vocabulary of the language, and if this includes only the predicates so far mentioned, DefN will include as numbers the whole series, both standard and non-standard. In such a case, the induction axioms that fail when quantification over numbers is represented by unrestricted individual quantification will still fail when the quantifiers are restricted by DefN. For a detailed presentation of the model-theoretic proof that not all instances of the induction scheme are derivable in a ramified theory like that of *Principia Mathematica*, see Myhill [1974]. We are forced to conclude that Poincaré was right (and Frege and Russell wrong), and that the principle of mathematical induction is one of the intuitive primitive principles of mathematical reasoning and not reducible to 'logic'.

From the indefinability of the natural numbers a variety of other indefinability results follow. Thus, for example, Boolos [1975], in a passage stressing the usefulness of impredicative second-order logic, lists several notions definable in second- but not in first-order logic. One is that of a *progression*, or series of the same order type as the series of natural numbers. The others include the notions of *well-ordering* and of the *ancestral* of a relation: but if either of these notions, or that of *isomorphism* or that of a property holding of *infinitely many* objects, were definable, that of a progression would be too. The system proved consistent in Fitch [1938] has a primitive operator expressing the notion of the ancestral of a relation. It is this that allows the development of arithmetic in the system and renders it stronger than that of *Principia Mathematica*.

Another of the notions on Boolos's list, that of *identity*, has a rather special status. Since the domain of a model for a first-order language may contain multiple elements which are not discriminated by the predicates (and other vocabulary) of the language, identity is not first-order definable. (To be sure, as has been pointed out in, e.g., Section 47 of Quine [1960], in any first-order language with a finite vocabulary we may define a dyadic predicate which we can *take*, consistently with any axioms expressed in the language, to express identity.) On the other hand, in an impredicative higher-order logic (impredicative second-order logic for identity of *individuals*) we may define two objects as identical if they are all their properties: $a = b$ if and only if

(DefId) $\forall F((F(a) \to F(b)) \wedge (F(b) \to F(a)))$.

The difficulty in transferring this definition to a ramified system is the familiar one. Like the variable in DefN, the bound predicate variable in DefId would have to be of some specific level. In the impredicative system, the fundamental rule of inference for identity – the rule of Identity Elimination, or Substitution of Identicals – can be derived from DefId by using Universal Instantiation, but in the ramified systems this would only licence Substitution of Identicals in formulas whose level does not exceed that of the bound predicate variable in DefId. Nor is Substitution of Identicals in higher level formulas derivable in some more roundabout way: Myhill [1974] contains a model-theoretic proof that substitution in formulas of any level higher than that of the quantified variable in DefId is not derivable.

Myhill's proof, however, has a peculiar feature which contrasts with the proofs outlined above. The terms *standard* and *non-standard model* are not in current use in discussions of ramified systems; it is not, in the general case, immediately obvious how they should be defined in order to accord with the predicative motivations of the systems. Still, a model in which the range of a higher level variable consists of the class of properties definable using quantification only over individuals and properties of lower level almost certainly deserves to be called standard. Models corresponding to substitution interpretations are standard in this sense, and the model appealed to in our argument for the non-derivability of induction from DefN (when the predicate variable in DefN is of lowest level). though non-standard as a model of *arithmetic*, was also a standard model of ramified logic. In contrast, in his argument that Substitution of Identicals in arbitrary formulas is not derivable from DefId, Myhill appeals to models in which the ranges of variables of levels higher than that of the variable in DefId contain infinitely many properties which are not so definable. Thus, though the non-derivability result is not in doubt, its intuitive significance is unclear – perhaps, one might wonder, Substitution of Identicals in higher level formulas is valid on the intended interpretation of ramified logics and should be adopted as an additional axiom scheme.

Thus, for example, material objects as conceived in classical physics and common sense are fully individuated by their spatio-temporal locations, and it would seem natural, if one thought of physical objects as individuals, to consider spatial and temporal relations as relations of the lowest level. It follows that, on such an interpretation, DefId with a quantified predicate variable of lowest order would, for individuals, define genuine indentity, and the premise that two individual terms denote the same physical object – the result, that is, of substituting the two terms for the free individual variables in DefId – will license their substitution for each other in any formula. It is, however, far from clear that this example is typical, or that Substitution of Identicals should be accepted as a general principle of logic when identity is defined by DefId. One worry is that it seems to be a logically contingent feature of physical objects that they can be individuated by their spatio-temporal relations (if, indeed, they can). A more serious problem is that the individuals might include entities of a very different sort. Thus, suppose that the individuals include *minds* or *spirits*, and that these are immaterial beings whose basic, psychological properties are not supervenient on, or at any rate not definable in terms of, anything physical. (There may, in fact, be no such entities, but the supposition that there are does not seem to be incoherent, and therefore it is undesirable for a very general logical or metaphysical theory to presuppose that there are none.)[35] Now, among the most important of the soul's properties are the *propositional attitudes*, such as belief, hope, wishing, etc. If we include minds among the individuals, we will almost certainly want to include predicates expressing these attitudes in our language: any reasonable interpretation of a ramified logic that includes spirits among the individuals will include belief among the properties. Belief, however, is a relation between souls and propositions; its level, therefore, will depend on that of the propositions to which it applies. If P and Q are propositions of mth and nth level, believing-that-P and believing-that-Q will be $m + $ 1st and $n + $ 1st level properties of individuals respectively.[36] It would seem possible for there to be two spirits, sharing all their non-psychological properties and believing (wishing true, desiring the truth of, etc.) exactly the same propositions at every level lower than some specified level, but differing in their attitudes toward some proposition of that level. If that *is* possible, then, since there is no highest level of propositions and so no highest level of propositional attitudes, there is no level such that the use of a bound predicate variable of that level in DefId will legitimize all instances of the Substitution of Identicals. A formal model inspired by this line of thought might well resemble Myhill's model in his proof of the indefinability of identity in the ramified theory of types.

5. REDUCIBILITY AND HETEROLOGICALITY[37]

The logicist program of reducing mathematics to logic is hopeless. Even when an axiom of infinity is assumed it is impossible to derive even first-order Peano arithmetic from a ramified logic, and even if arithmetic is assumed many of the classical proofs in analysis break down. Russell's response was to seek additional hypotheses, statable in the vocabulary of logic, from which the desired theorems would follow. One of these added assumptions was the Axiom of Infinity; another, a version of the set theoretic Axiom of Choice (the so-called Multiplicative Axioms) which, in view of the immense set theoretic literature, I shall not discuss. A third, which seems to have occasioned considerable confusion among later commentators, was the Axiom (or rather Axioms) of Reducibility, stating that to every higher level property there corresponds a coextensive property of lower level. More formally, in ramified second-order logic the axioms are, for every $n > 0$[38] and $m > 1$,

$$\forall F \exists G \forall x_1 \ldots \forall x_n [(Fx_1 \ldots x_n \to Gx_1 \ldots x_n) \wedge (Gx_1 \ldots x_n \to Fx_1 \ldots x_n)]$$

where F is an n-adic predicate variable of mth level, and G an n-adic predicate variable of the first, lowest, level. In a ramified logic of higher-order there would be similar axioms, with the place of F taken by a higher-order predicate variable of any level higher than needed to apply to its arguments, that of G by a variable of the same type and the lowest level compatible with its applying to its arguments, and those of the individual variables $x_1, \ldots, x_n$ with variables of appropriate type and highest possible level for the new predicate variables to apply to.

Russell [1908] said of these axioms that their

> assumption seems to be the essence of the usual assumption of classes; at any rate, it retains as much of the classes as we have any use for, and little enough to avoid the contradictions a less grudging admission of classes is apt to entail ...

and of the 'reduction' of higher or lower level properties postulated by the axioms of reducibility as being "what commonsense effects by the admission of classes".[39] In one direction, at least, it is easy to see what he meant. Suppose we assume that "there really are such things as classes" by allowing, as a new sort of individual, *sets* of our original individuals, with *membership* a primitive, lowest level, relation relating old and new individuals. On the realist conception of set implicit in standard set theory, any set of individuals, no matter how specified, will form a set, so, for any property of our original individuals, of whatever level, one of our new individuals will be the set of just those old ones having the property. Counting the set as an individual and membership as a first level relation, however, *membership in this set* will be a first level property coextensive with the original one. Thus the assumption of sets[40] as a kind of individual allows the derivation of Reducibility.

Given the Axioms of Reducibility, the classical development of mathematics proceeds as it does in the simple theory of types. This should not seem too surprising, for the versions of simple type theory usually considered in writings on the foundations of mathematics can be *interpreted* in ramified systems with reducibility. Simple type theory, as ordinarily encountered as a mathematical foundational system, has no non-logical constants, or, if the natural numbers are taken as individuals, only a few names for specific individuals and a few predicates of individuals. It does have variables of assorted types. Its syntax may, therefore, be identified with that of a fragment of a ramified system with the same names, and with the primitive predicates included as of lowest level. Of the variables of the ramified system, the fragment includes only those of the lowest level compatible with their arguments: more precisely, the fragment includes individual variables, second-order variables of lowest level, and at each higher order those variables taking as arguments terms of the types of variables already included in the fragment, and of lowest level compatible with taking those arguments. Since the rules for building formulas out of constants, variables, and logical symbols are the same for the two systems, and since there is a natural correlation between the types of variables in the fragment and those of the simple theory, every formula of the simple theory may be identified with a formula of this fragment of the language of the ramified system. Then, assuming that the two systems have the same axioms of infinity or other principles concerning individuals, it is easy to show that any formula of this fragment derivable in the simple theory is also derivable in the ramified system by use of Axioms of Reducibility. The basic propositional and quantificational logics are the same, with the only difference lying in the comprehension, or property existence, principles. In the simple theory we can infer the existence of a property expressed by a formula. In the ramified system, however, we can only infer the existence of higher level properties expressed by most quantified formulas of the fragment; these properties will not be in the ranges of the variables of the fragment. By the appropriate Axiom of Reducibility – itself not a formula of the fragment – we can then infer the existence of a coextensive lower level property that will be in the range of the variables of the fragment.

In this manner, detouring out of the fragment at each comprehension inference and then back into it by reducibility, any proof in the simple theory can be reproduced in the ramified system.[41]

What is perhaps harder to understand is Russell's claim that the adoption of the Axioms of Reducibility is somehow a weaker and less objectionable assumption than the postulation of classes. One tempting explanation is that, unlike Zermelo, he did not realize that a coherent notion of class, along the lines of the simple theory of types or its cumulative variant, the *iterative conception of set*, could be distilled out of the inconsistent mixture of notions implicit in Cantorian set theory. If coherence is interpreted as mere formal consistency, however, this is not fully satisfactory, for something like the interpretability argument sketched above was surely not beyond Russell's grasp. (As a thought experiment, try to imagine someone writing a three-volume work developing large parts of mathematics within ramified type theory with Axioms of Reducibility *without*, by the end, realizing how his proofs mimic those possible in simple type theory.) The problem was rather, perhaps, that Russell wouldn't have recognized the theories thus interpretable in ramified type theory with Reducibility as theories of *classes*. Cantor was not a logicist, and his explanations of the notion of a set make no reference to properties or concepts. Sets, on a Cantorian conception, have an existence totally independent of our concepts, and are thus entities of a completely different sort from the logician's 'propositional functions'. Even if a consistent theory of sets could be given, Russell might have asked, what right have we to suppose that they really exist in addition to propositional functions? And why is postulating their existence preferable to postulating numbers, for example, as primitive entities? The Axioms of Reducibility, in contrast, simply postulated more of a *kind* of entity he already believed in.

But what reason was there to believe in these extra properties? Much of the time Russell was willing to admit, as in Russell [1919], that the strongest argument in favor of the Axioms of Reducibility was the quasi-empirical one that they allowed the derivation of attractive theorems he saw no way of obtaining without them. This, however, smacks more of the pragmatic philosophy of mathematics espoused in, e.g., Quine [1960] and Putnam [1971] than it does of the Russell who scorned the advantages of theft over honest toil, and the outline of another defense can be extracted from Russell's work. The natural way of combining the epistemology of Russell [1912] with the logic of Russell [1908] is to take as the paradigm of a lowest-level property of individuals the having, by the individual, of some universal we know by acquaintance: if sense data are individuals, being bright red is a lowest level property of individuals. It is tempting to generalize this and say that any lowest level property is the having of some universal with which, in principle, some mind could be acquainted. (In practice, of course, there will be lots we aren't acquainted with, forcing us to appeal, not to the lowest level property of having a certain universal, but to some coextensive higher level property.) Now, it is at least somewhat plausible to say that, for some small finite collection of individuals, that we are acquainted with something they (and no others) have in common when we see them *as* a collection, when, that is, they form a *gestalt* for us (cf. Maddy [1980]). It is obviously impossible for us to perceive infinitely many objects as a unity, but Russell looked on the limitations of our finitude as logically contingent. An infinite intelligence could perhaps perceive infinitely many objects — even an infinite collection of objects that *we* could only characterize by citing some high level property they had in common — as a unity, and so be acquainted with a universal had by just the objects in the collection. The property of having this universal would then be a lowest level property coextensive with whatever higher-level property we might cite in specifying the collection. Such an appeal to the possibility of an infinite mind, however, is characteristic of the most extreme Platonist views in the philosophy of mathematics, and seems quite dubious. (If something like this *was* at the back of Russell's mind when he formulated the Axioms of Reducibility, he would have had to admit that *his* reasons for accepting the axioms were not as strong as those his Christian grandmother would have had!)

Since, from the point of view of deriving classical mathematics, a ramified logic with reducibility amounts to a somewhat roundabout way of getting the effect of the simple theory of types, a number of writers have been tempted to dismiss it as a mere historical curiosity. The dismissive tone of Quine [1936] (even more pronounced in the preface he wrote for the reprinting of Russell [1908] in Van Heijenoort [1967]) is typical. Such an attitude, however, is justified only if the goal in foundational studies is taken to be the formulation of an axiomatic system permitting the efficient derivation of large parts of classical mathematics without regard to the intuitive justification of the axioms. (Here Quine's dictum, that intuition, in matters connected with the logical paradoxes and the foundations of set theory, is bankrupt, expresses the most fundamental motivation for his pragmatic approach to the philosophy of mathematics.) In contrast, Russell, though admitting that axioms gained epistemic strength from plausible theorems as well as *vice versa*, was concerned with the intuitive justification of the axioms: the opening paragraph of Russell [1908] suggests that the ramified theory (with reducibility) is not

I.5: PREDICATIVE LOGICS

"wholly dependent on [the] indirect recommendation" of avoiding the familiar paradoxes, and claims for it "a certain consonance with common sense which makes it inherently credible". But more important than this statement of Russell's was his method. He required of his proposed system of the foundations of mathematics that it explain or 'solve' a variety of contradictions, some of them, like the Paradox of the Liar, of a decidedly nonmathematical flavor.

The concepts involved in the classical foundations of mathematics, and which are meant to be expressed by the predicates occurring in type theoretic foundational systems, have a very special kind of logical simplicity: they are *extensional*, in the sense that they do not distinguish between coextensive properties. This is trivially true of the concepts supposedly expressed by predicates of individuals: since they are not true of any properties, they do not distinguish among them. The properties of properties that can be expressed in a classical foundational system are the 'holds of' relations of various types, and whatever can be defined in terms of them using truth functions and quantification. It is possible to *express* the proposition that some property of properties is not extensional (just say that there are two properties holding of exactly the same things such that the higher-order property holds of one but not the other), but it is not possible to prove that there are such non-extensional properties,[42] and *a fortiori* no property of properties can be defined which can be shown to be non-extensional. Once one looks beyond the development of classical number theory and analysis to broader logical and epistemological concerns, however, many non-extensional properties of properties come to mind, and if one wants one's type theoretic system to be more than just a classical foundational system and to be adequate to the formalization of various sorts of non-mathematical reasoning, one will want to include primitive predicates expressing such properties.

One family of non-extensional properties is provided by the so-called *propositional attitudes*, belief, desire, hope, etc. These are naturally construed as relations between persons and propositions (or perhaps, cf. Lewis [1979], between persons and properties). As such, however, they are clearly non-extensional, for a person may believe one proposition and not another even if they have the same truth value, or want to have one property and not another even though the two properties are in fact coextensional. Note that, on the ramified account, each of the attitudes will divide into an infinite hierarchy: there will be a second level belief relation relating persons to first level propositions, and so on. It has been urged as an objection against the ramified theory that this splitting up of each attitude into a hierarchy is counterintuitive; there are, however, logical paradoxes concerning the attitudes,[43] and no other account that has been given of these paradoxes is clearly less counterintuitive than the ramified theory.

Another possible source of non-extensional properties is in the traditional distinction between relational and non-relational properties. Pittsburgh, Montreal, and Dublin are all cities of over 100 000 population; they are also all cities I have held jobs in: the first is a property the three cities have absolutely, whereas the second is one they have by relation to me. (Again, Montreal is further north than Pittsburgh and Dublin further north than Montreal; Montreal is also a city I first visited after my first visit to Pittsburgh, and Dublin one I first saw after Montreal: the 'further north than' relation is one that the city-pairs have absolutely, whereas the other is a relation they have by relation to me.)[44] It seems plausible to suppose that there are relational and non-relational properties that are coextensive with each other, but one will and the other will not have the higher-order property of being a relational property.

A third source of non-extensional notions is semantics. In sketching the proof that Axioms of Infinity allowed a ramified system to reproduce the derivations possible in simple type theory, I spoke loosely of properties being *expressed by* formulas as if this were an extensional property of properties: that a property was expressed by a formula just in case it was had by all and only the (sequences of) objects satisfying the formula. There is a stricter sense of the notion of expression on which it is not extensional in this way. The property of being (the Gödel number of) a provable formula of first-order logic is expressed by a number-theoretic formula, at least relative to a given arithmetization of syntax. The property of being a valid formula of first-order logic, on the other hand, is defined as truth on all interpretations, where an interpretation – a correlation assigning, at the very least, a domain to the variables and extensions to the predicates – is a higher type object and not a number. It, therefore, is *not* expressed by a number-theoretic formula, despite the fact that, by Gödel's completeness theorem, the two properties are coextensive.

If the language of a ramified system is capable of expressing any of these non-extensional notions, it cannot be assumed that the addition of Axioms of Reducibility will 'reduce' it to or 'collapse' it into the corresponding version of simple type theory. This can be seen by a consideration of the logical paradoxes. On the theory of properties embodied in simple type theory, these are genuine antinomies, and, when intuitively plausible axioms for the non-extensional notions involved are added to formal systems of simple type theory, lead to contradictions. The distinctions made by the ramified theory block the paradoxes, so that the corresponding formal systems of ramified type theory are consistent with the axioms for the nonextensional notions. Furthermore, the addition of Axioms of Reducibility does not reintroduce the contradictions, which would only follow if we further assumed that the lower level properties postulated by the Axioms of Reducibility were *identical*, and not merely coextensive, with their higher level 'originals'.

The Grelling Paradox, the paradox of heterologicality, provides a typical example, one discussed in, *inter alia*, Weyl [1918], Ramsey [1926] Church [1976] and Myhill [1979]. Call an adjective *heterological* just in case it does not itself have the property it expresses. Thus 'monosyllabic', being itself pentasyllabic rather than monosyllabic, is heterological, but (stipulating monosyllabicity as the criterion of shortness for words) 'short', being a short word, is not heterological. By definition, then, if the word 'heterological' is not heterological — if, that is, it lacks the property of heterologicality — then it *is* heterological. On the other hand, if it is heterological, then by definition it is *not* heterological, and, reasoning classically,[45] we have an explicit contradiction.

This argument can be formalized in a formal system of type theory with appropriate additional predicates and intuitive axioms for them. Let us first show how it produces a contradiction in simple type theory. (I shall state the extra axioms, etc., in English, but the translation into logical symbols is obvious.) The above informal exposition of the Grelling Paradox considered heterologicality as a property of *words* and introduced a newly defined adjective, 'heterological', but this is clearly inessential. In the formal version, heterologicality will be a property of (possibly quite complex) *formulas*: for definiteness, formulas containing, as their only free variable, the individual variable x.

The first requirement on our formal system of simple type theory, then, is that it has the vocabulary necessary to discuss, and the axioms needed to deduce the relevant propositions of, its own syntax, either *via* a scheme of arithmetization, as in Gödel [1931], or directly, as in Tarski [1936].[46] Next, we will want an *expresses* predicate, a two-place predicate taking as arguments terms for (in the first place) individuals and (in the second place) monadic properties of individuals. It will also be convenient to take identity, of properties as well as of individuals, as a primitive notion, with axioms allowing substitution of identicals in arbitrary contexts. The axioms for expression should guarantee three things. First, and trivially, they should provide that no individual which is not a formula containing a free occurrence of the variable x, or which is such a formula but contains free occurrences of other variables as well, expresses any property. Second, they should imply that every individual which *is* a formula of the specified kind expresses one and only one property. These latter implications should seem intuitively correct to anyone who finds the simple theory of types a credible theory of properties, for they amount to saying that the well-formed formulas of the formal system are both meaningful and unambiguous, and these seem like fairly minimal demands to make of an acceptable formalized theory. Note that the uniqueness of the property expressed by a formula licenses us to speak of *the* property expressed by that formula; in a fully formal presentation of the argument my uses of the definite article could be replaced by either universal or existential quantifications. Finally, they must imply all the 'disquotation' principles: given a formula, $\ldots x \ldots$, having x as its only free variable, if any individual x has the property expressed by the formula, then $\ldots x \ldots$, and if x does not have the property expressed by the formula, then it is not the case that $\ldots x \ldots$. To put this a bit more formally, for any such formula $\ldots x \ldots$, there will be a 'structural descriptive' (or, if we have chosen to arithmetize our syntax, arithmetical) formula $---y---$, saying that the individual y is the formula $\ldots x \ldots$. Then our third requirement for the expression axioms is that they imply all the *positive* disquotation principles, ⌜For all x and y, if $---y---$ and x has the property expressed by y, then $\ldots x \ldots$⌝, and also all the *negative* ones, ⌜For all x and y, if $---y---$ and x does not have the property expressed by y, then it is not the case that $\ldots x \ldots$⌝. Now let the formal analog of 'heterological' be the formula ⌜x does not have the property it itself expresses⌝. (For a system of simple type theory with classical logic: for a formal system based on one of the nonclassical logics mentioned in the earlier note, a different formula would have to be chosen.)

The contradiction may now be derived, essentially by a formalization of the informal argument given above. By the existence and uniqueness of properties expressed by formulas of the right sort, we know that we can speak of *the* property expressed by this formula; by a positive disquotation principle we can conclude that it doesn't have the property it expresses, and by a negative one that it does.

In a formal system of ramified type theory, the argument never gets started. We may assume the same treatment of syntax, and the primitive notion of identity (now for properties of all levels), but the expression relation must be assumed to have a level higher than the level of the properties it

I.5: PREDICATIVE LOGICS 373

relates to formulas. In our formal language, therefore, we should have a hierarchy of *expresses* predicates, a second level one relating formulas to first level properties, a third level one relating formulas to second level properties, and so on, and the intuitively justified axioms will be correspondingly weaker. Of the three things our expression axioms were to guarantee, the first (that only formulas of the right sort express anything) may be assumed for expression at any level. For the second (existence and uniqueness of expressed properties), it is plausible to assume that formulas of level $(n-1)$ or lower will *express* (in the sense of the *expresses* predicate of level n) unique properties of level $(n-1)$, but not that any formula *expresses* (in the sense of the *expresses* predicate of its own or any lower level) anything at all, or that it *expresses* (in any sense) a property of lower than its own level. Since the existence and uniqueness of expressed properties is no longer guaranteed, we must be more circumspect about the definite article. Treating it in accordance with Russell's Theory of Descriptions and using our earlier notation, a reasonable positive disquotation principle for the *expresses* predicate of level n will be ⌜For all x and y, if - - - y - - - and there is a unique property of level $(n-1)$ *expressed* by y and x has that property, then . . . x . . . ⌝, with the negative disquotation principles suffering similarly. Now the analogue of 'heterological' will be a formula of the same level as the *expresses* predicate it contains; in the sense of its own *expresses* predicate, it will not express anything. In order to derive the contradiction, we would need to use disquotation principles with *expresses* predicates of the same level in the antecedent and the consequent, but it follows from the last remark that such a disquotation principle will have a false antecedent, and so be true trivially, and unusable. An analogue of 'heterological' will, of course, express, in the sense of the next higher *expresses*, a property of the same level as its contained *expresses*, but the relevant disquotation principles here will have an *expresses* predicate in the antecedent one level higher than the one in the consequent, and so not lead to any contradiction.

If we assume the Axioms of Reducibility, we will be able to infer that the higher level property expressed by a given analogue of 'heterological' has a lower level copy coextensive with it, but only if we further assumed that copy and original are *identical* will we be able to substitute and rederive the contradiction from the disquotation principles.

The post-*Principia* history of discussion of the Axioms of Reducibility has been curiously muddled. Chwistek seems to have been simply confused about the distinctions connected with the Axioms of Reducibility; certainly the derivation of Richard's Paradox in Chwistek [1921] is fallacious.

Ramsey [1926] was not confused. He points out an error of Chwistek's and accurately summarizes the state of affairs with regard to the Grelling Paradox and the Axioms of Reducibility. Ramsey, however, was not in sympathy with the conceptualistic notion of property embodied in the ramified theory,[47] but rather believed that there really were such things as classes, and that his understanding of the concept of set (or, as he termed it, the 'extensional' concept of a propositional function) was clear enough to justify the adoption of the simple theory of types. As for the logical paradoxes, he distinguished between those (Russell's, Cantor's Paradox of the set of all sets, Burali-Forti's of the order type of the ordinals) which involve purely mathematical notions, and which seem to be blocked by the simple theory, and those on the other hand which involve 'linguistic' or (Ramsey's preferred formulation) 'epistemological' notions, and essentially recommended that students of the foundations of mathematics just ignore the latter.[48] By the 1930s the attention of most (classical minded) researchers in the foundations of mathematics had decisively shifted to simple type theory and the (formally simpler but conceptually very closely related) set theories of the Zermelian tradition, and the ramified theory as a whole, as well as the question of the Axioms of Reducibility in particular, was largely neglected. By the beginning of the second half of the century, despite a small revival of interest in predicative conceptions among mathematically minded logicians, it was possible for influential philosophers to dismiss the ramified theory of types as 'a philosophical mistake', and even to attribute it to a confusion of use and mention on Russell's part. (On the last point, Russell may not always have been unconfused, but much of what seems like use-mention confusion in, say, Russell [1908] will be seen by a sympathetic reader as a harmlessly loose mode of expression, allowing both more concision and a more relaxed and flowing style than is possible with a strict attention to the niceties of use-mention distinctions.) Only in the last few years, with the growth of interest in the logic of the attitudes and the revival, starting with Kripke [1975], of interest in the semantic paradoxes, is the tide perhaps beginning to turn.

The Axioms of Reducibility, however, seem likely to remain in disgrace. They are too closely related to the assumption of the real existence of classes and, though by appeal of explicitly set-theoretical principles one can assert the existence of models in which they are satisfied, the conceptualist notion of property seems to provide no grounds for belief in their truth. A convinced conceptualist or predicativist will want to consign the lowest level properties postulated by the Axioms of Reducibility to the same sort of Limbo as that inhabited by the infinitesimals of pre-Weierstrassian calculus. It is perhaps not inappropriate to paraphrase Bishop Berkeley and say that these postulated properties are the ghosts, not indeed of *departed* concepts, but of concepts that were never born.[49]

6. PREDICATIVE MATHEMATICS

The logicist program of reducing mathematics to logic is hopeless. If the predicative conception of pure logic, as embodied in the ramified theory of types, is accepted, something else besides logic is needed for the development of significant amounts of mathematics. One alternative is that discussed in the last section: that of accepting the impredicative reasoning of late 19th- and 20th-century mathematics on its own terms, of postulating (either directly or under the guise of the Axioms of Reducibility) the real existence of classes. In one form or another this has been the majority response among mathematicians, so much so that for a time the study of the foundations of (classical) mathematics could almost be identified with that of set theory.[50] From the point of view of the set-theoretic realist, ramified logics are of comparatively little interest. (In one respect, however, ramified type theory has been technically suggestive: the hierarchy of constructible sets of Gödel [1940] amounts to a ramified hierarchy in which the levels, and orders, go up through all the ordinals. Quine [1963] seems to recommend acceptance of the generalized continuum hypothesis, which holds on the assumption that all sets are constructible, on grounds of simplicity, but most set-theoretic realists tend to believe that it, and *a fortiori* the Axiom of Constructibility, is *false*. Cf. Gödel [1947].) A more conservative approach is to assume as somehow given, independently of logic, the notion of a *natural number* (or, equivalently, that of a *finite* set or sequence), and to build up the rest of mathematics by predicative means on this basis. In admitting, as objects of mathematical study, only the natural numbers and the (predicatively conceived) properties definable, ultimately, in terms of them, this latter approach is in the spirit of Kronecker's famous dictum, "God made the natural numbers; all the rest is human workmanship". Its natural formalization is a ramified system having the natural numbers as individuals, the familiar notions of elementary arithmetic, sum, product, etc., expressed by primitive predicates of level 1, and (recalling Poincaré's insistence that mathematical induction was a primitive, intuitive, form of inference, irreducible to logic) the principle of induction as a rule or axiom scheme for formulas of every level.

The inclusion of the principle of mathematical induction at higher levels makes such a system much stronger than the plain ramified logic, even when the latter is supplemented by an axiom of infinity. As pointed out in Section 2 above, *truth* for formulas of first order arithmetic can be defined in a predicative second-order system assuming the natural numbers as individuals. The only thing that keeps such a system from being able to prove the consistency of first-order arithmetic (or, *mutatis mutandis*, that keeps NBG from being able to prove the consistency of ZF) is the underivability of certain instances of the induction principle involving the notion of truth. With the aid of higher-level induction, it is possible in predicative second-order arithmetic to prove not only the consistency of first-order arithmetic, but also such further 'reflection principles' as that any sentence of first-order arithmetic provable in first-order arithmetic is true, and that any universal quantification (in the language of first-order arithmetic) all of whose numerical instances are provable is true. Proofs of these principles can be reproduced at higher levels as well: for each fragment of the full predicative system, containing quantifiers only over properties of some restricted hierarchy of levels (and orders), but admitting the principle of mathematical induction for formulas of each of its levels, it is possible to prove its reflection principles in the larger fragment containing quantifiers over properties of higher level and order, using, in the proof, mathematical induction for formulas containing these higher-level quantifiers. Higher *order* quantification is needed only because, on the natural interpretation, formulas containing second and higher order variables are satisfied by sequences having *properties* at some places. By appealing, in our proofs, to a substitutional interpretation of the second-order quantifiers, we can prove consistency and reflection principles in a system of ramified *second* order arithmetic: quantification over each new level, and induction for formulas containing the new quantifiers, allowing the proof of consistency and reflection principles for the system of second-order arithmetic of lower level.

Many classical proofs in analysis cannot be reconstructed within predicative systems of this sort, and not all of the important theorems of the subject can be proved to hold with full generality. The classical theorem that every set of real numbers bounded above has a least upper bound can serve as an example of the problems that arrive. Since the rational numbers can be coded by the natural numbers in various well-known ways, we can assume that the relationals are individuals of the system, and that the elementary theory of rational arithmetic can be developed without quantifying over properties.

I.5: PREDICATIVE LOGICS 377

Now, in the classical, set-theoretic development, real numbers may be taken as sets of rationals: one typical approach takes them to be *lower Dedekind cuts*, sets of rationals that do not contain all the rationals, but each of which contains every rational less than any rational it contains. The order of the reals is then easily definable in terms of that of the rationals – one real (i.e. one lower Dedekind cut) is greater than another just in case it contains some rational greater than any rational contained in the other. Now consider a set of such real numbers which is bounded above – that is, a set such that some real number not in the set is greater than any in the set. It is easy to see that the union of such a set of lower Dedekind cuts will itself fulfill the definition of a lower Dedekind cut, and that, considered as a real number, it is a least upper bound of the set, being, in the sense appropriate to real numbers, greater than or equal to any real in the set, but less than any other real which is greater than or equal to every real in the set. If we attempt to retrace the steps of this development in a predicative system, allowing properties to take the place of sets, we must specify *levels*. Suppose, then, we allow level n properties of rational numbers, for some n, to play the role of real numbers. (Since the condition on properties of being lower Dedekind cuts can be defined by quantifying only over the rationals themselves, we may let the reals be properties of any level we wish, and the property of being a real number will be a property of the lowest level taking properties of that level as arguments.) Corresponding to a set of real numbers will be a property of these properties, and the union of the set – the property of being a rational number in some members of the set – will be defined by quantifying over the properties having this property. Since it is defined by quantifying over real numbers, it will be of a level higher than the original reals, and thus cannot serve as least upper bound for a set of reals. The classical theory fails at one of its first steps.

Subtler proofs, however, allow the recovery of at least the most important, central, cases of many classical theorems. This has been established by the researches of a small number of mathematicians and logicians, starting a few years after the publication of *Principia Mathematica* with Weyl [1918]. Weyl considered ramified systems legitimate, but too complicated to work with practically, and so based his reconstruction of analysis on a system of second-order number theory quantifying only over properties of level 1. Despite this restriction, he was able to recover most of the familiar theory of *continuous* functions.

That such a recovery should be possible may at first seem surprising, but becomes more plausible when one recalls that a continuous function is determined by its values for *rational* arguments. It is therefore possible to represent continuous functions by dyadic relations of rational numbers (relating, e.g., each rational argument to all the rationals in the lower Dedekind cut corresponding to the value of the function for that argument), and thus to represent these functions as properties of the same level as that chosen to represent the real numbers themselves. Denumerable *sequences* of reals may similarly be represented by relations, the relation holding of a pair of (natural/rational) number arguments just in case the second is in the cut corresponding to the real number indexed by the first argument. The least upper bound of the reals in such a sequence, furthermore, may be defined by quantifying over the indexing *numbers* instead of over properties, and so may be of the same level as the sequence and as reals. What Weyl discovered was that, in place of the assumption of least upper bounds for arbitrary bounded *sets* of real numbers, it sufficed, for the development of the familiar theory of continuous functions, to have least upper bounds for bounded *sequences* of reals.

Further developments along the lines of Weyl [1918] have been Lorenzen [1965] and Zahn [1978] (which last I know only by the review Richman [1982]). The upshot would appear to be that, at the cost of some complication of the proofs, the basic mathematical methods used in the natural sciences can be justified by appeal to conceptualist and predicative principles. One implication for the philosophy of mathematics, then, is that whatever the theoretical validity of the so-called *indispensability argument* for Platonism – the argument that, since any acceptable systematization of physics involves quantifying over such entities as numbers, functions, or sets, such abstract entities *must* be accepted by any rational scientist[51] – it cannot, drawing on current science and mathematics for its premises, yield the strong form of Platonism embodied in impredicative set theory, but only the existence of such abstract entities as are quantified over in conceptualistic systems.

Weyl and his successors may be seen as giving 'lower bounds' to the extent of predicative mathematics by their positive results. At the opposite extreme there is the question of 'upper bounds' – the question of what parts of mathematics, if any, cannot be recovered at all in any predicative system, and of how the general notion of predicative provability may be characterized. Here we encounter conceptual or methodological difficulties. An interesting general characterization of the predicatively acceptable is unlikely to be predicatively acceptable itself. Thus, for example, in characterizing the predicatively acceptable, one will probably want to quantify over all predicatively definable properties, but if this kind of quantification is granted, *further* properties will probably be definable, whose definitions, *except* for the quantification over predicatively definable properties, will

I.5: PREDICATIVE LOGICS

meet predicative standards. If the quantification over all predicatively definable properties were itself predicatively acceptable, these further properties would be predicatively defined, *contra* our hypothesis that the original quantification was over *all* predicatively definable properties. The precise definitions of predicativity proposed by Feferman and others, therefore, are characterizations from the outside – as Feferman [1964] has put it, they are descriptions of "what the predicative 'universe' looks like to the impredicativist". Someone fully convinced of a conceptualist philosophy of mathematics that allowed only the predicative a place in genuine mathematics would dismiss them as mere myth, just as a believing intuitionist would reject the various classical interpretations of intuitionist logic.[52] Since I personally find the conceptualist philosophies of mathematics not unattractive, I suppose I should feel a twinge of guilt at presenting such characterizations: but then, predicativists, like intuitionists, are in the minority, and myths can have a heuristic value in speaking to the uninitiated.

If the hierarchy of levels is allowed, in the fashion of Gödel's [1940] hierarchy of constructible sets, to extend arbitrarily far into the transfinite, the ramified universe contains models of strong classical systems. The most famous result of this kind is, of course, Gödel's [1940] proof that the constructible sets themselves form a model of full ZF set theory with the Axiom of Choice.[53] Another is that there is a *denumerable* ordinal such that the axioms of full, impredicative, second-order arithmetic hold if its second-order variables are taken to range over all the properties in a ramified second-order hierarchy with natural numbers as individuals and properties of every level less than this ordinal. (Cf. Mostowski [1961] and Boyd *et al.* [1969].) The existence of such ordinals, however, is only established by reasoning no one regards as predicative, and the least such ordinal (known as β_0) is larger than any for which a recursive system of notation exists for the ordinals up to it, so there is no ramified *formal* system, of the sort described above, in which impredicative second-order arithmetic can be interpreted: there is no way to specify a system of level subscripts for our variables that would allow us to formulate a ramified system with levels up to β_0.

In characterizing the predicative, therefore, we must place stricter bounds on the levels admissible. Gödel [1944] suggests that "there would ... be no objection to extend [the hierarchy of levels] to such transfinite ordinals as can be constructed within the framework of finite orders". ('Orders' being so used that this implies finite level as well.) Wang [1954] makes the similar suggestion that we should start with a ramified system having levels up to some low transfinite level (low enough that the predicative character of the system is not in doubt), then allow as predicative ramified systems having levels up to any ordinal definable in the first system, and to count as a predicative system any system that can be obtained by repeated applications of this procedure. An alternative proposal, considered but dismissed in Wang [1954] and urged in Kreisel [1960] as giving an upper bound to the predicative levels, is to allow as levels all and only the *recursive* ordinals – the order types of well-ordered recursive relations of natural numbers. On this proposal, a level would be predicatively admissible just in case there was a formal system having variables for all levels up to it, and predicative mathematics could be identified with what is provable in some ramified formal system (having natural numbers as individuals and allowing mathematical induction for formulas of every level). (Note that there is no *one* formal system having variables of every recursive ordinal level – the order type of the recursive ordinals is not itself the order type of a recursive relation – so Gödel's incompleteness theorems would not apply to the corpus of predicative mathematics so defined. In fact Feferman [1962] has shown that every truth of first-order arithmetic is derivable in some formal system of the kind described.)

It follows from the result of Spector [1955] that (assuming the natural numbers as individuals and the usual elementary arithmetic predicates as of level 1) the Gödel–Wang proposal and the proposal to admit exactly the recursive ordinals as levels come to the same thing. The sets of natural numbers defined by formulas of ramified second-order arithmetic with recursive ordinals as levels – that is, the sets defined by predicative properties according to the second proposal – are exactly the *hyperarithmetic* sets, and by Spector's result, for any hyperarithmetic well-ordering, there is a recursive well-ordering of the same order type. (Cf. also Kleene [1955].)

Despite its naturalness, the identification of predicative mathematics with what is provable in ramified systems with effective notations for levels is subject to a serious objection. Consider how a ramified system with transfinite levels up to some recursive ordinal may be presented. There will have to be some sort of notation (a system of subscripts, say) marking the level of the variables, and with infinitely many levels the level symbols will have to be complex. The specification of the rules and axioms of the system must, therefore, in effect include an algorithm for recognizing which combinations of symbols are allowable level subscripts, and an algorithm for determining, of any two level subscripts, which is lower. If one is simply presented with an algorithm, however, there is no general way of telling whether it will give an output for every possible input (pair of inputs), and so whether it can be taken to define a set (relation), nor, given that an algorithm defines a relation, is there in general a way of telling whether or not this relation is a well-ordering. This implies that we may not be able to tell, of a purported presentation of a predicative system, whether or not it in fact is a presentation of a full system at all, nor, given that the presentation is of a well-defined formal system, whether that system is genuinely one of ramified second- (or higher-) order arithmetic, for on the one hand we may not have a recursive system of level subscripts, and on the other, even if we have a recursive system of subscripts, the supposed 'levels' may not be well-ordered.[54] Now consider something which is *in fact* a formal system of ramified second-order arithmetic, but with predicate variables of levels marked by some system of level subscripts so complex and unperspicuous that we have not been able to establish that they are well-ordered. Formulas in the language of such a system will in fact define properties of natural numbers, but we may hardly *assume*, in the course of our mathematical reasoning, that they do, just as a definite description of a natural number may *in fact* denote, though the fulfillment of its existence and uniqueness presuppositions may not be assumed in a proof unless it has been proved to denote. Theorems derivable in such a system will be true, but in the absence of some mathematical guarantee that the system is predicatively acceptable they cannot be thought of as having been proved.

What is needed, therefore, is a modification of the Gödel–Wang proposal that will allow an ordinal as a level of a predicatively acceptable system only if it can not only be *defined* within a system that has already been accepted as predicative, but can be shown to be an ordinal by means of principles available in the earlier system. Thus we come to the proposed analysis, studied notably by Feferman, of predicativity in terms of *autonomous* progressions of systems. (Cf. Feferman [1964, 1968, and 1978] and papers cited therein (notably Feferman's own more technical papers, and Kreisel [1960, 1960a]). Of these, Feferman [1964] in particular is a survey article, and perhaps the best introduction available to the whole field of predicative mathematics and its metatheory.) Very roughly, and without any hint of the variety of considerations Feferman adduces to support the naturalness of the proposal, we may describe it as follows. Call an ordinal a *provable* ordinal of a system just in case it is the order type of a relation which can be proven in the system to be a well-ordering, in the sense that a transfinite induction scheme is derivable for it. (Typically the provable ordinals of a system will be intermediate in magnitude between the level of the system and ordinals *definable* in the system. Thus, for example, first-order arithmetic is a system all of whose formulas are of level 1, in which all recursive well-orderings are definable, and has as provable ordinals just the ordinals less than ϵ_0.)

A (possibly infinite) series of systems of ramified second- (or higher-) order arithmetic, with later systems in the series having variables of higher level than earlier ones, is an *autonomous progression* if the levels of variables in a later system are all provable ordinals of earlier systems. Feferman's proposal, then, is to identify predicative mathematics with what can be proven in systems belonging to autonomous progressions (starting from systems of undoubted predicative character); he has shown that a particular recursive ordinal, called Γ_0, which unlike β_0, has a simple, nonmetamathematical, characterization in terms of classical notions of ordinal mathematics, is the limit of the provable ordinals of these systems. Predicative provability on Feferman's proposal, then, is more restricted than on the proposal allowing arbitrary recursive ordinals as levels: in particular, not all truths of first-order arithmetic are, on this conception, predicatively provable.

Delicate conceptual considerations are involved in any proposed general characterization of the predicative. One example of the delicacy: it is possible to specify a single formal system incorporating all of the predicative mathematics (on Feferman's proposal), essentially by formalizing the principle that theorems provable in a system with variables of a certain level may be asserted if the admissibility of that level has already been demonstrated by the derivation of an appropriate induction scheme. Gödel's incompleteness theorems will apply to this system, so its consistency (a sentence of first-order arithmetic) will not be provable in it. Surely, however, it is tempting to think that if *all* the theorems of a system are predicatively provable, a predicativist would be entitled to conclude that the system is consistent. 'Reflection inferences' of this kind, where the consistency of an accepted system is inferred, however, depend for their intuitive evidence on the notion of truth: we believe the system is consistent because we are confident that only truths are among its theorems. Now, if Feferman is right, the notion of truth for a language whose variables have levels bounded by some ordinal less than Γ_0 is a predicatively acceptable notion, but the general notion of truth for a system containing all of predicative mathematics is not; indeed, though every theorem of the system is a theorem of some predicatively acceptable system, it would not be possible, while maintaining a predicativist viewpoint, to recognize the whole system as correct. Feferman [1978] contains a discussion of this problem.

7. TARSKI AND SUCH

Since Russell formulated the ramified theory of types at least in part in response to what have since come to be termed the *semantic* paradoxes, it is of interest to compare the treatment of them that can be given on the basis of the ramified theory with more recent theories of truth. Recent theories, since Tarski [1936], have concentrated on the extensional semantic notions of *truth* and *satisfaction*. The natural semantic notion to take as primary, in the context of ramified type theory, is that of *expression*, or some comparable intensional relation holding between linguistic items (sentences, formulas, and so forth – which, for simplicity, we can assume to be individuals) and propositions or properties. Given such notions, it is possible to define the Tarskian semantic predicates in a simple and uniform way. Thus, for example, where x is an individual variable, p a propositional variable, S a (we may assume of level 1) monadic predicate of individuals (interpreted as *is a sentence*), and E a predicate of type (individual, proposition of the level of p) and of level one higher than the level of p (interpreted as *expresses*), we may define

as
$$x \text{ is true}$$

(T.D.) $\quad Sx \wedge \exists p(E(x,p) \wedge p).$

Notice, however, that this definition amounts to an infinite hierarchy of definitions, one for each level of propositional variable (and corresponding level of *expresses* predicate) used. Each definition will be adequate for sentences expressing propositions of low enough level to come within the range of the propositional variable used, but will fail when sentences are considered which express higher level propositions: almost automatically, we get a hierarchy of definitions of truth resembling that of Tarski [1936].

Church [1976] has shown that the resemblance is more than accidental, and that Tarski's [1936] hierarchies of truth predicates – the "levels of language" explanation of the semantic paradoxes – can be derived within a ramified theory. Start with a system of ramified type theory, comparable to that of *Principia Mathematica*, having variables of every finite level, the vocabulary needed to formulate its own syntax, and, for every natural number m and every positive integer n, an *expresses* predicate (which, if formulas are taken as individuals, may be assumed to be of level $n + 1$) relating formulas with m free variables to m-adic properties of level n. (Actually, depending on the details of the type structure of the system, the family of *expresses* predicates will usually be even more complicated.)

When truth and satisfaction predicates are defined in the obvious way in terms of expression, they will form a Tarskian hierarchy: there is an infinite series of truth and satisfaction predicates, each applying only to formulas containing no semantic predicates at all, or only semantic predicates defined in terms of lower level *expresses* predicates. Further, the principles of a Tarskian theory of truth are derivable within the ramified system from the definitions and plausible principles about expression. We obtain a Tarskian hierarchy of *languages* by considering those fragments of the language of the ramified system containing, for some n, only variables of level n or lower and predicates of level $n + 1$ or lower.

In one way, as Church points out, what this shows is that the Russellian treatment of the paradoxes can be seen as a 'special case' of the Tarskian: not all Tarskian hierarchies of languages and truth and satisfaction predicates arise in this manner from an underlying ramified type system. As Church also points out, however, the opposite perspective is possible. Ramified type theory, as a general theory of possible concepts, is more general than Tarski's theory of truth and satisfaction, and Tarski's hierarchies, which might otherwise seem somewhat *ad hoc*, are explained when they are derived from the more general theory.

Church [1976] also points out that the reduction by definition goes in only one direction: that, because of the intensional nature of the *expresses* predicate, it cannot be defined in terms of satisfaction. Where intensional considerations are unimportant, as in consistency proofs, however, the Tarskian and Russellian frameworks are on a par, for (along lines sketched in the appendix to Parsons [1971a]) a ramified type theory can be interpreted in a Tarskian system containing appropriate satisfaction predicates for fragments of its own language.

Clearly the restriction to finite levels is not essential.

The new theory of truth outlined in Kripke [1975] streamlines and improves Tarski's treatment in a number of respects. The mathematical generality and elegance of Kripke's theory (which is only hinted at in the published paper) need not concern us, but there are several points of conceptual interest which make the relation to the ramified theory of types, even with transfinite levels, far from obvious. First, although each occurrence of a truth or satisfaction predicate in a Kripkean language can, if grounded, be thought of as having an implicit level, the levels are not marked by any explicit notation. (In terms of the naturalness of the formal languages as models of our informal use of the semantic notions, this is clearly a point in Kripke's favor: for although, as Charles Parsons has pointed out, more things may be explicit in the 'deep structures' uncovered by syntactic and semantic analysis than are on the surface, at least superficially natural-language truth predicates have nothing comparable to level subscripts.) It is important to note that the omission of level notation is not simple ellipsis – it is not as if someone uttering a sentence of a Kripkean language *intends* each occurrence of a semantic predicate to have a level he does not explicitly specify, but rather that, as Kripke puts it, the semantic predicates *seek their own levels.* The believer in papal infallibility confidently asserts

(PI) Whatever the Pope says is true.

In order for this to express what the believer clearly intends, the level of the occurrence of the truth predicate in PI must be higher than that of any occurrence of a semantic predicate in any papal utterance. No level, therefore, can be specified simply in virtue of the form of PI, for whatever level one chose, it is at least possible that the Pope has at some time made a statement with a higher level. Nor could the believer, without an exhaustive examination of the whole corpus of papal utterances (and, if the Pope ever pronounces on the truth values of statements he does not quote, of their subject matter, and so on indefinitely), have intended his use of the truth predicate to have any specific level. Levels are assigned to occurrences of semantic predicates, and truth values to sentences containing them, by a transfinite recursion in which the levels, as well as the truth values, depend on the facts and not just on the form of the sentences. (It is thus tempting to compare Kripkean semantic predicates to pronouns and other indexical expressions: the level of an occurrence of a semantic predicate, like the referent of a pronoun, is not determined by the meaning of the expression, but depends on the situation in which it is uttered.) These levels may be as high as the ordinal of any well-ordering definable in the language containing the predicate. Thus, for example, suppose the language of first-order arithmetic is extended with a Kripkean truth predicate (assuming some scheme of Gödel numbering so that sentences of the language can be identified with numbers), and consider, for any formula A with two free variables and any number n, the sentence "Everything bearing the relation defined by A to n is true" (i.e. $\ulcorner \forall x(A(x, \bar{n}) \to T(x)) \urcorner$). If the relation defined by A is well-founded 'below' n, this sentence will be grounded, and will have as its level the 'height' of n in the ordering. Thus, for any recursive system of ordinal notation we might choose for level subscripts in a Tarskian language, there will be occurrences of the truth predicate in this Kripkean language of higher level than any whose level could be specified explicitly with those subscripts. Indeed, if something like Feferman's analysis of the extent of predicative provability (cf. Section 6 above) is right, it is possible to make truth attributions in this, relatively simple, Kripkean language having higher levels than any that could be explicitly specified in *any* Tarskian language whose system of level notation could be shown predicatively to be correct.

The other side of this flexibility, however, is risk. There is no way of ensuring, simply by looking at the syntactic structure of a sentence, that it will be grounded: indeed, much of the flexibility of Kripkean languages lies precisely in the fact that they allow sentences like PI which have levels and truth values if the facts are right, but will be ungrounded if the facts are unfavorable. As any Protestant logician will point out, PI may be ungrounded if the Pope ever asserts an ungrounded sentence, and the groundedness of the Pope's utterances will in turn depend on that of whatever statements by other people he choses to endorse or condemn as false. In order to allow for ungrounded and truthvalueless statements, Kripke considers a variety of three-valued valuation schemes ('logics'), favoring, at least in exposition and over some alternatives, that based on the so-called 'Strong Kleene' three-valued matrices.

In comparing Kripke's theory with one based on the ramified theory of types, it will be convenient to have versions of the two theories whose details are as similar as possible. To this end it would seem reasonably simple to formulate a three-valued variant of ramified type theory based on the Kleene matrices, with formulas getting the middle, undetermined, 'value' only if they contain property variables of undetermined level. On the other hand, it is probably more perspicuous to work with a relatively conventional ramified type theory, and reformulate Kripke's theory to make use of classical logic. Feferman [198?] contains just such a variant of Kripke's theory. In a Kripkean language based on the three-valued scheme, falsity is expressed by the negation of truth. The first step is to introduce a primitive falsity predicate (and, in the general case, a *nonsatisfaction*) predicate. Now, both in classical, two-valued, logic and on the Kleene matrices, negations can be driven inward, in the sense that every formula is demonstrably equivalent to one having only conjunction, disjunction, the quantifiers, and negation as logical constants (i.e. one in which the conditional does not occur), and in which the negation operator occurs only before atomic formulas. Feferman obtains what he calls the *positive approximant* of a sentence by first deriving its negations inward and then replacing negated occurrences of the truth (falsity) predicate with occurrences of the falsity (truth) predicate, so no semantic predicate stands in the scope of a negation operator. The *negative approximant* is defined as the positive approximant of the negation of the sentence. Feferman then assigns truth values (and levels) to sentences by a transfinite recursion like Kripke's, with a sentence being placed in the extension of the truth (falsity) predicate at a stage just in case its positive (negative) approximant is true at an earlier stage, with the level of a grounded sentence being determined by the stage at which it first goes into the extension of the truth or falsity predicate.

It would seem possible, then, to interpret this sort of variant of a Kripkean language in the framework of ramified type theory, using a definition very much like T.D. for truth with an *expresses*⁺ predicate holding between a sentence and a proposition just in case the positive approximant of the sentence expresses the proposition, and defining falsity in much the same way but with an *expresses*⁻ predicate that holds between a sentence and a proposition in case the negative approximant of the sentence expresses the proposition. (Clearly, expression⁺ and expression⁻ can be defined in terms of ordinary expression and standard syntactic notions.) Since, when Feferman's recursion is used instead of that based on the three-valued matrices, classical logic is appropriate in the language defined by the (minimal fixed point of the) recursion, there will be no problem about using a version of the ramified theory based on classical logic. What is a problem, and what will call for a departure from the conventional ramified formalisms, is the 'level seeking' feature of the Kripkean and Fefermanian languages. The language with level seeking semantic predicates will have to be interpreted, not in an ordinary ramified language with explicit level specifications for all its second- and higher-order variables, but in a language having an additional set of *unspecified* propositional variables (and in general, where satisfaction as well as truth is involved, property variables). The level of these unspecified variables may differ from one quantification to another, and their level in a particular quantification will be determined, not by syntactic features alone, but by the sorts of facts determining the levels of particular occurrences of the Kripkean truth and satisfaction predicates; in a semantic description of the language, the levels of unspecified variables would be defined by a recursion similar to those involved in the semantics of Kripkean languages. Unspecified variables, like level seeking semantic predicates, are risky: if the facts are unfavorable, the recursion will not give any level for the unspecified variables in a sentence. Thus, for example, a 'truth-teller' sentence ("This very sentence is true") will be translated along the lines of "There is a proposition p expressed⁺ by this sentence such that p". Its ungroundedness will take the form of a failure of the quantified unspecified variable to have any level; on the general ramified principle that a sentence does not express any determinate proposition if its quantified variables do not have levels, the traditional verdict on ungrounded sentences – that they fail to express propositions – follows.

But *is* this an interpretation of the Kripkean language within the framework of ramified type theory? Here we must distinguish between the ramified formal systems, on the one hand, and the ramified *theory* on the other. The essence of the ramified theory – it might more perspicuously be called the theory of ramified types, for the ramification is not in the structure of the theory but in the sorts of definitions of properties is admits – is the conceptualist or predicative conception of property existence, on which a property exists only in virtue of having a (non-circular) definition. Acceptance of this *theory* led Russell to propose a ramified formalism, but this formalism was always intended as an 'ideal language', by which he meant something like what Frege meant by *Begriffsschrift*, and not as a model of the colloquial language we use for communication. Its ideal feature was that the definition of the concept expressed by a formula in the language could be read off from the formula itself, something that is almost never the case with phrases of natural languages. (Thus, for example, Russell believed that the judgments we express using what are ordinarily called proper names would, in the ideal language, make use of complicated definite descriptions, and that just what definite description should replace a given name would in general depend on who was speaking, and could not be determined simply by examining the natural language sentence.) It seems to me entirely consistent with this conception of the nature of concepts and propositions that the language used for the ordinary, practical, purposes of communication should *not* be ideal in this sense, but should incorporate some such feature as the unspecified variables proposed here. Since the determination of the level of the unspecified variables in a given quantification, like that of the implicit level of an occurrence of a Kripkean semantic predicate, will, in general, depend on facts not known to the speaker asserting the sentence containing the quantification, the particular proposition (in the sense of the ramified theory) expressed by the sentence will not only not be recoverable from the syntax of the sentence, but will not even be determined by the speaker's intentions. This consequence of our attempt to capture the level seeking feature of Kripke's theory in the ramified framework might have been surprising to Russell, but seems entirely congruent with the philosophy of language developed in such papers as Putnam [1975]. "Meanings", as Putnam so elegantly phrases it, "just ain't in the head".

Even if this way of relating Kripke's and Feferman's theories to ramified types were worked out in detail, however, there would remain the question of what to say about ungrounded sentences – sentences to (the unspecified variables in) which the recursion assigns no level. Intuitively, one would want to say that such sentences do not express any propositions, but in saying this one would be quantifying over propositions, so that some level would have to be assigned to this analysis. Since a sentence saying of an ungrounded sentence that it does not express a proposition will itself be ungrounded, the level seeking 'mechanism' will not help here. Nor will any level assigned to any other sentence by the level seeking recursion be appropriate – for any level assigned to the unspecified variable in some sentence, there will be *grounded* sentences that will not express any proposition of that level. (Example: the sentence containing the variable itself.) In order to express the ungroundedness of a sentence by saying that it fails to express a proposition, it is necessary to quantify over propositions of a level (recall that we are construing levels as cumulative) higher than any assigned to any occurrence of the unspecified variables.

This problem becomes particularly pressing when we consider what are called *strengthened liar* arguments. These are the arguments, stressed in Parsons [1974a] and Burge [1979], in which we first establish the paradoxical quality of some sentence, then conclude *from that* that the sentence is neither true nor false, or expresses no proposition, and then realize that our conclusion is naturally expressed in the words of the original sentence! If, for example, the sentence "No proposition is expressed by this very sentence" is construed as quantifying an unspecified variable, it is clear that the recursive process assigning levels to level seeking locutions will not assign its variable a level, and so will not determine a proposition in the ramified hierarchy for it to express. Burge [1979] proposes that when the sentence is used to express the conclusion of a strengthened liar argument, it has a different interpretation from the one it has when it is uttered 'innocently' or considered as the subject matter in a strengthened liar argument. In our terms, the new interpretation would be characterized as involving the assignment as level to the unspecified variable of the *limit* of all the levels such variables could be assigned by the recursive process in environments other than strengthened liar arguments. Burge thus imputes an additional degree of indexicality to the level seeking locutions of a language beyond any involved in Kripke's analysis – the level of a sentence depends, not just on the facts about what sentences it refers to, and so on, but also on whether or not the speaker utters it as expressing the conclusion of a strengthened liar argument. (Compare the interpretation of 'bad' in "This is one of my bad days" as uttered by Job (i) thinking about his itching skin, or (ii) on the same day, but after having heard of the deaths of his sons and daughters.)

Kripke recognizes that his theory does not encompass the interpretation given to sentences like "This sentence is not true" when they express the conclusions of strengthened liar arguments, and suggests that his theory describes

> natural language at a stage before we reflect on the generation process associated with the concept of truth, the stage which continues in the daily life of nonphilosophical speakers (Kripke [1975] footnote 34).

This would seem to suggest that by reflecting on the recursive process giving levels to the level seeking locutions in our language we obtain a new concept: a conception of a certain ordinal number which is the limit of the levels assigned by that recursion to occurrences of these locutions in our pre-reflective use of the language, and that in saying, as the conclusion of a strengthened liar argument, that a certain sentence does not express any proposition, we intend this ordinal as the level of our propositional quantification. In the most general case, which would arise in the consideration of a language in which such reflective reasonings as strengthened liar arguments could themselves be formulated, this suggestion would appear to involve an impredicative mode of concept formation dangerously similar to that involved in paradoxes such as that of the least indefinable ordinal. (Kripke's wording, of course, explicitly excludes such cases.)

Commenting on Kripke's suggestion, Parsons[55] has claimed that the reflection embodied in a simple strengthened liar argument is more elementary and 'ordinary' than that implicit in the formulation and understanding of a grounded sentence of reasonably high transfinite level – an intuitively plausible claim which he uses in arguing for the plausibility of something like Burge's analysis. One respect in which the strengthened liar arguments are elementary is their logic – they do not employ specifically classical modes of reasoning, but could be formalized in a constructive logic incorporating *reductio ad absurdum* but not the classical form of indirect proof. Thus, it may be possible to interpret the conclusions of at least some strengthened liars as involving, not classical quantification over propositions of a new and higher level, but a sort of constructive quantification in a metalanguage surveying the entire classical ramified hierarchy. (Cf. discussion in Section 8 below.)

In the last few years (since, roughly, the publication of Kripke [1975]), research on the logical paradoxes has entered a new and fruitful stage. A number of theorists have proposed analyses promising deep insights, and the situation is still fluid: a variety of theories have probably not yet received their final forms, and the relations between them are still unclear. Further investigation along a number of lines is greatly needed. To cite only two:

first; although the insights of theories of truth and satisfaction such as Kripke's can probably be accommodated in the framework of ramified type theory, the additional complexity of the ramified theory's account of concepts will probably not seem to be 'paying its way' until it is applied to such (not explicitly *semantic*) paradoxes as are considered in Thomason [forthcoming] and, second, the relations of the ramified theory to the recent 'quasi-inductive' theories of Gupta and Herzberger are still extremely murky. (Cf. Gupta [1981, 1982], Herzberger [1982, 1982a], and Belnap [1982].)

8. CONCLUDING RUMINATIONS

Fitch [1946] charges the theory of types – since the argument doesn't depend on such technicalities, ramification isn't explicitly mentioned – with a kind of self-referential inconsistency. The claim was not that type-theoretic formal systems are inconsistent – the author of Fitch [1938] would hardly have believed that – but rather that the philosophical theory motivating these systems could not be advocated consistently. Russell's ramified theory of types is not a formal system (although the formal systems described in Section 2 above can, by transference, be called systems of ramified type theory) but a metaphysical or epistemological theory about the nature of concepts. Once again, on this theory concepts exist in virtue of, and only in virtue of, being defined; they therefore arrange themselves in a hierarchy of levels according to the order of their definitions, and because of this it is impossible for any concept to apply to all concepts whatever, and in particular for any proposition to be about absolutely all concepts, Russell, according to Fitch, was in a peculiar predicament. He wanted to assert that this theory was correct, but as soon as he had done so, he had also provided an example illustrating the falsity of the theory. Pick virtually any tenet of the theory of types – say, for example, *All properties have levels in the ramified type hierarchy.* In saying this, Russell would be asserting a proposition about all concepts whatsoever! Similarly, though the system of *Principia Mathematica* might be consistent, and though a description of it as an uninterpreted formal calculus could be given, it would not, on the ramified theory in the form in which Russell accepted it (i.e., without transfinite levels), have been possible to describe its intended interpretation, for if one said something along the lines of *The quantified variables of level n have as their range the properties of level n*, one would again have generalized about properties of all levels. If the theory of types can be asserted, and if the formal systems based on it can be described other than as formal calculi, it must be false and they must be of restricted application.

A concern with the needs of such general philosophical theories as the theory of types, then, led Fitch to reject the theory of types. It would be consistent with Fitch's position to regard ramified type theory as a true description of some special, restricted, domain of concepts, but it could not be the most general theory of concepts. As a corollary, although the sorts of reasoning formalized by type-theoretic formal systems might be appropriate in some applications, the logic of the language in which the most general foundational discussions are to be conducted cannot be type-theoretic. Some form of comprehension principle, however, is clearly appropriate in a theory which pretends to be about absolutely all properties and propositions – for such a theory to have restrictions on its comprehension principle would be the formal analogue of the self referential inconsistency of which Fitch accused Russell – so *something* has to be said about blocking the paradoxes. The conclusion of Fitch's criticism of the theory of types thus provides the motivation for his positive program of attempting to formulate a non-classical logic suitable for what, in the title of a later paper, he has called a "universal metalanguage for philosophy".

At this point there is an obvious candidate for the role of the special restricted class of concepts described by the ramified theory: they are the *classical* concepts, the concepts expressible in languages which can be assumed to be bivalent, for which classical logic is appropriate. A combined, Fitch-*cum*-Russell, system might then have the following features: classical logic would not hold in general, but the system would include a ramified type-theoretic subsystem within which classical logic would hold; there would be an unrestricted comprehension principle for what we may call (in order to distinguish them terminologically from the classical and ramified *concepts*) *notions*, and, given variables for the levels of the ramified hierarchy, a sort of restricted comprehension principle for concepts which would allow us to infer that, if the laws of classical logic can be shown to hold for a formula, then there is some concept at some level of the ramified hierarchy coextensive with the notion it expresses. (Cf. Hazen [1983].)

Such a system might not be as alien to Russell's thinking as it appears. Russell [1908] shows an awareness of the sort of consideration advanced against the theory of types later by Fitch, and sketches a reply that can be seen as a first step in the direction of Fitch's project of a universal metalanguage based on a non-classical logic. Unfortunately, his discussion of the matter, in Section 2 of Russell [1908], is very brief and somewhat confused, but (with a possibly tendentious change of terminology) we can describe the position as follows. In order to formulate general logical laws, it is necessary to use variables ranging over concepts – propositions or propositional functions – of all levels, but it is not possible to do this using bound variables, for a sentence with bound variables is only meaningful, only expresses a proposition, if the range of the variables is restricted to objects of a single level. It is, however, possible to have *free* variables ranging over concepts of all levels. A full formalization of Russell's theory, therefore, would be analogous to a formalization of primitive recursive arithmetic, having bound variables ranging only over restricted domains, but having unrestricted free variables. Such a formalism can be seen as embodying a non-classical logic much more restrictive than, say, the similar intuitionist logic. Intuitionist logic incorporates a negation that does not behave quite like classical negation, but the unrestricted generalizations of Russell's theory (or of primitive recursive arithmetic) do not, in the language of the theory, have any negations at all. The comparison with other non-classical logics may be more perspicuous if we think of logics, not as proposing different laws for the same logical operators, but as theories about domains of propositions that specify under what operations the domains are closed. Russell's implicit non-classical logic can then be compared with other theories according to which interesting domains of propositions are not closed under all classical logical operations. Mathematical intuitionism, in its logical aspect, can be seen as claiming that mathematical propositions do not in general have classical negations,[56] but other propositional operations are definable within intuitionistic mathematics which obey somewhat analogous formal laws, notably the familiar intuitionistic negation of Brouwer and Heyting and the 'constructible' negation of Nelson [1949]. Russell's theory is more restrictive in that unrestricted generalizations have no classical negations, and no substitute is proposed to fill, even partially, the gap.

Given the historical context of Russell's work, it is not surprising that he did not describe it in terms of non-classical logics (Brouwer's 'Unreliability of Logical Principles' appeared in the same year as Russell [1908]), but his lack of explicitness is unfortunate. If he had thought of his 'metalanguage' as based on one of a variety of possible non-classical logics, Russell might have given more thought to its justification. In fact, though the negationless metalanguage seems sufficient for the generalizations about concepts of all levels in Russell [1908], it is not sufficient for the things said in the introduction to *Principia Mathematica*. A free variable generalization is equivalent to a universally quantified statement whose initial quantifier includes the whole sentence in its scope. In the absence of negation, Russell's metalanguage provides no corresponding unrestricted *existential* quantification: no way of saying that *some* concept will satisfy a condition without specifying the level of the concept. At the end of the 'Theory of Logical Types' chapter in the introduction to *Principia Mathematica* there is a discussion of the fate of elementary arithmetic if the axiom of infinity is false. Given the definitions adopted in *Principia Mathematica*, the natural numbers can be construed as properties of any level (above a certain lower limit), and whatever level is chosen, some numerical equations will have the 'wrong' truth values if there are only finitely many individuals. However, Russell and Whitehead point out, even if there are only finitely many individuals, for any numerical equation *there is* a level such that the equation has the right truth value if the numbers are construed as properties of that or any higher level. Here an *existential* generalization is made without restriction as to level: something which would be impossible in the official 'metalanguage' of Russell [1908].

Clearly, *classical* logic is not appropriate for unrestricted generalizations about (classical) concepts, both because, with plausible comprehension principles, paradox would threaten, and because the use of classical logic would intuitively seem to imply that conditions formulated with such generality themselves expressed concepts, but something stronger than Russell's free variable calculus is intuitively plausible and apparently safe from the reappearance of the paradoxes. In fact, it appears possible to use full intuitionistic logic in connection with bound variables construed as ranging over all possible (classical) concepts of whatever level in the ramified hierarchy. (Curry's Paradox, however, shows that intuitionistic logic is not appropriate for an absolutely general theory of notions with an unrestricted comprehension principle.) In a way this is not surprising: the lesson of the paradoxes is surely that our grasp of the general notion of a (classical) concept, or of a possible classical language, is necessarily somewhat vague or incomplete, and the attempt to specify just what this vagueness or incompleteness comes to leads to formulations very similar to the statements of the intuitionistic position in, e.g. Dummett [1963].

Note, however, that the inappropriateness of classical logic when the quantified variables are interpreted as ranging over all (classical) concepts need not have, in a narrow sense of the word, *ontological* implications, as if, mysteriously, some strange entities would always escape our net whenever we reasoned classically. The problem with generalizing about all concepts is a conceptual one: that we are not able to define the notion of a concept, nor such correlative notions as that of a property's *holding* of an object.

On the predicative notion of properties as existing in virtue of their definitions, we might well identify (or at least correlate) concepts with certain quasi-syntactic entities (e.g. finite trees having individuals or basic properties of level 1 assigned to their terminal nodes), and in quantifying over all such objects we would be quantifying over a class of entities including (at least the correlates of) all concepts. If the possibility of quantifying classically over this more inclusive domain is accepted, it is the indefinability of the class of entities in that domain which are (correlated with) concepts that make it impossible to define (by simple restriction of quantifiers) quantification precisely over all concepts. The argument, in Section 3 of Russell [1908], that it is impossible to quantify over a proper superclass of a class over which it is impossible to quantify rests on an aspect of his view which is separable from, and less compelling than, the general predicative notion of a concept. (Cf. discussion in Parsons [1971a].)

A final question is suggested by the appearance of intuitionistic logic in connection with metalanguages for ramified type theory, and by historical connections: what connection is there between the various brands of conceptualism or constructivism in mathematics? Or: do the principles leading to predicativism also inevitably lead to Brouwerian intuitionism? Certainly there are similarities between the positions. Both reject the impredicative excesses of set-theoretic mathematics, and Gödel [1933] argued that this, and not the rejection of the law of excluded middle, was the fundamental difference between intuitionistic and classical mathematics. The history of the subject, moreover, suggests that the transition from classical predicativism to intuitionism is psychologically natural: Weyl, a few years after the publication of Weyl [1918], announced his conversion to intuitionism, and of course Poincaré (who was not explicit on logical matters) can be seen both as the founder of classical predicativism and as one of the forerunners of modern intuitionism. It does, however, seem consistent to reject impredicative set theory while maintaining a classical, realist, view of the natural numbers, and of properties predicatively conceived. Consistency, however, does not imply naturalness, and the failure of logicism may leave realism even about the natural numbers in an exposed position. The concept of a finite sequence, or of a natural number is not definable in ramified logic, and the existential requirements of arithmetic must be underwritten by a nonlogical postulate. In view of this it may seem hard to justify the assumption that all the infinitely many natural numbers actually exist, or that all the properties of some given level actually exist – assumptions implicit in the use of nonconstructive, classical, modes of reasoning about them.

Note, however, that the intuitionist standpoint is not properly more restrictive than the classical predicativist. Brouwer accepted as legitimate certain *inductive* definitions (e.g., of ordinals in the second number class) which cannot be justified predicatively, and introduced a fundamentally new notion, that of a *choice sequence*, which cannot be defined in (predicative) logical terms.

University of Melbourne

NOTES

[1] Some of Russell's papers from this period, and a bibliography, may be found in Lackey [1973]. Most of Poincaré's contributions to the debate were reprinted in the in the four books Poincaré [1902, 1905, 1908, 1913]. Anne-Françoise Schmid's study of Poincaré's thought contains a useful bibliography tracing the genealogy of the chapters of these books.

[2] Richard put it in terms of decimal expansions of real numbers, but clearly the logic is the same if we simply consider sets of naturals.

[3] Since there is no upper bound on the lengths of definitions, the sort of alphabetization used in ordinary dictionaries will not work, but a sort of 'cross-word puzzle dictionary' arrangement, in which the definitions of one length, in alphabetical order, appear before those one letter longer, and so on, will.

[4] Richard attributes the contradiction to the Richard set's not having been defined in a finite number of words, which, since he has just apparently defined it in a short paragraph, is a bit confusing. The definition depends in its details, however, on a specification of *which* expressions of the language really define sets of natural numbers, and there is no reason to believe that this set of expressions could be characterized.

[5] I shall, in what follows, use the term *property* as a generic term for conceptual entities, including, in addition to monadic properties, polyadic relations and '0-adic' properties, or propositions. The curious paper Russell [1906], which Russell withdrew from publication, presents an earlier attempted reduction related in interesting ways to the theory of judgment in Russell [1912] and the last chapter of Russell [1910]. Briefly, Russell thought of propositions as things that had *constituents*, which would include any individuals the proposition was about and also the universals denoted by the predicates that would be used in formulating the proposition. Quantification over, e.g., monadic properties was to be contextually defined in terms of a double quantification over propositions and some other kind of objects. Each pair of a proposition and one of its constituents was to represent the property, or the class, of objects such that the propositions resulting from the given one by replacing the given constituent with the new object was true. The system proposed was therefore rather like that of Fitch [1971]. Hylton [1980] argues that it was not an immediate ancestor of ramified type theory, and that, in the presence of auxiliary assumptions Russell would have accepted, it allowed the derivation of paradoxes.

I.5: PREDICATIVE LOGICS   397

[6] Cf. Gödel [1933], where the rejection of impredicative reasoning is seen as a more characteristic feature of intuitionistic mathematics than the rejection of the law of excluded middle.

[7] The words *order*, *type*, and *level* have all been used in a variety of senses in this area. Perhaps the most widely current term involving one of the words is *first-order logic* for a logic whose quantified variables range only over individuals – or rather, since theories of arbitrary kinds of objects can be formulated in first-order logic, a logic in which nothing in the formation rules or in the principles labelled *logical* reflects an intention that the variables should range over higher type entities. I shall, in what follows, use the terms *higher order* and *higher type* entities and theories to distinguish properties and theories of properties from individuals and theories of individuals. (*Individuals* is a bit of a misnomer, since in applications of these systems such objects as, e.g., the natural numbers may be taken as individuals. I am using the term *individual* to contrast with *property*; in applying one of the predicative logics it is assumed that the individuals are somehow more primitive than or given separately from such conceptual entitites as properties, but not necessarily that they are concrete *INDIVIDUALS* in any stronger metaphysical sense.) In my formal use I shall distinguish between the three words as follows, following approximately (but not exactly) Church's usage. *Type* has to do with the arguments an entity takes. Individuals form one type, propositions another. Two monadic properties are of the same type just in case they are defined for the same arguments, two relations just in case they take the same number of arguments, and the same kind of object at each argument place. Types of expressions in formal languages are defined accordingly, variables and constants having the types of the objects they range over or stand for. *Order* is an abstraction from this. Individuals are of first-order, propositions and properties of individuals of second, and in general properties are of the order one higher than the highest order of their arguments. Variables and constants again have the order of the entities they range over or stand for, and the order of a formula is the order of its highest-order bound variable, or one less than the order of its highest order constant or free variable, whichever is highest. (The logical connectives do not count as constants for this purpose.) This is in accordance with the common usage of the terms *first-order* and *second-order* as applied to variables and formulas. Finally, *level*, the key concept of the ramified theory, has to do with the hierarchy of definitions of properties. Individuals will be counted as of level 0; this is technically convenient and also accords well with the notion that they are not entities of the same *conceptual* sort as properties. We may assume there are at least some basic properties of individuals of level 1, and beyond that the levels of properties are determined by the levels of the objects they take as arguments or presuppose in their definition. Levels of objects, and the corresponding levels of expressions, will be more precisely defined in the text. Technically it is most convenient to allow levels to be *cumulative*, so that the objects and expressions of lower levels will be included among those of any higher-orders; informally it is occasionally convenient to use the terms exclusively. It will, I hope, be clear from the context when I mean, e.g., second-or-first and when I mean second-but-not-first. In reading other authors on the subject, it must be remembered that their usages of these terms are almost certainly not uniform with mine, or with each other's.

[8] It is a nice question whether we want to think of the domain of properties over which the second-order variables range as *exhausted* by those definable in the original first-order language. Cf. the next section.

[9] A precise statement can be found in Church [1956], where (in section 58), this kind of system is called a "predicative functional calculus of second order". This restrictive use of the term *predicative*, as applying to systems in which only properties of level 1 are quantified over, has some historical justification, but seems unfortunate: ramified logics in general are predicative in the sense of avoiding the assumption of impredicatively defined properties. The broader sense, in which ramified systems are predicative, is also current, as in the title of Feferman [1964].

[10] Well, you were warned that, in particular applications, the individuals might not be INDIVIDUALS in any strong metaphysical sense. Even in this application, however, the individuals are conceived of as entities of a different nature from properties, for Zermelo–Fraenkel set theory incorporates a strongly *Platonistic* conception of set, and not a conceptualistic one. Cf., for discussion on this point, Parsons [1974].

[11] Any system, of number theory, set theory, or *protosyntax* in the sense of Quine [1940], for which an analogue of Gödel's [1931] incompleteness proof can be given allows for enough.

[12] Exercise for the reader (solution in Section 6): why doesn't the definability of satisfaction for formulas of ZF allow us to prove the consistency of ZF in NBG? (In fact, NBG is a conservative extension of ZF.)

[13] For a formulation without abstracts, cf. Section 58 of Church [1956].

[14] Note that a formula or abstract can only be of level ω if it contains a free predicate variable of this level: this is what makes it convenient so to define limit levels that they contain no properties not already present at lower levels.

[15] The earliest explicit published proposal I know for a formal ramified logic with transfinite levels (and orders) is in Fitch [1938], where it is remarked that the consistency of such a system could be established by a straightforward extension of the means used to establish that of a system with finite levels. In the introduction to *Principia Mathematica*, p. 53, Whitehead and Russell reject transfinite levels (or orders) for a technical reason which reflects a deeper philosophical error. In their system, levels and orders were not construed as cumulative, so there is no obvious and natural way of extending it to the transfinite: "Since the orders of functions are only defined step by step, there can be no process of 'proceeding to the limit', and functions of an infinite order cannot occur". In their decision to keep the distinctions of order and level strict rather than cumulative, it seems likely that they were influenced by a confusion of two very different reasons for imposing a hierarchy of levels and orders. One was the predicative consideration that properties should not be presupposed (by being included in the range of quantified variables) in their own definition: this consideration motivates the ramified theory but receives its most natural formulation with cumulative levels. The other (influenced by Frege?) is a purely *grammatical* conception of types, on which they correspond to grammatical categories of words in a language: this conception indeed makes cumulativity seem unnatural, but has little to do with the epistemological analysis of the paradoxes that yields the ramified theory of types.

[16] This is not a fully rigorous statement. As is clear from Gödel's [1933] interpretation of classical in intuitionistic number theory, acceptance of the *formal* rules of classical quantifier logic for a certain language is not quite the same as assuming that formulas of that language have fully definite meanings in the classical sense. Cf. further discussion in Section 8 and Hazen [1983].

I.5: PREDICATIVE LOGICS 399

[17] Similar systems are described in Church [1976], Hatcher [1968], and Myhill [1979], the last influenced by the formulation of Schütte [1960]. Copi [1972] also contains an exposition of the ramified theory of types, but read Grover [1974] first.

[18] The question of the definability of identity (as possession of all the same properties) in ramified logics is taken up in Section 4.

[19] It is perhaps simpler to restrict this to identity between properties of the same level, for substitution of identicals could, if the terms of the identity were of different levels, lead from well-formed to ill-formed formulas. It is possible, however, to explain these ill-formed conclusions, in which predicates take arguments of higher level than themselves, as *façons de parler*, and so to assume coextensiveness as a general definition of property identity.

[20] Technically, since the specification of the type of a higher-order predicate includes a specification of the permitted levels of its arguments, adding new levels will generate new types: the types of predicates applying to properties with the new levels. The *purely* type theoretic structure — what remains when you ignore the levels of the arguments, the levels of the arguments those arguments themselves take, and so on — is not changed, however.

[21] With a number of extra axioms greatly increasing its mathematical strength: choice, infinity, and what might be called *ancestry*: at each level the existence at that level is asserted of the ancestrals of relations of that level. As Fitch remarks, the last may be thought of as a special case of Russell's axioms of reducibility.

[22] Quine recommends reference to the formal system as a 'strict' way of characterizing the conceptualist position, one that is 'free of metaphor' in contrast to the 'vague and metaphorical' intuitive formulation of the conceptualist viewpoint he has just given. In the next, concluding, section of 'Logic and the Reification of Universals', comparing the parts of mathematics acceptable to platonists and conceptualists, he says the conceptualist "tolerates elementary arithmetic and a good deal more". Since (cf. Section 4) the axioms of first-order Peano arithmetic are not derivable in Quine's system, even when an axiom of infinity is added to it, a better formulation of the conceptualist position what Quine calls the conceptualist, or *intuitionist* ("in a broad sense of the term"), position might take the natural numbers as individuals and add the principle of mathematical induction as a new primitive rule or axiom scheme. I shall not follow Quine's presentation in all its details; Professor Wang has informed me that formulations of ramified type theory similar to it were well known to logicians at the time.

[23] Quine elegantly construes as identities those formulas with terms of level 0 as predicates. He also, imitating the notation of set theory, writes the subject first, with a Greek letter between subject and predicate.

[24] Cf. also the article Chihara et al. [1975], which gives further details on the axiomatization of Σ_ω. In general, my aim in this section has been to suggest intuitively plausible axioms for ramified systems: particularly where transfinite *orders* are concerned, I make no claim that my axioms will be the *only* ones useful in practice. In particular, some sort of *induction* principle may be desirable in order to allow the derivation of theorems at limit levels summarizing lower theorems.

[25] One, that is, that does not postulate 'understood' quotation marks and satisfaction predicates, etc. Cf. Belnap and Grover [1973].

[26] This formulation is the natural one if we assume that the instances of a quantification are defined along the lines of Church [1956]. Other versions of the machinery of ramified logics will call for appropriate changes.

[27] Only Chihara, with the level ω variables of this system, and Parsons [1971], explicitly consider transfinite levels. As long as the notation for ordinal levels is effective, however, there is no problem in giving an effective definition of the substitution classes of the different sorts of variables, so there is no technical, and no apparent conceptual, difficulty in extending the substitution interpretations to ramified systems with transfinite levels.

[28] Other than, in the case of a property of individuals, those marking the arguments the property holds of.

[29] Assuming that X, Y, and Z were of level ω. In the case of axioms of the form ω_{Comp} for levels higher than ω, the formula replacing X must still be of higher level than those replacing the variables bound by the preceding quantifiers. In this case, too, we restrict our attention to instances in which the formula replacing X is not of limit level: a harmless restriction since (by, e.g., adding a logically true conjunct) any formula may be transformed into an equivalent one of higher level.

[30] *Quodlibet* is only the analogue of cut for certain special natural deduction formulations of classical logic, and even there can only be dispensed with in proofs of logical truths. In other natural deduction systems *modus ponens* can be seen as the culprit, and in yet others normal proofs are characterized by restrictions on the use of a variety of rules. That a common mathematical content is involved may be more obvious from a consideration of the corollaries that may be derived from these various versions of the *Hauptsatz* than it is from their statement!

[31] More precisely, we should (as can easily be done) define as the analogue of degree a function from formulas to ordinals such that, where A is a formula of level i,

$$\omega \cdot i \leq \text{degree}(A) < \omega \cdot (i + 1).$$

[32] Prawitz [1967] is a clearly written original paper, quite explicit on matters of motivation, and there is a thorough exposition in Takeuti [1975]. These proofs use methods related to the completeness proof in Henkin [1950]; there is apparently some sensitivity to details of formulation, for Schütte [1968] reports that an analogue of the *Hauptsatz* fails for another version of the simple theory of types.

[33] In its use of the notion of truth on a substitution interpretation, the proof outlined is based on that in Fitch [1938] (which, however, is for a stronger system and has correspondingly stronger presuppositions). As it stands, of course, it is not finitistic – the metalinguistic quantification over the infinitely many instances of an object language quantification is enough to disqualify it. It is, however, more elementary than it may look, for only the recursive characterization of truth is actually used, and it is not presupposed that this could be transformed into an explicit definition. (Cf. the distinction made in chapter 19 of Boolos and Jeffrey [1974] between the set of true sentences of arithmetic, which is not arithmetically definable, and the unit class of that set, which in an appropriate sense, is.) It should also be emphasized again that a strictly finitistic formulation, along the lines of Gentzen's proof of the consistency of arithmetic without induction, is possible.

[34] If we didn't have the constant 0 or the functor S, but only the predicate $G(x, y)$, and even if we had no usable primitive vocabulary, but only an existentially quantified version of Inf as an axiom of infinity, we would still, in a slightly more complicated way, be able to derive arithmetic in impredicative second-order logic. (This sort of treatment originates with Dedekind [1888].)

[35] It would also be quite un-Russellian to rule them out, for in Russell [1912], a work contemporary with *Principia Mathematica*, he counted the *self*, or Cartesian ego, as 'probably' among the things with which we were acquainted.

[36] The desirability of allowing for such primitive higher-level properties of individuals is one of the things that makes it difficult to formulate an interesting notion of a standard model of a ramified logic. Russell's tortuous theories of judgment and attempts at reductive analyses of propositions were perhaps in part motivated by a wish to avoid such higher level properties.

[37] On topics discussed in this section in particular, the exposition of Church [1976] cannot be too highly recommended. Another, very clear and careful discussion of the situation concerning the Axiom of Reducibility and the Grelling Paradox is in Myhill [1979].

[38] The corresponding principle with 0-adic – that is, propositional – variables being, of course, trivial.

[39] These passages were reproduced with minimal changes in *12 of *Principia Mathematica*.

[40] Or Fregean *Wertverläufe*. As a matter of historical speculation, it seems to me likely that Russell's conception of the distinction between classes and 'propositional functions' owed something to his reading of Frege.

[41] The simple theory of types as usually presented also includes Axioms of Extensionality. As Gandy [1956] showed, however, it is possible to interpret simple type theory *with* Extensionality in simple type theory *without*. (A similar but simpler proof, apparently arrived at independently, is in section 17 of Gallin [1975]; the fundamental idea seems to go back to Russell's 'No Class' theory.)

[42] This is a corollary of Gandy's [1956] proof that axioms of extensionality could be added consistently to a type theoretic foundational system. (Russell's proposal, in the second edition of *Principia Mathematica*, that axioms of extensionality be adopted, undoubtedly added to the confusion about the merits of the ramified theory and the Axioms of Reducibility.)

[43] Consider the Case of the Fortunate Fallibilist. He believes that he has some false beliefs, but he has never been deceived about anything else, and in fact all of his other beliefs are true.

[44] Another example: if there really are such things as classes, the first level property of belonging to the set of individuals with a certain higher level property is one the individuals have by relation to the set, but the higher level property might be non-relational.

[45] It is easy enough to modify the example so as to produce a contradiction employing only the (negation-free fragment of) intuitionistic logic. Thus, following the reasoning of the Curry Paradox, we might define an adjective as *p-heterological*, for some absurd proposition p, just in case p is implied by the supposition that the adjective has the property it expresses. If the adjective '*p*-heterological' is itself *p*-heterological, p follows by definition, so the supposition that '*p*-heterological' is *p*-heterological *does* imply p, from which conclusion p itself follows. Following the hints of Meyer *et al.* [1979], we could formulate a version assuming, as the logic of the conditional, an implicational logic much weaker than the intuitionistic.

[46] For simplicity I will follow those authors in treating symbols and formulas (or their Gödel numbers) as individuals, but given appropriate changes, including a change in the type of free variable in the formulas considered, treating expressions as entities of some higher type would not entail any essential modification in the argument.

[47] He doesn't seem to have had much sympathy for other constructivist tendencies either – in the conclusion of his discussion of the Axioms of Reducibility he writes, not only of "sav[ing] mathematics from the sceptics", but also of "preserv[ing] it from the Bolshevik menace of Brouwer and Weyl".

[48] Much of the subsequent literature restricts the term 'logical paradox' to Ramsey's first group. This seems unfortunate, suggesting a narrow construct of 'logic' which was not Russell's and is certainly inappropriate, given the growth of interest in intensional logic in the last few decades, to the contemporary scene. Ramsey's examples of 'epistemological' paradoxes are all semantic, but both the term and his description seem to encompass the similar contradictions involving the notions of the propositional attitudes.

[49] The claim of Leblanc [1975] to have provided a 'predicative' interpretation of a system containing Axioms of Reducibility bears comparison with Robinson's claim (in, e.g., Robinson [1967]) to have saved the notion of the infinitesimal with non-standard analysis. Both interpretations rest on classical model theory. (Leblanc's, based on ideas similar to those of Henkin [1950], claims a substitution interpretation, but the substitution instances are in a language containing new primitives foreign to the system interpreted, and their existence is proved non-constructively, and on the *assumption* of the consistency of the interpreted system.) Both assume at least the consistency of fairly strong classical systems, and neither would convince a philosopher with strong doubts about the usual, impredicative, standard formulations of analysis. It had long been obvious that the ramified systems with Reducibility were consistent if the simple theory of types was, and Leblanc seems to have provided no new insight on the consistency problem, whereas it was not initially obvious that very much of a theory of infinitesimals could be shown consistent even by set-theoretic means.

[50] It has been claimed that the majority of contemporary mathematicians are *realists* (who would accept set theory as a true description of really existing objects), a sizeable minority *formalists*, and only a very small minority *constructivists* of various stripes – but also suggested that the real majority are realists on weekdays, but respectably antimetaphysical formalists in their reflective moments. In any case, *predicativists*, as one stripe of constructivist, are a minority within a minority.

[51] *Loci classici* for the argument: Quine [1960], Putnam [1971]. Assorted criticisms: Chihara [1973], Van Fraassen [1975], Field [1980].

[52] Believing predicativists, if there are any, might echo the intuitionists in calling the assertions of classical mathematics *meaningless*. The accusation is not, of course, that classical mathematics is gibberish, or that they find it unintelligible in the way they would a completely unfamiliar foreign tongue – intuitionists and predicativists, after all, can *follow* classical proofs. The classical mathematician is rather, to a predicativist, like someone who, having access to all the usual sources of information about French politics, persists in using the term *the present king of France* as if France were a monarchy: with the added problems that his designationless terms are so necessarily rather than as a historical accident, and that they are predicates rather than individual terms.

[53] Gödel's opinion of the relation between his constructible sets and predicativity, stated in Gödel [1944], was that the original levels for which analogues of Russell's Axioms of Reducibility hold are "so great, that [they] presuppose [ ] impredicative totalities", but that when classical systems are interpreted in the constructible hierarchy "all impredicativities are reduced to one special kind, namely the existence of certain large ordinal numbers . . . and the validity of recursive reasoning for them".

[54] Note that a 'ramified' system whose levels are not well-ordered will not be predicatively acceptable, and its formulas will not seem, to a conceptualist, to define genuine properties of natural numbers, but it may not be *inconsistent*. Simple example: full classical second-order arithmetic, which may be thought of as 'ramified' system with a 'system' of one level subscript so ordered that the one subscript is lower than itself.

[55] In a postscript to Parsons [1974a] written for Parsons [1983]. I am indebted to Professor Parsons for showing me a manuscript of this postscript; it, and the paper to which it is a postscript, are as insightful as any examination of the logical paradoxes in the recent philosophical literature.

[56] Similarly, the point of quantum logic is that the classical negations, disjunctions, etc., of quantum mechanical observation propositions do not themselves belong to this domain. Cf. Van Fraassen [1975a], from which I learned this way of comparing logics.

REFERENCES

Belnap, N. D., Jr.: 1982, 'Gupta's rule of revision theory of truth', *J. Philos. Logic* **11**, 103–116.

Belnap, N. D. and Grover, D.: 1973, 'Quantifying in and out of quotes', in Leblanc [1973], pp. 17–47.

Benacerraf, P. and Putnam, H. (eds.) 1964, *Philosophy of Mathematics*, Prentice-Hall, Englewood Cliffs.

Bernays, P.: 1935, 'Sur le platonisme dans les mathematiques', *L'enseignement mathématique* **34**, 52–69. Eng. tr. 'On platonism in mathematics', in Benaceraff and Putnam [1964], pp. 274–286.

Boolos, G. S.: 1975, 'On second-order logic', *J. Philosophy* **72**, 509–527.

Boolos, G. S. and Jeffrey, R.: 1974, *Computability and Logic*, Cambridge University Press, Cambridge (second edition, 1980).

Boyd, R., Hensel, G., and Putnam, H.: 1979, 'A recursion-theoretic characterization of the ramified analytical hierarchy', *Trans. Amer. Math. Soc.* **141**, 37–62.

Burge, T.: 1979, 'Semantical paradox'. *J. Philosophy* **76**, 169–198.

Chihara, Ch. S.: 1973, *Ontology and the Vicious-circle Principle*, Cornell University Press, Ithaca.

Chihara, Ch. S., Lin, Y., and Schaffter, T.: 1975, 'A formalization of a nominalistic set theory', *J. Philos. Logic* **4**, 155–169.

Church, A.: 1956, *Introduction to Mathematical Logic*, Princeton University Press, Princeton.

Church, A.: 1976, 'Comparison of Russell's resolution of the semantical antimonies with that of Tarski', *J. Symbolic Logic* **41**, 747–760.

Chwistek, L.: 1921, 'Antynomje logiki formalnej', *Przeglad Filozoficzny* **24**, Engl. tr. 'Antinomies of formal logic', in Storrs McCall (ed.), *Polish Logic 1920–1939*, Clarendon Press, Oxford, 1967, pp. 338–345.

Coffa, A.: 1982, 'Kant, Bolzano, and the emergence of logicism', *J. Philosophy* **79**, 679–689.

Copi, I. M.: 1972, *The Theory of Logical Types*, Routledge and Kegan Paul, Boston and London.

Davis, M.: 1965, *The Undecidable*, Raven Press, Hewlett.
Dedekind, R.: 1888, *Was sind und was sollen die Zahlen?* Brunswick. Eng. tr. 'The nature and meaning of numbers', in Dedekind, *Essays on the Theory of Numbers*, Dover Publications, New York, 1963.
Dummett, M.: 1963, 'The philosophical significance of Gödel's theorem', *Ratio* 5, 140–155. Reprinted in M. Dummett, *Truth and Other Enigmas*, Duckworth, London, 1978; Ger. tr. 'Die philisophische Bedeutung von Gödels Theorem', *Ratio* 5, 124–137.
Feferman, S.: 1962, 'Transfinite recursive progressions of axiomatic theories', *J. Symbolic Logic* 27, 259–316.
Feferman, S.: 1964, 'Systems of predicative analysis', *J. Symbolic Logic* 29, 1–30. Reprinted in J. Hintikka (ed.), *The Philosophy of Mathematics*, Oxford University Press, London, 1969, pp. 95–127.
Feferman, S.: 1968, 'Autonomous transfinite progressions and the extent of predicative mathematics', in B. van Rootselaar and J. F. Staal (eds.), *Logic, Methodology and Philosophy of Science III*, North-Holland, Amsterdam.
Feferman, S.: 1978, 'A more perspicuous formal system for predicativity', in K. Lorenz (ed.), *Konstruktionen versus Positionen*, Walter de Gruyter, Berlin, pp. 68–93.
Feferman, S.: 198?, 'Toward useful free-type systems I', *J. Symbolic Logic* (to appear).
Field, H. H.: 1980, *Science Without Numbers*, Princeton University Press, Princeton.
Fisher, A.: 1982, *Formal Number Theory and Computability*, Clarendon Press, Oxford.
Fitch, F. B.: 1938, 'The consistency of the ramified *Principia*', *J. Symbolic Logic* 3, 140–149.
Fitch, F. B.: 1946, 'Self-reference in philosophy', *Mind* 55, 64–73. Reprinted in Fitch [1952].
Fitch, F. B.: 1952, *Symbolic Logic, An Introduction*, Ronald Tree Press, New York.
Fitch, F. B.: 1971, 'Propositions as the only realities', *American Philosophical Quarterly* 8, 99–103.
Gallin, D.: 1975, *Intensional and Higher-order Modal Logic*, North-Holland, Amsterdam.
Gandy, R. O.: 1956, 'On the axiom of extensionality, Part I', *J. Symbolic Logic* 21, 36–48.
Gentzen, G.: 1934, 'Untersuchungen über das logische Schliessen', *Mathematische Zeitschrift* 39, 176–210, 405–431; Fr. tr. *Recherches sur la deduction logic*, PUF, Paris, 1955 Eng. tr. 'Investigations into logical deduction', *American Philosophical Quarterly* 1 (1964), 288–306, 2 (1965), 204–218; Reprinted in Gentzen [1969].
Gentzen, G.: 1969, *Collected Papers*, North-Holland, Amsterdam.
Gödel, K.: 1931, 'Über formal unentscheidbare Sätze der Principa mathematica und verwandter Systeme I', *Monatshefte für Mathematik und Physik* 38, 173–198. Eng. tr. in Davis [1965] and Van Heijenoort [1967].
Gödel, K.: 1933, 'Zur intuitionistischen Arithmetik und Zahlentheorie', *Ergebnisse eines mathematischen Kolloquiums* 4, 34–38. Eng. tr. in Davis [1965].
Gödel, K.: 1940, *The Consistency of the Axiom of Choice and of the Generalized Continuum-hypothesis with the Axioms of Set Theory*, Princeton University Press, Princeton.
Gödel, K.: 1944, 'Russell's mathematical logic', in P. A. Schilpp (ed.), *The Philosophy of Bertrand Russell*, Tudor Publishing Company, New York. Reprinted in Benacerraf and Putnam [1964], pp. 211–232.
Gödel, K.: 1947, 'What is Cantor's continuum problem?' *Amer. Math. Monthly* 54, 515–525. Revised, expanded version in Benacerraf and Putnam [1964], pp. 258–273.

Gödel, K.: 1958, 'Über eine bisher noch nicht benüzte Erweiterung des finiten Standpunktes', *Dialectica* 12, 280–287. Eng. tr. 'On a hitherto unexploited extension of the finitary standpoint', *J. Philos. Logic* 9, 133–142 (1980).
Grover, D. L.: 1973, 'Propositional quantification and quotation contexts'. In Leblanc [1973], pp. 101–110.
Grover, D. L.: 1974, Review of Copi [1972], *Philosophical Rev.* 83, 281–283.
Gupta, A. K.: 1981, 'Truth and paradox', *J. Philosophy* 78, 735–736.
Gupta, A. K.: 1982, 'Truth and paradox', *J. Philos. Logic* 11, 1–60.
Hacking, I.: 1979, 'What is logic?' *J. Philosophy* 76, 285–319.
Hatcher, W. S.: 1968, *Foundations of Mathematics*, W. B. Saunders, Philadelphia. (Revised edition *The Logical Foundations of Mathematics*, Pergamon Press, Oxford, 1982.)
Hazen, A.: 1983, 'An example of a language with classical logic for which bivalence cannot be assumed', forthcoming in *Analysis*.
Henkin, L.: 1950, 'Completeness in the theory of types', *J. Symbolic Logic* 15, 81–91. Reprinted in J. Hintikka (ed.), *The Philosophy of Mathematics*, Oxford University Press, London, 1969.
Herzberger, H. G.: 1982, 'Notes on naive semantics', *J. Philos. Logic* 11, 61–102.
Herzberger, H. G.: 1982a, 'Naive semantics and the liar paradox', *J. Philosophy* 79, 479–497.
Hylton, P.: 1980, 'Russell's substitutional theory', *Synthese* 45, 1–31.
Kleene, S. C.: 1955, 'Hierarchies of number-theoretic predicates', *Bull. Amer. Math. Soc.* 67, 193–213.
Kreisel, G.: 1960, 'La predicativité', *Bulletin, Societé Mathématique de France* 88, 371–391.
Kreisel, G.: 1960a, 'Ordinal logics and the characterization of informal concepts of proof', in *Proceedings of the International Congress of Mathematicians, 14–21 August 1958*, Cambridge University Press, Cambridge, 1960.
Kripke, S.: 1975, 'Outline of a theory of truth', *J. Philosophy* 72, 690–715.
Lackey, D. (ed.): 1973, *Essays in Analysis by Bertrand Russell*, George Allen & Unwin, London.
Leblanc, H. (ed.): 1973, *Truth, Syntax and Modality*, North-Holland, Amsterdam.
Leblanc, H.: 1975, 'That Principia Mathematica, first edition, is predicative after all', *J. Philos. Logic* 4, 67–70.
Leblanc, H. and Weaver, G.: 1973, 'Truth-functionality and the ramified theory of types', in Leblanc [1973], pp. 148–167.
Lewis, D. K.: 1979, 'Attitudes *de dicto* and *de se*', *Philosophical Rev.* 88, 513–543.
Lorenzen, P.: 1965, *Differential and Integral*, Akad. Verlagsgesellschaft, Frankfurt/M. Eng. tr. *Differential and Integral*, University of Texas Press, Austin, 1971.
Maddy, P.: 1980, 'Perception and mathematical intuition', *Philosophical Rev.* 89, 163–196.
Meyer, R. K., Routley, R., and Dunn, J. M.: 1979, 'Curry's paradox', *Analysis* 39, 124–128.
Mostowski, A.: 1961, 'Formal system of analysis based on an infinitistic rule of proof', in *Infinitistic Methods: Proceedings of Symposium on Foundations of Mathematics*, PWN and Pergamon, Warsaw and Oxford. Reprinted in A. Mostowski, *Foundational Studies*, I, North-Holland and PWN, Amsterdam and Warsaw, 1979.

Myhill, J.: 1974, 'The undefinability of the set of natural numbers in the ramified *Principia*', in George Nakhnikian (ed.), *Bertrand Russell's Philosophy*, Harper and Row, New York, pp. 19-27.
Myhill, J.: 1979, 'A refutation of an unjustified attack on the axiom of reducibility', in G. W. Roberts (ed.), *Bertrand Russell Memorial Volume*, George Allen and Unwin, London, pp. 81-90.
Nelson, D.: 1949, 'Constructible falsity', *J. Symbolic Logic* **14**, 16-26.
Parsons, Ch. D.: 1971, 'A plea for substitutional quantification', *J. Philosophy* **68**, 231-237.
Parsons, Ch. D.: 1971a, 'Ontology and mathematics', *Philosophical Rev.* **80**, 151-176.
Parsons, Ch. D.: 1974, 'Sets and classes', *Noûs* **8**, 1-12.
Parsons, Ch. D.: 1974a, 'The liar paradox', *J. Philos. Logic* **3**, 381-412.
Parsons, Ch. D.: 1983, *Mathematics in Philosophy*. Cornell University Press, Ithaca.
Poincaré, H.: 1902, *La science et l'hypothèse*, Flammarion, Paris. Eng. tr. *Science and Hypothesis*, Dover Publications, New York, 1952.
Poincaré, H.: 1905, *La valeur de la science*, Flammarion, Paris. Eng. tr. *The Value of Science*, Dover Publications, New York, 1958.
Poincaré, H.: 1906, 'Les mathématiques et la logique', *Revue de métaphysique et de morale* **14**, 294-317.
Poincaré, H.: 1908, *Science et méthode*, Flammarion, Paris. Eng. tr. *Science and Method*, Dover Publications, New York, n.d.
Poincaré, H.: 1913, *Dernières pensées*, Flammarion, Paris. Eng. tr. *Mathematics and Science: Last Essays*, Dover Publications, New York, 1963.
Prawitz, D.: 1965, *Natural Deduction*, Almqvist and Wiksell, Stockholm.
Prawitz, D.: 1967, 'Completeness and Hauptsatz for second order logic', *Theoria* **67**, 246-258.
Prawitz, D.: 1971, 'Ideas and results in proof theory', in J. E. Fenstad (ed.), *Proceedings of the Second Scandinavian Logic Symposium*, North-Holland, Amsterdam, pp. 235-307.
Putnam, H.: 1971, *Philosophy of Logic*, Harper and Row, New York; reprinted in H. Putnam, *Mathematics, Matter and Method*, 2nd edn., CUP, Cambridge, 1979.
Putnam, H.: 1975, 'The meaning of "meaning" ', in H. Putnam (ed.), *Mind, Language and Reality: Philosophical Paper*, II, Cambridge University Press, Cambridge, pp. 215-271.
Quine, W. V. O.: 1936, 'On the axiom of reducibility', *Mind* **45**, 498-500.
Quine, W. V. O.: 1940, *Mathematical Logic*, Harper and Row, New York (revised edition, 1951).
Quine, W. V. O.: 1953, *From a Logical Point of View*, Harper and Row, New York (second edition, 1961).
Quine, W. V. O.: 1960, *Word and Object*, M.I.T. Press, Cambridge, Mass.
Quine, W. V. O.: 1963, *Set Theory and its Logic*, Belknap Press, Cambridge (revised edition, 1969).
Ramsey, F. P.: 1926, 'The foundations of mathematics', *Proceedings of the London Mathematical Society*, series 2, **25**, 338-384; reprinted in F. P. Ramsey *Foundations*, Routledge and Kegan Paul, London, 1978, pp. 152-212.
Richard, J.: 1905, 'Les principes des mathématiques et le probleme des ensembles', *Revue générale des sciences pures et appliquées* **16**, 541.
Richman, F.: 1982, Review of Zahn [1978], *J. Symbolic Logic* **47**, 703-705.

Robinson, A.: 1967, 'The metaphysics of the calculus', in J. Hintikka (ed.), *The Philosophy of Mathematics*, Oxford University Press, London, pp. 153–163.
Russell, B.: 1903, *Principles of Mathematics*, W. W. Norton, New York.
Russell, B.: 1905, 'On denoting'. *Mind* 14, 479–493. Reprinted in Lackey [1973].
Russell, B.: 1906, 'On the substitutional theory of classes and relations'. First published in Lackey [1973], pp. 165–189.
Russell, B.: 1908, 'Mathematical logic as based on the theory of types', *American Journal of Mathematics* 30, 222–262. Reprinted in Van Heijenoort [1962].
Russell, B.: 1910, *Philosophical Essays*, Simon and Schuster, New York, 1966.
Russell, B.: 1912, *The Problems of Philosophy*, Oxford University Press, London, 1959.
Russell, B.: 1919, *Introduction to Mathematical Philosophy*, George Allen and Unwin, London.
Schmid, Anne-Françoise.: 1978, *Une philosophie de savant: Henri Poincaré et la logique mathématique*, Maspero, Paris.
Schütte, K.: 1960, *Beweistheorie*, Springer-Verlag, Berlin.
Schütte, K.: 1968, 'On simple type theory with extensionality', in B. van Rootselaar and J. F. Staal (eds.), *Logic, Methodology and Philosophy of Sciences III*, North-Holland, Amsterdam, pp. 179–184.
Spector, C.: 1955, 'Recursive well-orderings', *J. Symbolic Logic* 20, 151–163.
Sundholm, G.: 1981, 'Hacking's logic', *J. Philosophy* 78, 160–168.
Tait, W. W.: 1981, 'Finitism', *J. Philosophy* 78, 524–546 *and* 618.
Takeuti, G.: 1975, *Proof Theory*, North-Holland, Amsterdam.
Tarski, A.: 1936, 'The concept of truth in formalized languages', in A. Tarski, *Logic, Semantics, Meta-mathematics*, Clarendon Press, Oxford, 1956. German tr. 'Der Wahrheitsbegriff in den formalisierten Sprachen', *Studia Philosophica* 1 (1936), 261–405.
Thomason, R. H.: 1970, *Symbolic Logic: An Introduction*, Macmillan, London.
Thomason, R. H.: 198?, 'Paradoxes of intentionality' (to appear).
Van Fraassen, B.: 1975, Review of Putnam [1971], *Canadian J. Philosophy* 4, 731–743.
Van Fraassen, B.: 1975a, 'The labyrinth of quantum logics', in C. A. Hooker (ed.) *The Logico-Algebraic Approach to Quantum Mechanics*, Vol. I, D. Reidel, Dordrecht.
Van Heijenoort, J.: 1967, *From Frege to Gödel*, Harvard University Press, Cambridge, Mass.
Wang, H.: 1952, 'Truth definitions and consistency proofs', *Transactions of the American Mathematical Society* 73, 243–275. Reprinted in H. Wang [1970].
Wang, H.: 1954, 'The formalization of mathematics', *J. Symbolic Logic* 19, 241–266. Reprinted in H. Wang [1970].
Wang, H.: 1970, *Logic, Computers, and Sets*, Chelsea Publishing Co., New York.
Weyl, H.: 1918, *Das Kontinuum*, De Gruyter, Leipzig; It. tr. *Il Continuo*, Bibliopolis, Naples, 1977.
Whitehead, A. N. and Russell, B.: 1910–1913, *Principia Mathematica* (3 vols.) Cambridge University Press, Cambridge. Partial paperback reissue *Principia Mathematica to *55*, Cambridge University Press, 1961.
Zahn, P.: 1978, *Ein konstruktiver Weg zur Masstheorie und Funktionalanalysis*, Wissenschaftliche Buchgesellschaft, Darmstadt.

CHAPTER I.6

ALGORITHMS AND DECISION PROBLEMS: A CRASH COURSE IN RECURSION THEORY

by *DIRK VAN DALEN*

0. Introduction	410
1. Primitive recursive functions	425
2. Partial recursive functions	435
3. Applications	454
Appendix	473
Historical notes	475
References	476

The justification for the inclusion of a chapter on the basic facts and techniques of recursion theory is admirably summed up by Wilhelm Busch in the lines:

> Also lautet ein Beschluß:
> Daß der Mensch was lernen muß –
> – Nicht allein das A-B-C
> Bringt den Menschen in die Höh;

An acquaintance with such topics as *diagonalization, arithmetization, self-reference, decidability, recursive enumerability* is indispensable for any student of logic. The mere knowledge of syntax (and semantics) is not sufficient to elevate him to the desired height.

The present chapter contains the bare necessities of recursion theory, supplemented by some heuristics and some applications to logic. The hard core of the chapter is formed by Sections 1 and 2 on primitive recursive functions and partial recursive functions – a reader who just wants the basic theory of recursivity can stick to those two sections. However, Section 0 provides a motivation for much that happens in Sections 1 and 2. In particular, it helps the reader to view recursive functions with a machine-oriented picture in mind. Section 3 contains a number of familiar applications, mainly to arithmetical theories.

The author does not claim any originality. There is a large number of texts on recursion theory (or computability) and the reader is urged to consult the literature for a more detailed treatment, or for alternative approaches. Our approach is aimed at a relatively complete treatment of some of the fundamental theorems, accompanied by a running commentary.

Drafts of this chapter have been read by a number of colleagues and students and I have received most helpful comments. I wish to thank all those who have kindly provided comments or criticism, but I would like to mention in particular the editors of the *Handbook* and Henk Barendregt, who tried out the first draft in a course, Karst Koymans and Erik Krabbe for their error detecting and Albert Visser for many helpful discussions.

0. INTRODUCTION

Algorithms have a long and respectable history. There are, e.g., Euclid's algorithm for determining the greatest common divisor of two numbers, Sturm's algorithm to find the number of zeros of a polynomial between given bounds.

Let us consider the example of Euclid's algorithm applied to 3900 and 5544.

After division of 5544 by 3900 the remainder is 1644
" " " 3900 " 1644 " " " 612
" " " 1644 " 612 " " " 420
" " " 612 " 420 " " " 192
" " " 420 " 192 " " " 36
" " " 192 " 36 " " " 12
" " " 36 " 12 " " " 0

Hence, the g.c.d. of 3900 and 5544 is 12.

There are three features in the above example:

(1) There is a proof that the algorithm does what it is asked to do. In this case, that 12 is actually the g.c.d., but in general that the outcome for any pair n, m is the g.c.d. (the reader will see the proof after a moment's reflection).

(2) The procedure is algorithmic, i.e. at each step it is clear what we have to do, and it can be done 'mechanically' by finite manipulations. This part is clear, assuming we know how to carry out the arithmetical operations on numbers given in decimal representation.

(3) The procedure stops after a finite number of steps. In a way (1) presupposes (3), but (1) might give the following result: if the procedure stops then the answer is correct, so we are still left with the burden of showing the halting of the procedure. In our example we observe that all entries in the

last column are positive and that each is smaller than the preceding one. So a (very) rough estimate tells us that we need at most 1644 steps.

Another example:

A palindrome is a word that reads the same forward or backwards, e.g. bob. Is there a decision method to test if a string of symbols is a palindrome? For short strings the answer seems obvious: you can see it a glance. However, a decision method must be universally applicable, e.g. also to strings of 2000 symbols. Here is a good method: compare the first and the last symbol and if they are equal, erase them. If not then the string is not a palindrome. Next repeat the procedure. After finitely many steps we have checked if the string is a palindrome. Here too, we can easily show that the procedure always terminates, and that the answer is correct.

The best-known example from logic is the decidability of classical propositional logic. The algorithm requires us to write down the truth table for a given proposition φ and check the entries in the last column if all of them are 1 (or T). If so then $\vdash \varphi$.

If φ has n atoms and m subformulas, then a truthtable with $m \cdot 2^n$ entries will do the job, so the process terminates. The truth tables for the basic connectives tell us that the process is effective and give us the completeness theorem.

The need for a notion of effectiveness entered logic in considerations on symbolic languages. Roughly speaking, syntax was assumed (or required) to be decidable, i.e. one either explicitly formulated the syntax in such a way that an algorithm for testing strings of symbols on syntactic correctness was seen to exist, or one postulated such an algorithm to exist, cf. Carnap [1937] or Fraenkel et al. [1973] p. 280 ff. Since then it is a generally recognized practice to work with a decidable syntax. This practice has vigorously been adopted in the area of computer languages.

The quest for algorithms has been stimulated by the formalist view of logic and mathematics, as being fields described by mechanical (effective) rules. Historically best-known is Hilbert's demand for a decision method for logic and arithmetic. A priori there are some philosophical arguments for decidability in a few matters. For example, the notion 'p is a proof of φ' should be decidable, i.e. we should be able to recognize effectively whether or not a given proof p proves a statement φ. Furthermore, it is a basic assumption for the usefulness of language that well-formedness should be effectively testable.

In the thirties, a number of proposals for the codification of the notion of 'algorithm' were presented. A very attractive and suggestive view was presented by Alan Turing, who defined effective procedures, or algorithms, as abstract machines of a certain kind (cf. Turing [1936], Kleene [1952], Davis [1958]).

Without aiming for utmost precision, we will consider these so-called Turing machines a bit closer. This will give the reader a better understanding of algorithms and, given a certain amount of practical experience, he will come to appreciate the ultimate claim that any algorithm can be carried out (or simulated) on a Turing machine. The reason for choosing this particular kind of machine and not, e.g., Markov algorithms or Register machines, is that there is a strong conceptual appeal to Turing machines. Turing has given a very attractive argument supporting the above claim – known as *Turing's Thesis*. We will return to the matter later.

A Turing machine can be thought of as an abstract machine (a black box) with a finite number of internal states, say $q_1, \ldots, q_n$, a reading and a printing device, and a (potentially infinite) tape. The tape is divided into squares and the machine can move one square at a time to the left or right (it may be more realistic to make the tape move, but realism is not our object). We suppose that a Turing machine can read and print a finite number of symbols $S_1, \ldots, S_n$. The actions of the machine are strictly local, it has a number of instructions of the form: *When reading S_j and being in state q_i print S_k, go into state q_l and move to the left (or right)*.

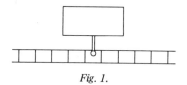

Fig. 1.

We collect this instruction into a convenient string $q_i S_j S_k q_l X$, where X is L or R. The machine is thus supposed to read a symbol, erase it, print a new one, and move left or right. It would not hurt to allow the machine to remain stationary, but it does not add to the algorithmic power of the class of Turing machines.

Of course we need some conventions or else a machine would just go on operating and we would never be able to speak of computations in a systematic way. Here are our main conventions: (1) we will always present the machine at the beginning with a tape whose squares, except for a finite number, are blank; (2) at the beginning of the process the machine scans the leftmost non-blank square; (3) the machine stops when it is in a state and

I.6: ALGORITHMS AND DECISION PROBLEMS

reads a symbol such that no instruction applies; (4) for any state and symbol read by the machine there is at most one instruction which applies (i.e. the machine is *deterministic*).

Another convention, which can be avoided at the cost of some complication, is that we always have a symbol B for 'blank' available. This helps us to locate the end of a given string, although even here there are some snags (e.g. suppose you move right until you get to a blank, how do you know that there may not be a non-blank square way out to the right?).

Now it is time for a few examples.

0.1. *The Palindrome Tester*

We use the ideas presented above. The machine runs back and forth checking the end symbols of the string, when it is a matching pair it erases them and proceeds to the next symbol. Let us stipulate that the machine erases all symbols when its finds a palindrome, and leaves at least one non-blank square if the result is negative. Thus, we can see at a glance the *yes* or *no* answer. We introduce the symbols a, b, B. During the process we will find out how many states we need. Consider the following example: the tape is of the form $\cdots BBaababaaBB \cdots$ and the machine reads the first a while being in the initial state q_0, we represent this by $\cdots B\underset{q_0}{a}ababaaB \cdots$.

We now want to move right while remembering that we scanned a first symbol a. We do that by changing to a new state q_a. We find out that we have passed the word when we meet our first B, so then we move back, read the symbol and check if it is an a, that is when we use our memory – i.e. the q_a.

Here are the necessary instructions:

$q_0 a a q_a R$ – in state q_0, read a, go to state q_a, move right,
$q_a a a q_a R$ – in state q_a, read a, do nothing, move right,
$q_a b b q_a R$ – in state q_a, read b, do nothing, move right,
$q_a B B q_1 L$ – in state q_a, read B, go to state q_1, move left,
$q_1 a B q_2 L$ – in state q_1, read a, erase a, go to state q_2, move left,
and now return to the front of the word.

We indicate the moves of the machine below:

$\underset{q_0}{B}aababaaB \to B\underset{q_a}{a}ababaaB \to \cdots \to BaababaaB\underset{q_a}{} \to$

$\to BaababaaB\underset{q_1}{} \to BaababaBB\underset{q_2}{} \to \cdots \to BaababaB\underset{q_2}{}$

We now move right, erase the first symbol, look for the next one and repeat the procedure.

More instructions:

move to the front
$\begin{cases} q_2 a a q_2 L \\ q_2 b b q_2 L \\ q_2 B B q_3 R \end{cases}$

erase the first symbol
$\begin{cases} q_3 a B q_0 R \\ q_3 b B q_0 R \end{cases}$

move right when you see a b and check the last symbol
$\begin{cases} q_0 b b q_b R \\ q_b b b q_b R \\ q_b a a q_b R \\ q_b B B q_4 L \\ q_4 b B q_2 L \end{cases}$

We indicate a few more steps in the computation:

$$B a a b a b a B \to B a a b a b a B \to B B a b a b a B \to \cdots \to$$
$$q_2 q_3 q_0$$

$$\to B b a b B \to B b a b B \to \cdots \to B b a b B \to B b a B \to \cdots \to$$
$$q_0 q_b q_4 q_2$$

$$\to B a B \to B a B \to B a B \to B B B \to B B B.$$
$$q_0 q_a q_1 q_2 q_3$$

Here the machine stops, there is no instruction beginning with $q_3 B$. The tape is blank, so the given word was a palindrome. If the word is not a palindrome, the machine stops at the end of the printed tape in state q_1 or q_4 and a non-blank tape is left.

One can also present the machine in the form of a graph (a kind of flow diagram). Circles represent states and arrows the action of the machine, e.g. $(q_i) \xrightarrow{S_j S_k X} (q_l)$ stands for the instruction $q_i S_j S_k q_l X$.

The graph for the above machine is given as Figure 2.

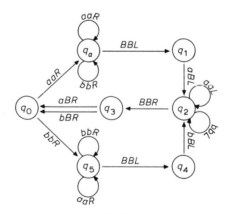

Fig. 2.

I.6: ALGORITHMS AND DECISION PROBLEMS

The expressions consisting of a finite part of the tape containing the nonblank part plus the state symbol indicating which symbol is being read, are called *state descriptions*. For better printing we place the state symbol in front of the scanned symbol instead of below it. A sequence of successive state descriptions is called a *computation*. Note that a computation may be infinite. In that case there is no output.

EXERCISES. Design Turing machines for the following tasks:
(1) check if a word (over the alphabet $\{a, b\}$) contains an a,
(2) check if a word (over the alphabet $\{a, b\}$) contains two a's,
(3) check if a word (over the alphabet $\{a, b\}$) has even length,
(4) interchange all a's and b's in a word,
(5) produce the mirror image of a word.

0.2. Some Arithmetical Operations

We represent natural numbers n by $n + 1$ strokes (so that 0 is taken along in accordance with modern usage). A pair of numbers is represented by two strings of strokes separated by a blank. We will denote the sequence of $n + 1$ strokes by $\bar{n}$.

The convention for reading the output is: count all the strokes that are on the tape when the machine stops. It is a simple exercise to convert the tape contents into the above unary representation, but it is not always required to make the conversion.

0.2.1. The identity function: $f(x) = x$. We just have to erase one stroke. Instructions:

$$q_0 \mid B\ q_0\ L.$$

Here is a computation

$$Bq_0 \mid \mid \mid .. \mid B \to q_0 BB \mid \mid .. \mid B.$$

0.2.2. The successor function: $f(x) = x + 1$. This is a very simple task: do nothing. So the machine has some dummy instruction, e.g. $q_0 BBq_0 R$. This machine stops right when it starts.

0.2.3. Addition: $f(x, y) = x + y$. Here we have to erase two strokes. It is tempting to erase both from the first string; however, the first string may contain only one $\mid$, so we have to be somewhat careful.

Here is the informal description: erase the first | and go into state q_1, move right until you meet the first |, erase it and stop. Instructions:

$$q_0 \mid B \, q_1 R$$
$$q_1 \mid B \, q_2 R$$
$$q_1 B B \, q_1 R.$$

Example:

$$q_0 \mid B \mid \mid \to B q_1 B \mid \mid \to B B q_1 \mid \mid \to B B B q_2 \mid$$

and

$$q_0 \mid \mid B \mid \to B q_1 \mid B \mid \to B B q_2 B \mid$$

0.2.4. *Subtraction:* $f(x, y) = x - y$. Observe that this is a *partial* function, for $x - y$ is defined only for $x \geqslant y$. The obvious procedure seems to erase alternatingly a stroke of y and one of x. If y is exhausted before x is, we stop. Instructions:

$$\left.\begin{array}{l} q_0 \mid \mid q_0 R \\ q_0 B B q_1 R \\ q_1 \mid \mid q_1 R \\ q_1 B B q_2 L \\ (*) \quad q_2 \mid a q_3 L \end{array}\right\} \text{move right and erase the last } \mid \text{ of } y, \text{ print a marker } a$$

$$\left.\begin{array}{l} q_3 \mid \mid q_3 L \\ q_3 B B q_4 L \\ (\dagger) \quad q_4 B B q_4 L \\ q_4 \mid B q_5 R \end{array}\right\} \text{move left until you meet the first } \mid \text{ of } x, \text{ erase it and move right again}$$

$$\left.\begin{array}{l} q_5 B B q_5 R \\ q_5 \mid \mid q_5 R \\ q_5 a a q_2 L \end{array}\right\} \text{move right until you meet the marker } a, \text{ erase the } \mid \text{ left of } a, \text{ cf. } (*)$$

For convenience we will write $\overset{*}{\to}$ to indicate a finite number of steps $\to$

Example.

$$B q_0 \mid \mid \mid B \mid \mid B \overset{*}{\to} B \mid \mid \mid B q_1 \mid \mid B \overset{*}{\to} B \mid \mid \mid B \mid \mid q_1 B \to$$
$$B \mid \mid \mid B \mid q_2 \mid B \to B \mid \mid \mid B q_3 \mid a B \to B \mid \mid \mid q_3 B \mid a B \to$$
$$B \mid \mid q_4 \mid B \mid a B \to B \mid \mid B q_5 B \mid a B \overset{*}{\to} B \mid \mid B B q_2 \mid a B \overset{*}{\to}$$
$$B \mid B B B q_5 a a B \to B \mid B B q_2 B a a B$$

If x is exhausted before y is, then by $(\dagger)$ the machine keeps moving left, i.e. it never stops. Hence for $x < y$ there is no output.

I.6: ALGORITHMS AND DECISION PROBLEMS

0.2.5. *The projection functions:* $U_i^n(x_0, \ldots, x_n) = x_i$ $(0 \leq i \leq n)$. The machine has to erase all the x_j's for $j \neq i$ and also to erase one | from x_i. Instructions:

$q_0 \mid B q_0 R$
$q_0 B B q_1 R$
$q_1 \mid B q_1 R$
$q_1 B B q_2 R$
$\vdots$
$q_i \mid B q_i' R$
$q_i' \mid\mid q_i' R$
$q_i' B B q_{i+1} R$
$\vdots$
$q_n \mid B q_n R$

By now the reader will have reached the point where he realizes that he is simply writing programs in a rather uncomfortable language for an imaginary machine. The awkwardness of the programming language is not accidental, we wanted to perform really atomic acts so that the evidence for the algorithmic character of Turing machines can immediately be read off from those acts. Of course, a high-level programming language is more convenient to handle, but it also stresses some features and neglects some other features, e.g. it might be perfect for numerical calculation and poor for string manipulations.

It is also about time to give a definition of the Turing machine, after all we have treated it so far as a *Gedankenexperiment*. Well, *a Turing machine is precisely a finite set of instructions*! For, given those instructions, we can perform all the computations we wish to perform. So, strictly speaking, adding or changing an instruction gives us a *new* machine. One can, in general, perform operations on Turing machines, e.g. for the purpose of presenting the output in a convenient way, or for creating a kind of memory for the purpose of recording the computation.

EXAMPLE. *Carrying out a computation between end markers.*
Let a machine M (i.e. a set of instructions) be given. We want to add two end markers, so that any computation of M has descriptions of the form $\$_1 - \$_2$, where the tape contains only blanks to the left of $\$_1$ and to the right of $\$_2$. We add two new symbols $\$_1$ and $\$_2$ to those of M and a number of instructions that take care of keeping the descriptions between $\$_1$ and $\$_2$.

For, in the course of a computation, one may need more space, so we have to build in a $-moving feature. The following instructions will do the job:

$$\left.\begin{array}{l} q_i\$_1 Bq_i'L \\ q_i'B\$_1 q_i R \end{array}\right\} \text{if } M \text{ reads } \$_1 \text{ print a blank, move one step left, print } \$_1, \text{ move back and go into the original state}$$

$$\left.\begin{array}{l} q_i\$_2 Bq_i''R \\ q_i''B\$_2 q_i L \end{array}\right\} \text{same action on the right hand side}$$

Here q' and q'' are new states not occurring in M.

The reader may try his hand at the following operations.

(1) Suppose that a computation has been carried out between end markers. Add instructions so that the output is presented in the form $\$_1 \bar{n} \$_2$ (sweeping up the strokes).

(2) Let M be given, add a terminal state to it, i.e. a new state q_t such that the new machine M' acts exactly like M, but when M stops M' makes one more step until it stops at the same description with the new q_t as state.

(3) Suppose a tape containing a word between end markers is given. Add instructions to a machine M such that during a computation M preserves the word intact, i.e. any time M reads, e.g. the left end marker, it moves the whole word one square to the right, and resumes its normal activity to the left of this marker.

The last exercise may serve to store, e.g., the input in the tape as memory, so that we can use it later.

We will now consider some operations on Turing machines, required for certain arithmetical procedures. The precise details of those operations can be found in the literature, e.g. Davis [1958] or Minsky [1967], we will present a rough sketch here.

0.2.6. *Substitution*. Suppose that machines M_1 and M_2 carry out the computations for the functions f and g. How can we compute $h(x) = f(g(x))$ by means of a Turing machine? To begin with, we make the sets of states of M_1 and M_2 disjoint. The idea is to carry out the computation of M_2 on input x, we add extra instructions so that M_2 moves into a terminal state q_t when it stops. Then we add some instructions that collect at the strokes into one string and make the machine scan the leftmost | in the initial state of M_1.

As simple as this sounds, it takes a certain amount of precaution to carry out the above plan, e.g. in order to sweep all the strokes together one has to

I.6: ALGORITHMS AND DECISION PROBLEMS   419

provide end markers so that one knows when all strokes have been counted, cf. the example above.

Schematically, we perform the following operations on the machines M_1, M_2: (1) change M_2 into a machine M_2' which carries out the same computations, but between end markers, (2) change M_2' into M_2'' which goes on to sweep all |'s together and stops in a terminal state q_t scanning the leftmost |, (3) renumber the states $q_0, \ldots, q_m$, of M_1 into $q_t, \ldots, q_{t+m}$, the resulting machine is M_1'. Then the instructions of M_2'' and M_1', joined together, define the required machine for h. Substitution with more variables is merely a more complicated variation of the above.

0.2.7. *Primitive recursion.* One of the standard techniques for defining new algorithms is that of recursion. We consider the simple parameterless case. If g is a given algorithm (and a total function) then so is f, with

$$\begin{cases} f(0) = n \\ f(x+1) = g(f(x), x). \end{cases}$$

We store n and x on the tape, say in the form $\$_1 \bar{n} \$_2 \bar{x} \$_3$. Next we erase one | from $\bar{x}$, if nothing is left between $\$_2$ and $\$_3$, then we erase one stroke of $\bar{n}$ and stop. If not, we transform the tape content into $\$_1 \bar{n} B \mid \$_2 \overline{x-1} \$_3 \bar{n} B \mid$ and let M_1 act on the part to the right of $\$_3$. The resulting strokes are swept up so that we get $\$_1 \bar{n} B \mid \$_2 \overline{x-1} \$_3 \bar{m}$ (where $m = f(1)$). Next, replace $\bar{n}$ by $\bar{m}$. Then we erase one more stroke between $\$_2$ and $\$_3$. If nothing is left we erase two strokes between $\$_1$ and $\$_2$ *and* all strokes to the right of $\$_2$, after which actions we stop. If *not*, then we add one stroke to the left of $\$_2$ and transform the tape content into $\$_1 \bar{m} B \mid\mid \$_2 \overline{x-2} \$_3 \bar{m} B \mid\mid$ and repeat the action.

The resulting machine eventually stops after x steps with $f(x)$ strokes on the tape. The addition of extra parameters is merely a matter of storing the parameters conveniently on the tape.

0.2.8. *Unbounded search or minimalization.* Suppose that we have a Turing machine M which computes a total function $g(x, y)$. Can we find a Turing machine M_1 that for a given y looks for the first x such that $g(x, y) = 0$?

Essentially, we will successively compute $g(0, y), g(1, y), g(2, y), \ldots$ and stop as soon as an output 0 has been produced. This is what we will do: (1) start with a tape of the form $\cdots B \$_1 \mid B\bar{y} \$_2 \$_3 B \cdots$ and read the first |, (2) copy the string between $\$_1$ and $\$_2$ between $\$_2$ and $\$_3$, (3) let M act on the string between $\$_2$ and $\$_3$, (4) add instructions that test if there is a | left between $\$_2$ and $\$_3$, if not erase $\bar{y}$ and one stroke to the right of $\$_1$ then stop, otherwise erase everything between $\$_2$ and $\$_3$ while shifting $\$_3$ to the left, then move left and add one | following $\$_1$, (5) repeat (2).

Clearly, if the new machine stops, then the tape content yields the desired output. The machine may, however, go on computing indefinitely. Contrary to the cases of substitution and recursion, the minimalization operation leads outside the domain of totally-defined algorithms!

The most striking feature of the family of Turing machines is that it contains a 'master' machine, that can mimic all Turing machines. This was established in Turing's very first paper on the subject. We will first give a loose and imperfect statement of this fact:

There is a Turing machine, such that if it is presented with a tape containing all the instructions of a Turing machine M plus an input, it will mimic the computations of M and yield the same output.

We will indicate the idea of the simulation process by means of a rough sketch of a simple case. Consider the addition-machine (0.2.3). On the tape we print the instructions plus the input separated by suitable symbols.

$$\$_1 q_0|Bq_1R*q_1|Bq_2R*q_1BBq_1R\,\$_2 q_0|B\,||\,\$_3$$

Note that the states of the addition-machine and its symbols, and the R and L have become symbols for the new machine. Now we start the machine reading the symbol to the right of $\$_2$, it moves one square to the right, stores $q_0|$ in its memory (i.e. it goes into a state that carries this information) and moves left looking for a pair $q_0|$ left of $\$_2$. When it finds such a pair, it looks at the three right-hand neighbours, stores them into its memory (again by means of an internal state), and moves right in order to replace the $q_0|$ following $\$_2$ by Bq_1. Then the machine repeats the procedure all over again. In this way the machine mimics the original computation.

$$\$_1 - \$_2 \bar{q}_0 q_0|B\,||\,\$_3 \overset{*}{\to} \$_1 - \$_2 B\bar{q}_k q_k B\,||\,\$_3 \overset{*}{\to} \cdots \overset{*}{\to}$$
$$\overset{*}{\to} \$_1 - \$_2 BB\bar{q}_i q_1\,||\,\$_3 \overset{*}{\to} \$_1 - \$_2 BBB\bar{q}_j q_2|\$_3 \overset{*}{\to}$$
$$\overset{*}{\to} \$_2 BBB\bar{q}_j q_2|\$_3.$$

The states of the machine have been indicated by barred q's. The final steps are to erase everything left of $\$_2$.

Of course, we have in a most irresponsible way suppressed all technical details, e.g. the search procedure, the 'memory' trick. But the worst sin is our oversimplification of the representation of the instructions. In fact we are dealing with an infinite collection of Turing machines and, hence, we have

I.6: ALGORITHMS AND DECISION PROBLEMS

to take care of infinitely many states q_i and symbols S_j. We solve this problem by a unary coding of the q_i's and S_j's, e.g. represent q_i by $qq \ldots q$ (i times) and S_j by $SS \ldots S$ ($j + 1$ times). This of course complicates the above schema, but not in an insurmountable way.

A more precise formulation of the theorem concerning the so-called *Universal Turing machine* is:

There is a Turing machine U such that for each Turing machine M it can simulate the computation of M with input x, when presented with an input consisting of a coded sequence of instructions of M and x. The output of U is identical with that of M (possibly up to some auxiliary symbols).

One can find proofs of this theorem in a number of places, e.g. Davis [1958], Minsky [1967], Turing [1936].

If the reader is willing to accept the above facts for the moment, he can draw some immediate consequences. We will give a few informal sketches.

Let us call the coded sequence e of instructions of a machine M its *index*, and let us denote the output of M with input x by $\varphi_e(x)$. Obviously the universal Turing machine has itself an index; up to some coding U can act on Turing machines (i.e. their indices), in particular, on itself. In a way we can view this as a kind of self-reference or self-application.

Since Turing machines are algorithmic, i.e. given an input they effectively go through a sequence of well-determined steps and hence, produce in an effective way an output when they stop, they can be used for decision procedures. Decision problems ask for effective yes-no answers, and Turing machines provide a particular framework for dealing with them. We can design a Turing machine that decides if a number is even, i.e. it produces a 1 if the input n is even and a 0 if n is odd.

If there is a Turing machine that produces in such a way 0–1 answers for a problem, we say that the problem is decidable. Question: are there undecidable problems? In a trivial way, yes. A problem can be thought of as a subset X of $\mathbb{N}$, and the question to be answered is: "Is n an element of X?" (In a way this exhausts all reasonably well-posed decision problems.) Since there are uncountably many subsets of $\mathbb{N}$ and countable many Turing machines, the negative answer is obvious. Let us therefore reformulate the question: are there interesting undecidable problems? Again the answer is *yes*, but the solution is not trivial; it makes use of Cantor's diagonal procedure.

It would be interesting to have a decision method for the question: does a Turing machine (with index e) eventually stop (and thus produce an output) on an input x? This is Turing's famous *Halting Problem*. We can make this precise in the following way: is there a Turing machine such that with input (e, x) it produces an output 1 if the machine with index e and input x eventually stops, and an output 0 otherwise. We will show (informally) that there is no such machine.

Suppose there is a machine M_0 with index e_0 such that

$$\varphi_{e_0}(e, x) = \begin{cases} 1 & \text{if } \varphi_e(x) \text{ exists,} \\ 0 & \text{if there is no such output for the machine with index } e \text{ on input } x. \end{cases}$$

We can change this machine M_0 slightly such that we get a new machine M_1 with index e_1 such that

$$\varphi_{e_1}(x) = 1 \quad \text{if } \varphi_{e_0}(x, x) = 0$$

and there is no output if $\varphi_{e_0}(x, x) = 1$. One can simply take the machine M_0 and change the output 0 into a 1, and send it indefinitely moving to the let if the output of M_0 was 1.

Now,

$$\varphi_{e_0}(e_1, e_1) = 0 \iff \varphi_{e_1}(e_1) = 1 \iff \varphi_{e_0}(e_1, e_1) = 1.$$

Contradiction. So the machine M_0 does not exist: the halting problem is undecidable.

Turing himself has put forward certain arguments to support the thesis that all algorithms (including the partial ones) can be carried out by means of Turing machines. Algorithms are here supposed to be of a 'mechanical' nature, i.e. they operate stepwise, each step is completely determined by the instructions and the given configurations (e.g. number symbols on paper, pebbles, or the memory content of a computer), and everything involved is strictly finite. Since computations have to be performed on (or in) some device (paper, strings of beads, magnetic tape, etc.) it will not essentially restrict the discussion if we consider computations on paper. In order to carry out the algorithm one (or a machine) has to act on the information provided by the configuration of symbols on the paper. The effectiveness of an algorithm requires that one uses an immediately recognizable portion of this information, so one can use only *local* information (we cannot even copy a number of 20 figures as a whole!), such as three numerals in a row or the letters attached to the vertices of a small-sized triangle. A Turing machine can only read one symbol at a time, but it can, e.g., scan three squares successively and use the information by using internal states for

I.6: ALGORITHMS AND DECISION PROBLEMS

memory purposes. So the limitations of the reading ability to one symbol at a time is not essential.

The finiteness conditions on Turing machines, i.e. both on the alphabet and on the number of states, is also a consequence of the effectiveness of algorithms. An infinite number of symbols that can be printed on a square would violate the principle of immediate recognizability, for a number of symbols would become so similar that they would drop below the recognizability threshold.

Taking into account the ability of Turing machines to simulate more complex processes by breaking them into small atomic acts, one realizes that any execution of an algorithm can be mimicked by a Turing machine, We will return to this matter when we discuss Church's Thesis.

There are many alternative but equivalent characterizations of algorithms – recursive functions, λ-calculable functions, Markov Algorithms, Register Machines, etc. – all of which have the discrete character in common. Each can be given by a finite description of some sort. Given this feature, it is a fundamental trick to code these machines, or functions, or whatever they may be, into natural numbers. The basic idea, introduced by Gödel, is simple: a description is given in a particular (finite) alphabet, code each of the symbols by fixed numbers and code the strings, e.g. by the prime-power-method.

EXAMPLE. Code a and b as 2 and 3. Then the strings aba, $aababb$, ... are coded as $2^2 \cdot 3^3 \cdot 5^2$, $2^2 \cdot 3^2 \cdot 5^3 \cdot 7^2 \cdot 11^3 \cdot 13^3$, ... Note that the coding is fully effective: we can find for each word its numerical code, and conversely, given a natural number, we simply factorize it and, by looking at the exponents, can check if it is a code of a word, and if so, of which word.

Our example is, of course, shockingly simple, but the reader can invent (or look up) more complicated and versatile codings.

The coding reduces the study of algorithms and decision methods to that of effective operations on natural numbers.

EXAMPLE. (1) We consider strings of a's and b's, and we want to test if such a string contains 15 consecutive b's.

First we code $a \to 1$, $b \to 2$ and next each string $x_1, x_2 \cdots x_n$ is coded as $p_1^{\bar{x}_1} \cdot p_2^{\bar{x}_2} \cdots p_n^{\bar{x}_n}$, where p_i is the ith prime and $\bar{x}_i$ the code of x_i ($x_i \in \{a, b\}$), e.g. $abba \to 2^1 \cdot 3^2 \cdot 5^2 \cdot 7^1 = 3150$, $bbbb \to 44100$.

Under this coding, the test for containing 15 consecutive b's is taken to be a test for a number to be divisible by 15 squares of consecutive primes, which is a purely number-theoretic test.

(2) We want an algorithm for the same set of strings that counts the number of a's. We use the same coding, then the algorithm is translated into a numerical algorithm: compute the prime factorization of n and count the number of primes with exponent 1.

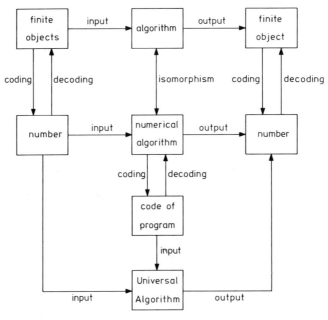

Fig. 3.

Figure 3 illustrates the use of codings. The lower half contains the so-called Universal Algorithm. Our working hypothesis is that there is a standard codification of algorithms that is specified in a certain language. By coding the linguistic expression for the algorithm in standard codification into a number, we obtain two inputs for a 'super'-algorithm that looks at the number that codes the algorithm and then proceeds to simulate the whole computation. We will meet this so-called universal algorithm in Section 2 under the disguise of clause R7.

We also can see now why the general form of a decision problem can be taken to be of the form '$n \in X$?', for a set of natural numbers X.

Say we want to decide if an object a has the property A; we consider a coding # of the class of objects under consideration into the natural numbers. Then A is coded as a predicate $A^{\#}(x)$ of natural numbers, which in

I.6: ALGORITHMS AND DECISION PROBLEMS

turn determines the set $A^\# = \{x \in \mathbb{N} | A^\#(x)\}$. So, we have reduced the question "Does a have the property A?" to "$\#(a) \in A^\#$?"

For theoretical purposes we can therefore restrict ourselves to the study of algorithms on natural numbers and to decision problems for sets of natural numbers.

We say that a set X is *decidable* if there is an algorithm F such that

$$F(n) = \begin{cases} 1 & \text{if } n \in X \\ 0 & \text{if } n \notin X. \end{cases}$$

We say that F tests for membership of X. In other words: X is *decidable* if its characteristic function is given by an algorithm.

1. PRIMITIVE RECURSIVE FUNCTIONS

Given the fact that numerical algorithms can simulate arbitrary algorithms, it stands to reason that a considerable amount of time and ingenuity has been invested in that specific area. A historically and methodologically important class of numerical algorithms is that of the *primitive recursive functions*. One obtains the primitive recursive functions by starting with a stock of acknowledged algorithms and constructing new algorithms by means of substitution and recursion. We have presented evidence that indeed recursion and substitution do lead from Turing machines to Turing machines, but the reader can easily provide intuitive arguments for the algorithmic character of functions defined by recursion from algorithmic functions. The primitive recursive functions are so absolutely basic and foundationally unproblematic (or rather, just as problematic as the natural number sequence), that they are generally accepted as a starting point for metamathematical research. Primitive recursive functions provide us with a surprisingly wide stock of algorithms, including codings of finite sequences of natural numbers as mentioned above, and one has to do some highly nontrivial tricks to get algorithms which are not primitive recursive.

The basic algorithms one departs from are extremely simple indeed: the successor function, the constant functions and the projection functions $(x_1, \ldots, x_n) \mapsto x_i$ $(i \leq n)$. The use of recursion was already known to Dedekind, and Landau made extensively use of it in his '*Foundations of Analysis*'. The study of primitive recursive functions was initiated in logic by Skolem, Herbrand, Gödel and others.

We will now proceed with a precise definition, which will be given in the form of an inductive defintion. First we present a list of initial functions of an unmistakably algorithmic nature, and then we specify how to get new algorithms from old ones.

The so-called *initial functions* are the *constant functions* C_m^k with $C_m^k(n_1, \ldots, n_k) = m$, the *successor function* S with $S(n) = n + 1$, and the *projection functions* P_i^k with $P_i^k(n_1, \ldots, n_k) = n_i$, $(i \leqslant k)$.

The recognized procedures are: *substitution* or *composition*, i.e. when $f(n_1, \ldots, n_k) = g(h_1(n_1, \ldots, n_k), \ldots, h_p(n_1, \ldots, n_k))$ then we say that f *is obtained by substitution from g and* $h_1, \ldots, h_p$, and *primitive recursion*, i.e. we say that f *is obtained by primitive recursion from g and h*

$$\text{if } \begin{cases} f(0, n_1, \ldots, n_k) = g(n_1, \ldots, n_k) \\ f(m + 1, n_1, \ldots, n_k) = h(f(m, n_1, \ldots, n_k), n_1, \ldots, n_k, m). \end{cases}$$

A class of functions is *closed under substitution or primitive recursion* if f belongs to it whenever it is obtained by substitution or primitive recursion from functions that already belong to that class.

1.1. DEFINITION. The class of *primitive recursive functions* is the smallest class containing the initial functions that is closed under substitution and primitive recursion.

Notation. For convenience we abbreviate sequences $n_1, \ldots, n_k$ as $\vec{n}$, whenever no confusion arises.

1.2. EXAMPLES. The following functions are primitive recursive.
(1) $x + y$

$$\begin{cases} x + 0 = x \\ x + (y + 1) = (x + y) + 1 \end{cases}$$

This definition can be put in a form that shows immediately that $+$ is primitive recursive

$$+(0, x) = P_1^1(x)$$
$$+(y + 1, x) = S(P_1^3(+(y, x), x, y)).$$

In accordance with tradition we write $x + y$ for $+(y, x)$. Note that we have given an h in the second line, that actually contains all the variables that the schema of recursion prescribes. This is not really necessary since the projection functions allow us to add dummy variables.

I.6: ALGORITHMS AND DECISION PROBLEMS

Example: let g contain only the variables x and y, then we can add the dummy variable z as follows $f(x, y, z) = g(P_1^3(x, y, z), P_2^3(x, y, z))$.

We will leave such refinements to the reader and proceed along traditional lines.

(2) $x \cdot y$

$$\begin{cases} x \cdot 0 = 0 \\ x \cdot (y + 1) = x \cdot y + x \quad \text{(we use (1))} \end{cases}$$

(3) x^y

$$\begin{cases} x^0 = 1 \\ x^{y+1} = x^y \cdot x \end{cases}$$

(4) the predecessor function $p(x) = \begin{cases} x - 1 & \text{if } x > 0 \\ 0 & \text{if } x = 0 \end{cases}$

$$\begin{cases} p(0) = 0 \\ p(x + 1) = x \end{cases}$$

(5) the cut-off subtraction (monus), $x \dotminus y$, where $x \dotminus y = x - y$ if $x \geq y$ and 0 else.

$$\begin{cases} x \dotminus 0 = x \\ x \dotminus (y + 1) = p((x \dotminus y)) \end{cases}$$

(6) the factorial function, $n! = 1 \cdot 2 \cdot 3 \cdots (n-1) \cdot n$.
(7) the signum function, $\text{sg}(x) = 0$ if $x = 0$, 1 otherwise.
(8) $\overline{\text{sg}}$, with $\overline{\text{sg}}(x) = 1$ if $x = 0$, 0 otherwise.
  Observe that $\overline{\text{sg}}(x) = 1 \dotminus \text{sg}(x)$.
(9) $|x - y|$, observe that $|x - y| = (x \dotminus y) + (y \dotminus x)$
(10) $f(\vec{x}, y) = \Sigma_{i=0}^{y} g(\vec{x}, i)$, where g is primitive recursive.
(11) $f(\vec{x}, y) = \Pi_{i=0}^{y} g(\vec{x}, i)$, idem.
(12) If f is primitive recursive and π is a permutation of the set $\{1, \ldots, n\}$, then g with $g(x_1, \ldots, x_n) = f(x_{\pi 1}, \ldots, x_{\pi n})$ is also primitive recursive.

Proof. $g(\vec{x}) = f(P_{\pi 1}^n(\vec{x}), \ldots, P_{\pi n}^n(\vec{x}))$. □

The reader may find it an amusing exercise to enlarge the stock for primitive recursive functions 'by hand'. We will, however, look for a more systematic way to obtain new primitive recursive functions.

1.3. DEFINITION. A relation R is primitive recursive if its characteristic function is so.

Note that this corresponds to the idea of testing R for membership: let K_R be the characteristic function of R then we know that $\vec{n} \in R$ iff

$$K_R(n_1, \ldots, n_k) = 1.$$

1.4. EXAMPLES. The following sets (relations) are primitive recursive
 (1) $\emptyset, K_\emptyset(x) = 0$
 (2) The set of even numbers, E.

$$\begin{cases} K_E(0) = 1 \\ K_E(x+1) = \overline{\text{sg}}(K_E(x)) \end{cases}$$

 (3) The equality relation: $K_=(x, y) = \overline{\text{sg}}(|x-y|)$
 (4) The order relation: $K_<(x, y) = \text{sg}(y \dot{-} x)$.

Note that relations are subsets of $\mathbb{N}^k$ for a suitable k; when dealing with operations or relations, we assume that we have the correct number of arguments, e.g. when we write $A \cap B$ we suppose that $A, B \subseteq \mathbb{N}^k$.

1.5. LEMMA. *The primitive recursive relations are closed under $\cup$, $\cap$, ^{c} and bounded quantification.*

Proof. Let $C = A \cap B$, then $x \in C \leftrightarrow x \in A \wedge x \in B$, so $K_C(x) = 1 \leftrightarrow K_A(x) = 1 \wedge K_B(x) = 1$. Therefore we put $K_C(x) = K_A(x) \cdot K_B(x)$. For union take $K_{A \cup B}(x) = \text{sg}(K_A(x) + K_B(x))$, and for the complement $K_{A^c}(x) = \overline{\text{sg}}(K_A(x))$.

We say that R is *obtained by bounded quantitication from S* if $R(n_1, \ldots, n_k, m) := Qx \leq m \, S(n_1, \ldots, n_k, x)$, where Q is one of the quantifiers $\forall, \exists$.

Consider the bounded existential quantification: $R(\vec{x}, n) := \exists y \leq n \, S(\vec{x}, y)$, then $K_R(\vec{x}, n) = \text{sg} \, \Sigma_{y \leq n} K_S(\vec{x}, y)$, so R is primitive recursive if S is so.

The $\forall$-case is very similar, and is left to the reader. □

1.6. LEMMA. *The primitive recursive relations are closed under primitive recursive substitutions, i.e. if $f_1, \ldots, f_n$ and R are primitive recursive, then so is $S(x_1, \ldots, x_k) := R(f_1(\vec{x}), \ldots, f_n(\vec{x}))$.*

Proof. $K_S(\vec{x}) = K_R(f_1(\vec{x}), \ldots, f_n(\vec{x}))$. □

1.7. LEMMA. *(definition by cases). Let $R_1, \ldots, R_p$ be mutually exclusive* *primitive recursive predicates, such that* $\forall \vec{x}(R_1(\vec{x}) \vee R_2(\vec{x}) \vee \cdots \vee R_p(\vec{x}))$ *and* $g_1, \ldots, g_p$ *primitive recursive functions, then f with*

$$f(\vec{x}) = \begin{cases} g_1(\vec{x}) & \text{if } R_1(\vec{x}) \\ g_2(\vec{x}) & \text{if } R_2(\vec{x}) \\ \vdots \\ g_p(\vec{x}) & \text{if } R_p(\vec{x}) \end{cases}$$

is primitive recursive.

Proof. If $K_{R_i}(\vec{x}) = 1$, then all the other characteristic functions yield 0, so we put $f(\vec{x}) = g_1(\vec{x}) \cdot K_{R_1}(\vec{x}) + \ldots + g_p(\vec{x}) \cdot K_{R_p}(\vec{x})$. □

The natural numbers have the fundamental and convenient property that each non-empty set has a least element ($\mathbb{N}$ is well-ordered). A natural question to pose is: can we effectively find this least element? In general the answer is negative, but if the set under consideration is non-empty and primitive recursive, then we can simply take the element that ensured its non-emptiness and test the smaller numbers one by one for membership.

Some notation: $(\mu y)R(\vec{x}, y)$ stands for the least number y such that $R(\vec{x}, y)$ if its exists. $(\mu y < m)R(\vec{x}, y)$ stands for the least number $y < m$ such that $R(\vec{x}, y)$ if such a number exists; if not, we simply take it to be m.

1.8. LEMMA. *If R is primitive recursive, then so is* $(\mu y < m)R(\vec{x}, y)$.

Proof. Consider the following table

R	$R(\vec{x}, 0)$	$R(\vec{x}, 1), \ldots, R(\vec{x}, i),$	$R(\vec{x}, i+1), \ldots, R(\vec{x}, m)$
K_R	0	0 ... 1	0 ... 1
g	0	0 ... 1	1 ... 1
h	1	1 ... 0	0 ... 0
f	1	2 ... i	i ... i

In the first line we write the values of $K_R(\vec{x}, i)$ for $0 \leq i \leq m$, in the second line we make the sequence monotone, e.g. take $g(\vec{x}, i) = \text{sg} \, \Sigma_{j=0}^{i} K_R(\vec{x}, i)$. Next we switch 0 and 1: $h(\vec{x}, i) = \overline{\text{sg}} \, g(\vec{x}, i)$ and finally we sum the h: $f(\vec{x}, i) = \Sigma_{j=0}^{i} h(\vec{x}, j)$. If $R(\vec{x}, j)$ holds for the first time in i, then $f(\vec{x}, m) = i$, and if $R(\vec{x}, j)$ does not hold for any $j < m$, then $f(\vec{x}, m) = m$. So $(\mu y < m)R(\vec{x}, y) = f(\vec{x}, m)$, and this bounded minimalization yields a primitive recursive function. □

We put $(\mu y \leq m)R(\vec{x}, y) := (\mu y < m+1)R(\vec{x}, y)$.

We now have sufficient equipment to establish the primitive recursiveness of a considerable number of functions and relations.

1.9. EXAMPLES. The following are primitive recursive.
(1) The set of primes: x is a prime $\leftrightarrow \forall yz \leqslant x(x = yz \rightarrow y = 1 \vee z = 1) \wedge x \neq 1$.
(2) The divisibility relation: $x|y \leftrightarrow \exists z \leqslant y(x \cdot z = y)$
(3) The exponent of the prime p in the factorization of x:
$$f(x) = (\mu y \leqslant x)(p^y|x \wedge \neg p^{y+1}|x)$$
(4) The 'nth prime' function:
$$\begin{cases} p_1 = 2 \\ p_{n+1} = (\mu x \leqslant \Pi_{i=1}^n p_i + 1)[x \text{ is prime} \wedge x > p_n]. \end{cases}$$

We can use the stock of primitive recursive functions that we built up so far to get a coding of finite sequences of natural numbers into natural numbers:
$$(n_1, \ldots, n_k) \mapsto 2^{n_1+1} \cdot 3^{n_2+1} \cdots p_i^{n_i+1} \cdots p_k^{n_k+1}.$$

Note that not all numbers figure as codes, e.g. 14 does not.

For convenience we add a code for the so-called 'empty sequence'.

Recall that, in the framework of set theory a sequence of length n is a mapping from $\{1, \ldots, n\}$ to $\mathbb{N}$, so we define the empty sequence as the unique sequence of length 0, i.e. the unique map from $\emptyset$ to $\mathbb{N}$, which is the empty function (set). The choice of the code is a matter of convenience, we put it 1. Following tradition, we write $1 = \langle \ \rangle$.

The reader should be aware that there are two traditions in counting sequences; some start with 0, some with 1. In mathematics and recursion theory the 0-tradition is observed. Since we will not develop recursion theory for its own sake we will stick to the 1-tradition.

The predicate Seq(n), 'n is a sequence number', is clearly primitive recursive, for it boils down to 'if a prime divides n, then each smaller prime divides it': $\forall p, q \leqslant n$ ('p is a prime' $\wedge$ 'q is a prime' $\wedge q < p \wedge p|n \rightarrow q|n) \wedge n \neq 0$. If n is a sequence number, say of $\langle a_1, \ldots, a_k \rangle$ we can find its 'length', i.e. k:
$$\text{lth}(n) := (\mu x \leqslant n + 1)[\neg p_x|n] \dotdiv 1.$$

Observe that lth(1) = 0 and we can 'decode' n:
$$(n)_i = (\text{the exponent of the }i\text{th prime in the factorization of }n) \dotdiv 1$$
(cf. Example 3 above).

I.6: ALGORITHMS AND DECISION PROBLEMS

Note that lth(n) and $(n)_i$ are primitive recursive. For a fixed k

$$(a_1, \ldots, a_k) \mapsto \Pi_{i=1}^{k} p_i^{a_i+1}$$

is primitive recursive. Notation: $\langle a_1, \ldots, a_k \rangle := \Pi_{i=1}^{k} p_i^{a_i+1}$

We can also code the 'concatenation' of two sequence numbers: $n * m$ is the code of $\langle a_1, \ldots, a_k, b_1, \ldots, b_p \rangle$ where n and m are the codes of $\langle a_1, \ldots, a_k \rangle$ and $\langle b_1, \ldots, b_p \rangle$. The definition of $*$ is as follows (but may be skipped):

$$n * m = n \cdot \Pi_{i=1}^{\text{lth}(m)} p_{\text{lth}(n)+i}^{(m)_i+1}.$$

There is one more form of recursion that will come in handy – the one where a value may depend on all preceding values. In order to make this precise we define for a function $f(y, \vec{x})$ its 'course of value' function $\bar{f}(y, \vec{x})$:

$$\begin{cases} \bar{f}(0, \vec{x}) = 1 \\ \bar{f}(y+1, \vec{x}) = \bar{f}(y \cdot \vec{x}) \cdot p_{y+1}^{f(y, \vec{x})+1}, \end{cases}$$

e.g. $f(0) = 1, f(1) = 0, f(2) = 7$, then

$$\bar{f}(0) = 1, \bar{f}(1) = 2^{1+1}, \bar{f}(2) = 2^{1+1} \cdot 3^1, \bar{f}(3) = 2^2 \cdot 3 \cdot 5^8.$$

Clearly, if f is primitive recursive, then so is $\bar{f}$. Since $\bar{f}(n+1)$ 'codes' so to speak all information on f up to the nth value, we can use $\bar{f}$ to formulate course-of-value recursion.

1.10. THEOREM. *If g is primitive recursive and $f(y, \vec{x}) = g(\bar{f}(y, \vec{x}), y, \vec{x})$, then so is f.*

Proof. We first define $\bar{f}$.

$$\bar{f}(0, \vec{x}) = 1$$

$$\bar{f}(y+1, \vec{x}) = \bar{f}(y, \vec{x}) * \langle g(\bar{f}(y, \vec{x}), y, \vec{x}) \rangle.$$

By primitive recursion $\bar{f}$ is primitive recursive. Now $f(y, \vec{x}) = (\bar{f}(y+1, \vec{x}))_y$, and so f is primitive recursive.

By now we have collected enough facts about the primitive recursive functions. We might ask if there are more algorithms than just the primitive recursive functions. The answer turns out to be yes. Consider the following construction: each primitive recursive function f is determined by its definition, which consists of a string of functions $f_0, f_1, \ldots, f_n = f$ such that each function is either an initial function, or obtained from earlier ones by substitution or primitive recursion.

It is a matter of dull routine to code the whole definition into a natural number such that all information can be effectively extracted from the code (see Grzegorczyk [1961] p. 41). The construction shows that we may define a function F such that $F(x, y) = f_x(y)$, where f_x is the primitive recursive function with code x. Now consider $D(x) = F(x, x) + 1$. Suppose that D is primitive recursive, so $D = f_n$ for a certain n, but then $D(n) = F(n, n) + 1 = f_n(n) + 1 \neq f_n(n)$. Contradiction.

Conclusion. We have 'diagonalized out' of the class of primitive recursive functions and yet preserved the algorithmic character. Hence, we have to consider a wider class of algorithms.

In case the reader should have qualms in accepting the above outlined argument, he may set his mind at ease. There are straightforward examples of algorithms that are not primitive recursive, e.g. Ackermann's function (cf. Section 2.4).

Since our class of primitive recursive functions evidently does not contain all algorithms, we will have to look for ways of creating new algorithms not covered by substitution or primitive recursion. There are various solutions to this problem. The most radical being a switch to a conceptually different framework, e.g. that of Turing machines. We want to stay, however, as close as possible to our mode of generating the primitive recursive functions.

One way out is to generalize the minimalization, e.g. if $g(\vec{x}, y)$ is an algorithm such that $\forall \vec{x} \exists y (g(\vec{x}, y) = 0)$ then $f(\vec{x}) = (\mu y)[g(\vec{x}, y) = 0]$ is an algorithm. This leads to the so-called μ-recursive functions.

Although we will ultimately adopt another approach that will quickly yield all the fundamental theorems of the field, we will dwell for a moment on the μ-recursive functions.

The operation of *minimalization* associates with each total function $g(\vec{x}, y)$ a partial function $f(\vec{x}) = \mu y [g(\vec{x}, y) = 0]$.

1.11. DEFINITION. The class of μ-recursive partial functions is the least set containing the initial functions P_i^k (projection), $+, \cdot, K_<$ (the characteristic function of 'less than') which is closed under substitution and minimalization.

Although the successor and the constant functions are obviously μ-recursive, we apparently have lost as much as we have won, for now we no longer have closure under recursion. One can, fortunately, show that the class of μ-recursive (partial) functions is closed under recursion. The prove rests on the presence of a coding of finite sequences of numbers, for a computation associated with a function defined by recursion proceeds by computing

I.6: ALGORITHMS AND DECISION PROBLEMS

successively $f(0), f(1), \ldots, f(x)$. Although we cannot in any obvious way use the coding via the prime factorization – since we cannot make use of the exponential function – we can get an alternative coding. The main tool here is *Gödel's β-function*:

1.12. THEOREM. *There is a μ-recursive function β such that $\beta(n, i) \leqslant n \dotdiv 1$ and for any sequence $a_0, a_1, \ldots, a_{n-1}$ there is an a with $\beta(a, i) = a_i$ for $i < n$.*
For a proof, cf. Shoenfield [1967] p. 115.

One then defines the coding of $a_0, \ldots, a_{n-1}$ as $\mu a [\bigwedge_{i<n} \beta(a, i) = a_i]$. Here we have skipped the traditional lemma's on μ-recursive functions and relations (in particular the closure properties), cf. Shoenfield [1967] or Davis [1958].

If we denote this particular coding temporarily by $[a_0, \ldots, a_{n-1}]$, then we can get closure under recursion as follows:
Let

$$\begin{cases} f(0, \vec{x}) = g(\vec{x}) \\ f(y+1, \vec{x}) = h(f(y, \vec{x}), \vec{x}, y) \end{cases}$$

put

$$f'(y, \vec{x}) = [f(0, \vec{x}), \ldots, f(y, \vec{x})]$$

then

$$f'(y, \vec{x}) = \mu z [\text{Seq}(z) \wedge \forall i < y ([z]_0 = g(\vec{x}) \wedge [z]_{i+1} = h([z]_i, \vec{x}, i)]$$

Here Seq is the obvious predicate which states that z is a coded sequence and $[\ ]_i$ is the decoding function belonging to $[\ ]$. Taking the closure properties for granted we see that $f'(y, \vec{x})$ is μ-recursive. But then so is f, since $f(y, \vec{x}) = [f'(y, \vec{x})]_{\text{lth}(y)}$, where lth is the proper length function.

The definition of recursiveness via minimalization has the advantage that it does not ask for fancy apparatus, just two innocent closure operations. One has, however, to work harder to obtain the fundamental theorems that concern the properties of algorithms as finite, discrete, structured objects.

The sketch of Turing-machine computability that we have presented should, however, make it clear that all (partial) μ-recursive functions can be simulated by Turing machines. The converse is also correct: every function that can be computed by a Turing machine is μ-recursive (cf. Davis [1958]).

The approach to the partial recursive functions that we will use is that of Kleene using indices of recursive functions in the definition. The most striking aspect of the approach is that we postulate right away the existence of a universal function for each class of (partial) recursive functions of n arguments. The system has, so to speak, its diagonalization built in. Because of this we cannot have total functions only, for suppose that we have a universal recursive function $g(x, y)$ for the class of all unary recursive functions, i.e. for each f in the class there is a y such that $f(x) = g(x, y)$. Taking for granted that the recursive functions are closed under identification of variables, we get a unary recursive function $g(x, x)$. Evidently $g(x, x) + 1$ is also recursive, so $g(x, x) + 1 = g(x, y)$ for some y. For this particular y, we get $g(y, y) + 1 = g(y, y)$. *Contradiction*. Since $g(x, y)$ was taken to be recursive, we cannot conclude to have diagonalized out of the class of recursive functions. Instead, we conclude that $g(y, y)$ is undefined. So not all recursive functions are total!

Surprising as it may seem, we thus escape a diagonalization paradox for recursion theory.

Before we start our definition of the recursive functions in earnest, it may be helpful to the reader to stress an analogy with the theory of Turing machines.

We have seen that there is a universal Turing machine that operates on suitably coded strings of instructions. Calling such a coded string the *index* of the machine that is being simulated by the universal Turing machine, we introduced the notation $\varphi_e(x) = y$ for 'the machine with index e yields output y on input x'. We can now refer to the Turing machines by their indices, e.g. the existence of the universal Turing machine comes to: there is an index u such that for all indices e $\varphi_u(e, x) \simeq \varphi_e(x)$. The last expression has to be read as 'both sides are undefined, or they are defined and identical'.

Whereas in the case of Turing machines there is quite a lot of work to be done before one gets the universal machine, we will take the easy road and give the 'universal' recursive functions by one of the closure properties (clause R7 in Definition 2.1).

One final remark: matters of terminology in recursion are somewhat loosely observed. One should always speak of partial recursive functions, and add the predicate *total* when such a function is defined for all arguments. However, the total partial recursive functions are called just 'recursive'. Moreover, some authors simply drop the adjective 'partial' and always speak of 'recursive functions'. We will steer a middle course and add whatever adjectives that may help the reader. Nonetheless, the reader should beware!

I.6: ALGORITHMS AND DECISION PROBLEMS

2. PARTIAL RECURSIVE FUNCTIONS

We will now extend the class of algorithms as indicated above. This extension will yield new algorithms *and* it will automatically widen the class to partial functions.

In our context functions have natural domains, i.e. sets of the form $\mathbb{N}^n$ $(= \{(m_1, \ldots, m_n) | m_i \in \mathbb{N}\}$, so called Cartesian products), a partial function has a domain that is a subset of $\mathbb{N}^n$. If the domain is all of $\mathbb{N}^n$, then we call the function *total*.

Example: $f(x) = x^2$ is total, $g(x) = \mu y\, [y^2 = x]$ is partial and not total, ($g(x)$ is the square root of x if it exists).

The algorithms that we are going to introduce are called *partial recursive functions*; maybe recursive, partial functions would have been a better name, anyway this is the name that has come to be generally accepted. The particular technique for defining partial recursive functions that we employ here goes back to Kleene. As before, we use an inductive definition; apart from clause R7 below, we could have used a formulation almost identical to that of the definition of the primitive recursive functions. Since we want a built-in universal function, we have to employ a more refined technique that allows explicit reference to the various algorithms. The trick is not esoteric at all, we simply give each algorithm a code number, what we call its *index*. We fix these indices in advance so that we can speak of 'the algorithm with index e yields output y on input $(x_1, \ldots, x_n)$', symbolically represented as $\{e\}(x_1, \ldots, x_n) \simeq y$.

Note that we do not know in advance that the result is a partial function, i.e. that for each input there is at most one output. However plausible that is, it has to be shown. Kleene has introduced the symbol $\simeq$ for equality in the context of undefined terms. A proper treatment would be by means of the existence predicate and $\simeq$ would be the $\equiv$ of Van Dalen [Chapter III.5]. The convention ruling $\simeq$ is: if $t \simeq s$ then t and s are simultaneously defined and identical, or they are simultaneously undefined, Kleene [1952] p. 327.

2.1. DEFINITION. The relation $\{e\}(\vec{x}) \simeq y$ is inductively defined by

R1 $\quad \{\langle 0, n, q\rangle\}(m_1, \ldots, m_n) \simeq q$

R2 $\quad \{\langle 1, n, i\rangle\}(m_1, \ldots, m_n) \simeq m_i \quad \text{for } 1 \leq i \leq n$

R3 $\quad \{\langle 2, n, i\rangle\}(m_1, \ldots, m_n) \simeq m_i + 1 \quad \text{for } 1 \leq i \leq n$

R4 $\{\langle 3, n+4 \rangle\}(p, q, r, s, m_1, \ldots, m_n) \simeq p$ if $r = s$

       $\{\langle 3, n+4 \rangle\}(p, q, r, s, m_1, \ldots, m_n) \simeq q$ if $r \neq s$

R5 $\{\langle 4, n, b, c_1, \ldots, c_k \rangle\}(m_1, \ldots, m_n) \simeq p$ if there are $q_1, \ldots, q_k$ such that $\{c_i\}(m_1, \ldots, m_n) \simeq q_i$ $(1 \leq i \leq k)$ and $\{b\}(q_1, \ldots, q_k) \simeq p$

R6 $\{\langle 5, n+2 \rangle\}(p, q, m_1, \ldots, m_n) = S_n^1(p, q)$

R7 $\{\langle 6, n+1 \rangle\}(b, m_1, \ldots, m_n) \simeq p$ if $\{b\}(m_1, \ldots, m_n) \simeq p$.

The function S_n^1 from R6 will be specified in the S_n^m theorem. It is a pure technicality, slipped in to simplify the proof of the normal form theorem. We will comment on it below.

Keeping the above reading of $\{e\}(\vec{x})$ in mind, we can paraphrase the schema's as follows:

R1   the machine with index $\langle 0, n, q \rangle$ yields for input $(m_1, \ldots, m_n)$ output q (the *constant function*),

R2   the machine with index $\langle 1, n, i \rangle$ yields for input $\vec{m}$ output m_i (the *projection function* P_i^n),

R3   the machine with index $\langle 2, n, i \rangle$ yields for input $\vec{m}$ output $m_i + 1$ (the *successor function* on the ith argument),

R4   the machine with index $\langle 3, n+4 \rangle$ tests the equality of the third and fourth argument of the input and puts out the first or second argument accordingly (the *discriminator function*),

R5   the machine with index $\langle 4, n, b, c_1, \ldots, c_k \rangle$ first simulates the machines with index $c_1, \ldots, c_k$ with input $\vec{m}$, then uses the output sequence $(q_1, \ldots, q_k)$ as input and simulates the machine with index b (*substitution*)

R7   the machine with index $\langle 6, n+1 \rangle$ simulates for a given input $b, m_1, \ldots, m_n$, the machine with index b and input $m_1, \ldots, m_n$ (*reflection*).

The machine with index $\langle 6, n+1 \rangle$ acts as a *universal machine* for all machines with n-argument inputs.

Remarks. (1) The index of a machine contains all relevant information, the first coordinate tells us which clause to use, the second coordinate always gives the number of arguments. The remaining coordinates contain the specific information.

(2) R7 is very powerful, it yields an enumeration of all machines with a fixed number of arguments. Exactly the kind of machine we needed above

for the diagonalization. Intuitively, the existence of such a machine seems quite reasonable. If one can effectively recognize the indices of machines, then a machine should be able to do so, and thus to simulate each single machine.

The scrupulous might call R7 a case of cheating, since it does away with all the hard work one has to do in order to obtain a universal machine, e.g. in the case of Turing machines.

The relation $\{e\}(\vec{x}) \simeq y$ is functional, i.e. we can show

Fact. $\{e\}(\vec{x}) \simeq y, \{e\}(\vec{x}) \simeq z \Rightarrow y = z$.

Proof. Use induction on the definition of $\{e\}$. □

The above definition tells us implicitly what we have to consider a computation: to compute $\{e\}(\vec{x})$ we look at e, if the first 'entry' of e is 0, 1, 2, then we compute the output via the corresponding initial function. If the first 'entry' is 3, then we determine the output 'by cases'. First 'entry' 5 is handled as indicated in the S_n^m theorem. If the first entry is 4, then we first carry out the subcomputations with indices $c_1, \ldots, c_k$, followed by the subcomputation with index b, and find the output according to R5. At first 'entry' 6, we jump to the subcomputation with index b (cf. R7).

In the presence of R7 we are no longer guaranteed that the process will stop; indeed, we may run into a loop, as the following simple example shows.

By R7 there exists an e such that $\{e\}(x) = \{x\}(x)$.

To compute $\{e\}$ for the argument e we pass, according to R7, onto the right-hand side, i.e. we must compute $\{e\}(e)$, since e was introduced by R7, we must repeat the transitions to the right hand side, etc. Evidently our procedure does not get us anywhere!

Loops and non-terminating computations account for algorithms being undefined at some inputs.

There could also be a trivial reason for not producing outputs, e.g. $\{0\}(\vec{x}) \simeq y$ holds for no y, since 0 is not an index at all, so $\{0\}$ is the empty function.

Some terminology:

(i) If for a partial function φ $\exists y(\varphi(\vec{x}) = y)$, then we say that φ *converges* at $\vec{x}$, otherwise φ *diverges* at $\vec{x}$.

(ii) If a partial function converges for all inputs, it is called *total*.

(iii) A total partial recursive function (sic!) will be called a *recursive function*.

(iv) A set (relation) is called *recursive* if its characteristic function is recursive.

The definition of $\{e\}(\vec{x}) = y$ has the consequence that a partial recursive function diverges if one of its arguments diverges. This is an important feature, not shared for example by λ-calculus or combinatory logic. It tells us that we *have* to carry out all subcomputations. We could, for instance, not assert that $\{e\}x - \{e\}x = 0$ for all e and x, we must know that $\{e\}x$ converges!

This feature is sometimes inconvenient and slightly paradoxical, e.g. in direct applications of the discriminator scheme R4 $\{\langle 3, 4\rangle\}(\varphi(x), \psi(x), 0, 0)$ is undefined when the (seemingly irrelevant) function $\psi(x)$ is undefined. With a bit of extra work, we can get an index for a partial recursive function that does *definition by cases* on partial recursive functions:

$$\{e\}(\vec{x}) = \begin{cases} \{e_1\}(\vec{x}) & \text{if } g_1(\vec{x}) = g_2(\vec{x}) \\ \{e_2\}(\vec{x}) & \text{if } g_1(\vec{x}) \neq g_2(\vec{x}) \end{cases}$$

for recursive g_1, g_2.

Define

$$\varphi(\vec{x}) = \begin{cases} e_1 & \text{if } g_1(\vec{x}) = g_2(\vec{x}) \\ e_2 & \text{if } g_2(\vec{x}) \neq g_2(\vec{x}) \end{cases}$$

by $\varphi(\vec{x}) = \{\langle 3, 4\rangle\}(e_1, e_2, g_1(\vec{x}), g_2(\vec{x}))$, use R5. Then an application of R7 and R5 to $\{\varphi(\vec{x})\}(\vec{x})$ yields the desired $\{e\}(\vec{x})$.

We will adopt the following notational convention after Rogers' [1967] book: partial recursive functions will be denoted by $\varphi, \psi, \ldots$, and the total ones by $f, g, h, \ldots$. From now on we will indiscriminatingly use '=' for '$\simeq$' and for the ordinary equality.

After some preliminary work, we will show that all primitive recursive functions are recursive. We could forget about the primitive recursive functions and just discuss partial recursive ones. However, the primitive recursive functions form a very natural class, and they play an important role in metamathematics.

The following important theorem has a neat machine motivation. Consider a machine with index e operating on two arguments x and y. Keeping x fixed, we have a machine operating on y. So we get a sequence of machines, one for each x. Does the index of each such machine depend in a decent way on x? The plausible answer seems 'yes'. The following theorem confirms this.

2.2. THE S_n^m THEOREM. *For every m, n such that $0 < m < n$ there exists a primitive recursive function S_n^m such that $\{S_n^m(e, x_1, \ldots, x_m)\}(x_{m+1}, \ldots, x_n) = \{e\}(\vec{x})$.*

I.6: ALGORITHMS AND DECISION PROBLEMS

Proof. The first function S_n^1 is given by R6, we write down its explicit definition:

$$S_n^1(e, y) = \langle 4, (e)_2 \dotminus 1, e, \langle 0, (e)_2 \dotminus 1, y\rangle, \langle 1, (e)_2 \dotminus 1, 1\rangle, \dots,$$
$$\langle 1, (e)_2 \dotminus 1, n \dotminus 1\rangle\rangle.$$

Then $\{S_n^1(e, y)\}(\vec{x}) = z \longleftrightarrow \exists q_1 \cdots q_n [\{\langle 0, (e)_2 \dotminus 1, y\rangle\}(\vec{x}) = q_1 \wedge \{\langle 1, (e)_2 \dotminus 1, 1\rangle\}(\vec{x}) = q_2 \wedge \cdots \wedge \{\langle 1, (e)_2 \dotminus 1, n \dotminus 1\rangle\}(\vec{x}) = q_n \wedge \{e\}(q_1, \dots, q_n) = z]$.

By the clauses R1 and R2 we get $q_1 = y$ and $q_{i+1} = x_i$, so $\{S_n^1(e, y)\}(\vec{x}) = \{e\}(y, \vec{x})$. Clearly S_n^1 is primitive recursive.

S_n^m is obtained by applying S_n^1 m times. Note that S_n^m is partial recursive. □

The S_n^m theorem expresses a *uniformity* property of the partial recursive functions. It is obvious indeed that, say for a partial recursive function $\varphi(x, y)$, each individual $\varphi(n, y)$ is partial recursive (substitute the constant n function for x), but this does not yet show that the index of $\lambda y \cdot \varphi(x, y)$ is in a systematic, uniform way computable from the index of φ and x.

There are numerous applications, we will just give one: define $\varphi(x) = \{e\}(x) + \{f\}(x)$, then by 2.8 φ is partial recursive and we would like to express the index of φ as a function of e and f. Consider $\psi(e, f, x) = \{e\}(x) + \{f\}(x)$. ψ is partial recursive, so it has an index n, i.e. $\{n\}(e, f, x) = \{e\}(x) + \{f\}(x)$. By the S_n^m theorem there is a primitive recursive function h such that $\{n\}(e, f, x) = \{h(n, e, f)\}(x)$. Therefore, $g(e, f) = h(n, e, f)$ is the required function.

Next we will prove a fundamental theorem about partial recursive functions that allows us to introduce partial recursive functions by inductive definitions, or by implicit definition. We have see that we can define a primitive recursive function by using all (or some) of the preceding values to get a value in n. We might, however, just as well make the value depend on future values, only then we can no longer guarantee that the resulting function is total (let alone primitive recursive!).

EXAMPLE:

$$\varphi(n) = \begin{cases} 0 & \text{if } n \text{ is a prime, or 0, or 1} \\ \varphi(2n + 1) + 1 & \text{otherwise.} \end{cases}$$

Then $\varphi(0) = \varphi(1) = \varphi(2) = \varphi(3) = 0$, but $\varphi(4) = \varphi(9) + 1 = \varphi(19) + 2 = 2$, $\varphi(5) = 0$, and, e.g. $\varphi(85) = 6$. Prima facie, we cannot say much about such a sequence. The following Theorem of Kleene shows that we can always find a partial recursive solution to such an equation for φ.

2.3. THE RECURSION THEOREM. *There exists a primitive recursive function* rc *such that* $\{rc(e)\}(\vec{x}) = \{e\}(rc(e), \vec{x})$.

Before we prove the theorem let us convince ourselves that it solves our problem. We want a partial recursive φ such that $\varphi(\vec{x}) = \{e\}(\ldots \varphi \ldots \vec{x})$ (where the notation is meant to indicate that φ occurs on the right-hand side). To ask for a partial recursive function is to ask for an index for it, so replace φ by $\{z\}$, where z is the unknown index:

$$\{z\}(\vec{x}) = \{e\}(\ldots \{z\} \ldots, \vec{x}) = \{e'\}(z, \vec{x}).$$

Now it is clear that rc(e') gives us the required index for φ.

Proof. Let $\varphi(m, e, \vec{x}) = \{e\}(S^2_{n+2}(m, m, e), \vec{x})$ and let p be an index of φ. Put rc$(e) = S^2_{n+2}(p, p, e)$, then

$$\{rc(e)\}(\vec{x}) = \{S^2_{n+2}(p, p, e)\}(\vec{x}) = \{p\}(p, e, \vec{x}) = \varphi(p, e, \vec{x})$$
$$= \{e\}(S^2_{n+2}(p, p, e), \vec{x}) = \{e\}(rc(e), \vec{x}). \quad \square$$

As a special case we get the

COROLLARY. *For each e there exists an n such that* $\{n\}(\vec{x}) = \{e\}(n, \vec{x})$.

Remark. Although we have not yet shown that the class of partial recursive functions contains all primitive recursive functions, we know what primitive recursive functions are and what their closure properties are. In particular, if $\{e\}$ should happen to be primitive recursive, then by $\{rc(e)\}(\vec{x}) = \{e\}(rc(e), \vec{x})$ $\{rc(e)\}$ is also primitive recursive.

2.4. EXAMPLES. (1) There is a partial recursive function φ such that $\varphi(n) = (\varphi(n+1) + 1)^2$: Consider $\{z\}(n) = \{e\}(z, n) = (\{z\}(n+1) + 1)^2$. By the recursion theorem there is a solution rc(e), hence φ exists. A simple argument shows that φ cannot be defined for any n, so the solution is the empty function (the machine that never gives an output).

(2) *The Ackermann function.* Consider the following sequence of functions.

I.6: ALGORITHMS AND DECISION PROBLEMS

$$\varphi_0(m, n) = n + m$$
$$\varphi_1(m, n) = n \cdot m$$
$$\varphi_2(m, n) = n^m$$
$$\vdots$$
$$\begin{cases} \varphi_{k+1}(0, n) = n \\ \varphi_{k+1}(m+1, n) = \varphi_k(\varphi_{k+1}(m, n), n) \quad (k \geqslant 2) \end{cases}$$

This sequence consists of faster and faster growing functions. We can lump all those functions together in one function

$$\varphi(k, m, n) = \varphi_k(m, n).$$

The above equations can be summarized as

$$\begin{cases} \varphi(0, m, n) = n + m \\ \varphi(k+1, 0, n) = \begin{cases} 0 & \text{if } k = 0 \\ 1 & \text{if } k = 1 \\ n & \text{else} \end{cases} \\ \varphi(k+1, m+1, n) = \varphi(k, \varphi(k+1, m, n), n). \end{cases}$$

Note that the second equation has to distinguish cases according to the φ_{k+1} being the multiplication, exponentiation, or the general case ($k \geqslant 2$).

Using the fact that all primitive recursive functions are recursive (Corollary 2.8) we rewrite the three cases into one equation of the form $\{e\}(k, m, n) = f(e, k, m, n)$ for a suitable recursive f. Hence, by the recursion theorem there exists a recursive function with index e that satisfies the equations above. Ackermann has shown that the function $\varphi(n, n, n)$ grows faster than any primitive recursive function. The recursion theorem can also be used for inductive definitions of sets or relations, this is seen by changing over to characteristic functions, e.g. suppose we want a relation $R(x, y)$ such that

$$R(x, y) \leftrightarrow (x = 0 \wedge y \neq 0) \vee (x \neq 0 \wedge y \neq 0 \wedge R(x \dotdiv 1, y \dotdiv 1)).$$

Then we write

$$K_R(x, y) = \text{sg}(\overline{\text{sg}}(x)) \cdot \text{sg}(y) + \text{sg}(x) \cdot \text{sg}(y) \cdot K_R(x \dotdiv 1, y \dotdiv 1)),$$

so there is an e such that

$$K_R(x, y) = \{e\}(K_R(x \dotdiv 1, y \dotdiv 1), x, y).$$

Now if K_R has index z then we have

$$\{z\}(x, y) = \{e'\}(z, x, y).$$

The solution n provided by the recursion theorem is the required characteristic function. One immediately sees that R is the relation 'less than', so $\{n\}$ is recursive and hence R too (cf. 1.4), note that by the remark following the recursion theorem we even get the primitive recursiveness of R. The partial recursive functions are a rather wild lot, they have an enormous variety of definition (in terms of R1–R7). We can, however, obtain them in a uniform way by one minimalization from a fixed predicate.

2.5. NORMAL FORM THEOREM. *There is a primitive recursive predicate T such that* $\{e\}(\vec{x}) = ((\mu z)T(e, \langle\vec{x}\rangle, z))_1$.

Proof. See the Appendix. □

The predicate T formalizes the statement 'z is the computation of the partial recursive function (machine) with index e operating on input $\langle\vec{x}\rangle$', where 'computation' has been defined such that the first projection is the output.

For applications the precise structure of T is not important. One can obtain the well-known undecidability results from the S_n^m theorem, the recursion theorem and the normal form theorem.

The partial recursive functions are closed under a general form of minimalization, sometimes called *unbounded search*, which for a given recursive function $f(y, \vec{x})$ and arguments $\vec{x}$ runs through the values of y and looks for the first one that makes $f(y, \vec{x})$ equal to zero.

2.6. THEOREM. *Let f be a recursive function, then $\varphi(\vec{x}) = \mu y[f(y, \vec{x}) = 0]$ is partial recursive.*

Proof. Our strategy consists of testing successively all values of y until we find the first y such that $f(y, \vec{x}) = 0$. We want a function ψ such that $\psi(y, \vec{x})$ produces a 0 if $f(y, \vec{x}) = 0$ and moves on to the next y while counting the steps if $f(y, \vec{x}) \neq 0$. Let this function have index e. We introduce auxiliary functions ψ_1, ψ_2 with indices b and c such that $\psi_1(e, y, \vec{x}) = 0$ and $\psi_2(e, y, \vec{x}) = \psi(y + 1, \vec{x}) + 1 = \{e\}(y + 1, \vec{x}) + 1$. If $f(y, \vec{x}) = 0$ then we consider ψ_1, if not, ψ_2. So we introduce, by clause R4, a new function χ_0:

I.6: ALGORITHMS AND DECISION PROBLEMS

$$\chi_0(e, y, \vec{x}) = \begin{cases} b & \text{if } f(y, x) = 0 \\ c & \text{else} \end{cases}$$

and we put $\chi(e, y, \vec{x}) = \{\chi_0(e, y, \vec{x})\}(e, y, \vec{x})$.

The recursion theorem provides us with an index e_0 such that $\chi(e_0, y, \vec{x}) = \{e_0\}(y, \vec{x})$.

We claim that $\{e_0\}(0, \vec{x})$ yields the desired value, if its exists at all, i.e. e_0 is the index of the ψ we were looking for.

For, if $f(y, \vec{x}) \neq 0$ then $\chi(e_0, y, \vec{x}) = \{c\}(e_0, y, \vec{x}) = \psi_2(e_0, y, \vec{x}) = \psi(y + 1, \vec{x}) + 1$, and if $f(y, \vec{x}) = 0$ then $\chi(e_0, y, \vec{x}) = \{b\}(e_0, y, \vec{x}) = 0$.

So suppose that y_0 is the first value y such that $f(y, \vec{x}) = 0$, then

$$\psi(0, \vec{x}) = \psi(1, \vec{x}) + 1 = \psi(2, \vec{x}) + 2 = \cdots = \Psi(y_0, \vec{x}) + y_0 = y_0.$$
□

Note that the given function need not be recursive, and that the above argument also works for partial recursive f. We then have to reformulate $\mu y[f(x, \vec{y}) = 0]$ as the y such that $f(y, \vec{x}) = 0$ and for all $z < y$ $f(z, \vec{x})$ is defined and positive.

We need minimalization in our approach to obtain closure under primitive recursion. We could just as well have thrown in an extra clause for primitive recursion, (and deleted R4 and R6), but that would have obscured the power of the reflection clause R7. Observe that in order to get closure under primitive recursion, we need a simple consequence of it, namely R6.

It is easy to see that the predecessor function, $x \dotminus 1$, can be obtained:

$$\text{define } x \dotminus 1 = \begin{cases} 0 & \text{if } x = 0 \\ \mu y[y + 1 = x] & \text{else} \end{cases}$$

where $\mu y[y + 1 = x] = \mu y[f(y, x) = 0]$ with

$$f(y, x) = \begin{cases} 0 & \text{if } y + 1 = x \\ 1 & \text{else} \end{cases}$$

2.7. THEOREM. *The recursive functions are closed under primitive recursion.*

Proof. We want to show that if g and h are recursive, then so is f, defined by

$$\begin{cases} f(0, \vec{x}) = g(\vec{x}) \\ f(y + 1, \vec{x}) = h(f(y, \vec{x}), \vec{x}, y). \end{cases}$$

We rewrite the schema as

$$f(y, \vec{x}) = \begin{cases} g(\vec{x}) & \text{if } y = 0 \\ h(f(y \dotdiv 1, \vec{x}), \vec{x}, y \dotdiv 1) & \text{otherwise.} \end{cases}$$

Since the predecessor is recursive, an application of definition by cases yields the following equation for an index of the required function f: $\{e\}(y, \vec{x}) = \{a\}(y, \vec{x}, e)$ (where a can be computed from the indices of g, h and the predecessor). By the recursion theorem the equation has a solution e_0. One shows by induction on y that $\{e_0\}$ is total, so f is a recursive function. □

We now get the obligatory

2.8. COROLLARY. *All primitive recursive functions are recursive.*

2.9. DEFINITION.
    (i) A set (or relation) is (recursively) *decidable* if it is recursive.
    (ii) A set is *recursively enumerable* (RE) if it is the domain of a partial recursive function.
    (iii) $W_e^k = \{\vec{x} \in \mathbb{N}^k \mid \exists y(\{e\}(\vec{x}) = y)\}$, i.e. the domain of the partial recursive function $\{e\}$. We call e the RE index of W_e^k. If no confusion arises we will delete the superscript.

We write $\varphi(\vec{x})\downarrow$ (resp $\varphi(\vec{x})\uparrow$) for $\varphi(\vec{x})$ converges (resp. φ diverges).

One can think of a recursively enumerable set as a set that is accepted by an abstract machine; one successively offers the natural numbers, 0, 1, 2, ..., and when the machine produces an output the input is 'accepted'.

The next theorem states that we could also have defined RE sets as those produced by a machine.

It is good heuristics to think of RE sets as being accepted by machines, e.g. if A_i is accepted by machine M_i ($i = 0, 1$), then we make a new machine that simulates M_0 and M_1 running parallel, and so n is accepted by M if it is accepted by M_0 or M_1. Hence the union of two RE sets is also RE.

2.10. EXAMPLES OF RE SETS.
    (i) $\mathbb{N}$ = the domain of the constant function.
    (ii) $\emptyset$ = the domain of the empty function. This function is partial recursive, as we have already seen.
    (iii) Every recursive set is RE. Let A be recursive, put

I.6: ALGORITHMS AND DECISION PROBLEMS

$$\psi(\vec{x}) = \begin{cases} 0 & \text{if } \vec{x} \in A \\ \varphi(\vec{x}) & \text{if } \vec{x} \notin A, \text{ where } \varphi \text{ is the empty function.} \end{cases}$$

Then Dom $\psi = A$.

The recursively enumerable sets derive their importance from the fact that they are effectively given, in the sense that they are produced by partial recursive functions, i.e. they are presented by an algorithm. Furthermore it is the case that the majority of important relations (sets) in logic are RE. For example the set of provable sentences of arithmetic or predicate logic is RE. The RE sets represent the first step beyond the decidable sets, as we will show below.

2.11. THEOREM. *The following statements are equivalent,* $(A \subseteq \mathbb{N})$:
  (i) $A = \text{Dom}(\varphi)$ *for some partial recursive* φ,
  (ii) $A = \text{Ran}(\varphi)$ *for some partial recursive* φ,
  (iii) $A = \{x \mid \exists y R(x, y)\}$ *for some recursive* R.

Proof. (i) $\Rightarrow$ (ii). Define $\psi(x) = x \cdot \text{sg}(\varphi(x) + 1)$. If $x \in \text{Dom}\varphi$, then $\psi(x) = x$, so $x \in \text{Ran}\psi$, and if $x \in \text{Ran}\psi$, then $\varphi(x)\downarrow$, so $x \in \text{Dom}\varphi$.

(ii) $\Rightarrow$ (iii). Let $A = \text{Ran}\{g\}$ then

$$x \in A \longleftrightarrow \exists w [T(g, (w)_1, (w)_2) \wedge x = (w)_{2,1}].$$

The relation in the scope of the quantifier is recursive.

Note that w acts as a pair: first coordinate – input, second coordinate – computation.

(iii) $\Rightarrow$ (i). Define $\varphi(x) = \mu y R(x, y)$.

φ is partial recursive and $\text{Dom}\varphi = A$.

Observe that (i) $\Rightarrow$ (iii) also holds for $A \subseteq \mathbb{N}^k$. □

Since we have defined recursive sets by means of characteristic functions, and since we have established closure under primitive recursion, we can copy all the closure properties of primitive recursive sets (and relations) for the recursive sets (and relations).

Next we list a number of closure properties of RE-sets.

2.12. THEOREM.
  (i) If A and B are RE, then so are $A \cup B$ and $A \cap B$
  (ii) If $R(x, \vec{y})$ is RE, then so is $\exists x R(x, \vec{y})$

(iii) If $R(x, \vec{y})$ is RE and φ partial recursive, then $R(\varphi(\vec{y}, \vec{z}), \vec{y})$ is RE
(iv) If $R(x, \vec{y})$ is RE, then so are $\forall x < z R(x, \vec{y})$ and $\exists x < z R(x, \vec{y})$

Proof. (i) There are recursive R and S such that

$$A\vec{y} \longleftrightarrow \exists x R(x, \vec{y}), \ B\vec{y} \longleftrightarrow \exists x S(x, \vec{y}).$$

Then

$$A\vec{y} \wedge B\vec{y} \longleftrightarrow \exists x_1 x_2 (R(x_1, \vec{y}) \wedge S(x_2, \vec{y}))$$
$$\longleftrightarrow \exists z (R((z)_1, \vec{y}) \wedge S((z)_2, \vec{y})).$$

The relation in the scope of the quantifier is recursive, so $A \cap B$ is RE. A similar argument establishes the recursive enumerability of $A \cup B$. The trick of replacing x_1 and x_2 by $(z)_1$ and $(z)_2$ and $\exists x_1 x_2$ by $\exists z$ is called *contraction of quantifiers*.

(ii) Let $R(x, \vec{y}) \longleftrightarrow \exists z S(z, x, \vec{y})$ for a recursive S, then $\exists x R(x, \vec{y}) \longleftrightarrow \exists x \exists z S(z, x, \vec{y}) \longleftrightarrow \exists u S((u)_1, (u)_2, \vec{y})$. So the *projection* $\exists x R(x, \vec{y})$ of R is RE.

$\exists x R(x, \vec{y})$ is indeed a projection. Consider the two-dimensional case (Figure 4)

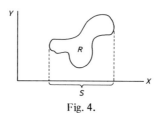

Fig. 4.

The vertical projection S of R is given by $Sx \longleftrightarrow \exists y R(x, y)$.

(iii) Let R be the domain of a partial recursive ψ, then $R(\varphi(\vec{y}, \vec{z}), \vec{y})$ is the domain of $\psi(\varphi(\vec{y}, \vec{z}), \vec{y})$.

(iv) Left to the reader.

2.13. We observe that the graph of a partial function is RE iff the function is partial recursive.

Proof. $G = \{(\vec{x}, y) | y = \{e\}(\vec{x})\}$ is the graph of $\{e\}$. Now $(\vec{x}, y) \in G \Longleftrightarrow \exists z (T(e, \langle \vec{x} \rangle, z) \wedge y = (z)_1)$, so G is RE. Conversely, if G is RE, then $G(\vec{x}, y) \Longleftrightarrow \exists z R(\vec{x}, y, z)$ for some recursive R. Hence $\varphi(\vec{x}) = (\mu w R(\vec{x}, (w)_1, (w)_2))_1$, so φ is partial recursive. □

I.6: ALGORITHMS AND DECISION PROBLEMS

We can also characterize recursive sets in terms of RE-sets. Suppose both A and A^c are RE, then (heuristically) we have two machines enumerating A and A^c. Now the test for membership of A is simple: turn both machines on and wait for n to turn up as output of the first or second machine. This must necessarily occur in finitely many steps since $n \in A$ or $n \in A^c$ (principle of the excluded third!). Hence, we have an effective test. We formalize the above:

2.14. THEOREM. *A is recursive $\iff A$ and A^c are RE.*

Proof. $\Rightarrow$ is trivial, $A(\vec{x}) \leftrightarrow \exists y A(\vec{x})$, where y is a dummy variable. Similarly for A^c. $\Leftarrow$ Let $A(\vec{x}) \leftrightarrow \exists y R(\vec{x}, y)$, $\neg A(\vec{x}) \leftrightarrow \exists z S(\vec{x}, z)$. Since $\forall \vec{x}(A(\vec{x}) \vee \neg A(\vec{x}))$, we have $\forall \vec{x} \exists y (R(\vec{x}, y) \vee S(\vec{x}, y))$, so $f(\vec{x}) = \mu y [R(\vec{x}, y) \vee S(\vec{x}, y)]$ is recursive and if we plug the y that we found in $R(\vec{x}, y)$, then we know that if $R(\vec{x}, f(\vec{x}))$ is true, the $\vec{x}$ belongs to A. So $A(\vec{x}) \leftrightarrow R(\vec{x}, f(\vec{x}))$, i.e. A is recursive.

For partial recursive functions we have a strong form of definition by cases:

2.15. THEOREM. *Let $\psi_1, \ldots, \psi_k$ be partial recursive, $R_1, \ldots, R_k$ mutually disjoint RE-relations, then*

$$\varphi(\vec{x}) = \begin{cases} \psi_1(\vec{x}) & \text{if } R_1(\vec{x}) \\ \psi_2(\vec{x}) & \text{if } R_2(\vec{x}) \\ \vdots \\ \psi_k(\vec{x}) & \text{if } R_k(\vec{x}) \\ \uparrow & \text{else} \end{cases}$$

is partial recursive.

Proof. We consider the graph of the function φ.

$$G(\vec{x}, y) \leftrightarrow (R_1(\vec{x}) \wedge y = \psi_1(\vec{x})) \vee \cdots \vee (R_k(\vec{x}) \wedge y = \psi_k(\vec{x})).$$

By the properties of RE-sets $G(\vec{x}, y)$ is RE and, hence, $\varphi(\vec{x})$ is partial recursive. □

Note that the last case in the definition is just a bit of decoration.
Now we can show the existence of undecidable RE sets.

2.16 (1). THE HALTING PROBLEM (A. TURING). Consider $K = \{x | \exists z T(x, x, z)\}$. K is the projection of a recursive relation, so it is RE. Suppose that K^c is also RE, then $x \in K^c \leftrightarrow \exists z T(e, x, z)$ for some index e. Now $e \in K \leftrightarrow \exists z T(e, e, z) \leftrightarrow e \in K^c$. Contradiction. Hence K is not recursive by the above theorem.

The decision problem for K is called the halting problem, because it can be paraphrased as "decide if the machine with index x performs a computation that halts after a finite number of steps when presented with x as input". Note that it is *ipso facto* undecidable if 'the machine with index x eventually halts on input y'.

We will exhibit a few more examples of undecidable problems.

(2). It is not decidable if $\{x\}$ is a total function.

Suppose it were decidable, then we would have a recursive function f such that $f(x) = 0 \leftrightarrow \{x\}$ is total.

Now consider

$$\varphi(x, y) := \begin{cases} 0 & \text{if } x \in K \\ \uparrow & \text{else} \end{cases} \quad \text{(note that } \varphi(x, y) = 0 \cdot \{x\}(x)\text{)}.$$

By the S_n^m theorem there is a recursive h such that $\{h(x)\}(y) = \varphi(x, y)$. Now $\{h(x)\}$ is total $\leftrightarrow x \in K$, so $f(h(x)) = 0 \leftrightarrow x \in K$, i.e. we have a recursive characteristic function $\text{sg}(f(h(x)))$ for K. Contradiction. Hence such an f does not exist, that is $\{x | \{x\}$ is total$\}$ is not recursive.

(3). The problem 'W_e is finite' is not recursively solvable. Suppose that there was a recursive function f such that $f(e) = 0 \leftrightarrow W_e$ is finite.

Consider the $h(x)$ defined in example (2). Clearly $W_{h(x)} = \text{Dom}\{h(x)\} = \emptyset \leftrightarrow x \notin K$, and $W_{h(x)}$ is infinite for $x \in K$. $f(h(x)) = 0 \leftrightarrow x \notin K$, and hence $\text{sg}(f(h(x)))$ is a recursive characteristic function for K. Contradiction.

Remark. $x \in K \leftrightarrow \{x\}x\downarrow$, so we can reformulate the above solutions as follows: in (2) take $\varphi(x, y) = 0.\{x\}(x)$ and in (3) $\varphi(x, y) = \{x\}(x)$.

(4). The equality of RE sets is undecidable, i.e. $\{(x, y) | W_x = W_y\}$ is not recursive. We reduce the problem to the solution of (3) by choosing $W_y = \emptyset$.

(5). It is not decidable if W_e is recursive. Put $\varphi(x, y) = \{x\}(x) \cdot \{y\}(y)$, then $\varphi(x, y) = \{h(x)\}(y)$ for a certain recursive h, and

$$\text{Dom}\{h(x)\} = \begin{cases} K & \text{if } x \in K \\ \emptyset & \text{otherwise.} \end{cases}$$

I.6: ALGORITHMS AND DECISION PROBLEMS 449

Suppose there were a recursive function f such that $f(x) = 0 \leftrightarrow W_x$ is recursive, then $f(h(x)) = 0 \leftrightarrow x \notin K$ and, hence, K would be recursive. Contradiction. □

There are several more techniques for establishing undecidability. We will consider the method of inseparability.

2.17. DEFINITION. Two disjoint RE-sets W_m and W_n are *recursively separable* (Figure 5) if there is a recursive set A such that $W_n \subseteq A$ and $W_m \subseteq A^c$. Disjoint sets A and B are *effectively inseparable* if there is a partial recursive φ such that for every m, n with $A \subseteq W_m$, $B \subseteq W_n$, $W_m \cap W_n = \emptyset$ we have $\varphi(m, n)\downarrow$ and $\varphi(m, n) \notin W_m \cup W_n$.

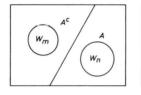
recursively separable sets

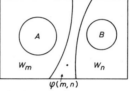
effectively inseparable sets

Fig. 5.

We immediately see that effectively inseparable RE sets are recursively inseparable, i.e. not recursively separable.

2.18. THEOREM. *There exist effectively inseparable RE sets.*

Proof. Define $A = \{x \mid \{x\}(x) = 0\}$, $B = \{x \mid \{x\}(x) = 1\}$. Clearly $A \cap B = \emptyset$ and both are RE.

Let $W_m \cap W_n = \emptyset$ and $A \subseteq W_m$, $B \subseteq W_n$. To define φ we start testing $x \in W_m$ or $x \in W_n$, if we first find $x \in W_m$, then we put an auxiliary function $\sigma(x)$ equal to 1, if x turns up first in W_n then we put $\sigma(x) = 0$.

Formally

$$\sigma(m, n, x) = \begin{cases} 1 & \text{if } \exists z(T(m, x, z) \text{ and } \forall y < z \,\neg T(n, x, y)) \\ 0 & \text{if } \exists z(T(n, x, z) \text{ and } \forall y \leq z \,\neg T(m, x, y)) \\ \uparrow & \text{else.} \end{cases}$$

By the S_n^m theorem $\{h(m, n)\}(x) = \sigma(m, n, x)$ for some recursive h.

Now

$$h(m, n) \in W_m \Rightarrow h(m, n) \notin W_n \Rightarrow \exists z (T(m, h(m, n), z) \text{ and}$$
$$\forall y < z \neg T(n, h(m, n), y))$$
$$\Rightarrow o(m, n, h(m, n)) = 1 \Rightarrow \{h(m, n)\}(h(m, n)) = 1$$
$$\Rightarrow h(m, n) \in B \Rightarrow h(m, n) \in W_n.$$

Contradiction. Hence $h(m, n) \notin W_m$. Similarly $h(m, n) \notin W_n$. Thus h is the required φ. □

As a corollary we find that A^c is *productive*, i.e. there is a partial recursive ψ such that for each $W_k \subseteq A^c$ we have $\psi(k) \in A^c - W_k$. Simply take in the above proof $W_{m_0} = A$ and $W_n = B \cup W_k$.

Using the simple fact that there is a recursive f such that $W_x \cup W_y = W_{f(x, y)}$, we find a recursive g such that $B \cup W_k = W_{g(k)}$. Putting $\psi(k) = \varphi(m_0, g(k))$ (φ as defined in 2.18), we find the desired production function: $\psi(k) \in A^c - W_k$.

Such a productive set is in a strong sense not RE: if one tries to fit in an RE set then one can uniformly and effectively indicate a point that eludes this RE set.

2.19. *Relative Recursiveness*

Following Turing we can widen the scope of computability (recursiveness) a bit, by allowing one (or finitely many) functions to be added to the initial functions, e.g. let f be such an extra initial function, then Definition 2.1 yields a wider class of partial functions. We call these functions '*recursive in f*', and we may think of them as being computable when f is given beforehand as an *oracle*. The theory of this section can be carried through for the new concept, just replace 'recursive' or 'RE' by 'recursive in f' or 'RE in f'.

The notion of 'recursive in' is particularly interesting when applied to (characteristic functions of) sets. We say that A is Turing reducible to B (notation $A \leq_T B$) if K_A is recursive in K_B. K_A stands for a membership test for A, so $A \leq_T B$ means that there is an algorithm such that we can test $n \in A$ by applying the algorithm to the (given) membership test for B (i.e. K_B).

By assigning an index to K_B, we can write this as $K_A(x) = \{e\}^B(x)$, where e is computed as before. The superscript B (or in general f) is added to indicate the dependence on B (or f).

It is not terribly difficult to show that in computing $\{e\}^B(x)$ we can only use a finite part of the function K_B. Heuristically this means that in order to test $n \in A$ we carry out an algorithm while during the computation we may ask finitely many questions to the oracle K_B (or B).

It is easily seen that $\leqslant_T$ is transitive, but not a partial order. Since $A \leqslant_T B$ means roughly that A is, from a recursive viewpoint, less complicated than B, $A \leqslant_T B \wedge B \leqslant_T A$ means that A and B are equally complicated. $A =_T B :=$ $A \leqslant_T B \wedge B \leqslant_T A$ can be shown to be an equivalence relation, the equivalence classes are called *degrees of unsolvability* or *Turing degrees*, cf. Shoenfield, [1971].

2.20. *Church's Thesis*

Are there more algorithms than just the recursive ones? This question has never been settled, partly due to the nature of the problem. The same question for the primitive recursive functions has been answered positively. We have been able to 'diagonalize out of the class of primitive recursive functions' in an effective way. The same procedure does not work for the recursive functions, since there is no effective (i.e. recursive) way to enumerate them.

If one accepts the fact that the initial functions are algorithmic, and that the closure under substitution and reflection leads from algorithms to algorithms, then there is an inductive proof that all recursive functions are algorithms. Or, if one takes partial functions into consideration, that all partial recursive functions are algorithmic.

The converse poses the real hard question: are all algorithms recursive, or in a negative form: are there any non-recursive algorithms?

The exhibition of a non-recursive algorithm would settle the problem in the negative. A positive solution would require an exact characterization of the class of algorithms, something that is lacking. Actually the partial recursive functions have been introduced precisely for this purpose. To put it succinctly: an algorithm is a function that we recognize as effectively computable. So there is on the one hand the mathematically precise notion of a *partial recursive function* and on the other hand the anthropological, subjective notion of an algorithm.

In 1936 Alonzo Church proposed to identify the two notions, a proposal that since has become known as *Church's Thesis*: A (number theoretic) function is algorithmic if and only if it is recursive. Similar proposals have been made by Turing and Markov.

There are a number of arguments that support Church's Thesis.

(1) *A pragmatic argument: all known algorithms are recursive.* As a matter of fact, the search for non-recursive algorithms has not yielded any result. The long experience in the subject has led to acceptance for all practical purposes of the thesis by all who have practiced the art of recursion theory. This has led to a tradition of 'proof by Church's Thesis', cf. Rogers [1967], which takes the following form: one convinces oneself by any means whatsoever that a certain function is computable and then jumps to the conclusion that it is (partial) recursive. Similarly, for 'effectively enumerable' and 'RE'.

We will demonstrate a 'proof by Church's Thesis' in the following

EXAMPLE. Each infinite RE set contains an infinite recursive set.

Proof. Let A be infinite RE. We list the elements of A effectively, $n_0, n_1, n_2, n_3, \ldots$

From this list we extract an increasing sublist: put $m_0 = n_0$, after finitely many steps we find an n_k such that $n_k > n_0$, put $m_1 = n_k$. We repeat this procedure to find $m_2 > m_1$, etc. This yields an effective listing of the subset $B = \{m_0, m_1, m_2, \ldots\}$ of A, with the property $m_i < m_{i+1}$.

Claim. B is decidable. For, in order to test $k \in B$ we must check if $k = m_i$ for some i. Since the sequence of m_i's is increasing we have to produce at most $k + 1$ elements of the list and compare them with k. If none of them is equal to k, then $k \notin B$. Since this test is effective, B is decidable and, by Church's Thesis, recursive. □

This practice is not quite above the board, but it is very convenient, and most experienced recursion theorists adhere to it.

(2) A conceptual analysis of the *notion of computability.* An impressive specimen is to be found in Alan Turing's fundamental paper [1936], also cf. Kleene [1952]. Turing has broken down the human computational procedures in elementary steps formulated in terms of abstract computers, the so-called *Turing machines.*

Robin Gandy has pursued the line of Turing's analysis in his paper 'Church's thesis and principles for mechanisms' (Gandy [1980]), which contains a list of four principles that underly, so to speak, the conceptual justification of Church's Thesis.

(3) A stability argument: all the codifications of the notion of computability that have been put forward (by, e.g., Gödel-Herbrand, Church, Curry, Turing, Markov, Post, Minski, Shepherdson-Sturgis) have been shown to be equivalent. Although – as Kreisel pointed out – this does not rule out

I.6: ALGORITHMS AND DECISION PROBLEMS

a systematic mistake, it does carry some weight as a heuristic argument: the existence of a large number of independent but equivalent formulations of the same notion tends to underline the naturalness of the notion.

The algorithms referred to in Church's Thesis must be 'mechanical' in nature, i.e. they should not require any creativity or inventiveness on the part of the human performer. The point is to be kept in mind; in the chapter on intuitionistic logic [Chapter III.5) we will return to it.

One should also keep in mind that the notion of computability that is under discussion here is an abstract one. Matters of feasibility are not relevant to Church's Thesis, but they *are* of basic interest to theoretical computer scientists. In particular, the time (or tape) complexity has become a subject of considerable importance. Computations in 'polynomial time' are still acceptable from a practical point of view. Unfortunately, many important decision methods (algorithms) require exponential time (or worse), cf. Schnorr [1974].

There is a constructive and a non-constructive approach to the notion of recursiveness. There seems little doubt that the proper framework for a theory of (abstract) computability is the constructive one. Let us illustrate an anomaly of the non-constructive approach: there is a partial recursive function with at most one output, that is the Gödel number of the name of the President of American in office on the first of January of the year two thousand, if there is such a president, and which diverges otherwise. This may seem surprising; is the future fully determined? A moments reflection shows that the above statement is a cheap, magician's trick: consider the empty function and all constant functions (we can even bound by the number by putting a bound on the possible length of the name of the future president). Exactly one of those partial (recursive) functions is the required one, we don't know *which one*, but a repeated application of the principle of the excluded third proves the statement. Here is another one: consider some unsolved problem P (e.g. Fermat's last problem or the Riemann hypothesis) – there is a recursive function f such that f has (constant) output 1 if P holds and 0 if P is false. Solution: consider the constant 0 and 1 functions f_0 and f_1. Since $P \vee \neg P$ holds (classically) f_1 or f_0 is the required recursive function.

Constructively viewed, the above examples are defective, since the principle of the excluded third is constructively false (cf. the Chapter on intuitionistic logic [Chapter III.5]). The constructive reading of "there exists a partial recursive function φ" is: we can effectively compute an index e.Ròsza Pèter has used the constructive reading of recursion theory as an argument for the circularity of the notion of recursiveness, when based on Church's Thesis, Péter [1959]. The circularity is, however, specious. Recursive functions are not used for computing single numbers, but to yield outputs for given inputs. The computation of isolated discrete objects precedes the manipulations of recursive functions, it is one of the basic activities of constructivism. So the computation of an index of a partial recursive function does itself not need recursive functions.

The notion of a recursive function has received much attention. Historically speaking, its emergence is an event of the first order. To quote Gödel:

> It seems to me that this importance [i.e. of the notion of recursive function] is largely due to the fact that with this concept one has for the first time succeeded in giving an absolute definition of an interesting epistemological notion, i.e. one not depending on the formalism chosen. Gödel [1946], Wang [1974] p. 81.

3. APPLICATIONS

It is no exaggeration to say that recursion theory was conceived for the sake of the study of arithmetic. Gödel used the machinery of recursion theory to show that theories containing a sufficient portion of arithmetic are incomplete. Subsequent research showed that arithmetic is undecidable, and many more theories to boot.

We will briefly sketch some of the methods and results.

3.1. The first-order theory of arithmetic, **PA** (Peano's arithmetic), has a language with $S, +, \cdot$ and 0.

Its axioms are

$$Sx \neq 0 \qquad x + Sy = S(x + y)$$
$$Sx = Sy \to x = y \qquad x \cdot 0 = 0$$
$$x + 0 = x \qquad x \cdot Sy = x \cdot y + x$$
$$\varphi(0) \wedge \forall x(\varphi(x) \to \varphi(Sx)) \to \forall x \varphi(x).$$

(the induction schema).

In **PA** we can define the order relation: $x < y := \exists z(x + Sz = y)$ and we can prove its properties: $\neg(x < 0)$, $x < Sy \leftrightarrow x < y \vee x = y$, $x < y \vee x = y \vee y < x$, $x < y \wedge y < z \to x < z$, by induction. The individual natural number symbols are defined by

I.6: ALGORITHMS AND DECISION PROBLEMS

$1 = S0, \quad 2 = SS0, \quad 3 = SSS0, \ldots$

R. Robinson introduced a finitely axiomatized subsystem **Q** of **PA** with the schema of induction replaced by one axiom:

$x \neq 0 \to \exists y(x = Sy).$

Shoenfield also introduced a finitely axiomatized subsystem **N** of **PA**. This system has $<$ as a primitive symbol, and the schema of induction is replaced by the axioms $\neg(x < 0), x < Sy \to x < y \lor x = y, x < y \lor x = y \lor y < x.$

3.2. Arithmetization

One can code the expressions of arithmetic as natural numbers in such a way that the relevant syntactical properties become primitive recursive predicates of the codes.

There are many ways to carry out the actual coding. Unfortunately this part of the theory is strongly 'coordinate dependent', i.e. it depends on the underlying coding of finite sequences of natural numbers. Canonical codings have been proposed at various points, cf. Jeroslow [1972], but there has always remained a residue of arbitrariness. We will sketch a coding based on the coding of Examples 1.9. Following the tradition, we will call the codes *Gödel numbers*.

(1) We assign Gödel numbers to the symbols of the alphabet.

$x_i \mapsto 2_i, 0 \mapsto 1, \lor \mapsto 3, \neg \mapsto 5, \exists \mapsto 7, S \mapsto 9, + \mapsto 11, \cdot \mapsto 13,$

$= \mapsto 15, (< \mapsto 17, \text{ if considering } \mathbf{N})$

(2) The Gödel numbers of terms are defined by

$\ulcorner x_i \urcorner = \langle 2_i \rangle, \ulcorner 0 \urcorner = \langle 1 \rangle, \ulcorner St \urcorner = \langle 9, \ulcorner t \urcorner \rangle, \ulcorner (t + s) \urcorner = \langle 11, \ulcorner t \urcorner, \ulcorner s \urcorner \rangle,$

$\ulcorner t \cdot s \urcorner = \langle 13, \ulcorner t \urcorner, \ulcorner s \urcorner \rangle$

(3) The Gödel numbers of formulas are defined by

$\ulcorner (t = s) \urcorner = \langle 15, \ulcorner t \urcorner, \ulcorner s \urcorner \rangle, \ulcorner (\varphi \lor \psi) \urcorner = \langle 3, \ulcorner \varphi \urcorner, \ulcorner \psi \urcorner \rangle, \ulcorner \neg \varphi \urcorner = \langle 5, \ulcorner \varphi \urcorner \rangle,$

$\ulcorner (\exists x_i \varphi) \urcorner = \langle 7, \ulcorner x_i \urcorner, \ulcorner \varphi \urcorner \rangle.$

The above functions that assign Gödel numbers to expressions are defined by recursion, and one would like to know if, e.g., the set of Gödel numbers of formulas is decidable. The following lemmas provide answers.

(4) LEMMA. *The following predicates are primitive recursive*
  (a) *n is the Gödel number of a variable*
  (b) *n is the Gödel number of a term*
  (c) *n is the Gödel number of a formula*

Proof. We will write down the predicate in a suitable form such that by means of the usual closure properties the reader can immediately conclude the primitive recursiveness.

(a) $\text{Vble}(n) \leftrightarrow \exists x \leqslant n \, (n = \langle 2x \rangle)$

(b) $\text{Term}(n) \leftrightarrow [\text{Vble}(n) \lor n = \langle 1 \rangle \lor \exists x < n \, (n = \langle 9, x \rangle) \land \text{Term}(x)) \lor \exists x, y < n \, (n = \langle 11, x, y \rangle \land \text{Term}(x) \land \text{Term}(y)) \lor \exists xy < n \, (n = \langle 13, x, y \rangle \land \text{Term}(x) \land \text{Term}(y))$

(c) $\text{Form}(n) \leftrightarrow$
  $\exists xy < n \, (n = \langle 15, x, y \rangle \land \text{Term}(x) \land \text{Term}(y)) \lor$
  $\exists xy < n \, (n = \langle 3, x, y \rangle \land \text{Form}(x) \land \text{Form}(y)) \lor$
  $\exists x < n \, (n = \langle 5, x \rangle \land \text{Form}(x)) \lor$
  $\exists xy < n \, (n = \langle 7, x, y \rangle \land \text{Vble}(x) \land \text{Form}(y)).$

Note that the lemma is established by an appeal to the primitive recursive version of the recursion theorem (cf. 2.4); by switching to characteristic functions, one can make use of course of value recursion. Another standard procedure, e.g. for the coding of terms, is to arithmetize the condition "there is a finite sequence of which each member is either a variable or 0, or obtained from earlier ones by applying S, $+$ or $\cdot$". This immediately establishes the recursiveness, for primitive recursiveness one only has to indicate a bound on the code of this sequence. □

(5) At certain places it is important to keep track of the free variables in terms and formulas, e.g. in $\forall x \varphi(x) \to \varphi(t)$. The following predicate expresses that 'x is free in A' (where A is a term or a formula – we handle them simultaneously): 'x is A itself or A is built from two parts not by means of $\exists$ and x is free in at least one of those, or built from two parts by means of $\exists$ and x is not the first part and free in the second part, etc.".

So put

$\text{Fr}(m, n) \leftrightarrow (m = n \land \text{Vble}(m)) \lor \exists xy < n \, (n = \langle 3, x, y \rangle \land (\text{Fr}(m, x) \lor \text{Fr}(m, y))) \lor (\exists xy < n (n = \langle 11, \ldots \rangle \ldots) \lor$
$(\ldots n = \langle 13, \ldots \rangle \ldots) \lor (\ldots n = \langle 15, \ldots \rangle \ldots) \lor \ldots$
$\ldots \exists xy < n \, (n = \langle 7, x, y \rangle \land m \neq x \land \text{Fr}(m, y))$

Again Fr is primitive recursive.

I.6: ALGORITHMS AND DECISION PROBLEMS

(6) We need a number-theoretic function that mimicks the substitution operation, i.e. that from the Gödel numbers of φ, x and t computes the Gödel number of $\varphi[t/x]$.
The function Sub must satisfy $\text{Sub}(\ulcorner A \urcorner, \ulcorner x \urcorner, \ulcorner t \urcorner) = \ulcorner A[t/x] \urcorner$, where A is a term or a formula.
So put

$$\text{Sub}(m, n, k) = \begin{cases} k & \text{if Vble}(m) \wedge m = n \\ \langle 9, \text{Sub}(p, n, k) \rangle & \text{if } m = \langle 9, p \rangle \\ \langle 5, \text{Sub}(p, n, k) \rangle & \text{if } m = \langle 5, p \rangle \\ \langle a, \text{Sub}(p, n, k), \text{Sub}(q, n, k) \rangle & \text{if } m = \langle a, p, q \rangle \\ \quad \text{for } a = 3, 11, 13, 15 \\ \langle 7, p, \text{Sub}(q, n, k) \rangle & \text{if } m = \langle 7, p, q \rangle \wedge p \neq x, \\ m & \text{otherwise.} \end{cases}$$

Sub is primitive recursive.

(7) Depending on the logical basis of **PA** (e.g. a Hilbert type system, or sequent calculus, etc.) one can find a primitive recursive predicate $\text{Prov}(m, n)$ which expresses that n is the Gödel number of a proof of the formula with Gödel number m. Every textbook will give the details, but by now the reader will be able to concoct such a predicate by himself.

In order to avoid cumbersome manipulations, one usually assumes the set of (Gödel numbers of) axioms to be recursive. For **PA**, **Q** and **N** this evidently is correct.

(8) The predicate $\text{Thm}(m) := \exists x \, \text{Prov}(m, x)$ expresses that m is the Gödel number of a theorem of **PA**. The existential quantifier makes Thm recursively enumerable, and we will show that it is *not* recursive.

(9) Superficially speaking, natural numbers lead a double life; they occur in the theory of arithmetic and in the real world. In the first case they occur as symbols, the so-called *numerals*. The numeral for the number n is denoted by $\bar{n}$. To be specific $\bar{0} = 0$ (recall that **PA** had 0 as a constant symbol, we could have used a bold face zero, but the reader will not be confused), $\overline{n+1} = S(\bar{n})$. Now numerals are symbols, so they have Gödel numbers. We define the function Num which associates with each number n the Gödel number of $\bar{n}$:

$$\text{Num}(0) = \langle 1 \rangle,$$
$$\text{Num}(n + 1) = \langle 9, \text{Num}(n) \rangle.$$

So $\text{Num}(n) = \ulcorner \bar{n} \urcorner$.

(10) In order to avoid needless restrictions, we quote the following theorem of Craig:

THEOREM *Every axiomatizable theory can be axiomatized by a recursive set of axioms.*

Here a theory is called axiomatizable if the set of its axioms is RE. The following informal argument may suffice.

Let $\varphi_0, \varphi_1, \varphi_2, \ldots$ be an effective enumeration of the axioms of T (it is no restriction to assume an infinite list). Then $\varphi_0, \varphi_0 \wedge \varphi_1, \varphi_0 \wedge \varphi_1 \wedge \varphi_2, \ldots$ also axiomatizes T, and we can effectively test whether a given sentence σ belongs to this set. For the length of the axioms is strictly increasing so after a finite number of those axioms have been listed we know that if σ has not yet occurred, it will not occur at all. In the literature axiomatizable arithmetical theories are also called RE theories. Non-axiomatizable theories are *ipso facto* indecidable, for even their class of theorems is not RE.

3.3. *Representability of the Recursive Functions and Predicates*

In the preceding section we have reduced, so to speak, logic and arithmetic to recursion theory. Now we will reduce recursion theory to the formal theory of arithmetic. We will show that inside **PA** (or **Q**, or **N**, for that matter) we can speak about recursive functions and relations, be it in a rather complicated way.

For convenience we will treat the three theories of arithmetic above on an equal footing by considering a theory T which is an axiomatizable extension of **Q** or **N**. If we need special features of T (e.g. $T = $ **PA**), then we will explicitly list them.

3.3.1. DEFINITION. An n-ary function f is *represented* by a formula φ with $n + 1$ variables $x_1, \ldots, x_n, y$ if for all $k_1, \ldots, k_n, l$

$$f(k_1, \ldots, k_n) = l \Longleftrightarrow T \vdash \varphi(\bar{k}_1, \ldots, \bar{k}_n, y) \leftrightarrow \bar{l} = y.$$

An n-ary relation R is *represented* by a formula φ with n free variables if $R(k_1, \ldots, k_n) \Rightarrow T \vdash \varphi(\bar{k}_1, \ldots, \bar{k}_n)$ and *not-*$R(k_1, \ldots, k_n) \Rightarrow T \vdash \neg\varphi(\bar{k}_1, \ldots, \bar{k}_n)$.

The basic theorem states that

THEOREM. *All recursive functions and predicates are representable in any of the systems* **PA**, **Q**, **N** *(and hence in any T)*

I.6: ALGORITHMS AND DECISION PROBLEMS

It is a simple exercise to show that a relation is representable iff its characteristic function is so. Therefore it suffices to prove the theorem for recursive functions. For this purpose it is most convenient to use the characterization of recursive functions by means of μ-recursion.

The proof of the theorem is not very difficult but rather clerical, the reader may look it up in, e.g., Davis [1958], Kleene [1952] or Shoenfield [1967].

As a consequence we now have available formulas in T for the predicates that we introduced in Section 3.2. In particular, there is a formula that represents Prov. We will, for convenience, use the same symbol for the representing formulas. It will always appear from the context which reading must be used.

The soundness theorem for predicate logic tells us that the theorems of a theory T are true in all models of T. In particular, are all provable sentences of T true in the standard model, following tradition we call sentences true in $\mathbb{N}$ simply *true*.

Until the late Twenties it was hoped and expected that, conversely, all true sentences would also be provable in **PA**. Gödel destroyed that hope in 1931. Nonetheless, one might hope to establish this converse for a significant class of sentences.

The following theorem provides such a class.

Convention. We call a formula $\Sigma_1^0(\Pi_1^0)$ if it is (equivalent to) a prenex formula with a prefix of existential (universal) quantifiers followed by a formula containing only bounded quantifiers.

3.3.2. THEOREM. (Σ_1^0-*completeness*)

$$\varphi \Rightarrow \mathbf{PA} \vdash \varphi \text{ for } \Sigma_1^0 \text{ sentences } \varphi.$$

The proof proceeds by induction on the structure of φ, cf. Shoenfield [1967] p. 211. Note that the premiss is 'φ is true', or to spell it out '$\mathbb{N} \models \varphi$'.

For Π_1^0 sentences truth does not imply provability as we will show below.

Before we proceed to the incompleteness theorem, let us have another look at our formal system. The language of arithmetic contains function symbols for $+$, $\cdot$, S, should we leave it at that? After all, exponentiation, the square, the factorial, etc. belong to the daily routine of arithmetic, so why should we not introduce them into the language? Well, there is no harm in doing so since primitive recursive functions are given by defining equations which can be formulated in arithmetical language. To be specific, for each primitive recursive function f we can find the representing formula $\varphi(\vec{x}, y)$

such that not only $f(\vec{m}) = n \iff \mathbf{PA} \vdash \varphi(\vec{m}, y) \leftrightarrow y = n$, but also $\mathbf{PA} \vdash \forall \vec{x}\, \exists! y \varphi(\vec{x}, y)$. Note that here one essentially needs induction.

Now it is a well-known fact of elementary logic that one can add a function symbol F to the language, and an axiom $\forall \vec{x}\, \varphi(\vec{x}, F(\vec{x}))$ without essentially strengthening the theory. To be precise: the extended theory T' is *conservative* over the original theory T, for formulas ψ not containing F we have $T \vdash \psi \iff T' \vdash \psi$.

In this case we even have a translation ^{0} that eliminates the symbol F so that $T' \vdash \psi \iff \psi^0$ and $T' \vdash \psi \iff T \vdash \psi^0$ (cf. Shoenfield [1967] p. 55 ff., Van Dalen [1981] p. 144 ff.).

Summing up, we can conservatively add function symbols and defining axioms for all primitive recursive functions to **PA**.

The proof of this fact runs parallel to the proof of the representability of the primitive recursive functions (and makes use of Gödel's β-function), cf. Shoenfield [1967] p. 206 ff.

Observe that if F is related to the primitive recursive f in the above manner, then F represents f:

$$f(\vec{m}) = n \iff \mathbf{PA} \vdash F(\vec{m}) = \bar{n}.$$

Note that we can also add function symbols for recursive functions since a recursive f is represented by a formula φ. For put $\psi(\vec{x}, y) = [\exists z \varphi(\vec{x}, z) \to \varphi(\vec{x}, y) \land \forall u < y \neg \varphi(\vec{x}, y)] \land [\neg \exists z \varphi(\vec{x}, z) \to y = 0]$, then clearly ψ also represents f and $\mathbf{PA} \vdash \forall \vec{x}\, \exists! y \psi(\vec{x}, y)$.

For metamathematical purposes it usually does not harm to consider the conservative extension by primitive recursive functions. E.g. with respect to decidability and completeness T' and T behave exactly alike. T' is decidable (complete) $\iff T$ is decidable (complete).

Since the presence of function symbols for primitive recursive functions considerably streamlines the presentation of certain results, we will freely make use of the above *definitional* extension of arithmetic, which we, by abuse of notation, also call **PA**.

3.4. *The First Incompleteness Theorem and the Undecidability of* **PA**.

Gödel formulated a sentence which had a certain analogy to the famous Liar Paradox. Paraphrased in everyday language, it states: "I am not provable in **PA**". This kind of self-referential sentence makes full use of the ability of **PA** (and related systems) to formulate its own syntax and derivability relation. A convenient expedient is the

3.4.1. FIXED POINT THEOREM. *Let* $\varphi(x_0)$ *have only the free variable* x_0 *then there is a sentence* ψ *such that* **PA** $\vdash \psi \leftrightarrow \varphi(\ulcorner\psi\urcorner)$.

Proof. We will use the substitution function. Put $s(m, n) = \text{Sub}(m, \ulcorner x_0 \urcorner,$ $\text{Num}(n))$, and let $k := \ulcorner \varphi(s(x_0, x_0)) \urcorner$, then $\psi := \varphi(s(\bar{k}, \bar{k}))$ will do the trick.

For $s(k, k) = s(\ulcorner \varphi(s(x_0, x_0)) \urcorner, k) = \ulcorner \varphi(s(\bar{k}, \bar{k})) \urcorner$; so by the representability of s, **PA** $\vdash \psi \leftrightarrow \varphi(\ulcorner\psi\urcorner)$. □

Observe that Theorem 3.4.1 holds for any axiomatizable extension of **PA**.

We now quickly get the incompleteness result, by applying the Fixed Point Theorem to $\neg \exists y \, \text{Prov}(x, y)$. **PA** $\vdash \psi \leftrightarrow \neg \exists y \, \text{Prov}(\ulcorner\psi\urcorner, y)$ for a certain ψ.

Now **PA** $\vdash \psi \Rightarrow$ **PA** $\vdash \text{Prov}(\ulcorner\psi\urcorner, \bar{k})$ where k is the Gödel number of the actual proof of ψ. So **PA** $\vdash \exists y \, \text{Prov}(\ulcorner\psi\urcorner, y)$. But also **PA** $\vdash \neg \exists y \, \text{Prov}(\ulcorner\psi\urcorner, y)$, so **PA** $\vdash \bot$. Assuming consistency we get **PA** $\nvdash \psi$.

If **PA** $\vdash \neg \psi$, then **PA** $\vdash \neg \exists y \, \text{Prov}(\ulcorner\neg\psi\urcorner, y)$, and also **PA** $\vdash \exists y \, \text{Prov}(\ulcorner\neg\psi\urcorner, y)$. Without any additional assumptions we only get **PA** $\vdash \exists y \, \text{Prov}(\ulcorner\bot\urcorner, y)$ (we may take $0 = 1$ for $\bot$). If we assume, e.g., ω-consistency, i.e. it is not the case that **PA** $\vdash \sigma(n)$, for all n, and **PA** $\vdash \exists x \neg \sigma(x)$, for arbitrary σ, then we may derive a contradiction and conclude **PA** $\nvdash \neg \psi$. Since ω-consistency implies consistency (immediate), we also get **PA** $\nvdash \psi$. So neither ψ nor $\neg \psi$ are theorems of **PA**. Hence, **PA** is incomplete.

By appealing to the fact that $\mathbb{N}$ is a model of **PA** we can avoid those technicalities mentioning consistency or ω-consistency: **PA** $\vdash \exists y \, \text{Prov}(\ulcorner\psi\urcorner, y)$ $\Rightarrow \mathbb{N} \models \exists y \, \text{Prov}(\ulcorner\psi\urcorner, y)$, so there is an n such that $\text{Prov}(\ulcorner\psi\urcorner, \bar{n})$ holds, but this implies that **PA** $\vdash \psi$. *Contradiction.* Note that the above Gödel sentence ψ is Π_1^0.

3.4.2. *Remarks.* (1) Since **PA** extends **Q** and **N** we also have established the incompleteness of **Q** and **N** (although for just incompleteness a simple model-theoretic argument would suffice, e.g., in **Q** we cannot even prove addition to be commutative). By means of a careful rewording of the Gödel sentences, such that no use is made of primitive recursive functions *in* the system one can present a similar argument which shows that all consistent axiomatizable extensions T of **Q** (or **N**) are incomplete. Cf. Mendelson [1979].

(2) Rosser eliminated the appeal to ω-consistency by applying the fixed-point theorem to another formula: there is a proof of my negation and no proof of me precedes it.

In symbols: $\exists y [\text{Prov}(\text{neg} \, x, y) \land \forall z < y \, \neg \text{Prov}(x, z)]$. Here neg is a primitive recursive function such that $\text{neg}(\ulcorner\varphi\urcorner) = \ulcorner\neg\varphi\urcorner$. The whole formula is formulated in the language of a conservative extension of **PA**.

Call this formula $R(x)$ and apply the fixed point theorem: $\mathbf{PA} \vdash \psi \leftrightarrow R(\ulcorner\psi\urcorner)$. For better readability we suppress the reference to $\mathbf{PA}$.

$$\vdash \psi \leftrightarrow \exists y (\text{Prov}(\ulcorner\neg\psi\urcorner, y) \wedge \forall z < y \neg\text{Prov}(\ulcorner\psi\urcorner, z)) \qquad (1)$$

Suppose $\vdash \psi$, then $\text{Prov}(\ulcorner\psi\urcorner, \bar{n})$ is true for some n.

So $\vdash \text{Prov}(\ulcorner\psi\urcorner, \bar{n})$ \hfill (2)

Also $\vdash \exists y (\text{Prov}(\ulcorner\neg\psi\urcorner, y) \wedge \forall z < y \neg\text{Prov}(\ulcorner\psi\urcorner, z))$ \hfill (3)

(2), (3) $\Rightarrow \vdash \exists y < \bar{n} \, \text{Prov}(\ulcorner\neg\psi\urcorner, y)$ \hfill (4)

Using $\vdash y < \bar{n} \leftrightarrow y = \bar{0} \vee y = \bar{1} \vee \cdots \vee y = \overline{n-1}$ we find $\vdash \text{Prov}(\ulcorner\neg\psi\urcorner, \bar{0}) \vee \cdots \vee \text{Prov}(\ulcorner\neg\psi\urcorner, \overline{n-1})$. Now one easily establishes $\vdash \sigma \vee \tau \Rightarrow \vdash \sigma$ or $\vdash \tau$ for sentences σ and τ with only bounded quantifiers, this yields $\vdash \text{Prov}(\ulcorner\neg\psi\urcorner, \bar{m})$ for some $m < n$.

Hence, $\vdash \neg\psi$. But this contradicts the consistency of $\mathbf{PA}$.

Suppose now $\vdash \neg\psi$ then $\not\vdash \psi$, i.e. $\vdash \neg\text{Prov}(\ulcorner\psi\urcorner, \bar{n})$ for all n, \hfill (5)

From $\vdash \neg\psi$ we get $\vdash \forall y (\text{Prov}(\ulcorner\neg\psi\urcorner, y) \rightarrow \exists z < y \, \text{Prov}(\ulcorner\psi\urcorner, z))$ and $\vdash \text{Prov}(\ulcorner\neg\psi\urcorner, \bar{m})$ for some m.

Hence $\vdash \exists z < \bar{m} \, \text{Prov}(\ulcorner\psi\urcorner, z)$, so as before we get $\vdash \text{Prov}(\ulcorner\psi\urcorner, \bar{k})$ for some $k < m$. Together with (5) this contradicts the consistency of $\mathbf{PA}$.

Conclusion. $\not\vdash \psi$ and $\not\vdash \neg\psi$.

(3) Interpreting the Gödel sentence ψ in the standard model $\mathbb{N}$ we see that it is true. So the Gödel sentence provides an instance of a true but unprovable sentence.

(4) From the above incompleteness theorem we can easily conclude that $\{\ulcorner\varphi\urcorner | \mathbb{N} \models \varphi\}$, i.e. the set of (Gödel numbers of) true sentences of arithmetic is not decidable (i.e. recursive). For suppose it were, then we could use the true sentences as axioms and repeat the incompleteness theorem for this system $\mathbf{Tr}$: there is a φ such that $\mathbf{Tr} \not\vdash \varphi$ and $\mathbf{Tr} \not\vdash \neg\varphi$. But this is impossible since the true and false sentences exhaust all sentences. Hence, $\mathbf{Tr}$ is not recursive.

(5) By means of the provability predicate one can immediately show that in axiomatizable extensions of $\mathbf{Q}$ or $\mathbf{N}$ representable predicates and functions are recursive. To be precise, in any axiomatizable theory with 0 and S such that $\vdash \bar{m} = \bar{n} \Rightarrow n = m$ the representable predicates and functions are recursive. This provides another characterization of the recursive functions: the class of recursive functions coincides with that of the functions representable in $\mathbf{PA}$. Theories satisfying the above conditions are called *numerical*.

(6) The technique of arithmetization and the representability of recursive functions allows us to prove the undecidability of extensions of $\mathbf{Q}$ and $\mathbf{N}$ (and hence $\mathbf{PA}$) (3.4.4). We say that $\mathbf{Q}$ and $\mathbf{N}$ are *essentially undecidable*.

Exactly the same method allows us to prove the general result: numerical theories in which all recursive functions are representable are undecidable. We can also derive the following theorem of Tarski on the undefinability of truth in arithmetic. We say that a formula τ is a *truth definition* in a numerical theory T if $T \vdash \varphi \leftrightarrow \tau(\ulcorner\varphi\urcorner)$.

3.4.3. TARSKI'S THEOREM. *No consistent axiomatizable extension of* **Q** *or* **N** *has a truth definition.*

Proof. Suppose a truth definition τ exists, then we can formulate the Liar Paradox.

By the fixed point theorem there is a sentence λ such that $T \vdash \lambda \leftrightarrow \neg\tau(\ulcorner\lambda\urcorner)$. Since τ is a truth definition, we get $T \vdash \lambda \leftrightarrow \neg\lambda$. This conflicts with the consistency of T. □

For the undecidability of **Q**, **N** or **PA** we use the set K from 2.16. Or, what comes to the same thing, we diagonalize once more.

Consider the set Thm and an RE-set A disjoint from Thm (Figure 6), for any axiomatizable extension T of **Q** (or **N**).

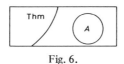

Fig. 6.

Define

$$\varphi(k) = \begin{cases} 0 & \text{if } \vdash \exists z T(\bar{k}, \bar{k}, z) \wedge Uz \neq 0 \\ 1 & \text{if } \ulcorner(\exists z T(\bar{k}, \bar{k}, z) \wedge Uz \neq 0)\urcorner \in A \\ \uparrow & \text{otherwise.} \end{cases}$$

Clearly φ is a partial recursive function, say with index e.

(i) Then $\{e\}e = 0 \Rightarrow T(e, e, n) \wedge U\bar{n} = 0$ for some $n \Rightarrow$

$\vdash T(\bar{e}, \bar{e}, \bar{n}) \wedge U\bar{n} = \bar{0} \Rightarrow \vdash \exists z(T(\bar{e}, \bar{e}, z) \wedge Uz = 0),$ \hfill (1')

Also $\{e\}e = 0 \Rightarrow \vdash \exists z(T(\bar{e}, \bar{e}, z) \wedge Uz \neq 0).$ \hfill (2)

From the definition the T-predicate it easily follows that

$\vdash T(\bar{e}, \bar{e}, z) \wedge T(\bar{e}, \bar{e}, z') \rightarrow z = z'$ \hfill (3)

So (1), (2), (3) imply that T is inconsistent.

(ii) $\{e\}e = 1 \Rightarrow T(e, e, n) \wedge Un = 1$ for some $n \Rightarrow \vdash T(\bar{e}, \bar{e}, \bar{n}) \wedge U\bar{n} = 1 \Rightarrow$

$\vdash \exists z T(\bar{e}, \bar{e}, z) \wedge Uz \neq \bar{0}$. Also $\{e\}e = 1 \Rightarrow \ulcorner \exists z T(\bar{e}, \bar{e}, z) \wedge Uz \neq 0 \urcorner \in A$, but that contradicts the disjointness of Thm and A.

So we have shown that Thm is productive (cf. 2.18) and, hence, undecidable. A slight adaptation of the argument shows that the set of theorems and the set of refutable sentences (i.e. σ with $T \vdash \neg \sigma$) are effectively inseparable. The above proof established

3.4.4. THEOREM. *If T is a consistent extension of* **Q** (*or* **N**) *then T is undecidable.*

As a corollary we get

3.4.5. THEOREM (Church). *First-order predicate logic is undecidable.*

Proof. Consider a first-order language containing at least the language of arithmetic. Since **Q** is finitely axiomatized we can write $\mathbf{Q} \vdash \varphi \Leftrightarrow \vdash \alpha \rightarrow \varphi$, where α is the conjunction of the axioms of **Q**. Clearly any decision procedure for first-order predicate logic would also provide a decision procedure for **Q**. □

The undecidability of **PA** yields another proof of its incompleteness, for there is a fairly obvious theorem of model theory that states

THEOREM. *A complete, axiomatizable theory is decidable* (Vaught).

Observe that it is standard practice to identify 'decidable' and 'recursive', i.e. to rely on Church's Thesis. In a more cautious approach one would, of course, use phrases such as "the set of Gödel numbers of theorems of **PA** is not recursive".

There are various paths that lead from recursion theory to the area of decidable and undecidable theories. Usually one starts from a suitable class of undecidable (sometimes also called *unsolvable*) problems in one of the many approaches to the notion of algorithm, e.g. the Halting problem for Turing machines or register machines, or Post's correspondence problem and somehow 'interpretes' them in a convenient logical form. The actual role of recursion theory by then is modest or often even nil. Most of the techniques in the particular field of undecidable theories are of a logical nature, e.g. the construction of suitable translations.

3.5. Decidable and Undecidable Theories

In a large number of cases there are reductions of the decision problem of certain theories to well-known theories.

We list a few undecidable theories below:
(1) Peano's arithmetic,
(2) Theory of rings (Tarski (1949)),
(3) Ordered fields (J. Robinson (1949)),
(4) Theory of lattices (Tarski (1949)),
(5) Theory of a binary predicate (Kalmár (1936)),
(6) Theory of a symmetric binary predicate (Church, Quine (1952)),
(7) Theory of partial order (Tarski (1949)),
(8) Theory of two equivalence relations (Rogers (1956)),
(9) Theory of groups,
(10) Theory of semi groups,
(11) Theory of integral domains.

One should note that the underlying logic does play a role in decidability results, e.g. whereas classical monadic predicate logic is decidable, intuitionistic monadic predicate logic is *not* (Kripke (1968), cf. Gabbay [1981] p. 234).

The decidability aspects of predicate logic have been widely studied. One of the oldest results is the decidability of monadic predicate calculus (Löwenheim (1915), Behmann (1922)), and the high-water mark is the undecidability of predicate logic (Church (1936)). In this particular area there has been much research into solvable and unsolvable cases of the decision problem.

There are special classes Γ of formulas, given by syntactical criteria, for which undecidability has been established by a reduction procedure that effectively associates to each φ a formula φ^* of Γ such that $\vdash \varphi \Longleftrightarrow \vdash \varphi^*$. Such a class Γ is called a *reduction class with respect to provability* (or validity). Analogously one has reduction classes with respect to satisfiability.

Among the syntactic criteria that are used we distinguish, e.g., (i) the number of arguments of the predicates, (ii) the number of predicates, (iii) the length of the quantifier prefix for the prenex normal form, (iv) the number of quantifier changes in the same.

For convenience we introduce some notation: $Q_1^{n_1} \cdots Q_m^{n_m}$ stands for the class of all prenex formulas with prefixes of n_1 quantifiers Q_1, n_2 quantifiers $Q_2, \ldots, n_m$ quantifiers Q_m. A superscript ∞ indicates a quantifier block of arbitrary finite length.

Restrictions on the nature of predicate symbols is indicated by finite sequences, e.g. (0, 2, 1) indicates a predicate logic language with no unary predicates, two binary and one ternary predicate. Combining the two notations we get the obvious classes, such as $\forall^\infty \exists^2 (0, 1)$, $\exists^2 \forall^1 \exists^3 (2, 1)$, etc. Here too, ∞ means 'all finite n'.

An immediate simple but not trivial example is furnished by the *Skolem normal form* for satisfiability: the class of $\forall^\infty \exists^\infty$ formulas is a reduction class for satisfiability. So this class consists of prenex formulas with universal quantifiers followed by existential quantifiers.

Of course, this can be improved upon since the undecidability proofs for predicate logic provide more information.

We list some examples of reduction classes. There will be no function symbols or constants involved.

$\exists \forall \exists \forall$ (0, 3)	(Büchi (1962))
$\forall \exists \forall$ (0, ∞)	(Kahr, Moore, Wang (1962))
$\forall^3 \exists$ (0, ∞)	(Gödel (1933))
$\exists \forall \exists \forall$ (∞, 1)	(Rödding (1969))
$\forall \exists \forall$ (∞, 1)	(Kahr)
$\exists^\infty \forall^2 \exists^2 \forall^\infty$ (0, 1)	(Kalmar (1932))
$\forall^\infty \exists$ (0, 1)	(Kalmar, Surányi (1950))
$\forall \exists \forall^\infty$ (0, 1)	(Denton (1963))
$\forall \exists \forall \exists^\infty$ (0, 1)	(Gurevich (1966))

One can also consider prenex formulas with the matrix in conjunctive or disjunctive normal form, and place restrictions on the number of disjuncts, conjuncts, etc.

$\mathcal{D}_n$ is the class of disjunctive normal forms with at most n disjuncts: $\mathcal{C}_n$ is the class of conjunctive normal forms with at most n disjuncts per conjunct. *Krom formulas* are those in $\mathcal{D}_2$ or $\mathcal{C}_2$; *Horn formulas* those in $\mathcal{C}_\infty$ with at most one negated disjunct per conjunct.

Prefix classes of Krom and Horn formulas have been investigated by Aandera, Dreben, Börger, Goldfarb, Lewis, Maslov and others. For a thorough treatment of the subject the reader is referred to Lewis [1979].

The reader should not get the impression that logicians deal exclusively with undecidable theories. There is a considerable lore of decidable theories, but the actual decision methods usually employ little or no recursion theory. This is in accordance with the time-honored practice: one recognizes a decision method when one sees one.

The single most important decision method for showing the decidability

I.6: ALGORITHMS AND DECISION PROBLEMS

of theories is that of *quantifier elimination*. Briefly, a theory T is said to have (or allow) quantifier elimination if for each formula $\varphi(x_1, \ldots, x_n)$, with all free variables shown, there is an open (i.e. quantifier free) formula $\psi(x_1, \ldots, x_n)$ such that $T \vdash \varphi(x_1, \ldots, x_n) \longleftrightarrow \psi(x_1, \ldots, x_n)$. So for theories with quantifier elimination one has only to check derivability for open formulas. This problem usually is much simpler than the full derivability problem, and it yields a decision procedure in a number of familiar cases.

An early, spectacular result was that of Presburger (1930), who showed that the theory of arithmetic with only successor and addition has quantifier elimination and is decidable. However, additive number theory is not the most impressive theory in the world, so Presburger's result was seen as a curiosity (moreover, people hoped at that time that full arithmetic was decidable, so this was seen as an encouraging first step).

The result that really made an impression was Tarski's famous *Decision Method for Elementary Algebra and Geometry* (1948), which consisted of a quantifier elimination for real closed (and algebraically closed) fields.

Since then quantifier elimination has been established for a long list of theories, among which are *linear dense ordering, abelian groups* (Szmielev (1955)), *p-adic fields* (Cohen (1969)), *Boolean algebras* (Tarski, (1949)), *equality* (Hilbert-Bernays, (1934)). For a survey of decidable theories, including some complexity aspects, cf. Rabin [1977].

We will give a quick sketch of the method for the theory of equality.

(1) Since one wants to eliminate step by step the quantifier in front of the matrix of a prenex formula it clearly suffices to consider formulas of the form $\exists y \varphi(y, x_1, \ldots, x_n)$.

(2) We may suppose φ to be in disjunctive normal form $\mathbf{W}_i \varphi_i$ and, hence, we can distribute the $\exists$-quantifier. So it suffices to consider formulas of the form $\exists y \psi(y, x_1, \ldots, x_n)$ where ψ is a conjunction of atoms and negations of atoms.

(3) After a bit of rearranging, and eliminating trivial parts (e.g. $x_i = x_i$ or $\neg y = y$), we are left with $\exists y (y = x_{i_1} \wedge \cdots \wedge y = x_{i_k} \wedge y \neq x_{j_1} \wedge \cdots \wedge y \neq x_{j_l} \wedge \delta)$, where δ does not contain y. By ordinary logic we reduce this to $\exists y(\text{\underline{\hspace{1em}}}) \wedge \delta$. Now the formula $\exists y(\text{\underline{\hspace{1em}}})$ is logically equivalent to $(x_{i_1} = x_{i_2} \wedge \cdots \wedge x_{i_1} = x_{i_k} \wedge x_{i_1} \neq x_{j_1} \wedge \cdots \wedge x_{i_1} \neq x_{j_l})$.

So we eliminated one quantifier. However, there are a number of special cases to be considered, e.g. there are no atoms in the range of $\exists y$. Then we cannot eliminate the quantifier. So we simply introduce a constant c and replace $\exists y(y \neq x_{j_a})$ by $c \neq x_{j_a}$ (the Henkin constants, or witnesses), i.e. we consider conservative extensions of the theory.

In this way we finally end up with an open formula containing new constants. If we started with a sentence then the result is a Boolean combination of sentences which express conditions on the constants, namely which ones should be unequal. Such conditions can be read as conditions of cardinality: there are at least n elements.

From the form of the resulting sentence we can immediately see whether it is derivable or not. Moreover, the method shows what completions the theory of identity has (cf. Chang and Keisler [1973] p. 55).

Whereas the actual stepwise elimination of quantifiers is clearly algorithmic, there are a number of decidability results that have no clear algorithmic content. Most of these are based on the fact that a set is recursive iff it and its complement are recursive enumerable. An example is the theorem: if T is axiomatizable and complete then T is decidable.

3.6. *The Arithmetical Hierarchy*

We have seen that decision problems can be considered to be of the form "does n belong to a set X?", so – up to a coding – the powerset of $\mathbb{N}$ presents us with all possible decision problems. Put in this way, we get too many unrealistic decision problems. For, a reasonable decision problem is usually presented in the form: test to find out if an element of a certain effectively generated (or described) set is an element of another such set, e.g. if a formula is a theorem. So among the subsets of $\mathbb{N}$, we are interested in certain sets that are somehow effectively described.

We have already met such sets, e.g. the primitive recursive, the recursive, and the recursive enumerable sets. It seems plausible to consider sets of natural numbers definable in a theory or in a structure. The very first candidate that comes to mind is the standard model of arithmetic, $\mathbb{N}$. Sets definable in $\mathbb{N}$ are called *arithmetical*, they are of the form $\{n \mid \mathbb{N} \models \varphi(\bar{n})\}$.

We already know that recursive sets and RE sets are arithmetical, and we know at least one particular set that is not arithmetical: the set of (Gödel numbers of) true sentences of *arithmetic* (Tarski's theorem).

It is natural to ask if the collection of arithmetical sets has some structure, for example with respect to a certain measure of complexity. The answer, provided by Kleene and Mostowski, was 'yes'. We can classify the arithmetical sets according to their syntactic form. The classification, called the *arithmetical hierarchy*, is based on the prenex normal forms of the defining formulas.

The hierarchy is defined as follows:

I.6: ALGORITHMS AND DECISION PROBLEMS 469

a Σ_1^0-formula is of the form $\exists \vec{y} \varphi(\vec{x}, \vec{y})$,
a Π_1^0-formula is of the form $\forall \vec{y} \varphi(\vec{x}, \vec{y})$,

where φ contains only bounded quantifiers,

if $\varphi(\vec{x}, \vec{y})$ is a Σ_n^0 formula, then $\forall \vec{y} \varphi(\vec{x}, \vec{y})$ is a Π_{n+1}^0 formula,
if $\varphi(\vec{x}, \vec{y})$ is a Π_n^0 formula, then $\exists \vec{y} \varphi(\vec{x}, \vec{y})$ is a Σ_{n+1}^0 formula,

Σ_n^0 and Π_n^0 sets are defined as extensions of corresponding formulas, and a set is Δ_n^0 if it is both Σ_n^0 and Π_n^0.

The following inclusions are easily established

$$\begin{array}{ccccccc}
 & \Sigma_1^0 & & \Sigma_2^0 & & \Sigma_n^0 & & \Sigma_{n+1}^0 \\
 \subseteq & & \subseteq & & \subseteq \cdots \subseteq & & \subseteq & \subseteq \\
\Delta_1^0 & & \Delta_2^0 & & & \Delta_{n+1}^0 & & \\
 \subseteq & & \subseteq & & \subseteq \cdots \subseteq & & \subseteq & \subseteq \\
 & \Pi_1^0 & & \Pi_2^0 & & \Pi_n^0 & & \Pi_{n+1}^0
\end{array}$$

e.g. if $X \in \Sigma_1^0$, then it is defined by a formula $\exists \vec{y} \varphi(x, \vec{y})$, by adding dummy variables and quantifiers we can get an equivalent Π_2^0-formula $\forall z \exists \vec{y} \varphi(x, \vec{y})$.

The above classification is based on syntactic criteria; there is, however, a 'parallel' classification in terms of recursion theory. At the lower end of the hierarchy we have the correspondence

Δ_1^0 recursive sets,

Σ_1^0 RE sets,

Π_1^0 complements of RE sets.

We know, by Theorem 2.14, that $\Sigma_1^0 \cap \Pi_1^0 = \Delta_1^0$.

The predicate $\exists z T(x, y, z)$ is universal for the RE sets in the sense that each RE set is of the form $\exists z T(e, y, z)$ for some e. Via the above correspondence, $\exists z T(x, y, z)$ is universal for the class Σ_1^0. Likewise $\neg \exists z T(x, y, z)$ is universal for Π_1^0 (for proofs of the facts stated here see any of the standard texts in the bibliography). To progress beyond the RE sets and their complements, we use an operation that can be considered as an infinite union, let $U_n(x, y)$ be universal for Π_n^0, then $\exists z U_n(x, \langle z, y \rangle)$ is universal for Σ_{n+1}^0. The basic technique here is in complete analogy to that in the so-called *hierarchy of Borel sets*, one can make this precise and show that the arithmetical hierarchy is the finite part of the effective Borel hierarchy (cf. Shoenfield [1967], Hinman [1978]).

Since one goes from Π_n^0 to Σ_{n+1}^0 by adding an existential quantifier in front of the defining Π_n^0 formula, we can also express the relation between Π_n^0 and Σ_{n+1}^0 in another recursion-theoretic way.

We have already defined the notion of relative recursiveness: f is recursive in g if $f(x) = \{e\}^g(x)$ for some e (where the superscript g indicates that g is an extra initial function), or A is recursive in B if $K_A(x) = \{e\}^B(x)$ for some e.

We use this definition to introduce the notion: A is *recursively enumerable in B*.

A is RE in B if A is the domain of $\{e\}^B$ for some e. Generalizing the T-predicate in an obvious way to a relativized version we get A is RE in B if $n \in A \iff \exists z T^B(e, n, z)$ for some e.

Now the relation between Π_n^0 and Σ_{n+1}^0 can be stated as

$$A \in \Sigma_{n+1}^0 \iff A \text{ is RE in some } B \in \Pi_n^0$$

There is another way to go up in the hierarchy. The operation that one can use for this purpose is the *jump*. The *jump of a set A*, denoted by A', is defined as

$$\{x \mid \{x\}^A(x)\downarrow\}, \text{ or } \{x \mid \exists z T^A(x, x, z)\},$$

e.g. the jump $\emptyset'$ of $\emptyset$ is K.

The successive jumps of $\emptyset$: $\emptyset', \emptyset'', \ldots \emptyset^{(n)}, \ldots$ are the paradigmatic examples of Σ_n^0 sets, in the sense that

$$A \in \Sigma_{n+1}^0 \iff A \text{ is RE in } \emptyset^{(n)}$$

What happens if we take a set that is recursive in $\emptyset^{(n)}$? The answer is given by *Post's theorem*:

$$A \in \Delta_{n+1}^0 \iff A \leqslant_T \emptyset^{(n)}$$

An important special case is: $A \in \Delta_2^0 \iff A$ is recursive in the Halting Problem (i.e. K).

A diagonal argument, similar to that of 2.16, shows that $\Sigma_n^0 - \Pi_n^0 \neq \emptyset$ and $\Pi_n^0 - \Sigma_n^0 \neq \emptyset$, hence all the inclusions in the diagram above are proper (this is called the *Hierarchy Theorem*) and the hierarchy is infinite.

For arithmetical predicates one can simply compute the place in the arithmetical hierarchy by applying the reduction to (almost) prenex form, i.e. a string of quantifiers followed by a recursive matrix.

Example: "e is the index of a total recursive function". We can reformulate this as "$\forall x \exists y T(e, x, y)$" and this is a Π_2^0 expression.

The more difficult problem is to find the lowest possible place in the hierarchy. So in the example: could $\{e \mid \{e\} \text{ is total}\}$ be Σ_1^0?

I.6: ALGORITHMS AND DECISION PROBLEMS

There are various ways to show that one has got the best possible result. One such technique uses *reducibility* arguments.

DEFINITION. Let $A, B \subseteq \mathbb{N}$. A is many-one (one-one) reducible to B if there is a (one-one) recursive function f such that $\forall x(x \in A \iff f(x) \in B)$. Notation: $A \leqslant_m B (A \leqslant_1 B)$:

EXAMPLE: (i) $W_e \leqslant_1 \{\langle x, y \rangle | x \in W_y\} = \{\langle x, y \rangle | \{y\}x\downarrow\} = \{\langle x, y \rangle | \exists z T(y, x, z)\}$. Define $f(n) = \langle n, e \rangle$, then clearly $n \in W_e \iff \langle n, e \rangle \in \{\langle x, y \rangle | x \in W_y\}$ and f is injective.

(ii) $\{\langle x, y \rangle | x \in W_y\} \leqslant_m K = \{x | x \in W_x\} = \{x | \{x\}x\downarrow\}$

Consider the function $\{(z)_1\}(z)_0$, and define

$$\{e\}(z, x) = \begin{cases} 1 & \text{if } \{(z)_1\}(z)_0 \downarrow \\ \text{divergent otherwise} \end{cases}$$

By the S_n^m theorem $\{e\}(z, x) = \{f(z)\}(x)$ and

$$a \in W_b \iff \{b\}a\downarrow \iff \{(\langle a, b \rangle)_1\}(\langle a, b \rangle)_0 \downarrow \iff$$
$$\iff \{f\langle a, b\rangle\}f\langle a, b\rangle \downarrow \iff f\langle a, b\rangle \in K.$$

An extra argument shows that $\leqslant_m$ can be replaced by $\leqslant_1$.

Now we call $A \in \Sigma_n^0(\Pi_n^0)$ Σ_n^0 complete (Π_n^0 complete) if for all $B \in \Sigma_n^0(\Pi_n^0)$ $B \leqslant_1 A$.

One can show that the nth jump of $\emptyset$, $\emptyset^{(n)}$, is Σ_n^0 complete.

Therefore, we use the sets $\emptyset^{(n)}$ to establish, e.g., that $A \in \Sigma_n^0$ does not belong to Π_{n-1}^0 or Σ_{n-1}^0 by showing $\emptyset^{(n)} \leqslant_m A$. For, otherwise $\emptyset^{(n)} \leqslant_m A \leqslant_1 \emptyset^{(n-1)}$ (or its complement), which is not the case. For more applications and examples see Rogers [1967] Section 14.8.

The complexity of models can also be measured by means of the arithmetical hierarchy, i.e. we call a model Σ_n^0 (Π_n^0 or Δ_n^0) if its universe is the set of all natural numbers and the relations are Σ_n^0 (Π_n^0 or Δ_n^0).

We mention several results:
  (i) *The Hilbert-Bernays completeness theorem.* If a theory (in a countable language) has an RE set of axioms, and it is consistent then it has a Δ_2^0 model (Kleene, Hasenjaeger).
  (ii) A decidable theory has a recursive model.
  (iii) The only recursive model of arithmetic is the standard model (Tennenbaum).

In 1970, Matijasevič gave a negative solution to Hilbert's tenth problem: *there is no algorithm that decides if any given Diophantine-equation* (i.e. of the form $p(x_1, \ldots, x_n) = 0$, for a polynomial p with integer coefficients) *has a solution*. To be precise, he established that every RE subset of $\mathbb{N}$ is Diophantine, that is for each RE set A we can find a polynomial p with integer coefficients such that $A = \{n \mid \exists x_1 \ldots x_n p(n, x_1, \ldots, x_n) = 0\}$.

Using results of Davis, Julia Robinson and Putnam, Matijasevič solved the problem by a purely number-theoretic argument, Matijasevič [1973].

Basis theorem, or do decent sets have decent elements? Let us agree for a moment that decent subsets of the Euclidean plane are open sets and that decent points are points with rational coordinates, then each non-empty decent set contains a decent point. For, a non-empty open set A is a union of open discs, so if $p \in A$ then $p \in C \subseteq A$, where C is an open disc. Now a bit of geometry tells us that C contains decent points. We express this by saying that the rational points form a basis for the open sets. Had we taken points with integer coordinates, then we would not have found a basis.

The problem of finding a basis for a given family of sets is of general interest in logic. One usually considers families of subsets of $\mathbb{N}$, or of functions from $\mathbb{N}$ to $\mathbb{N}$. 'Decency' is mostly expressed in terms of the arithmetical (or analytical) hierarchy.

Since most basis-theorems deal with classes of functions, we will classify functions by means of their graphs. To be precise, a function is Σ_n^0 (Π_n^0 or Δ_n^0) if its graph is so.

Unfortunately the majority of basis theorems deal with the analytical hierarchy instead of the arithmetical hierarchy, so we restrict ourselves to a theorem that fits into the arithmetical hierarchy, and also has nice applications in logic:

KREISEL'S BASIS THEOREM. *The Δ_2^0 functions form a basis of the Π_1^0 sets of characteristic functions.*

Here the notions of Σ_n^0, Π_n^0 generalize in a natural way to classes of functions or sets. Consider a language for arithmetic that also contains set variables X_i and a relation symbol $\in$ (for 'element of'), i.e. a language for second-order arithmetic. For this language we define the notions Σ_n^0 and Π_n^0 exactly as in first-order arithmetic. Now a class A of sets is Σ_n^0 if $A = \{X \mid \varphi(X)\}$ for a Σ_n^0 formula φ. One may just as well consider a language for second-order arithmetic with function variables.

For a detailed treatment of basis theorems, cf. Shoenfield [1967] and Hinman [1978].

APPENDIX

Proof of the Normal Form Theorem

The normal form theorem states, roughly speaking, that there is a primitive recursive relation $T(e, u, z)$ that formalizes the heuristic statement 'z is a (coded) computation that is performed by a partial recursive function with index e on input u (i.e. $\langle \vec{x} \rangle$). The 'computation' has been arranged in such a way that its first projection is its output.

The proof is a matter of clerical perseverence – not difficult but not exciting either.

We have tried to arrange the proof in a readable manner by providing a running commentary.

We have displayed below the ingredients for, and conditions on, computations. The index contains the information given in the clauses R_i. The computation codes the following items:

(1) the output,
(2) the input,
(3) the index,
(4, 5) subcomputations.

	Index e	Input u	Step z	Conditions on subcomputations
R_1	$\langle 0, n, q \rangle$	$\langle \vec{x} \rangle$	$\langle q, u, e \rangle$	
R_2	$\langle 1, n, i \rangle$	$\langle \vec{x} \rangle$	$\langle x_i, u, e \rangle$	
R_3	$\langle 2, n, i \rangle$	$\langle \vec{x} \rangle$	$\langle x_i + 1, u, e \rangle$	
R_4	$\langle 3, n + 4 \rangle$	$\langle p, q, r, s, \vec{x} \rangle$	$\langle p, u, e \rangle$ if $r = s$ $\langle q, u, e \rangle$ if $r \neq s$	
R_5	$\langle 4, n, b, c_1, \ldots \\ \ldots, c_k \rangle$	$\langle \vec{x} \rangle$	$\langle (z')_1, u, e, z', \\ \langle z_1'', \ldots, z_k'' \rangle \rangle$	$z', z_1'', \ldots, z_k''$ are computations with indices $b, c_1, \ldots, c_k$. z_i'' has input u and z' has input $\langle (z_1'')_1, \ldots, (z_k'')_1 \rangle$.
R_6	$\langle 5, n + 2 \rangle$	$\langle p, q, \vec{x} \rangle$	$\langle s, u, e \rangle$	(cf. 2.2)
R_7	$\langle 6, n + 1 \rangle$	$\langle b, \vec{x} \rangle$	$\langle (z')_1, u, e, z' \rangle$	z' is a computation with input $\langle \vec{x} \rangle$ and index b.

Note that z is the 'master number', i.e. we can read off the remaining data from z, e.g. $e = (z)_3$, $\text{lth}(u) = (z)_{3,2}$, but in particular the 'master numbers' of the subomputations. So, by decoding the code for a computation, we can effectively find the codes for the subcomputations, etc. This suggests a primitive recursive algorithm for the extraction of the total 'history' of a computation from its code. As a matter of fact, that is essentially the content of the normal form theorem.

We will now proceed in a (slightly) more formal manner, by defining a predicate $C(z)$ (for z is a computation), using the information of the preceding table. For convenience, we assume that in the clauses below, sequences u (in seq(u)) have positive length.

$C(z)$ is defined by cases as follows:

$$C(z) := \begin{cases} \exists q, u, e < z[z = \langle q, u, e \rangle \wedge \text{Seq}(u) \wedge e = \langle 0, \text{lth}(u), q \rangle] & (1) \\ \text{or} \\ \exists u, e, i < z[z = \langle (u)_i, u, e \rangle \wedge \text{Seq}(u) \wedge e = \langle 1, \text{lth}(u), i \rangle] & (2) \\ \text{or} \\ \exists u, e, i < z[z = \langle (u)_i + 1, u, e \rangle \wedge \text{Seq}(u) \wedge e = \langle 2, \text{lth}(u), i \rangle] & (3) \\ \text{or} \\ \exists u, e < z[\text{Seq}(u) \wedge e = \langle 3, \text{lth}(u) \rangle \wedge \text{lth}(u) > 4 \wedge ([z = \langle (u)_1, u, e \rangle \wedge \\ \quad \wedge (u)_3 = (u)_4] \vee [z = \langle (u)_2, u, e \rangle \wedge (u)_3 \neq (u)_4])] & (4) \\ \text{or} \\ \text{Seq}(z) \wedge \text{lth}(z) = 5 \wedge \text{Seq}((z)_3) \wedge \text{Seq}((z)_5) \wedge \text{lth}((z)_3) = \\ \quad = 3 + \text{lth}((z)_5) \wedge (z)_{3,1} = 4 \wedge C((z)_4) \wedge (z)_{4,1} = (z)_1 \wedge (z)_{4,2} = \\ \quad = \langle (z)_{5,1,1}, \ldots, (z)_{5,\text{lth}((z)_5),1} \rangle \wedge (z)_{4,3} = (z)_{3,3} \wedge \\ \quad \wedge \bigwedge_{i=1}^{\text{lth}((z)_5)} [C((z)_{5,i} \wedge (z)_{5,i,3} = (z)_{1,3+i} \wedge (z)_{5,i,2} = (z)_2] & (5) \\ \text{or} \\ \exists s, u, e < z[z = \langle s, u, e \rangle \wedge \text{Seq}(u) \wedge e = \langle 5, \text{lth}(u) \rangle \wedge \\ \quad s = \langle 4, (u)_{1,2} \dotdiv 1, (u)_1, \langle 0, (u)_{1,2} \dotdiv 1, (u)_2 \rangle, \langle 1, (u)_{1,2} \dotdiv 1, 1 \rangle, \ldots \\ \quad \ldots, \langle 1, (u)_{1,2} \dotdiv 1, (e)_2 \dotdiv 2 \rangle \rangle], & (6) \\ \text{or} \\ \exists u, e, w < z[\text{Seq}(u) \wedge e = \langle 6, \text{lth}(y) \rangle \wedge z = \langle (w)_1, u, e, w \rangle \wedge C(w) \wedge \\ \quad \wedge (w)_3 = (u)_1 \wedge (w)_2 = \langle (u)_2, \ldots, (u)_{\text{lth}(u)} \rangle] & (7) \end{cases}$$

I.6: ALGORITHMS AND DECISION PROBLEMS

Now observe that each clause of the disjunction only refers to C for smaller numbers. Furthermore, each disjunct clearly is primitive recursive. By changing to the characteristic function of C, we see that (1)–(7) present us (via definition by cases) with a course of value recursion. Hence, K_C is primitive recursive, and so is C.

This brings us to the end of the proof; define $T(e, u, z) := C(z) \wedge u = (z)_2 \wedge e = (z)_3$ (i.e. "z is a computation with index e and input u), then T is primitive recursive, and $\{e\}(u) = (\mu z T(e, u, z))_1$. □

HISTORICAL NOTES

Recursion theory was the offspring of Gödel's famous investigation into the completeness of arithmetic (1931). In the course of his proof of the incompleteness theorem he introduced the primitive recursive functions (under the name 'recursive'). In a subsequent paper (1934) he introduced, following a suggestion of Herbrand, a wider class of the so-called Herbrand–Gödel recursive functions. In the following years a number of approaches to the theory of algorithms were worked out: λ-*calculus* – Church, Kleene, cf. Barendregt [1981]; *Turing machines* – Turing [1936], cf. Davis [1965]; *Combinatory Logic* – Schönfinkel [1924], Curry [1929], cf. Barendregt [1981]; *Post Systems* [1943], cf. Hopcroft and Ullman [1969], *Register machines* – Minsky, Shepherdson and Sturgis [1961, 1962], cf. Schnorr [1974], Minsky [1967], *Markov algorithms* [1951], cf. Mendelson [1979].

The theory of recursive functions as a discipline in its own right was developed by Kleene, who proved all the basic theorems: the S_n^m theorem, the recursion theorem, the normal form theorem, and many others. Turing developed the theory of Turing machines, constructed a Universal Turing machine and showed the unsolvability of the Halting Problem. After Gödel's pioneering work, Rosser modified the independent statement so that only the consistency of arithmetic was required for the proof [1936], cf. Smoryński [1977].

Following Post [1944], a study of subsets of ℕ was undertaken, leading to notions as *creative, productive, simple*, etc. At the same time Post initiated the classification of sets of natural numbers in terms of 'Turing reducibility', i.e. a set A is Turing reducible to a set B if the characteristic function of A is recursive when the characteristic function of B is given as one of the initial functions – in popular terms if we can test membership of A given a test for membership of B. The theory of *degrees of unsolvability* has grown out of this notion, cf. Shoenfield [1971].

The applications to logic are of various sorts. In the first place there is the refinement of (un)decidability results of Gödel, Rosser, Turing and others, now known as the theory of *reduction types*, in which syntactical classes with unsolvable decision problem are studied, cf. Lewis [1979], on the other hand a study of solvable cases of the decision problem has been carried out, cf. Dreben and Goldfarb [1979].

The theory of (arithmetical and other) *hierarchies* started with papers of Kleene [1943] and Mostowski [1947], the subject has been extensively explored and generalized in many directions, cf. Hinman [1978]. Generalizations of recursion theory to objects other than natural numbers have been studied extensively.

To mention a few approaches: *Recursion of higher types*, Kleene [1959]; *Recursion on ordinals* Kripke, Platek; *Admissible Sets*, Kripke, Barwise, a.o.; *Definability Theory*, Moschovakis a.o.; *Axiomatic recursion theory*, Wagner-Strong, Friedman, a.o.; *Recursion on continuous functionals* Kleene, Kreisel. The reader is referred to Barwise [1975], Hinman [1978], Fenstad [1980], Moschovakis [1974], Norman [1980].

The connection between recursion theory and intuitionism was first established by Kleene [1945], since then the subject has proliferated, cf. Troelstra [1973]. For a historical account of the subject cf. Kleene [1976], [1981], Mostowski [1966], Van Heijenoort [1967].

University of Utrecht

REFERENCES

Barendregt, H. P.: 1981, *Lambda Calculus: Its Syntax and Semantics*, North-Holland, Amsterdam.
Barwise, J.: 1975, *Admissible Sets and Structures*, Springer-Verlag, Berlin.
Barwise, J. (ed.): 1977, *Handbook of Mathematical Logic*, North-Holland, Amsterdam.
Carnap, R.: 1937, *The Logic Syntax of Language*, Routledge and Kegan Paul, London.
Chang, C. C. and Keisler, H. J.: 1973, *Model Theory*, North-Holland, Amsterdam.
Davis, M.: 1958, *Computability and Unsolvability*, McGraw-Hill, New York.
Davis, M. (ed.): 1965, *The Undecidable*, Raven Press, New York.
Davis, M.: 1977, 'Unsolvable problems', in Barwise [1977] pp. 567–594.
Dreben, B. and Goldfarb, W. D.: 1979, *The Decision Problem. Solvable Classes of Quantificational Formulas*, Addison-Wesley, Reading, Mass.
Fenstad, J. E.: 1980, *Generalized Recursion Theory: An Axiomatic Approach*, Springer-Verlag, Berlin.
Fraenkel, A. A., Bar-Hillel, Y., Levy, A., and Van Dalen, D.: 1973, *Foundations of Set Theory*, North-Holland, Amsterdam.

I.6: ALGORITHMS AND DECISION PROBLEMS

Gabbay, D. M.: 1981, *Semantical Investigations in Heyting's Intuitionistic Logic*, D. Reidel, Dordrecht.
Gandy, R.: 1980, 'Church's thesis and principles for mechanisms', in J. Barwise, H. J. Keisler and K. Kunen (eds.), *Kleene Symposium*, North-Holland, Amsterdam.
Gödel, K.: 1946, 'Remarks before the Princeton Bicentennial Conference', in Davis [1965].
Grzegorczyk, A.: 1961, *Fonctions Récursives*, Gauthier-Villars, Paris.
Hinman, P. G.: 1978, *Recursion-Theoretic Hierarchies*, Springer-Verlag, Berlin.
Hopcroft, J. E. and Ullman, J. D.: 1969, *Formal Languages and Their Relations to Automata*, Addison-Wesley, Reading, Mass.
Jeroslow, R. G.: 1972, *On the Encodings Used in the Arithmetization of Metamathematics*, University of Minnesota.
Kleene, S. C.: 1952, *Introduction to Metamathematics*, North-Holland, Amsterdam.
Kleene, S. C.: 1959, 'Recursive functionals and quantifiers of finite type 1', *Trans. Am. Math. Soc.* 91, 1-52.
Kleene, S. C.: 1976, 'The work of Kurt Gödel', *J. Symbolic Logic* 41, 761-778.
Kleene, S. C.: 1981, 'Origins of recursive function theory', *Ann. Hist. Comp. Sci.* 3, 52-67.
Landau, E.: 1966, *Foundations of Analysis*, 3rd edn, Chelsea Publishing Co., New York.
Lewis, H. R.: 1979, *Unsolvable Classes of Quantificational Formulas*, Addison-Wesley, Reading, Mass.
Matijasevič, Y.: 1973, 'Hilbert's tenth problem', in P. Suppes, L. Henkin, A. Joyal and G. C. Moisil (eds.), *Logic, Methodology and Philosophy of Science*, North-Holland, Amsterdam, pp. 89-110.
Mendelson, E.: 1979, *Introduction to Mathematical Logic*, Van Nostrand, New York.
Minsky, M.: 1967, *Computation: Finite and Infinite Machines*, Prentice-Hall, Englewood Cliffs, N.J.
Monk, D.: 1976, *Mathematical Logic*, Springer-Verlag, Berlin.
Moschovakis, Y.: 1974, *Elementary Induction on Abstract Structures*, North-Holland, Amsterdam.
Mostowski, A.: 1966, *Thirty Years of Foundational Studies*, Blackwell, Oxford.
Norman, D.: 1980, *Recursion on the Countable Functionals*, Springer-Verlag, Berlin.
Pèter, R.: 1959, 'Rekursivität und Konstruktivität', in A. Heyting (ed.), *Constructivity in Mathematics*, North-Holland, Amsterdam.
Rabin, M. O.: 1977, 'Decidable theories', in Barwise [1977], pp. 595-629.
Rogers, H.: 1967, *Theory of Recursive Functions and Effective Computability*, McGraw-Hill, New York.
Schnorr, C. P.: 1974, *Rekursive Funktionen und ihre Komplexität*, Teubner, Stuttgart.
Shoenfield, J. R.: 1967, *Mathematical Logic*, Addison-Wesley, Reading, Mass.
Shoenfield, J. R.: 1971, *Degrees of Unsolvability*, North-Holland, Amsterdam.
Smoryński, C.: 1977, 'The incompleteness theorems', in Barwise [1977], pp. 821-866.
Smoryński, C.: 1981, 'Fifty years of self-reference', *Notre Dame J. Formal Logic* 22, 357-374.
Tarski, A.: 1951, *A Decision Method for Elementary Algebra and Geometry*, 2nd rev. edn. Berkeley.
Tarski, A., Mostowski, A., and Robinson, R. M.: 1953, *Undecidable Theories*, North-Holland, Amsterdam.

Troelstra, A. S.: 1973, *Metamathematical Investigations of Intuitionistic Arithmetic and Analysis*, Springer-Verlag, Berlin.
Turing, A.: 1936, 'On computable numbers, with an application to the Enscheidungsproblem', *Proc. London Math. Soc.* **42**, 230–265 (Also in Davis [1965]).
Van Dalen, D.: 1981, *Logic and Structure*, Springer-Verlag, Berlin.
Van Heijenoort, J.: 1967, *From Frege to Gödel. A Source Book in Mathematical Logic 1879–1931*, Harvard Univ. Press, Cambridge, Mass.
Wang, Hao: 1974, *From Mathematics to Philosophy*, Routledge and Kegan Paul, London.

NAME INDEX

Aandera, S. 466
Ackermann, W. 4, 82, 294, 441
Ajtai, M. 294
Altham, J. 39
Anderson, J. M. 31
Ax, J. 84, 85

Barendregt, H. 325
Bar-Hillel, Y. 411
Barnes, R. F. 263
Barwise, J. 39, 53, 83, 87, 100, 101, 103, 105, 279, 280, 281, 285, 291, 295, 298, 301, 304, 308, 318, 476
Behmann, H. 82, 465
Bell, J. L. 20, 24, 51, 65, 68, 74, 122, 180, 302
Belnap, N. D. 18, 30, 31, 35, 190, 195, 209, 210, 213, 214, 266, 391, 399
Bencivenga, E. 270
Bendall, K. 264, 267, 268, 270
Bernays, P. 20, 60, 71, 74, 117, 122, 134, 137, 147, 179, 265, 336, 467
Beth, E. 23, 25, 65, 106, 107, 180, 209, 213, 260, 263
Bocheński, I. M. 57
Bolzano, B. 45, 46, 56, 57, 58
Bode, G. 3
Boolos, G. 92, 362, 400
Börger, E. 466
Boyd, R. 379
Brouwer, L. E. J. 92, 393, 395, 396, 402
Burge, T. 389. 390

Cantor, G. 99
Carnap, R. 48, 55, 76, 213, 215, 225, 245, 262, 264, 267, 268, 411
Chang, C. C. 68, 74, 87, 107, 122, 284, 296, 468

Chastain, C. 44
Cherlin, G. 68
Chihara, Ch. S. 351, 352, 399, 400, 402
Church, A. 65, 78, 136, 143, 260, 311, 312, 331, 339, 343, 357, 360, 371, 383, 384, 397, 398, 399, 401, 402, 451, 452, 464, 465, 475
Chwistek, L. 373, 374
Coffa, A. 334
Cohen, L. J. 7
Cohen, P. 85, 467
Cook, S. A. 26
Cooper, R. 39, 83, 281
Copi, I. 399
Corcoran, J. 270
Craig, W. 20, 107, 321
Curry, H. 452, 475

Davidson, D. 262
Davis, M. 412, 418, 421, 433, 459, 472, 475
Dedekind, R. 75, 111, 333, 400, 425
De Morgan, A. 3
Dowty, D. 4
Drake, F. R. 297
Dreben, B. 466, 476
Dummett, M. 6, 35, 162, 166, 173, 394
Dunn, M. 18, 35, 190, 195, 209, 210, 213, 214, 266, 401

Ehrenfeucht, A. 102, 103
Ellis, B. 267
Enderton, H. 142, 299, 321
Evans, G. 44

Feferman, S. 38, 59, 87, 148, 331, 379, 380, 381, 382, 385, 386, 387, 388, 398

479

Fenstad, J. E. 476
Field, H. 264, 265, 402
Fisher, A. 358
Fitch, F. B. 31, 193, 347, 352, 355, 362, 391, 392, 396, 398, 400
Fitting, M. 180, 270
Flum, J. 101, 105
van Fraassen, B. 264, 265, 268, 402, 403
Fraenkel, A. 86, 117, 292, 411
Fraissé, R. 99, 102, 276, 286, 287, 288
Frege, G. 3, 6, 13, 32, 37, 38, 39, 48, 49, 74, 114, 116, 134, 260, 262, 265, 275, 289, 333, 337, 362, 388, 398
Friedman, M. 476

Gabbay, D. M. 465
Gaifman, H. 225, 264, 266, 267, 268
Gallin, D. 313, 315, 317, 326, 401
Gandy, R. O. 86, 401, 452
Garey, M. R. 26
Garson, J. W. 263
Gentzen, G. 3, 23, 35, 38, 39, 31, 32, 59, 134, 144, 157, 158, 159, 166, 168, 171, 172, 173, 332, 355, 357, 358, 359, 362
Gödel, K. 3, 4, 60, 68, 75, 86, 87, 90, 91, 92, 114, 115, 122, 311, 351, 360, 371, 375, 379, 395, 397, 398, 402, 423, 425, 452, 454, 459, 475
Goldblatt, R. 4
Goldfarb, W. D. 39, 265, 266, 466, 476
Gottlieb, D. 263
Greenbaum, S. 84
Grzegorczyk, A. 432
Grice, P. 7
Grover, D. L. 352, 399
Guenthner, F. 123
Gumb, R. D. 263, 265, 269, 270
Gupta, A. 391

Hacking, I. 357
Hajinal, A. 4
Hanf, W. 279
Havel, D. 4
Harper, W. L. 247, 248, 264, 265 269
Hatcher, W. S. 399

Hasenjaeger, G. 66, 194, 471
Hausdorff, F. 99
Hazen, A. 392, 398
Henkin, L. 60, 68, 94, 194, 197, 260, 262, 266, 306, 312, 315, 321, 325, 326, 400, 402
Hensel, G. 379
Herbrand, J. 3, 4, 12, 28, 59, 137, 425, 452
Herzberger, H. 391
Heyting, A. 393
Hilbert, D. 3, 4, 12, 20, 32, 48, 49, 56, 60, 71, 74, 75, 90, 92, 134, 137, 144, 147, 262, 356, 411, 467
Hinman, P. G. 469, 473, 476
Hintikka, J. 23, 48, 61, 103, 180, 209, 217, 218, 219, 224, 260, 267, 286, 299, 308, 327
Hodges, W. 11, 57,88, 186
Hopcroft, J. E. 475
Huntington, E. V. 264
Hylton, P. 396

Jeffrey, R. C. 24, 65, 186, 225, 245, 267, 400
Jennings, R. E. 265
Jeroslow, R. G. 455
Johnson, D. S. 26
Johnstone, H. W. 31
Johnstone, P. T. 87

Kalish, D. 11, 15, 31, 82, 142
Kalmár, L. 33, 465
Kamp, H. 40, 123, 317
Kanger, S. 180
Kaplan, D. 82
Karp, C. 99, 100
Kearns, J. T. 263
Keisler, H. J. 68, 74, 87, 167, 122, 276, 279, 282, 284, 287, 296, 467
Kemeny, J. 308
Keynes, J. M. 245
Kleene, S. 13, 20, 28, 33, 66, 82, 90, 109, 147, 165, 173, 177, 180, 298, 311, 380, 412, 433, 435, 440, 452, 459, 468, 472, 475, 476
Klenk, V. 88

NAME INDEX

Kneale, W. 31
Kochen, S. 84, 85
Kolmogorov, A. N. 225, 226, 245, 264, 267
Koymans, K. 410
Krabbe, E. 410
Kreisel, G. 67, 96, 296, 380, 381, 452, 472, 476
Kripke, S. 35, 143, 263, 332, 374, 384, 385, 386, 388, 389, 390, 391, 465, 476
Krivine, J. L. 96
Kronecker, L. 60, 375
Kunen, K. 295

Lackey, D. 396
Ladd, Ch. 37
Lazeski, E. 270
Lakoff, G. 4
Landau, E. 425
Langford, C. H. 3
Leblanc, H. 190, 191, 192, 193, 195, 197, 199, 210, 211, 213, 220, 225, 226, 246, 263, 264, 265, 266, 267, 268, 269, 352, 357, 402
Leisenring, A. C. 60
Leivant, D. 151, 166
Lemmon, E. J. 31
Levin, M. 268
Levy, A. 86, 90, 117, 411
Lewis, D. 369, 466, 476
Lindenbaum, A. 195
Lindström, P. 85, 97, 100, 101, 103, 276, 278, 280, 287
Lorenzen, P. 94, 378
Łos, J. 122, 283, 284
Löwenheim, L. 3, 68, 465
Łukasiewicz, J. 17, 32, 59

McArthur, R. P. 263
McCarthy, T. 263
McCawlwy, J. D. 4
Machover, M. 20, 24, 51, 65, 180
Maddy, P. 368
Magidor, M. 292, 295
Mal'cev, A. 68
Malitz, J. 292

Manna, Z. 4
Marcus, R. B. 262
Markov, A. A. 451, 452, 475
Martin-Löf, P. 164
Maslov, S. 466
Mates, B. 11, 32
Matijasevič, Y. 472
Mendelson, E. 33, 65, 122, 143, 178, 461, 475
Menger, K. 68
Meyer, R. K. 401
Minsky, M. 418, 421, 452, 475
Mirimanoff, D. 117
Mitchell, O. H. 37
Monk, D. 142, 278
Montague, R. 4, 11, 15, 31, 83, 142, 305, 312, 313, 317
Moore, G. E. 262
Morgan, C. G. 264, 265, 268
Moschovakis, Y. 311, 476
Mostowski, A. 68, 311, 379, 468, 476
Mycielski, E. 123
Mycielski, J. 123
Myhill, J. 297, 362, 363, 364, 371, 399, 401

Nelson, D. 393
Nemeti, I. 4
Norman, D. 476

Orenstein, A. 270
Orey, S. 317

Parsons, Ch. 263, 341, 352, 355, 384, 389, 390, 395, 398, 403
Partee, B. 44
Peano, G. 3, 111, 114
Perry, J. 49
Pèter, R. 453, 454
Pierce, C. S. 3, 6, 32, 37, 38, 82
Platek, R. 476
Poincaré, H. 333, 334, 337, 362, 375, 395, 396
Popper, K. R. 31, 190, 225, 226, 245, 246, 264, 267, 268, 269
Post, E. 4, 12, 19, 452, 475
Pottinger, G. 173

NAME INDEX

Prawitz, D. 30, 31, 151, 161, 162, 163, 165, 172, 180, 355, 400
Presburger, M. 467
Prior, A. 17, 31
Putnam, H. 34, 88, 336, 337, 379, 388, 402, 472

Quine, W. V. 8, 25, 31, 40, 78, 84, 193, 215, 260, 263, 267, 276, 332, 340, 341, 348, 362, 267, 368, 375, 398, 399, 402, 465
Quirk, R. 84

Rabin, M. O. 294, 467
Ramsey, F. P. 262, 337, 371, 374
Rasiowa, H. 20, 60
Reichenbach, H. 245
Rényi, A. 245
Ressayre, J. P. 304
Richard, J. 335, 396
Robinson, A. 85, 209, 262, 267, 402
Robinson, J. 465, 472
Robinson, R. 455
Rogers, H. 452, 465, 471
Rosser, J. B. 193, 461, 476
Routley, R. 401
Russell, B. 3, 4, 80, 81, 116, 260, 262, 275, 276, 209, 331, 333, 335, 336, 337, 338, 351, 354, 360, 362, 365, 367, 368, 369, 373, 374, 382, 388, 391, 392, 393, 394, 396, 398, 401

Sacks, G. E. 68
Schlipf, J. 304
Schmid, A. 396
Schmidt, H. A. 59
Schnorr, C. P. 453, 475
Schönfinkel, M. 475
Schotch, P. K. 265
Schröder, E. 3, 37, 57
Schütte, K. 28, 61, 178, 180, 209, 213, 260, 263, 399, 400
Schwemmer, O. 94
Schwichtenberg, H. 179
Scott, D. 180, 186, 297
Scott, R. S. 72
Seager, W. 265

Shelah, S. 276, 284
Shepherdson, J. C. 452, 475
Shoenfield, J. R. 209, 262, 267, 433, 451, 459, 460, 469, 473, 475
Shoesmith, D. J. 31
Sikorski, R. 20, 60
Skolem, T. 3, 59, 60, 63, 68, 76, 82, 87, 88, 94, 96, 114, 117, 208, 260, 292, 425
Slomson, A. B. 68, 74, 122, 302
Smiley, T. J. 31
Smoryński, C. 148, 475
Smullyan, R. 24, 65, 180, 183, 194, 215, 217, 224
Sneed, J. D. 74, 107
Spector, C. 380
Stalnaker, R. 264, 265
Stegmüller, W. 74
Steiner, M. 12
Stevenson, L. 35, 265
Strong, M. R. 476
Sturgis, H. E. 452, 475
Sundholm, G. 357
Suppes, P. 31, 86, 107, 117
Svenonius, L. 301
Szmielev, W. 467

Takeuti, G. 173, 178, 400
Tait, W. W. 178, 356
Tarski, A. 3, 32, 55, 56, 57, 58, 59, 68, 74, 90, 91, 137, 142, 162, 195, 262, 266, 332, 371, 382, 383, 384, 465, 467
Tennenbaum, S. 471
Thomason, R. H. 11, 31, 266, 355
Tennant, N. 31, 39, 155, 167
Troelstra, A. S. 165, 476
Turing, A. 311, 412, 420, 421, 422, 448, 450, 451, 452, 475, 476
Ullman, J. D. 475

Vaught, R. L. 4, 60, 68, 82, 464
Van Benthem, J. 281, 300, 301, 303, 317, 323, 327
Van Dalen, D. 31, 123 167, 411, 460
Van Heijenoort, J. 12, 368, 476
Visser, A. 410

Von Neumann, J. 86, 117, 351
Von Wright, H. 245, 269

Wagner, E. G. 476
Wang, H. 3, 4, 59, 60, 341, 379, 380, 399, 454
Ward, A. 270
Weaver, G. 352, 357
Weyl, H. 371, 377, 378, 385, 402

Whitehead, A. 3, 4, 80, 260, 262, 351, 394, 398
Wiredu, J. E. 59
Wisdom, W. A. 197, 199, 210, 220, 225, 266, 267, 270
Wittgenstein, L. 6, 260, 262, 265, 267

Zahn, P. 378
Zermolo, E. 86, 116, 292
Zucker, J. 173

SUBJECT INDEX

Abbreviation 15, 51
Abstract model theory 278f,
Abstraction 340
Absurdity 7
Ackermann function 432, 440
Addition 415
Algorithm 410, 451
   universal 424
Almost all 291
Ambiguous symbol 49
Analytic hierarchy 311, 476
   theorem 311
Anaphora 44
Argument 2
   schema 11, 45
   validity of 10, 11
Arithmetical hierarchy 310, 468f, 470, 476
Arithmetical set 468
Assumption 29, 136
   class 151
Arithmetization 165, 455
   see also Gödel number
Assignment
   in propositional logic 11
   in predicate logic 41
   suitable 41
Axiom 32, 135
   schema 135
Axiomatic system 74
Axiomatic theory 72f
Axiom of choice 120, 298, 321
Axiom of constructibility 295, 375
Axiom of descriptiveness 322
Axiom of infinity 357, 361
Axiom of plenitude 323
Axiom of reducibility 365, 373
Axioms 32, 72

complete set of 76
   first-order Peano arithmetic 114
   naive arithmetic 114
   predicate logic 108
   for propositional logic 108
   set theory 117f

Back-and-forth equivalence 97f, 284
β-function 433
Beth's definability theorem 74, 106
Bi-conditional 7
Bolzano-valid 46
Borel set 469
Brackets 15
Branching quantifier, *see* Quantifier

Calculus
   sound 27
   complete 27, 58, 60
   natural deduction 29
   see also Proof system, deductive system
Cardinal 120, 295
   inaccessible 293, 295
   measurable 295
   supercompact 296
Cardinality 120
Categorial grammer 312
Church-Rosser theorem 325
Church's theorem 67, 92, 464
Church's thesis 67, 451f
Closure 113
Coding 423
Commutative groups 74
Compactness theorem 67, 76, 101, 104, 121, 278
Complete theory 76

Completeness theorem 60f, 293
    weak 203
    strong 205
    in model-set semantics 223f
    in probabilistic semantics 240f
    in standard semantics 203f
    in substitutional semantics 207f
    in truth-value semantics 213f
    general 318f
Complexity 453
    of a formula 12, 50
    hierarchy 308
    of a model 471
Completeness proof 60
Composition 426
Comprehension axiom 321, 339, 349, 351, 353
Computation 415
Concept 336
Conclusion 26, 45
Conjunction 7
Conjunctive normal form 20
Connective 9
    definition of 325
    see also Truth functor
Conservative extension 79
Consistent theory 60
Constant 45
    ambiguous 45
    individual 46
    non-referring 57
    relation 57
    see also Witness
Constant domain schema 177
Constant function 426, 436
Constructibility 379
Contextual definition 335
Contextual restriction 38
Continuum hypothesis 296
Continuum problem 296
Counterexample 24
Contraction rule 173
Course-of-value recursion 431
Craig's theorem 458
Cumulative hierarchy 292
Cut elimination theorem 28
Cut rule 28, 171

Decidability 411, 425
    see also Decision method
Decidable 425, 444
    set 444
Decision method 21, 421
    for propositional logic 21, 411
Decision problem 424
Dedekind completeness 290
Deduction theorem 18, 109, 137f
Deductive system
    Hilbert-Frege style 134f
    natural deduction 148f
    sequent calculus 168f
Deep structure 33
Definability 73
    first-order 73, 277, 285, 318
    generalized first-order 73
    hierarchy 302
    higher-order 310f
    in arithmetic 92
    Σ_1^1-sentence 300
    Π_1^1-sentence 302, 303
Definite descriptions 80f
    see also Russell's theory of descriptions
Definition 107
    by cases 428, 338
    inductive 13, 113
    recursive 112
    and axioms 72
    Δ-elementary 302
    Σ-elementary 302
Δ_1-relation 90
Degrees of unsolvability 451
Derivability 135
    relation 165
Derivation 109
    rule 32, 109
    see also Proof
Description 335
Deterministic 413
Diagonalization 432, 434
Discriminator function 436
Disjunctive normal form 19, 57
Disjunction 7
Disquotation 372
Domain 196
    empty 56

non-empty 57
see also Quantification
Double negation 148
Double rewrite 204

Ehrenfeucht-Fraïssé game 99, 102
Elementary equivalence 96
Elimination rule 29, 58, 148
see also Natural deduction
Empiricism 354
Empty domain 56, 110
see also Domain
Encoding 86f, 115f
Entailment 17
see also Logical implication
ϵ-calculus 60
Equivalence relation 73
Express-predicate 373
Expression relation 383
Expressive power
of first-order logic 83f
limits of 276f
of second-order logic 289f
Extension 47
Extension of first-order logic 97, 277, 279
Extensionality 341, 346, 369
axiom of 77, 319, 349
axiom in set theory 117

Falsehood 6
Falsum, *see* Absurdity
Finite intersection property 122
Finiteness 276, 291
Finite set 121, 350
Finitist methods 356
First-order definability 281, 338
see also Definability
First-order languages 119
First-order logic, *see* Logic
First-order Peano arithmetic 92, 97, 454f
undecidability of 460f
Fixed point theorem 461
Formal system 26
see also Deductive system
Formal vs. informal 67
Formation tree 8
Formula 8, 50

arithmetical 92
atomic 9, 50
basic 13, 57
closed 50
compound 9
derivable 26
quantificational complexity 309
signed 180
subformula 50
universal 95
well-formed 51
Δ^n_m 310
Π^0_n 469
Π^1_1 297f
Π^n_m 297
Σ^0_n 469
Σ^1_1 297f
Σ^n_m 309
see also Complexity
Function 71, 119
measure 244
two-valued 241
Function constant 71
Functional application 312

Game 92, 93f, 301
General completeness 318f
General model, *see* Model
Generalization 140, 143
rule 142
Generalized quantifier, *see* Quantifier
Generative semantics 4
Gentzen's Hauptsatz 171f, 355f
'genuine logic' 84
Gödel number 91, 455
Grelling paradox 371

Hanf number 295
Halting problem 422, 448
Henkin extension 194
Henkin interpretation 197
see also Model
Henkin set 216
Heyting Arithmetic 164
Higher-order logic 305f
reduction to first-order logic 315
see Logic, Second-order logic, Type

theory
Hilbert-Bernays completeness theorem 471
Hilbert's ε-calculus 59
Hilbert's tenth problem 472
Hilbert-style system 107f
Hintikka sentence 103, 104
Hintikka set 61
Horn formula 466
Hyperarithmetic 380

Identity 68f, 78
   of indiscernibles 288
   in ramified type theory 346, 362f
   reflexivity of 69
   standard 68
   see Leibniz' law
Identify function 415
Implication (logical) 15, 56
Implication (material) 7
Impredicative definition 336
Incompleteness theorem 360, 380, 460
Index 421, 434, 435
Indexical 49
Individuals 122
Induction 361
Induction axiom 111
   see also Peano arithmetic
Inductive definition 115
Inference rule 135, 139, 143, 148
   see also Derivation rule, (rule of) Proof
Infinitary logic 96, 101
Infinity axiom 118
Initial segment 111
Inseparable 449, 464
Intended interpretation 75
Interpolation theorem 280, 304, 305
   in propositional logic 20
   in predicate logic 106
Introduction rule 29, 58, 148
   see Natural deduction
Intuitionism 379, 394, 395
   see also Logic
Isomorphism 98
   partial 284
   n-partial 286

Jump 470

Keisler's theorem 282
König's tree lemma 63
Kreisel's basis theorem 472
Kripke's theory of truth 384f
Krom formula 466

Lambda-abstraction 312f
   see also Abstraction
Lambda-calculus 324f
Lambda-conversion 324
Language 5, 49
   of propositional logic 5
   of predicate logic 50
   many-sorted 55
   set- 67
Leibniz' law 70
Level 343, 345, 397
   cumulative 345
   transfinite 347
Liar Paradox 460
Lindström's theorem 85f, 101f, 281, 287f
Logic 97
   first-order 4, 33f
   higher-order 305f
   infinitary 279
   intuitionist 30, 144
   many-sorted 38
   minimal 145, 155
   modal 139
   propositional 5f
   relevant 151
   second-order 76, 288f
Logical consequence 56
   see Logical implication
Logical equivalence 18
Logical form 4
Logicist program 333, 337, 362
Łoś equivalence 283
Łoś-Tarski theorem 303
Löwenheim number 295
Löwenheim-Skolem theorem 97, 208, 277, 278
   downward 68, 100, 104
   upward 68, 101, 104

SUBJECT INDEX

Many-sorted language 59
Mathematical induction 11
    course-of-values 11
Meaning postulates 76
Metalanguage 7
Metavariables 8, 9, 50
Minimal logic 145
Minimalization 419, 432
Modal logic
    S4 139
    axioms and rules 139, 140
Model 14, 53, 193
    construction of 60f, 65f
    elementary 284
    equivalent 98
    finite 276, 279
    Henkin 197
    isomorphic 98
    Henkin model of second-order logic 77
    intended (or standard) 75, 77
    non-standard 76, 277
    general 316f
    recursive 471
    resplendent 303
    of a theory 56
    see Structure
Model associate 201
Model counterpart 211
Model set 217
Model-theorist 33, 34, 85
Modus ponens 29, 109
Montague Grammar 4, 299, 305
most 277, 285
μ-recursive function 432f

n-approximation 113
n-equivalent 285
    elementary 286
Naive arithmetic 110
Name 39
    and satisfaction 40
Natural deduction 29, 148f
    higher-order logic 355f
    intuitionist logic 149f
    modal logic 163f
    predicate logic 58f
    propositional logic 29f

    relation to Hilbert-style system 155
    see also Calculus
Natural number 118
Necessitation 139
Negation 7
Non-axiomatizability 293
Non-constructive dilemma 159
Non-empty domain 57
    see Domain
Non-standard model of arithmetic 76
Normal form theorem 442, 473f
Normalization theorem 162
Noun phrases 78f
Null-set axiom 117
Numeral 114, 457

Objectual interpretation 197
    consistency 461
Ontology 394
Oracle 450
Order 397
Ordering
    partial 73
    total 73
Ordinal 118, 380
    provable 381
Overspill methods 85

Pair-set axiom 118
Palindrome tester 413
Paradox 335, 337, 365, 374, 382f, 391f
Parameter 134, 143, 144
    eigen-parameter 146
    proper 145
Partial recursive function 435, 451
    converges 437
    diverges 437
    total 434
Partial truth definition 341
Peano arithmetic 66, 75f, 110f
    induction axiom 289
    second-order 293
Phenomenalism 354
Platonism 334
Polish notation 17
Polynomial 25
    function 26

solvable in polynomial time 26
Ponential 193
Post's theorem 470
Power-set axiom 118
Pragmatism 334, 367
Predicate 37
    n-place 41
    relativization 38
Predicative conception of property 338
Predicative definition 337, 338, 379
Premise 26, 45
Prenex form 58, 309
Prenex complexity hierarchy 308f
Prenex normal form theorem 309
Preservation 317
Primitive recursion 419, 426
Primitive recursive function 425
Principal type structure 316
Probabilistic semantics 225f
Probability associate 254
Probability function 226
Productive set 450, 464
Progression of systems 381
Projection function 417, 436
Projective relation 311
Pronouns 45
Proof 26, 150, 194
    from assumptions 137
    cut-free sequent 27
    indirect 159
    rule of 135, 139, 143
    see also Deductive system
Proof system 107f
    higher-order logic 355f
    see also Deductive system
Proof-theorist 33, 84
Proper class 341
Property 336, 338, 344
    definition of 344
    existence 344f, 351
    non-extensional 370
    non-relational 370
    relational 352
Propositional attitude 364, 369
Propositional function 336
Propositional logic
    see Logic

Provability 134
    see also Deductive system
Pseudo-wff 144
Pure set 123
Pure set structure 66

Quantification 35f
    domain of 36, 47
    restricted 36, 51
    substitutional 35
    vacuous 193
Quantifier 35f
    branching 299
    contraction 446
    existential 43
    generalized 279, 291f
    and games 92
    most 277
    restricted 90
    sortal 43
    universal 43
    two-thirds
    word 37
Quantifier-depth 285
Quantifier-elimination 467

Rabin's theorem 294
Ramified analysis 318
Ramified functional calculus of second-order 343
Ramified higher-order systems 345
Real number 333, 376f
Recursion theorem 440
Recursive definition 115
Recursive function, *see also* Partial recursive function, primitive recursive function
Recursive saturation 304
Recursive set 437
Recursively enumerable in 470
Recursively enumerable set 444
Recursively separable 449
Reducibility argument 471
Reductio ad absurdum 30
Reduction 161
Reduction class 465
Reduction type 476

SUBJECT INDEX

Reflection 436
Regularity axiom 117
Relation 47
   recursive 310
   recursively enumerable 310
   hyperarithmetical 311
   primitive recursive 428
   projective 311
   see also Arithmetic hierarchy, analytic hierarchy, formula
Relative recursiveness 450f
Relativization (of a logic) 98
Replacement axiom 120
Representative 458
Restriction term 38
Richard's paradox 335
Robinson's non-standard arithmetic 85
Russell's theory of descriptions 335

Satisfaction 39f, 46, 53
Satisfaction class 342
Scope 79
Second-order arithmetic 116
Second-order condition 88
Second-order logic 288, 238f
   non-axiomatizability 293
   set-theoretic aspects 294f
   see also Logic
Self-reference 421
Semantic tableaux 23, 65, 180f
Semantics
   linguistic 281
   probabilistic 225f
   standard first-order 191f
   substitutional 198f
   truth-value 209f
   see also Truth definition
Sentence 5
   letter 8
   open 41
   schema 45
   see also Formula
Separation Axiom 117, 351
Sequence number 430
Sequent 15, 56, 165
   arrow 134
   calculus 28, 168f, 175

   derivable 26
   generalized 25
   normal 172
   proof 25
Set 86f, 116f
   see also Recursive set, Recursively enumerable set
Set-class theory of Gödel–Bernays 122
Set language 67
Set-theoretic foundations of mathematics 86
Set theory 86f, 296, 367
   non-standard models of 88
   ZF set theory 117f
   axioms of ZF set theory 117f
   second-order 292f
Σ_1^0-complete 459
Σ_n^0-complete 471
Similarity type 12, 49
Simple theory of types 366
Situation 5, 45
   see Model, structure
Skolem–Behmann theorem 83
Skolem function 92, 95
Skolem normal form for satisfiability 95, 96
Skolem's paradox 87, 208
S_n^m-function 438
S_n^m-theorem 438
sound 58
Soundness theorem 201
   in model-set semantics 222f
   in probabilistic semantics 238f
   in standard semantics 201f
   in substitutional semantics 207f
   in truth-value semantics 212f
Standard model 75
   see Model
State description 415
Stationary logic 291
Strategy 93f
   see Game
Strengthened liar arguments 389
Structure 13, 34, 47, 52
   classification of 78
   domain of 52
   L-structure 52

SUBJECT INDEX

n-equivalent 102
truth in 13
see Model
Sub-formula property 177
Subset 117
Substatement 192
Substitution 418, 436
Substitution of terms 141
Substitution lemma 198, 202
Substitution instance 192
Substitution rule 136
Substitutional quantification 198f
   in higher-order logic 351f
   see Substitutional semantics
Substitutional semantics 35, 198f
Subtraction 416
Successor function 415, 436
Summation 349
Svenonius' theorem 30
Syllogism 2
Syntax 49
   see Language

Tableaux, see Semantic Tableaux
Tarski's undefinability theorem 90
   see also Truth definition
Tautology 15
Term 50
   substitutable 141
   substitution 141
Term extension 195
Tertium non datur 159
Theorem 135
   schema 136
Theoremhood 135
Theory 56, 194
   axiomatizable 458
   consistent 60, 194
   complete 76
   decidable 465
   maximally consistent 194
   numerical 462
   undecidable 462, 465
   ω-complete 194
Thinning 166, 173
tonk 31
Topos 87

Traditional Logician 33
Transfer methods 85
Tree diagram 23
Truth 6, 53
   see Truth definition
Truth condition 197
Truth definition 341f, 382, 463
   hierarchies 383
   propositional logic 14
   predicate logic 52f
Truth set 214, 267
Truth table 6f, 10
Truth trees 23
Truth value 6, 54
Truth-value assignment 210
Truth-value associate 211
Truth-value counterpart 211
Truth-value semantics 209f
Truth-value function 241
Turing degrees 451
Turing machine 412
   universal 421
Turing reducible 450
Turing's thesis 412
Turnstile 15, 26
Type (in higher-order logic) 306, 345, 347, 397
   level 345
   order of 306
Type structure 316
Type theory 311f
   functional 312
   intensional 313
   ramified 345f
   simple 366
   reduction of 316
   transfinite extension of 343, 347, 350f

Ultrafilter 121
   principal 121
   regular 121
Ultraproduct 74, 122
Unbounded search 419, 442
Undecidability 462f
   see also Church's theorem
Undecidable problems 448

SUBJECT INDEX

Undefinability of truth, see Tarski's theorem
Union axiom 118
Universal algorithm 424
Universality of language 391
Universe (in type theory) 307

Validity 66
Valuation 52
Variable 36, 50
    assignment 52
    binding of 43
    clash of 51
    free 41
    free occurrence 43
    sortal 38
    suitable 53
Virtual class 340
Von Neuman–Bernays–Gödel set theory 341
Von Neumann ordinal 118

Well-founded relation 290
Well-ordering axiom 120
Wff, see (well-formed) Formula
Witness 61, 326

Zermelo set theory 119, 341

CPC 135
CQC 140
CS 176
H 108
HCPC 135
HCQC 141
HIQC 151
HS4 164
IQC 149
IS 170
IS' 174
IS$^+$ 171
$L_{\omega,\omega}$ 96, 279
$L_{\infty,\omega}$ 98
L_n 307
L_ω 308
L^+ 316
N 455
NBG 341
NIQC 149
NS4 164
P 114
PA 454
Q 455
QC 143
S4 139
T 182
ZF 117
ZFC 86

TABLE OF CONTENTS TO VOLUME II

ACKNOWLEDGEMENTS

PREFACE

A NOTE ON NOTATION

II.1. ROBERT A. BULL and KRISTER SEGERBERG / Basic Modal Logic
II.2. JOHN BURGESS / Basic Tense Logic
II.3. RICHMOND H. THOMASON / Combinations of Tense and Modality
II.4. JOHAN VAN BENTHEM / Correspondence Theory
II.5. JAMES W. GARSON / Quantification in Modal Logic
II.6. NINO B. COCCHIARELLA / Philosophical Perspectives on Quantification in Tense and Modal Logic
II.7. C. ANTHONY ANDERSON / General Intensional Logic
II.8. DONALD NUTE / Conditional Logic
II.9. CRAIG SMORYŃSKI / Modal Logic and Self-reference
II.10. DAVID HAREL / Dynamic Logic
II.11. LENNART ÅQVIST / Deontic Logic
II.12. DAVID HARRAH / The Logic of Questions

NAME INDEX

SUBJECT INDEX

TABLE OF CONTENTS TO VOLUMES I, III, AND IV

TABLE OF CONTENTS TO VOLUME III

ACKNOWLEDGEMENTS

PREFACE

A NOTE ON NOTATION

III.1. STEVEN BLAMEY / Partial Logic
III.2. ALISDAIR URQUHART / Many-valued Logic
III.3. MICHAEL DUNN / Relevance Logic and Entailment
III.4. DIRK VAN DALEN / Intuitionistic Logic
III.5. WALTER FELSCHER / Dialogues as a Foundation for Intuitionistic Logic
III.6. ERMANNO BENCIVENGA / Free Logics
III.7. MARIA-LUISA DALLA CHIARA / Quantum Logic
III.8. GÖRAN SUNDHOLM / Proof Theory and Meaning

NAME INDEX

SUBJECT INDEX

TABLE OF CONTENTS TO VOLUMES I, II, AND IV

TABLE OF CONTENTS TO VOLUME IV

ACKNOWLEDGEMENTS

PREFACE

A NOTE ON NOTATION

IV.1. BARBARA PARTEE / Approaches to the Semantics of Natural Language
IV.2. DAG WESTERSTÅHL / Quantifiers in Formal and Natural Languages
IV.3. UWE MOENNICH and GEORGE BEALER / Property Theories
IV.4. NINO B. COCCHIARELLA / Philosophical Perspectives on Formal Theories of Predication
IV.5. HANS KAMP / Predication and Vagueness
IV.6. JEFFREY PELLETIER and LEN SCHUBERT / Mass Terms
IV.7. NATHAN SALMON / Reference and Information Content: Names and Descriptions
IV 8. STEVEN KUHN / Indexicals
IV.9. RAINER BAEUERLE and MAX J. CRESSWELL / Propositional Attitudes
IV.10. ESA SAARINEN / Quantifiers and Modality
IV.11. STEVEN KUHN / Tense, Aspect, Time and Events
IV.12. WILLIAM LADUSAW / Non-indicative Speech Acts
IV.13. SCOTT SOAMES / Presupposition
IV.14. ALBERT VISSER / Semantics and the Liar Paradox

NAME INDEX

SUBJECT INDEX

TABLE OF CONTENTS TO VOLUMES I, II, AND III